Birkhäuser

## Trends in Mathematics

*Trends in Mathematics* is a series devoted to the publication of volumes arising from conferences and lecture series focusing on a particular topic from any area of mathematics. Its aim is to make current developments available to the community as rapidly as possible without compromise to quality and to archive these for reference.

Proposals for volumes can be submitted using the Online Book Project Submission Form at our website www.birkhauser-science.com.

Material submitted for publication must be screened and prepared as follows:

All contributions should undergo a reviewing process similar to that carried out by journals and be checked for correct use of language which, as a rule, is English. Articles without proofs, or which do not contain any significantly new results, should be rejected. High quality survey papers, however, are welcome.

We expect the organizers to deliver manuscripts in a form that is essentially ready for direct reproduction. Any version of TEX is acceptable, but the entire collection of files must be in one particular dialect of TEX and unified according to simple instructions available from Birkhäuser.

Furthermore, in order to guarantee the timely appearance of the proceedings it is essential that the final version of the entire material be submitted no later than one year after the conference.

More information about this series at http://www.springer.com/series/4961

Javier Fernández de Bobadilla • Tamás László •
András Stipsicz
Editors

# Singularities and Their Interaction with Geometry and Low Dimensional Topology

## In Honor of András Némethi

Birkhäuser

*Editors*
Javier Fernández de Bobadilla
Basque Center for Applied Mathematics
Ikerbasque (Basque Foundation for Science), Bilbao, Spain

Tamás László
Faculty of Mathematics and Computer Science
Babeş-Bolyai University
Cluj-Napoca, Romania

András Stipsicz
Alfréd Rényi Institute of Mathematics
Budapest, Hungary

ISSN 2297-0215 ISSN 2297-024X (electronic)
Trends in Mathematics
ISBN 978-3-030-61957-2 ISBN 978-3-030-61958-9 (eBook)
https://doi.org/10.1007/978-3-030-61958-9



This book is published under the imprint Birkhäuser, www.birkhauser-science.com, by the registered company Springer Nature Switzerland AG.
The registered company address is: Gewerbestrasse 11, 6330 Cham, Switzerland

*Dedicated to our friend and colleague*
*András Némethi*
*on the occasion of his 60th birthday.*

András Némethi, Budapest 2019

# Preface

This volume is based on the conference entitled "Némethi60: Geometry and Topology of Singularities", which was organized in Budapest from 27 to 31 May 2019 in honour of András Némethi's 60th birthday. The aim of the conference was to bring together researchers from algebraic geometry, singularity theory and low dimensional topology in order to present and discuss new results and perspectives for recent and old problems in the fusion of these fields. At the same time, the event provided a great opportunity to celebrate the fundamental contributions of András, which play a decisive role in the interactions between the fields mentioned above, a fact that also motivated the title of the present volume.

The volume consists of 14 articles covering a wide range of active subjects in the crossroads of singularity theory, algebraic geometry and low dimensional topology, such as hypersurfaces with isolated singularities, complex surface singularities and their smoothings, rational cuspidal curves, fundamental groups of algebraic varieties, Hilbert schemes of points, semicontinuity of invariants, Schubert calculus, motivic Chern classes, knot theory and Blanchfield forms, lattices and Heegaard Floer correction terms. The articles have been peer reviewed by renowned experts on the respective subjects. We would like to thank both the authors and the reviewers for the work they have done for the publication of this collection of papers.

Last but not the least, we would like to acknowledge the generous support of the Foundation Compositio Mathematica, the National Science Foundation (USA), the NKFIH-National Research, Development and Innovation Office (Hungary), the Hungarian Academy of Sciences, the Rényi Institute of Mathematics and Birkhäuser, which made the realization of the conference and the publication of the volume possible. Special thanks go to Sarah Goob and Sabrina Hoecklin for their kind help during the preparation of this volume.

We recommend this book to all interested readers in the hope that it will become a useful source for both senior researchers and the younger generations.

Bilbao, Spain
Cluj-Napoca, Romania
Budapest, Hungary
July 2020

Javier Fernández de Bobadilla
Tamás László
András Stipsicz

# The Scientific Life and Work of András Némethi

In this short essay, we attempt to give a summary about the scientific life and work of András Némethi. Undoubtedly, such a summary is always very subjective and incomplete, yet we would like to frame the main moments and achievements that have greatly determined his scientific life so far.

András Némethi was born on 29 May 1959 in a small Transylvanian town, called Erdőszentgyörgy (Sângeorgiu de Pădure, Romania). He completed the high school studies close to his hometown in Sighişoara, where, thanks to the guidance of his teacher Miklós Farkas, he decided to choose mathematics as a career. After high school, he became a student of mathematics at the University of Bucharest. After graduating from the university in 1984 with a (Romanian) diploma degree, as a usual procedure at that time, he worked as a school teacher in Harghita county. After 1 year, he returned to Bucharest and became a young researcher at the mathematical section of the National Institute for Science and Technical Research (INCREST; today, this institute is called "Simion Stoilow" Institute of Mathematics of the Romanian Academy).

He was admitted to the Doctoral School of Mathematics at the University of Bucharest in 1988, and in 1990, he defended his doctoral dissertation entitled "Topology of Algebraic and Analytic Singularities" under the supervision of Professor Lucian Bădescu. Taking advantage of the opening of the borders following the Romanian revolution in 1989, he visited the Mathematical Institute of the Hungarian Academy of Sciences in Budapest and then in the fall of 1990, he spent several months at the universities of Utrecht and Nijmegen at the invitation of Dirk Siersma and Joseph Steenbrink. During the years in Bucharest, he achieved various fundamental results related to the Milnor fibre and zeta function of different types of singularities. His main collaborators were A. Zaharia and M. Dădărlat. With the latter, he has several publications in operator theory. His most significant articles from this period are the following:

- *Lefschetz Theory for Complex Affine Varieties*, in Rev. Roumaine Math. Pures Appl. 33 nr. 3 (1988) 233–250;

- *Shape theory and (connective) K-theory* (joint with M. Dadarlat) in J. Operator Theory 23 (1990) 207–291;
- *On the bifurcation set of a polynomial function and Newton boundary* (joint with A. Zaharia), in Publ. RIMS. Kyoto Univ. 26 (1990) 681–689;
- *The Milnor fiber and the zeta function of the singularities of type $f = P(h, g)$*, in Compositio Math. 79 (1991) 63–97;
- *Generalized local and global Sebastiani-Thom type theorems*, in Compositio Mathematica 80 (1991) 1–14.

In 1991, András moved to the USA at the invitation of Professor Henri Moscovici from the Ohio State University and began his career as a PhD student under Moscovici's guidance. Throughout his stay in the USA, he worked at the Ohio State University, going through all university positions—instructor (1991–1995), assistant professor (1995–1998) and associate professor (1998–2002)—until he was appointed as a full professor in 2002. During this time, Némethi's work and professional development was extremely influenced and helped by Professor Moscovici and Dan Burghelea, as well as the seminars and work environment they led. To this day, he looks back to the years they spent together at the Ohio State University with the deepest gratitude.

During this period, he had several shorter and longer scientific visits, which resulted in several decisive working relationships and fundamental contributions. Just to mention the most significant ones, he visited Piere Millman at the University of Toronto, worked with Joseph Steenbrink during his 1-year stay in Nijmegen, spent 3 months in Paris collaborating with Claude Sabbah and visited Alexandru Dimca at the University of Nice-Sophia Antipolis as well as János Kollár at the Princeton University. Amongst his achievements during this period, it is worth mentioning the continuation of the Artin-Laufer theory, results regarding mixed Hodge structures and variation structures, and about the monodromy of polynomial maps and the topology of non-isolated hypersurface singularities. The most important articles can be summarized as follows:

- *Extending Hodge bundles for Abelian Variations* (joint with J. Steenbrink), in Annals of Mathematics 143 (1996) 131–148;
- *Dedekind sums and the signature of $f(x, y) + z^N$*, in Selecta Mathematica New series 4 (1998) 361–376;
- *"Weakly" Elliptic Gorenstein singularities of surfaces*, in Inventiones math. 137 (1999) 145–167;
- *Five lectures on normal surface singularities; lectures delivered at the Summer School in "Low dimensional topology", Budapest, Hungary 1998*, in Proc. of the Summer School, Bolyai Society Mathematical Studies 8, Low Dimensional Topology, (1999) 269–351;
- *On the monodromy of complex polynomials* (joint with A. Dimca), in Duke Math. Journal 108 Nr. 2 (2001) 199–209.

In the early 2000s, he started a collaboration with Liviu Nicolaescu from the University of Notre Dame about the study of the link of normal surface singularities with three-dimensional Seiberg-Witten theory. They formulated the so-called Seiberg-Witten invariant conjecture, which attempted to connect the geometric genus of the singularity with the Seiberg-Witten invariant of its link. Their work not only provided new methods and directions for studying the link of a singularity but also produced fundamental examples for the general understanding of the Seiberg-Witten invariants. This resulted in three fundamental articles with the following titles:

- *Seiberg-Witten invariants and surface singularities*, in Geometry and Topology Vol. 6 (2002) 269–328;
- *Seiberg-Witten invariants and surface singularities II (singularities with good $\mathbb{C}^*$-action)* in Journal of London Math. Soc. (2) 69 (2004) 593–607;
- *Seiberg-Witten invariants and surface singularities: Splicings and cyclic covers*, in Selecta Mathematica New series, Vol. 11, Nr. 3–4 (2005) 399–451.

He also established a fruitful collaboration with the Spanish node of singularity theory—I. Luengo, A. Melle-Hernández and J. Fernández de Bobadilla—regarding new results and conjectures about the theory of rational cuspidal curves. This resulted in a number of articles, amongst which we would mention the following:

- *Links and analytic invariants of superisolated singularities* (joint with I. Luengo and A. Melle-Hernández), in Journal of Algebraic Geometry 14 (2005) 543–565;
- *On rational cuspidal projective plane curves* (joint with J. F. de Bobadilla, I. Luengo and A. Melle-Hernández), in Proc. of London Math. Soc. 92 (3) (2006) 99–138.

The continuation of András' work in the study of the topology of links was strongly influenced by Peter Ozsváth and Zoltán Szabó with the appearance of their famous Heegaard Floer homology as a categorification of the Seiberg-Witten invariants. In 2004, he published the article:

- *On the Ozsváth-Szabó invariant of negative definite plumbed 3-manifolds* in Geometry and Topology 9 (2005) 991–1042,

as the first main contribution concerning the intimate relationship between the theory of normal surface singularities and Heegaard Floer homology. One of the outstanding achievements in this relationship is to apply and emphasize the power of the algebro-geometric methods to low dimensional topology. Around 2006, this contribution led to the birth of the theory of his *lattice cohomology*. This tool shed new light on some old problems regarding the topology of links of normal surface singularities and created many new directions in the crossroads of singularity theory, algebraic geometry and low dimensional topology. Lattice cohomology is still one of the main areas of his research, with many interesting problems and conjectures, such as its isomorphism with Heegaard Floer homology in case of singularity links. In the sequel, we list his most significant articles on this topic:

- *Lattice cohomology of normal surface singularities*, in Publ. RIMS. Kyoto Univ. 44 (2008) 507–543;

- *The Seiberg–Witten invariants of negative definite plumbed 3-manifolds*, in Journal of EMS 13(4) (2011) 959–974;
- *Reduction theorem for lattice cohomology* (joint with T. László), in IMRN, Issue 11 (2015) 2938–2985;
- *Lattice and Heegaard Floer Homologies of Algebraic Links* (joint with E. Gorsky), in IMRN (2015) 12737–12780;
- *The geometric genus of hypersurface singularities* (joint with B. Sigurðsson), in Journal of EMS 18 (2016) 825–851;
- *Lattice cohomology and rational cuspidal curves* (joint with J. Bodnár), in Math. Research Letters 23 (2016) no:2 339–375;
- *Links of rational singularities, L-spaces and LO fundamental groups*, in Inventiones mathematicae 210(1) (2017) 69–83;
- *On the set of L–space surgeries for links* (joint with E. Gorsky), in Adv. in Math. 333 (2018) 386–422.

Strengthening the relationship with the Hungarian mathematical community, he worked as a visiting professor at the Rényi Institute of Mathematics in the academic year 1999–2000. In 2004, he moved to Budapest, and as Full Professor at the Rényi Institute, he founded a new group of "algebraic geometry and differential topology". It was his mission to build a research community of topologists and algebraic geometers. Starting that year, many former and new colleagues, PhD students and postdocs from all around the world visited András at the Rényi Institute in Budapest. His most representative publications (apart from those mentioned in the previous context) starting with this period are the following:

- *The Seiberg-Witten invariant conjecture for splice-quotients* (joint with T. Okuma), in Journal LMS 28 (2008) 143–154;
- *On the Casson Invariant Conjecture of Neumann-Wahl* (joint with T. Okuma), in Journal of Algebraic Geometry 18 (2009) 135–149;
- *Generalized monodromy conjecture in dimension two* (joint with W. Veys), in Geometry and Topology 16(1) (2012) 155–217;
- *The cohomology of line bundles of splice-quotient singularities*, in Advances in Math. 229(4) 2503–2524 (2012);
- *The 'corrected Durfee's inequality' for homogeneous complete intersections* (joint with D. Kerner), in Math. Zeitschrift 274, Issue 3–4 (2013) 1385–1400;
- *Ehrhart theory of polytopes and Seiberg-Witten invariants of plumbed 3-manifolds* (joint with T. László), in Geometry and Topology 18 (2014) 717–778;
- *Holomorphic arcs on analytic spaces* (joint with J. Kollár), in Inventiones Math. 200 issue 1 (2015) 97–147;
- *Euler characteristics of Hilbert schemes of points on surfaces with simple singularities* (joint with A. Gyenge and B. Szendrői), in IMRN Issue 13 (2017) 4152–4159;
- *Durfee's conjecture on the signature of smoothings of surface singularities* (joint with J. Kollár, with an appendix by T. de Fernex), in Annales Scient. de l'Ecole Norm. Sup. 50 (2017) 787–798;

- *Combinatorial duality for Poincaré series, polytopes and invariants of plumbed 3-manifolds* (joint with T. László and J. Nagy), in Selecta Math. New Series 25, art.nr: 21 (2019);
- *The Abel map for surface singularities I. Generalities and examples.* (joint with J. Nagy), in Math. Annalen 375 (2019) 1427–1487;
- *The Abel map for surface singularities II. Generic analytic structure.* (joint with J. Nagy), to appear in Adv. of Math.

From 2008, he became a professor in the Department of Geometry at ELTE (University of Budapest), taking a large part in educating the new generation of mathematicians in Hungary. András had nine PhD students, two of whom graduated in Ohio and seven in Budapest. At this moment, he is supervising three more students. In recognition of his outstanding research, teaching and school-creating activities in Hungary, he received first the Rényi Prize of the Rényi Institute of Mathematics (2007), then the Prize of the Hungarian Academy of Sciences (2010), and in 2017, the most prestigious Hungarian state award given to scientists, the Széchenyi Prize. He was elected as a corresponding member of the Hungarian Academy of Sciences in 2019.

András Némethi is characterized by his strong motivation, humility and by his pure, very natural thinking, which is deeply present in his articles, in his teaching, as well as in his relationships with his colleagues, students and anyone around him. We would like to thank him for this and wish him many more active years in the fascination of the geometry and topology of singularities, and joy with his family, friends and colleagues.

# Contents

# Versality, Bounds of Global Tjurina Numbers and Logarithmic Vector Fields Along Hypersurfaces with Isolated Singularities

**Alexandru Dimca**

*To András Némethi on the occasion of his 60th birthday*

**Abstract** We recall first the relations between the syzygies of the Jacobian ideal of the defining equation for a projective hypersurface $V$ with isolated singularities and the versality properties of $V$, as studied by du Plessis and Wall. Then we show how the bounds on the global Tjurina number of $V$ obtained by du Plessis and Wall lead to substantial improvements of our previous results on the stability of the reflexive sheaf $T\langle V\rangle$ of logarithmic vector fields along $V$, and on the Torelli property in the sense of Dolgachev-Kapranov of $V$.

**Keywords** Projective hypersurfaces · Syzygies · Logarithmic vector fields · Stable reflexive sheaves · Torelli properties

**Subject Classifications** Primary 14C34; Secondary 14H50, 32S05

## 1 Introduction

Let $V : f = 0$ be a degree $d$ singular hypersurface in the complex projective space $\mathbb{P}^n$, having only isolated singularities. Let $S = \mathbb{C}[x_0, \ldots, x_n]$ be the graded polynomial ring, and consider the graded $S$-module of *Jacobian syzygies or Jacobian relations* of $f$ defined by

$$AR(f) = \{(a_0, \ldots, a_n) \in S^{n+1} : a_0 f_0 + \cdots + a_n f_n = 0\}, \tag{1.1}$$

A. Dimca (✉)
Université Côte d'Azur, CNRS, LJAD, Nice, France

Simion Stoilow Institute of Mathematics, Bucharest, Romania
e-mail: dimca@unice.fr



J. Fernández de Bobadilla et al. (eds.), *Singularities and Their Interaction with Geometry and Low Dimensional Topology*, Trends in Mathematics,
https://doi.org/10.1007/978-3-030-61958-9_1

where $f_j$ denotes the partial derivative of the polynomial $f$ with respect to $x_j$ for $j = 0, \dots, n$. This module has a natural $S$-graded submodule $KR(f)$, the module of *Koszul syzygies or Koszul relations* of $f$, defined as the $S$-submodule spanned by the obvious relations of the type $f_j f_i + (-f_i) f_j = 0$. Note that the syzygies in $AR(f)$ are regarded as vector fields annihilating $f$ in the papers by A. du Plessis and C.T.C. Wall, while the Koszul syzygies are called Hamiltonian vector fields. The quotient

$$ER(f) = AR(f)/KR(f) \tag{1.2}$$

is the graded module of *essential Jacobian relations*. One has the following description in terms of global polynomial forms on $\mathbb{C}^{n+1}$. If we denote by $\Omega^j$ the graded $S$-module of such forms of exterior degree $j$, then

1. $\Omega^{n+1}$ is a free $S$-module of rank one generated by $\omega = dx_0 \wedge dx_1 \wedge \ldots \wedge dx_n$.
2. $\Omega^n$ is a free $S$-module of rank $n+1$ generated by $\omega_j$ for $j = 0, \dots, n$ where $\omega_j$ is given by the same product as $\omega$ but omitting $dx_j$.
3. The kernel of the wedge product $df \wedge : \Omega^n \to \Omega^{n+1}$ can be identified up to a shift in degree to the module $AR(f)$. To see this, it is enough to use the formula

$$df \wedge \left( \sum_{j=0,n} (-1)^j a_j \omega_j \right) = \left( \sum_{j=0,n} a_j f_j \right) \omega.$$

4. $\Omega^{n-1}$ is a free $S$-module of rank $\binom{n+1}{2}$ generated by $\omega_{ij}$ for $0 \le i < j \le n$ where $\omega_{ij}$ is given by the same product as $\omega$ but omitting $dx_i$ and $dx_j$.
5. The image of the wedge product $df \wedge : \Omega^{n-1} \to \Omega^n$ can be identified up to a shift in degree to the submodule $KR(f)$. To see this, it is enough to use the formula

$$df \wedge \omega_{ij} = f_i \omega_j - f_j \omega_i.$$

In conclusion, it follows that one has

$$ER(f)_m = H^n(K^*(f))_{m+n} \tag{1.3}$$

for any $m \in \mathbb{N}$, where $K^*(f)$ is the Koszul complex of $f_0, \dots, f_n$ with the natural grading $|x_j| = |dx_j| = 1$ defined by

$$0 \to \Omega^0 \to \Omega^1 \to \ldots \to \Omega^{n+1} \to 0 \tag{1.4}$$

with all the arrows given by the wedge product by $df = f_0 dx_0 + f_1 dx_1 + \ldots + f_n dx_n$. In other words, one has

$$ER(f) = H^n(K^*(f))(-n). \tag{1.5}$$

Using these graded $S$-modules of Jacobian syzygies, we introduce two numerical invariants for the hypersurface $V : f = 0$ as follows. The integer

$$mdr(f) = \text{indeg}(AR(f)) = \min\{k \ : \ AR(f)_k \neq 0\} \tag{1.6}$$

is called the *minimal degree of a relation* for $f$, while the integer

$$mder(f) = \text{indeg}(ER(f)) = \min\{k \ : \ ER(f)_k \neq 0\} \tag{1.7}$$

is called the *minimal degree of an essential relation* for $f$. From the definition, it is clear that $mdr(f) \leq mder(f)$ with equality if $mdr(f) < d - 1$. Note also that $0 \leq mdr(f) \leq d - 1$ and $0 \leq mder(f) \leq n(d - 2)$, where the last inequality follows from [3, Corollary 11], see also Theorem 2.2 below. It is also clear that $mdr(f) = 0$ if and only if $V$ is a cone over a hypersurface in $\mathbb{P}^{n-1}$, case excluded in our discussion from now on.

Let $\alpha_V$ be the Arnold exponent of the hypersurface $V$, which is by definition the minimum of the Arnold exponents of the singular points of $V$, see [8]. Using Hodge theory, one can prove that

$$mder(f) \geq \alpha_V d - n, \tag{1.8}$$

under the *additional hypothesis that all the singularities of V are weighted homogeneous*, see [8]. This inequality is the best possible in general, as one can see by considering hypersurfaces with a lot of singularities, see [10]. However, for situations where the hypersurface $V$ has a small number of singularities this result is far from optimal, and in such cases one has the following inequality

$$mder(f) > n(d - 2) - \tau(V), \tag{1.9}$$

where $\tau(V)$, the Tjurina number of $V$, is the sum of the Tjurina numbers of all the singularities of $V$, see [7].

Jacobian syzygies and these two numerical invariants $mdr(f)$ and $mder(f)$ occur in a number of fundamental results due to A. du Plessis and C.T.C. Wall, see [18–21], some of which we recall briefly below. The first type of their results is devoted to the versality properties of projective hypersurfaces. These results are recalled in Sect. 2, where we explain that [18, Theorem 1.1], which is stated as Theorem 2.1 below, is essentially equivalent to the first part of [4, Theorem 1], which is stated as Theorem 2.2 below for the reader's convenience.

The second type of results by A. du Plessis and C.T.C. Wall are related to finding lower and upper bounds for the global Tjurina number $\tau(V)$. Their main result in this direction is [21, Theorem 5.3], which is stated as Theorem 3.1 below. We show that this result can be used to greatly strengthen two of our main results in [7], one on the stability of the reflexive sheaf $T\langle V\rangle$ of logarithmic vector fields along a surface $V$ in $\mathbb{P}^3$, and the other on the Torelli property in the sense of Dolgachev-Kapranov of the hypersurface $V$, see Theorems 3.4 and 3.7 below. Since the proofs of our results given in [7] are rather long and technical, we present here only the minor changes in these proofs, possible in view of du Plessis and Wall's result in Theorem 3.1, and leading to much stronger claims, as explained in Remarks 3.5 and 3.8.

## 2 Versality of Hypersurfaces with Isolated Singularities

Fix an integer $a \geq 0$, and call the hypersurface $V : f = 0$ in $\mathbb{P}^n$ *a-versal*, resp. *topologically a-versal*, if the singularities of $V$ can be simultaneously versally, resp. topologically versally, deformed by deforming the equation $f$, in the affine chart of $\mathbb{P}^n$ given by $\ell \neq 0$, with $\ell \in S_1$, containing all the singularities of $V$, by the addition of all monomials of degree $n(d-2)-1-a$. Otherwise, we say that $V$ is *(topologically) a-non-versal*. With the above notation, one has the following result proved by A. du Plessis, see [18, Theorem 1.1].

**Theorem 2.1** *For a positive integer $a \geq 0$, the hypersurface $V : f = 0$ is a-non-versal if and only if*

$$a \geq mder(f).$$

*Equivalently, for a positive integer $b \geq 0$, the hypersurface $V : f = 0$ is b-versal if and only if*

$$b < mder(f).$$

Let $\Sigma$ be the singular subscheme of $V$, defined by the Jacobian ideal of $f$ given by

$$J_f = (f_0, \ldots, f_n) \subset S.$$

Then, for $p$ a singular point of $V$, one has an isomorphism of Artinian $\mathbb{C}$-algebras

$$\mathcal{O}_{\Sigma,p} = T(g),$$

where $g = 0$ is a local equation for the germ $(V, p)$ and $T(g)$ is the Tjurina algebra of $g$, which is also the base space of the miniversal deformation of the isolated singularity $(V, p)$, see [4, Lemma 1]. More precisely, one has

$$T(g) = \frac{\mathbb{C}\{y_1, \ldots, y_n\}}{J_g + (g)}, \tag{2.1}$$

where $(y_1, \ldots, y_n)$ is a local coordinate system centered at $p$ and $J_g$ is the Jacobian ideal of $g$ in the local ring $\mathcal{O}_{\mathbb{P}^n,p} = \mathbb{C}\{y_1, \ldots, y_n\}$. Note that, for any integer $k$, one can consider the natural evaluation morphism

$$eval_k : S_k \to \oplus_{p \in \Sigma} \mathcal{O}_{\Sigma,p}, \ \ h \mapsto ([h/\ell^k]_p)_{p \in \Sigma},$$

computed in the chart $\ell \neq 0$. Alternatively, $eval_k$ is the morphism

$$H^0(\mathbb{P}^n, \mathcal{O}_{\mathbb{P}^n}(k)) \to H^0(\mathbb{P}^n, \mathcal{O}_\Sigma(k)) = H^0(\mathbb{P}^n, \mathcal{O}_\Sigma),$$

induced by the exact sequence

$$0 \to \mathcal{I}_\Sigma \to \mathcal{O}_{\mathbb{P}^n} \to \mathcal{O}_\Sigma \to 0, \tag{2.2}$$

where $\mathcal{I}_\Sigma$ is the ideal sheaf defining the singular subscheme $\Sigma$. We set

$$\mathrm{def}_k(\Sigma) = \dim(\mathrm{coker}\, eval_k) = \dim H^1(\mathbb{P}^n, \mathcal{I}_\Sigma(k)), \tag{2.3}$$

the *defect of* $\Sigma$ *with respect to homogeneous polynomials of degree* $k$. It follows that the hypersurface $V : f = 0$ is $a$-versal if and only if the corresponding evaluation morphism $eval_{n(d-2)-1-a}$ is surjective, i.e. the defect $\mathrm{def}_{n(d-2)-1-a}(\Sigma)$ vanishes. We see in this way that Theorem 2.1 is essentially equivalent to the first part of [4, Theorem 1], which we state now.

**Theorem 2.2** *With the above notation, one has*

$$\dim ER(f)_k = \mathrm{def}_{n(d-2)-1-k}(\Sigma)$$

*for* $0 \leq k \leq n(d-2) - 1$ *and* $\dim ER(f)_j = \tau(V)$ *for* $j \geq n(d-2)$.

The proofs of both Theorems 2.1 and 2.2 use the Cayley-Bacharach Theorem, as discussed for instance in [22].

*Example 2.3* If we take $a = n(d-2) - d - 1$, then the hypersurface $V : f = 0$ is $a$-versal if and only if the family of all hypersurfaces of degree $d$ in $\mathbb{P}^n$ versally deform all the singularities of $V$, a property called *T-condition* or *T-smoothness* in [28, 31, 32]. This property holds if and only if

$$n(d-2) - d - 1 < mder(f). \tag{2.4}$$

For instance, in the case of a plane curve, $n = 2$ and the condition becomes

$$d - 5 < mder(f).$$

The inequality (1.8) implies that this condition holds for any nodal plane curve $V$, i.e. a curve in $\mathbb{P}^2$ having only $A_1$-singularities, since in this case one has $\alpha_V = 1$. The inequality (1.9) implies that the condition (2.4) holds if the hypersurface $V$

satisfies $\tau(V) \leq d-1$. In fact, for $d \geq 5$, it is known that the condition (2.4) holds if $\tau(V) < 4(d-1)$, see [27, 31] for the case $n = 2$, and [20, 32] for the case $n \geq 2$.

One has also the following result, see [18, Theorem 2.1], which we recall for the completeness of our presentation, and to illustrate the concept of topological versality.

**Theorem 2.4** *With the above notation, we suppose that* $\dim ER(f)_a = 1$, *and* $\rho = (a_0, \ldots, a_n) \in ER(f)_a$ *is a non-zero element. If there is a non-simple singular point* $p \in V$ *such that* $\rho(p) \neq (0, \ldots, 0)$, *then* $V$ *is topologically a-versal.*

*Example 2.5* Let $n = 2$ and $V : f = x_0^d + x_1^{d-1}x_2$, with $d \geq 5$. Then $V$ has a non-simple singularity at $p = (0 : 0 : 1)$ and $\rho = (0, x_1, -(d-1)x_2) \in ER(f)_1$ does not vanish at $p$. It follows that $V$ is topologically 1-versal, i.e. by the addition of the monomials of degree $2d-6$ in $x, y, z$, the singularity at $p$ is topologically versally unfolded.

## 3 Bounds on the Global Tjurina Number, Stability and Torelli Properties

One has the following result, see [21, Theorem 5.3].

**Theorem 3.1** *With the above notation, we set* $r = mdr(f)$. *Then*

$$(d-r-1)(d-1)^{n-1} \leq \tau(V) \leq (d-1)^n - r(d-r-1)(d-1)^{n-2}.$$

For $n = 2$ this result was obtained in [19], and played a key role in the understanding of free curves. Indeed, when $n = 2$, the reduced curve $V$ is free if and only if

$$\tau(V) = (d-1)^2 - r(d-r-1),$$

i.e. the upper bound is attained, see [6, 23] for related results. When $n > 2$, a free hypersurface $V$ has non-isolated singularities, and so freeness must be related to other invariants, see for instance [5].

*Remark 3.2* The lower bound in Theorem 3.1 is attained for any pair $(d, r)$. Indeed, it is enough to find a degree $d$, reduced curve $C : f'(x_0, x_1, x_2) = 0$ such that $r = mdr(f')$ and

$$\tau(C) = (d-r-1)(d-1),$$

and then take $V : f = 0$, with

$$f(x_0, \ldots, x_n) = f'(x_0, x_1, x_2) + x_3^d + \cdots + x_n^d.$$

This formula for $f$ implies that $mdr(f) = mdr(f')$. The existence of curves $C$ as above is shown in [15, Example 4.5] and a complete characterization of them is given in [15, Theorem 3.5 (1)].

*Remark 3.3* The upper bound in Theorem 3.1 is attained for any pair $(d, r)$ with $2r < d$, since for such pairs $(d, r)$ the existence of free plane curves $C : f' = 0$ of degree $d$ and with $r = mdr(f')$ is shown in [11] and then one constructs the hypersurface $V$ as in Remark 3.2 above. Note that the free curves, which are characterized by the equality

$$\tau(C) = (d-1)^2 - r(d-r-1),$$

and the nearly free curves, which are characterized by the equality

$$\tau(C) = (d-1)^2 - r(d-r-1) - 1,$$

are closely related to the *rational cuspidal curves* in $\mathbb{P}^2$, see [2, 6, 12–14, 25, 26]. It is an interesting *open question* to improve the upper bound in Theorem 3.1 when $2r \geq d$. The better upper bound for such pairs is known when $n = 2$, see [19], and is given by the stronger inequality

$$\tau(C) \leq (d-1)^2 - r(d-r-1) - \binom{2r+2-d}{2}.$$

Moreover, it is conjectured and verified in many cases that this inequality is the best possible for any pair $(d, r)$ with $2r \leq d$ when $n = 2$, see [1, 16].

If we start with a degree $d$, reduced curve $C : f'(x_0, x_1, x_2) = 0$ in $\mathbb{P}^2$ such that $r = mdr(f') \geq d/2$ and take $V : f = 0$ the corresponding hypersurface in $\mathbb{P}^n$, with

$$f(x_0, \ldots, x_n) = f'(x_0, x_1, x_2) + x_3^d + \cdots + x_n^d,$$

then clearly $mdr(f) = mdr(f') \geq d/2$ and

$$\tau(V) \leq (d-1)^n - r(d-r-1)(d-1)^{n-2} - \binom{2r+2-d}{2}(d-1)^{n-2}. \quad (3.1)$$

However, this stronger inequality fails for hypersurfaces not constructed as suspensions of plane curves. To have an example, consider Cayley surface

$$V : f = xyz + xyw + xzw + yzw = 0$$

in $\mathbb{P}^3$ having four $A_1$-singularities. Then $d = 3$, $r = mdr(f) = 2 > d/2$ and $\tau(V) = 4$. The inequalities in Theorem 3.1 are in this case

$$0 \leq \tau(V) \leq 8,$$

while the bound given by the inequality (3.1) is 2, which is clearly not correct.

The exact sequence of coherent sheaves on $X = \mathbb{P}^n$ given by

$$0 \to T\langle V\rangle \to \mathcal{O}_X(1)^{n+1} \to \mathcal{I}_\Sigma(d) \to 0, \tag{3.2}$$

where the last non-zero morphism is induced by $(a_0, \ldots, a_n) \mapsto a_0 f_0 + \ldots + a_n f_n$ and $\mathcal{I}_\Sigma$ is, as above, the ideal sheaf defining the singular subscheme $\Sigma$, can be used to define the sheaf $T\langle V\rangle$ of logarithmic vector fields along $V$, see [24, 29, 30, 33]. This is a reflexive sheaf, in particular a locally free sheaf $T\langle V\rangle$ in the case $n = 2$. The above exact sequence clearly yields

$$AR(f)_m = H^0(X, T\langle V\rangle(m-1)), \tag{3.3}$$

for any integer $m$. This equality can be used to show the reflexive sheaf $T\langle V\rangle$ is stable in many cases. This was done already in the case $n = 2$ in [9] and in the case $n = 3$ in [7]. The next result is a substantial improvement of [7, Theorem 1.3].

**Theorem 3.4** *Assume that the surface $V : f = 0$ in $\mathbb{P}^3$ of degree $d = 3d' + \epsilon \geq 2$, with $d' \in \mathbb{N}$ and $\epsilon \in \{1, 2, 3\}$, has only isolated singularities and satisfies*

$$\tau(V) < (d - d' - 1)(d - 1)^2.$$

*Then $F = T\langle V\rangle(d' - 1)$ is a normalized stable rank 3 reflexive sheaf on $\mathbb{P}^3$, with first Chern class $c_1(F) = 1 - \epsilon \in \{0, -1, -2\}$. This reflexive sheaf is locally free if and only if $V$ is smooth.*

***Proof*** Checking the proof of [7, Theorem 1.3], we see that the only point to be explained is the vanishing of $H^0(\mathbb{P}^3, F)$. Using the formula (3.3), it follows that we have to show that $AR(f)_{d'} = 0$ or, equivalently, that $r = mdr(f) > d'$. Using Theorem 3.1, we see that $r \leq d'$ implies $\tau(V) \geq (d - d' - 1)(d - 1)^2$. This ends the proof of the vanishing $AR(f)_{d'} = 0$. □

*Remark 3.5* The hypothesis in [7, Theorem 1.3] is

$$\tau(V) \leq 8d' + 3(\epsilon - 2),$$

hence the upper bound for $\tau(V)$ is, as a function of $d$, equivalent to $\frac{8}{3}d$. On the other hand, the upper bound for $\tau(V)$ in Theorem 3.4 is, as a function of $d$, equivalent to $\frac{2}{3}d^3$, hence it has a much faster growth when $d$ increases. For example, when $d = 12$ we have with the above notation $d' = 3 = \epsilon$, and hence the bound from [7, Theorem 1.3] is

$$\tau(V) \leq 24 + 3 = 27.$$

Our new bound, given by Theorem 3.4 is

$$\tau(V) < 8 \cdot 11^2 = 968,$$

which makes the result applicable in much more cases.

Recall the following notion.

**Definition 3.6** A reduced hypersurface $V \subset \mathbb{P}^n$ is called *DK-Torelli* (where DK stands for Dolgachev-Kapranov) if the hypersurface $V$ can be reconstructed as a subset of $\mathbb{P}^n$ from the sheaf $T\langle V\rangle$.

For a discussion of this notion and various examples and results we refer to the papers [7, 9, 17, 33]. The following result uses Theorem 3.1 to improve [7, Theorem 1.5] when $n \geq 3$. More precisely we prove the following.

**Theorem 3.7** *Let $V : f = 0$ be a degree $d \geq 4$ hypersurface in $\mathbb{P}^n$, having only isolated singularities. Set $m = \lfloor \frac{d-2}{2} \rfloor$ and assume*

$$\tau(V) < \binom{m+n-1}{n-1}.$$

*Then one of the following holds.*

1. *$V$ is DK-Torelli;*
2. *$V$ is of Sebastiani-Thom type, i.e. in some linear coordinate system $(x_0, \ldots, x_n)$ on $\mathbb{P}^n$, the defining polynomial $f$ for $V$ is written as a sum $f = f' + f''$, with $f'$ (resp. $f''$) a homogeneous polynomial of degree $d$ involving only $x_0, \ldots, x_r$ (resp. $x_{r+1}, \ldots, x_n$) for some integer $r$ satisfying $0 \leq r < n$.*

***Proof*** We indicate only the changes to be made in the proof of [7, Theorem 1.5]. Let $I$ be the saturation of the ideal $J_f$ with respect to the maximal ideal $(x_0, \ldots, x_n)$. The first step in the proof is to show the existence of two polynomials $h_1, h_2 \in I_m$ having no common factor. As explained in the proof of [7, Theorem 1.5], to get this it is enough to assume

$$\tau(V) < \binom{m+n-1}{n-1},$$

which is exactly our assumption now. The second step is to show that $r = mdr(f) > d-2 \geq 2m$. If we assume $r \leq d-2$, it follows from Theorem 3.1 that

$$\tau(V) \geq (d-r-1)(d-1)^{n-1} \geq (d-1)^{n-1}.$$

But this is impossible, since

$$\binom{m+n-1}{n-1} < (d-1)^{n-1}.$$

To see this, it is enough to check that

$$\frac{m+k}{k} < d-1$$

for $k = 1, \ldots, n-1$ which is obvious using the definition of $m$ and the fact that $d \geq 4$.

Using Wang's result in [34, Theorem 1.1], to complete our proof it is enough to show that $V$ cannot have a singular point $p$ of multiplicity $d-1$. Assume such a point $p$ exists and let $g = 0$ be a local equation for the singularity $(V, p)$. Since all the elements in $J_g + (g)$ have order at least $d-2$ and since

$$m \leq d-3,$$

the definition of $T(g)$ in (2.1) shows that the monomials in $y_j$'s of degree $m$ are linearly independent in $T(g)$. It follows that

$$\tau(V) \geq \tau(V, p) = \dim T(g) \geq \binom{m+n-1}{n-1},$$

a contradiction. □

*Remark 3.8* The hypothesis in [7, Theorem 1.5] is

$$\tau(V) \leq \frac{(n-1)(d-4)}{2} + 1,$$

which is the same as the hypothesis above for $n = 2$, but much more restrictive for $n \geq 3$. For instance, for $n = 3$ and $d = 2d'$ even, the assumption in Theorem 3.7 is

$$\tau(V) < \binom{d'+1}{2},$$

while the assumption in [7, Theorem 1.5] is $\tau(V) \leq 2d' - 3$. As an example, for $n = 3$ and $d = 12$, the hypothesis in Theorem 3.7 is

$$\tau(V) \leq \binom{7}{2} = 21,$$

while the hypothesis in [7, Theorem 1.5] is $\tau(V) \leq 9$, which is much more restrictive.

**Acknowledgments** This work has been partially supported by the French government, through the UCA$^{\mathrm{JEDI}}$ Investments in the Future project managed by the National Research Agency (ANR) with the reference number ANR-15-IDEX-01 and by the Romanian Ministry of Research and Innovation, CNCS—UEFISCDI, grant PN-III-P4-ID-PCE-2016-0030, within PNCDI III.

## References


1. Abe, T., Dimca, A., Sticlaru, G.: Addition-deletion results for the minimal degree of logarithmic derivations of arrangements. J. Algebraic Combin. (2019). https://doi.org/10.1007/s10801-020-00986-9
2. Bodnár, J., Némethi, A.: Lattice cohomology and rational cuspidal curves. Math. Res. Lett. **23**, 339–375 (2016)
3. Choudary, A.D.R., Dimca, A.: Koszul complexes and hypersurface singularities. Proc. Am. Math. Soc. **121**, 1009–1016 (1994)
4. Dimca, A.: Syzygies of Jacobian ideals and defects of linear systems. Bull. Math. Soc. Sci. Math. Roumanie Tome **56(104)**, 191–203 (2013)
5. Dimca, A.: Freeness versus maximal degree of the singular subscheme for surfaces in $\mathbb{P}^3$. Geom. Dedicata **183**, 101–112 (2016)
6. Dimca, A.: Freeness versus maximal global Tjurina number for plane curves. Math. Proc. Camb. Philos. Soc. **163**, 161–172 (2017)
7. Dimca, A.: Jacobian syzygies, stable reflexive sheaves, and Torelli properties for projective hypersurfaces with isolated singularities. Algebr. Geom. **4**, 290–303 (2017)
8. Dimca, A., Saito, M.: Generalization of theorems of Griffiths and Steenbrink to hypersurfaces with ordinary double points. Bull. Math. Soc. Sci. Math. Roum. **60(108)**, 351–371 (2017)
9. Dimca, A., Sernesi, E.: Syzygies and logarithmic vector fields along plane curves. J. Ecole Polytech. Math. **1**, 247–267 (2014)
10. Dimca, A., Sticlaru, G.: On the syzygies and Alexander polynomials of nodal hypersurfaces. Math. Nachr. **285**, 2120–2128 (2012)
11. Dimca, A., Sticlaru, G.: On the exponents of free and nearly free projective plane curves. Rev. Mat. Complut. **30**, 259–268 (2017)
12. Dimca, A., Sticlaru, G.: Free divisors and rational cuspidal plane curves. Math. Res. Lett. **24**, 1023–1042 (2017)
13. Dimca, A., Sticlaru, G.: Free and nearly free curves vs. rational cuspidal plane curves. Publ. RIMS Kyoto Univ. **54**, 163–179 (2018)
14. Dimca, A., Sticlaru, G.: On the freeness of rational cuspidal plane curves. Moscow Math. J. **18**, 659–666 (2018)
15. Dimca, A., Sticlaru, G.: Plane curves with three syzygies, minimal Tjurina curves, and nearly cuspidal curves. Geom. Dedicata. **207**, 29–49 (2020). https://doi.org/10.1007/s10711-019-00485-7
16. Dimca, A., Sticlaru, G.: Plane curves with maximal global Tjurina numbers. arXiv: 1901.05915
17. Dolgachev, I., Kapranov, M.: Arrangements of hyperplanes and vector bundles on $\mathbb{P}^n$. Duke Math. J. **71**, 633–664 (1993)
18. du Plessis, A.A.: Versality properties of projective hypersurfaces. In: Real and Complex Singularities, Trends in Mathematics, pp. 289–298. Birkhäuser, Basel (2006)
19. du Plessis, A.A., Wall, C.T.C.: Application of the theory of the discriminant to highly singular plane curves. Math. Proc. Camb. Philos. Soc. **126**, 256–266 (1999)
20. du Plessis, A.A., Wall, C.T.C.: Singular hypersurfaces, versality and Gorenstein algebras. J. Algebr. Geom. **9**, 309–322 (2000)
21. du Plessis, A.A., Wall, C.T.C.: Discriminants, Vector Fields and Singular Hypersurfaces. New Developments in Singularity Theory (Cambridge, 2000), pp. 351–377, NATO Sci. Ser. II Math. Phys. Chem., vol. 21. Kluwer, Dordrecht (2001)
22. Eisenbud, D., Green, M., Harris, J.: Cayley Bacharach theorems and conjectures. Bull. Am. Math. Soc. **33**, 295–324 (1996)
23. Ellia, Ph.: Quasi complete intersections and global Tjurina number of plane curves. J. Pure Appl. Algebra **224**, 423–431 (2020)
24. Faenzi, D., Vallès, J.: Logarithmic bundles and line arrangements, an approach via the standard construction. J. Lond. Math. Soc. **90**, 675–694 (2014)

25. Fernandez de Bobadilla, J., Luengo, I., Melle-Hernandez, A., Némethi, A.: On rational cuspidal projective plane curves. Proc. Lond. Math. Soc. (3) **92**, 99–138 (2006)
26. Fernandez de Bobadilla, J., Luengo, I., Melle-Hernandez, A., Némethi, A.: Classification of rational unicuspidal projective curves whose singularities have one Puiseux pair. In: Real and Complex Singularities Sao Carlos 2004, Trends in Mathematics, pp. 31–45. Birkhauser, Basel (2006)
27. Greuel, G.-M., Lossen, C.: Equianalytic and equisingular families of curves on surfaces. Manuscripta Math. **91**, 323–342 (1996)
28. Greuel, G.-M., Lossen, C., Shustin, E.: Singular Algebraic Curves, with an Appendix by O. Viro. Springer, Berlin (2018)
29. Marchesi, S., Vallès, J.: Nearly free curves and arrangements: a vector bundle point of view. Math. Proc. Camb. Philos. Soc. (2019). https://doi.org/10.1017/S0305004119000318
30. Sernesi, E.: The local cohomology of the Jacobian ring. Doc. Math. **19**, 541–565 (2014)
31. Shustin, E.: Versal deformations in the space of planar curves of fixed degree. Funct. Anal. Appl. **21**, 82–84 (1987)
32. Shustin, E., Tyomkin, I.: Versal deformation of algebraic hypersurfaces with isolated singularities. Math. Ann. **313**, 297–314 (1999)
33. Ueda, K., Yoshinaga, M.: Logarithmic vector fields along smooth divisors in projective spaces. Hokkaido Math. J. **38**, 409–415 (2009)
34. Wang, Z.: On homogeneous polynomials determined by their Jacobian ideal. Manuscripta Math. **146**, 559–574 (2015)

# On Ideal Filtrations for Newton Nondegenerate Surface Singularities

**Baldur Sigurðsson**

*Dedicated to András Némethi on his 60th birthday*

**Abstract** We compare three naturally occurring multi-indexed filtrations of ideals on the local ring of a Newton nondegenerate hypersurface surface singularity with rational homology sphere, which in many cases are all distinct. These are the divisorial, the order, and the image filtrations. These filtrations are indexed by the lattice associated with a toric partial resolution of the singularity, or equivalently, the free Abelian group generated by the compact facets of the Newton polyhedron.

We prove that there exists a top dimensional cone contained in the Lipman cone having the property that the three ideals indexed by order vectors from this cone coincide. As a corollary, if a periodic constant can be associated with the Hilbert series associated with these filtrations on the Lipman cone, then they coincide.

In some cases, the Poincaré series associated with one of these filtrations has been shown to coincide with a zeta function associated with the topological type of the singularity. In the end of the article, we show that this is the case for all three filtrations considered in the case of a Newton nondegenerate suspension singularity. As a corollary, in this case, the zeta function provides a direct method of determining the Newton diagram from the link.

**Keywords** Newton nondegenerate hypersurface singularity · Divisorial filtration · Order filtration · Image filtration · Poincaré series · Newton nondegenerate suspension singularity

B. Sigurðsson (✉)
Universidad Nacional Autónoma de México, Instituto de Matemáticas, Unidad Cuernavaca, Mexico City, Mexico
e-mail: baldur@im.unam.mx



J. Fernández de Bobadilla et al. (eds.), *Singularities and Their Interaction with Geometry and Low Dimensional Topology*, Trends in Mathematics,
https://doi.org/10.1007/978-3-030-61958-9_2

# 1 Introduction

Let $(X, 0) \subset (\mathbb{C}^3, 0)$ be a hypersurface singularity given as the vanishing set of a function $f \in \mathcal{O}_{\mathbb{C}^3,0}$ with Newton nondegenerate principal part. Assume further that the link is a rational homology sphere. Let $\bar{G}$ be the dual graph to the compact Newton boundary of $f$. That is, the vertex set $\mathcal{N}$ indexes the compact facets of $\Gamma_+(f)$ so that for $n \in \mathcal{N}$ we have a face $F_n = F_n(f)$, and two vertices are joined by an edge if and only if the corresponding faces intersect in a segment. There is a corresponding toric modification of $\mathbb{C}^3$ which yields a $V$ resolution $\pi : \tilde{X} \to X$. To each $n \in \mathcal{N}$ there corresponds an irreducible component of the exceptional $\pi^{-1}(0)$, say $E_n$. This correspondence is bijective.

For each $n \in \mathcal{N}$ we denote by $\mathrm{div}_n$ the valuation on $\mathcal{O}_{X,0}$ associated with the divisor $E_n$. Furthermore, the positive primitive normal vector to the face $F_n$ provides a valuation $\hat{\mathrm{wt}}_n$ on $\mathcal{O}_{\mathbb{C}^3,0}$ which induces the order function $\mathrm{wt}_n$ on $\mathcal{O}_{X,0}$ via

$$\mathrm{wt}_n(g) = \max\left\{\hat{\mathrm{wt}}_n(h) \,\middle|\, h|_X = g\right\}.$$

For $g \in \mathcal{O}_{X,0}$ we set $\mathrm{div}\, g = (\mathrm{div}_n\, g)_{n\in\mathcal{N}}$ and $\mathrm{wt}\, g = (\mathrm{wt}_n\, g)_{n\in\mathcal{N}}$. For $k \in \mathbb{Z}^{\mathcal{N}}$ we define

$$\mathcal{F}(k) = \left\{g \in \mathcal{O}_{X,o} \middle| \mathrm{div}\, g \geq k\right\}, \quad \mathcal{G}(k) = \left\{g \in \mathcal{O}_{X,o} \middle| \mathrm{wt}\, g \geq k\right\}.$$

Similarly, let $\hat{\mathcal{G}}$ be the divisorial filtration on $\mathcal{O}_{\mathbb{C}^3,0}$ associated with the valuations $\hat{\mathrm{wt}}_n$, $n \in \mathcal{N}$. We define $\mathcal{I}(k)$ as the image of $\hat{\mathcal{G}}(k)$ under the natural projection $\mathcal{O}_{\mathbb{C}^3,0} \to \mathcal{O}_{X,0}$.

It follows from these definition that for all $k \in \mathbb{Z}^{\mathcal{N}}$ we have inclusions

$$\mathcal{I}(k) \subset \mathcal{G}(k) \subset \mathcal{F}(k). \tag{1}$$

In general, we may not expect equality here. In [5], Lemahieu shows that the $\mathcal{I}$ and $\mathcal{G}$ coincide if and only if the Newton diagram of $f$ is bi-stellar, i.e. every pair of compact facets of $\Gamma_+(f)$ shares a point. In Example 7.6 of [8], Némethi provides an example of a Newton nondegenerate singularity whose diagram contains only two compact faces (in particular, it is bi-stellar) for which the inclusion $\mathcal{G} \subset \mathcal{F}$ is shown to be proper.

The following theorem is proved in Sect. 6.

**Theorem 1.1** *Let $(X, 0)$ be a Newton nondegenerate hypersurface singularity in $(\mathbb{C}^3, 0)$ with a rational homology sphere link. Then there exists an $|\mathcal{N}|$ dimensional polyhedral cone $C \subset \mathcal{S}_{\mathbb{R}}$ (see Definitions 5.1 and 5.2 for $C$ and $\mathcal{S}_{\mathbb{R}}$) satisfying*

$$\forall k \in C \cap \mathbb{Z}^{\mathcal{N}} : \mathcal{F}(k) = \mathcal{G}(k) = \mathcal{I}(k).$$

In Sect. 7 we define the zeta function and prove the following.

**Theorem 1.2** *If* $(X, 0)$ *is a Newton nondegenerate suspension singularity with rational homology sphere link, then* $\mathcal{I}, \mathcal{G}, \mathcal{F}$ *all coincide. Furthermore, the associated Poincaré series coincides with the reduced zeta function* $Z_0^{\mathcal{N}}(t)$ *with respect to nodes (see Definition 7.8), which is given by the formula*

$$\frac{1 - t^{\hat{\mathrm{wt}}\, f}}{(1 - t^{\hat{\mathrm{wt}}\, x})(1 - t^{\hat{\mathrm{wt}}\, y})(1 - t^{\hat{\mathrm{wt}}\, z})}. \tag{2}$$

## 2 Associated Power Series and the Search for an Equation

For a better understanding of these filtrations, the associated *Hilbert* and *Poincaré* series are introduced:

$$H^{\mathcal{F}}(t) = \sum_{k \in \mathbb{Z}^{\mathcal{N}}} h_k^{\mathcal{F}} t^k, \quad P^{\mathcal{F}}(t) = \sum_{k \in \mathbb{Z}^{\mathcal{N}}} p_k^{\mathcal{F}} t^k = -H^{\mathcal{F}}(t) \prod_{n \in \mathcal{N}} (1 - t_n^{-1}),$$

where $h_k^{\mathcal{F}} = \dim_{\mathbb{C}} \mathcal{O}_{X,0}/\mathcal{F}(k)$. Similar definitions are made for the other filtrations.

These series provide very strong numerical invariants of the analytic structure of the singularity. Two leading questions in the theory of surface singularities are, on one hand, whether numerical analytic invariants such as these can be characterized by the topology of $(X, 0)$, and on the other, whether numerical invariants can be used to construct variables and equations realizing singularities with a given topology.

The divisorial filtration $\mathcal{F}$ is intrinsic to the singularity $(X, 0)$, and therefore one may hope for it to have the most direct relation to the link, whether or not the singularity $(X, 0)$ is a hypersurface. Indeed, in [8], Némethi provides a topological invariant, the zeta function, which coincides with $P^{\mathcal{F}}$ in many cases, e.g. for rational singularities and minimally elliptic singularities whose minimal resolution is good. These are examples of classes of singularities whose intrinsic analytic structure has restrictions. The main identity in [8] is not true for arbitrary singularities, but has been proved for singularities of splice-quotient type [9].

On the other hand, the filtrations $\mathcal{I}$ and $\mathcal{G}$ are given in terms of the embedding of the singularity $(X, 0) \subset (\mathbb{C}^3, 0)$. It is not clear how to relate the topology of $(X, 0)$, or its embedded type to the Hilbert or Poincaré series associated with these filtration. On the other hand, as we shall see, there are cases when the knowledge of the Poincaré series can be used to rebuild the singularity, or a similar one.

There are no relations between the monomials of the ring $\mathcal{O}_{\mathbb{C}^3,0}$, and the filtration $\hat{\mathcal{G}}$ is given by a grading of these monomials. As a result, one computes easily (see also Proposition 1 of [2]):

$$P^{\hat{\mathcal{G}}}(t) = \frac{1}{(1 - t^{\hat{\mathrm{wt}}\, x})(1 - t^{\hat{\mathrm{wt}}\, y})(1 - t^{\hat{\mathrm{wt}}\, z})}.$$

By a result of Lemahieu [5], this gives

$$P^{\mathcal{I}}(t) = (1 - t^{\hat{\mathrm{wt}}\, f}) P^{\hat{\mathcal{G}}} = \frac{1 - t^{\hat{\mathrm{wt}}\, f}}{(1 - t^{\hat{\mathrm{wt}}\, x})(1 - t^{\hat{\mathrm{wt}}\, y})(1 - t^{\hat{\mathrm{wt}}\, z})}.$$

If we assume that $f$ has a *convenient* Newton diagram (meaning in our case that $f(x, y, z)$ contains monomials of the form $x^a$, $y^b$, $z^c$ with nonzero coefficients), then the arguments of section 5 of [5] show that this series in fact determines the Newton polyhedron (it is also determined by it). In particular, if this series can be computed using only the topological type of $(X, 0)$, then one obtains a method of determining from only the topology of $(X, 0)$ an equation for a singularity with that topological type. We shall see in Sect. 7 that this program actually runs in the case of suspension singularities with rational homology sphere.

In fact, in [1], Braun and Némethi found, using totally different methods, that when the link of a Newton nondegenerate hypersurface singularity is a rational homology sphere, then the link determines the Newton diagram, up to permutation of the coordinates. Nonetheless, the above route identifies a more conceptual way of finding an equation determining a given topology.

## 3 Newton Nondegeneracy

In this section we define the Newton polyhedron and its normal fan. We do not subdivide the normal fan to obtain a smooth variety. As a result, we obtain a partial resolution of $(X, 0)$ which has at most cyclic quotient singularities. This construction is described in details in [11].

Let $f$ be a convergent power series in three variables given as $f(x) = \sum_{u \in \mathbb{N}^3} a_u x^u$. We define the *support* of $f$ as

$$\mathrm{supp}(f) = \left\{ u \in \mathbb{N}^3 \,\middle|\, a_u \neq 0 \right\}$$

and the *Newton polyhedron* of $f$ as

$$\Gamma_+(f) = \mathrm{conv}(\mathrm{supp}(f) + \mathbb{R}^3_{\geq 0}).$$

A *facet* of $\Gamma_+(f)$ is a face of dimension 2. We index the compact facets of $\Gamma_+(f)$ by a set $\mathcal{N}$, which we take as the vertex set of a graph $\bar{G}$ as in the introduction. We define the graph $\bar{G}^*$ similarly, but we allow in this case noncompact facets as well. We denote the vertex set of $\bar{G}^*$ by $\mathcal{N}^*$.

To a vertex $n \in \mathcal{N}^*$, there corresponds a facet $F_n \subset \Gamma_+(f)$. To each such $n$ there corresponds a unique primitive integral linear functional $\ell_n : \mathbb{R}^n \to \mathbb{R}$ having $F_n$ as its minimal set in $\Gamma_+(f)$.

We identify the set of integral linear functionals $\ell : \mathbb{Z}^3 \to \mathbb{Z}$ taking nonnegative values on $\mathbb{N}^3$ with $\mathbb{N}^3$ via the standard intersection product. Thus, for each $n \in \mathcal{N}$, the functional $\ell_n$ corresponds to the primitive normal vector to $F_n$ pointing into $\Gamma_+(f)$. For any face $F \subset \Gamma_+(f)$ (of any dimension) denote by

$$f_F = \sum \left\{ a_u x^u \,\middle|\, u \in F \cap \operatorname{supp}(f) \right\}.$$

**Definition 3.1** The function $f$ is *Newton nondegenerate* if for any compact face $F \subset \Gamma_+(f)$, the affine scheme

$$\left\{ x \in (\mathbb{C}^*)^3 \,\middle|\, f_F(x) = 0 \right\}$$

is smooth.

**Definition 3.2** The *normal fan*, denoted by $\Delta_f$ of the polyhedron $\Gamma_+(f)$ subdivides the positive octant $\mathbb{R}_{\geq 0}$ as follows.

- The one dimensional cones are generated by $\ell_n$ for $n \in \mathcal{N}^*$.
- A two dimensional cone in the normal fan is generated by two vectors $\ell_n$ and $\ell_{n'}$ where $n, n'$ are adjacent in $\bar{G}^*$. Equivalently, for any segment $S = F_n \cap F'_{n'}$, with $\dim S = 1$, there is a cone consisting of those functionals whose minimal value on $\Gamma_+(f)$ is taken on all of $S$.
- The above construction splits the positive octant into chambers, whose closures are the three dimensional cones in the normal fan. Equivalently, to each vertex $u \in \Gamma_+(f)$, there is a three dimensional cone in the normal fan consisting of those linear functions whose minimum on $\Gamma_+(f)$ is realized at the point $u$.

Denote by $Y_f$ the toric variety associated with $\Delta_f$. Then we have a canonical morphism $Y_f \to \mathbb{C}^3$. Denote by $\bar{X} \subset Y_f$ the strict transform of $X$. Denote by $O_n$ the orbit in $Y_f$ corresponding to the cone generated by $\ell_n$, and by $E_n$ the closure of $O_n \cap \bar{X}$.

## 4 The Intersection Lattice

If $f$ is Newton nondegenerate, then the strict transform $\bar{X}$ has transverse intersections with all orbits in $Y_f$, meaning that, if $O$ is an orbit, then the scheme theoretic intersection $\bar{X} \cap O$ is smooth. Furthermore, the divisors $E_n$ are irreducible [11].

We will identify the lattice $\bar{L} = \mathbb{Z}^{\mathcal{N}}$ with the set of divisors on $\bar{X}$ supported on the exceptional divisor, that is, the Abelian group freely generated by the irreducible divisors $E_n$ for $n \in \mathcal{N}$. An intersection product is obtained on this lattice as follows. Take a resolution $\phi : \tilde{X} \to \bar{X}$ which is an isomorphism outside the singular set $\bar{X}_{\text{sing}}$ (note that $\bar{X}$ has isolated singularities). In particular, there is a well defined intersection theory on $\tilde{X}$. For any curve $C \subset \bar{X}$, the pullback $\phi^* E$ is defined as

$\tilde{C} + C_{\text{exc}}$, where $\tilde{C}$ is the strict transform of $C$, and $C_{\text{exc}}$ is the unique rational divisor supported on $\phi^{-1}(\bar{X}_{\text{sing}})$, satisfying $(E, \tilde{C} + C_{\text{exc}}) = 0$ for any divisor $E$ supported on $\phi^{-1}(\bar{X}_{\text{sing}})$. This equation has a unique solution for $C_{\text{exc}}$ since the lattice of divisors supported on the exceptional divisor of the resolution $\phi$ is negative definite, in particular, nonsingular. We then set $(C, C') = (\phi^* C, \phi^* C')$. Note that for integral cycles $C, C' \in \bar{L}$, the intersection number $(C, C')$ is rational, but not necessarily integral.

**Definition 4.1** We refer to $\bar{L}$ with the intersection form defined above as the *intersection lattice*. Elements of $\bar{L}$, or $\bar{L}_{\mathbb{R}} = \bar{L} \otimes \mathbb{R}$ are referred to as *cycles*. Let $n \in \mathcal{N}$ and $n' \in \mathcal{N}^*$. We set $e_n = E_n^2 = (E_n, E_n)$. Furthermore

- Denote by $t_{n,n'}$ the length of the segment $F_n \cap F_{n'}$, that is, the number of relative interior integral points on this segment. In particular, $t_{n,n'} = 0$ if and only if $n, n'$ are not adjacent.
- Denote by $\alpha_{n,n'}$ the index of the lattice generated by $\ell_n$ and $\ell_{n'}$ in its saturation in $\text{Hom}(\bar{L}, \mathbb{Z})$.

**Proposition 4.2** *The intersection lattice is negative definite. In particular, we have* $e_n < 0$. *Let* $n, n' \in \mathcal{N}$ *be adjacent in* $\bar{G}$. *Then* $(E_n, E_{n'}) = t_{n,n'}/\alpha_{n,n'}$. *Furthermore, for any* $n \in \mathcal{N}$, *we have*

$$e_n \ell_n + \sum_{n'} \frac{t_{n,n'}}{\alpha_{n,n'}} \ell_{n'} = 0.$$

***Proof*** The intersection lattice can be seen as a subspace of the intersection lattice associated with a resolution of $(X, 0)$, which is negative definite, see e.g. [7]. The rest follows from [11], see also [1]. □

## 5 Cycles, Newton Diagrams and the Cone

In this section we define the cone $C$ which appears in Theorem 1.1. This requires some analysis of the geometry of Newton diagrams associated with arbitrary cycles. Lemma 5.3 shows that $C$ has the right properties, that is, it is a top dimensional rational cone contained in the Lipman cone. Lemma 5.5 is a workhorse used in the proof of Theorem 1.1.

**Definition 5.1** The *Lipman cone* $\mathcal{S}_{\mathbb{R}}$ is the set of vectors $Z \in \bar{L}_{\mathbb{R}}$ satisfying $(Z, E) \leq 0$ for any effective cycle $E$.

It is well known that the Lipman cone is an $|\mathcal{N}|$-dimensional simplicial cone generated by elements with all coordinates positive.

We associate to a cycle $Z \in \bar{L}_{\mathbb{R}}$ the *Newton polyhedron*

$$\Gamma_+(Z) = \left\{ u \in \mathbb{R}^3_{\geq 0} \middle| \forall n \in \mathcal{N},\ \ell_n(u) \geq m_n(Z) \right\}$$

where the $m_n$ are defined by $Z = \sum_{n \in \mathcal{N}} m_n(Z) E_n$. For a subgraph $A$ of $\bar{G}$ (or a subset of $\mathcal{N}$) let $\mathcal{N}_A$ be the set of vertices either in $A$ or connected to a vertex in $A$. For a cycle $Z$ let

$$\Gamma_+^A(Z) = \left\{ u \in \mathbb{R}^3_{\geq 0} \middle| \forall n \in \mathcal{N}_A,\ \ell_n(u) \geq m_n(Z) \right\}$$

and for $a \in A$, denote by $F_a^A(Z)$ the corresponding face of this polyhedron, given by

$$F_a^A(Z) = \left\{ u \in \Gamma_+^A(Z) \,\middle|\, \ell_a(u) = m_n(Z) \right\}.$$

Note that we may have $F_a^A(Z) = \emptyset$.

**Definition 5.2** Let $C$ be the set of divisors $Z \in \bar{L}$ satisfying

- $\emptyset \neq F_n^{\{n\}}(Z) = F_n(Z)$ for all $n \in \mathcal{N}$.
- If $n, n' \in \mathcal{N}$ are adjacent in $\bar{G}$ and $\rho(F_n(f) \cap F_{n'}(f)) + u \subset F_n(Z) \cap F_{n'}(Z)$ for some $\rho \geq 0$ and $u \in \mathbb{R}^3$ then $\rho F_n(f) + u \subset F_n(Z)$.

**Lemma 5.3** *$C$ is a top dimensional polyhedral cone contained in the Lipman cone $\mathcal{S}_{\mathbb{R}}$.*

***Proof*** The definition of $C$ is equivalent to a finite number of rational inequalities, and so the set $C$ is a rational polyhedron. Furthermore, assume that $\lambda \in \mathbb{R}_{\geq 0}$ and $Z, Z' \in C$. Then $F_n^A(\lambda Z) = \lambda F_n^A(Z)$ for any $A \subset \mathcal{N}$, which shows $\lambda Z \in C$. Furthermore, $F_n^A(Z + Z') = F_n^A(Z) + F_n^A(Z')$. Thus, if $n, n' \in \mathcal{N}$ are adjacent in $\bar{G}$, and

$$\rho > 0, \quad u \in \mathbb{R}^3, \quad \rho(F_n(f) \cap F_{n'}(f)) + u \subset F_n(Z + Z') \cap F_{n'}(Z + Z'),$$

then there are $\rho_1, \rho_2 > 0, u_1, u_2 \in \mathbb{R}^3$ so that

$$\rho_1(F_n(f) \cap F_{n'}(f)) + u_1 \subset F_n(Z) \cap F_{n'}(Z),$$
$$\rho_2(F_n(f) \cap F_{n'}(f)) + u_2 \subset F_n(Z') \cap F_{n'}(Z'),$$

and we get

$$\rho F_n(f) + u = (\rho_1 F_n(f) + u_1) + (\rho_2 F_n(f) + u_2) \subset F_n(Z) + F_{n'}(Z) = F_n(Z + Z').$$

As a result, we find $Z, Z' \in C$, and so $C$ is a cone.

Next, we prove $C \subset \mathcal{S}_{\mathbb{R}}$. Let $n \in \mathcal{N}$ and choose an $u \in F_n^{\{n\}}$, which is nonempty by assumption. We find

$$(E_n, Z) = e_n m_n(Z) + \sum_{n' \in \mathcal{N}_n} \frac{t_{n,n'} m_{n'}(Z)}{\alpha_{n,n'}} \leq e_n \ell_n(u) + \sum_{n' \in \mathcal{N}_n} \frac{t_{n,n'} \ell_{n'}(u)}{\alpha_{n,n'}} = 0.$$

Finally, we prove that $C$ has dimension $|\mathcal{N}|$. We will use the terminology introduced in [1], in particular, central faces and edges, arms and hands. Let $n_0 \in \mathcal{N}$ be a vertex so that $F_{n_0}(f)$ intersects all the coordinate planes. Then the complement $\mathcal{N} \setminus n_0$ is a disjoint union of parts of arms. Let the vertices of the $k$-th partial arm have vertices $n_{k,j}$ in such a way that $n_{k,1}$ is adjacent to $n_0$, and for $j \geq 2$, $n_{k,j}$ is adjacent to $n_{k,j-1}$. We also set $n_{k,0} = n_0$ for any $k$.

Define $Z \in \bar{L}_{\mathbb{R}}$ recursively as follows. Start by choosing $\varepsilon > 0$ very small and set $m_{n_0}(Z) = \hat{\mathrm{wt}}_{n_0} f$ and $m_{n_{k,1}}(Z) = \hat{\mathrm{wt}}_{n_{k,1}}(f) - \varepsilon$. Note that at this point we have a well defined facet

$$F_{n_0}^{\{n_0\}}(Z) = \left\{ u \in \mathbb{R}^3_{\geq 0} \,\middle|\, \ell_{n_0} = m_{n_0}(Z), \quad \forall k : \ell_{n_{k,1}} \geq m_{n_{k,1}}(Z) \right\}$$

and it follows from this construction that this face intersects each coordinate hyperplane in a segment of positive length.

Next, assume that we have defined $m_{n_{k,j}}$ for $0 < j \leq j_0$ for some $j_0 > 0$. In particular, the facet $F_{n_{k,j_0-1}}^{\{n_{k,j_0-1}\}}(Z)$ is well defined similarly as above. Unless $n_{k,j_0}$ is a hand, define

$$m_{n_{k,j_0+1}}(Z) = \min \left\{ \ell_{n_{k,j_0+1}}(u) \,\middle|\, u \in F_{n_{k,j_0-1}}^{\{n_{k,j_0-1}\}}(Z) \right\} - \varepsilon.$$

In particular, the face $F_{n_{k,j_0}}^{\{n_{k,j_0}\}}(Z)$ is now well defined.

Note now that if $n$ is a node, and $F_n^{\{n\}}(Z)$ is already well defined, then the value $m_{n_{k,j_0+1}}(Z)$ is smaller than the minimal value of $\ell_{n_{k,j_0+1}}$ on $F_n^{\{n\}}(Z)$. Therefore, we find

$$\forall n \in \mathcal{N} : F_n^{\{n\}}(Z) = F_n(Z),$$

proving the first condition for $Z \in C$. The second condition follows similarly.

Finally note that at every step in the definition of $Z$, we may as well have used different epsilons, meaning that a generic small perturbation of $Z$ is also in $C$. It follows that $C$ contains an open subset of $\bar{L}_{\mathbb{R}}$, and so has highest dimension possible, $|\mathcal{N}|$. □

*Remark 5.4* By the above lemma, if $Z \in C$, then either $Z = 0$, or all coordinates of $Z$ are positive, that is, $m_n(Z) > 0$ for all $n \in \mathcal{N}$, since this holds for any element of the Lipman cone.

**Lemma 5.5** *Let $Z \in C$, $\rho \in \mathbb{R}_{>0}$ and $u \in \mathbb{R}^3$ satisfying $\rho F_n(f) + u \subset F_n(Z)$ for some $n \in \mathcal{N}$. Then $\rho \Gamma_+(f) + u \subset \Gamma_+(Z)$.*

***Proof*** For $A \subset \mathcal{N}$ a subset inducing a connected subgraph of $\bar{G}$ containing $n$, let $P_A(Z)$ be the following condition:

(i) We have $\rho F_k(f) + u \subset F_k^A(Z)$ for all $k \in A$.

(ii) For any $l \in \mathcal{N} \setminus A$, $l' \in \mathcal{N}_l$ and dilation $\phi : \mathbb{R}^3 \to \mathbb{R}^3$, $x \to \rho' x + u'$ so that $\phi(F_l(f) \cap F_{l'}(f)) \subset F_l^B(Z) \cap F_{l'}^B(Z)$ where $B$ is the connected component of $\bar{G} \setminus A$ containing $l$, we have $\phi(F_l(f)) \subset F_l^B(Z)$.

The assumptions of the lemma imply $P_{\{n\}}(Z)$. Assuming there is a $Z' \in \bar{L}$ with $Z' \geq Z$ so that $P_{\mathcal{N}}(Z')$ holds, we find $\rho\Gamma_+(f)+u \subset \Gamma_+(Z') \subset \Gamma_+(Z)$, proving the lemma. Thus, it is enough to prove that given an $n \in A \subset \mathcal{N}$ inducing a connected subgraph of $\bar{G}$, and a $Z' \geq Z$ so that $P_A(Z')$ holds, and an $i \in \mathcal{N}_A \setminus A$, there is a $Z'' \geq Z'$ so that $P_{A\cup\{i\}}(Z'')$ holds.

So, let such an $i$ be given, assume that it is adjacent in $\bar{G}$ to a $j \in A$. Since $\rho F_j(f) + u \subset F_j^A(Z)$ we have $m_i(Z) \leq \rho\, \hat{\mathrm{wt}}_i(f) + \ell_i(u)$. Let $s = \frac{\rho\, \hat{\mathrm{wt}}_i(f)+\ell_i(u)}{m_i(Z)}$. Note that the denominator here is nonzero by Remark 5.4. Then $s \geq 1$. Let $B$ be the connected component of $\bar{G} \setminus A$ containing $i$ and define the cycle $Z''$ by

$$m_k(Z'') = \begin{cases} s m_k(Z) & \text{if } k \in B, \\ m_k(Z) & \text{else.} \end{cases}$$

Then $Z'' \geq Z'$. We start by noting that condition $P_{A\cup\{i\}}(Z'')$(ii) follows immediately from $P_A(Z')$(ii).

We are left with proving $P_{A\cup\{i\}}(Z'')$(i). We must show that for $k \in A \cup \{i\}$ and $l \in \mathcal{N}_{A\cup\{i\}}$ we have

$$m_l(Z'') \leq \min_{\rho F_k(f)+u} \ell_l, \tag{3}$$

with equality in the case $k = l$.

If $k \in A$ and $l \neq i$, then this is clear from $P_A(Z')$(i).

The minimum of $\ell_i$ on $\cup_{k\in A}\rho F_k + u$ is taken on $(\rho F_i + u) \cap (\rho F_j + u)$, and so by definition of $m_i(Z'')$, Eq. (3) holds also for $l = i$ and any $k \in A$.

Equation (3) is also clear when $k = i$ and $l$ is either $i$ or $j$.

Finally, we prove Eq. (3) in the case $k \in A \cup \{i\}$ and $l \neq j$. Similarly as above, the function $\ell_l$ restricted to $\cup_{k\in A\cup\{i\}}\rho F_k + u$ takes its minimal value on $(\rho F_i + u) \cap (\rho F_l + u)$, and so it suffices to consider the case $k = i$.

Let $\rho' > 0$ and $u' \in \mathbb{R}^3$ be such that $\rho'(F_i(f) \cap F_j(f)) + u' = F_i(Z') \cap F_j(Z')$. By $P_A(Z')$(ii), we have $\rho' F_i(f) + u' \subset F_i(Z')$. By the definition of $Z''$, we find $s \cdot F_i(Z') \subset F_i(Z'')$. As a result,

$$s(\rho' F_i(f) + u') \subset s F_i(Z') \subset F_i(Z'').$$

An application of Lemma 5.6 now shows that if $\rho'' > 0$ and $u''$ are such that $\rho''(F_i(f) \cap F_j(f)) + u'' = F_i(Z'') \cap F_j(Z'')$, then $\rho'' F_i(f) + u'' \subset F_i(Z'')$. Now, we get

$$\rho(F_i(f) \cap F_j(f)) + u \subset \rho''(F_i(f) \cap F_j(f)) + u''.$$

which then implies

$$\rho F_i(f) + u \subset \rho'' F_i(f) + u'' \subset F_i(Z''),$$

which is $P_{A\cup\{i\}}(Z'')$(i) for $k = i$. □

**Lemma 5.6** *Let $A \cong \mathbb{R}^2$ be an affine plane, $\ell_i : A \to \mathbb{R}$ affine functions for $i = 0, \ldots, s$. Assume that $P, Q \subset A$ are polygons given by inequalities $\ell_i \geq p_i$ and $\ell_i \geq q_i$ respectively, in such a way that $p_i = \min_P \ell_i$ and $q_i = \min_Q \ell_i$. Let $P_i$ and $Q_i$ be the minimal sets of $\ell_i$ on $P$ and $Q$ respectively. We assume that $Q \subset P$ and that $Q_0 = P_0$ is a segment of positive length.*

*Take a $p_0' < p_0$ in such a way that we have a polygon $P'$ defined by inequalities $\ell_0 \leq p_0'$ and $\ell_i \leq p_i$ for $i > 0$, and $p_0' = \min_{P'} \ell_0$, and define $P_i'$ as the minimal set of $\ell_i$ on $P'$. Assume that $P_0'$ is a segment of positive length. Let $\phi : A \to A$ be the unique affine isomorphism which preserves directions (i.e. if $L \subset A$ is a line, then $L$ and $\phi(L)$ are parallel) so that $\phi(P_0) = P_0'$. Then $\phi(Q) \subset P'$.*

***Proof*** We can assume that $P_1'$ and $P_s'$ are adjacent to $P_0'$. Consider three cases.

The first case is when the lines spanned by the segments $P_1'$ and $P_s'$ are not parallel, and their intersection point $a$ satisfies $\ell_0(a) < p_0$. In this case, $\phi$ is a homothety with center $a$ and ratio $< 1$. As a result, if we define $P^1$ as the convex hull of $P$ and $a$, then $\phi(P) \subset P^1$. In particular, $\phi(Q) \subset P^1$. The polygon $P^1$ can be defined by the inequalities $\ell_i \geq c_i$ for $i > 0$. It is clear that $\phi(Q)$ also satisfies $\ell_0 \geq c_0'$. As a result, $\phi(Q) \subset P'$ (Fig. 1).

In the second case, assume that the segments $P_1'$ and $P_s'$ are parallel. In this case, $\phi$ is a translation preserving the lines spanned by $P_1'$ and $P_s'$, and $P'$ is the convex hull of $P$ and $\phi(P)$. In particular, $Q \subset P'$.

In the third case, the lines spanned by $P_1'$ and $P_s'$ are not parallel, and their intersection point $a$ satisfies $\ell_0(a) > c_0$. In this case, $\phi$ is a homothety with center $a$ and ratio $> 1$ and similarly as in the second case, $P'$ is the convex hull of $P$ and $\phi(P)$, and so $\phi(Q) \subset P'$. □

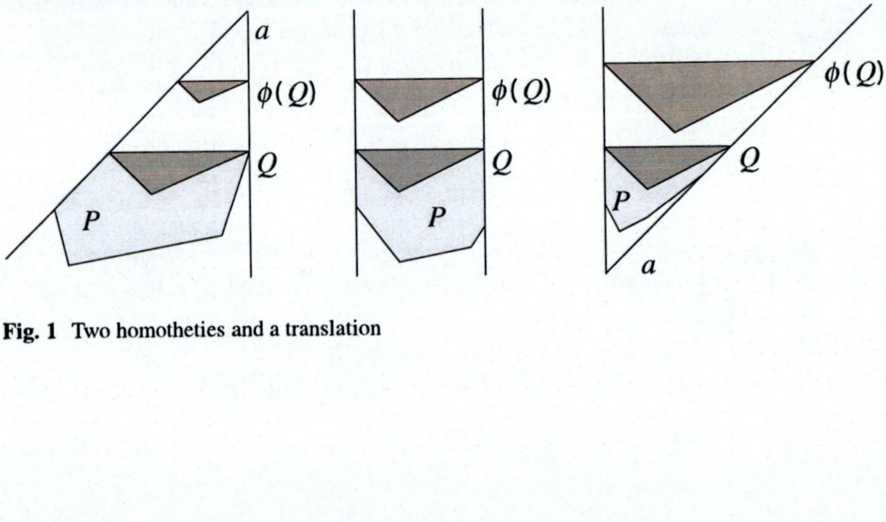


**Fig. 1** Two homotheties and a translation

## 6 Equality Between Ideals

For $g \in \mathcal{O}_{\mathbb{C}^3,0}$ denote by $g_n$ the principal part of $g$ with respect to the weight function $\ell_n$. For $i = 1, 2, 3$ and $\hat{g} \in \mathcal{O}_{\mathbb{C}^3,0}$, we denote by $\hat{\mathrm{wt}}_i$ the weight of $g$ with respect to the $i$-th natural basis vector, i.e. $\hat{\mathrm{wt}}_i(x_j) = \delta_{i,j}$.

**Lemma 6.1** *Let $g \in \mathcal{O}_{\mathbb{C}^3,0}$. Then $\hat{\mathrm{wt}}_n\, g \leq \mathrm{div}_n\, g|_X$ with sharp inequality if and only if $f_n$ divides $g_n$ over the ring of Laurent polynomials.*

***Proof*** See e.g. the proof of Proposition 1 of [3]. □

**Lemma 6.2** *Let $g \in \mathcal{O}_{\mathbb{C}^3,0}$ and assume $\hat{\mathrm{wt}}_n\, g < \mathrm{div}_n\, g$ for some $n \in \mathcal{N}$. Let $h = g_n/f_n$ (a Laurent polynomial by Lemma 6.1). Writing $\{1, 2, 3\} = \{i, j, k\}$, if $F_n(f)$ intersects the $x_j x_k$ coordinate plane, then $\hat{\mathrm{wt}}_i(h) \geq 0$.*

***Proof*** Assume that $h$ contains a monomial with a negative power of $x_i$. Then the same would hold for $g_n = hf_n$, since $f_n$ contains monomials with no power of $x_i$. □

***Proof of Theorem 1.1*** We want to show that for any $Z \in C$, we have $\mathcal{F}(Z) = \mathcal{G}(Z) = \mathcal{I}(Z)$. In light of Eq. (1), it suffices to show that $\mathcal{F}(Z)$ contains $\mathcal{I}(Z)$, that is, if $g \in \mathcal{F}(Z)$, then there exists a $\hat{g} \in \mathcal{O}_{\mathbb{C}^3,0}$ restricting to $g$ with $\hat{\mathrm{wt}}\, \hat{g} \geq m_n(Z)$ for all $n \in \mathcal{N}$.

We use the classification in [1] to set up an induction on the vertices of $\bar{G}$. Assume that $n_0$ is a vertex which intersects all the coordinate axes. This can be done by Proposition 2.3.9 of [1] by choosing $F_{n_0}$ either as a central facet or containing a central edge. We define the partial ordering $\leq$ on $\mathcal{N}$ by setting $n_1 \leq n_2$ if $n_1$ lies on the geodesic connecting $n_0$ and $n_2$. Note that $\bar{G}$ has well defined geodesics since it is a tree.

We prove inductively the statement $P(A)$ that for a subset $A \subset \mathcal{N}$ satisfying

$$n \in A, \quad n' \leq n, \quad \Rightarrow \quad n' \in A,$$

there exists a $\hat{g} \in \mathcal{O}_{\mathbb{C}^3,0}$ satisfying $\hat{g}|_X = g$ and $\hat{\mathrm{wt}}_n\, g \geq m_n(Z)$ for any $n \in A$.

The initial case $P(\emptyset)$ is clear, but we prove $P(\{n_0\})$ as well. Take any $\hat{g} \in \mathcal{O}_{\mathbb{C}^3,0}$ restricting to $g$. If $\mathrm{wt}_{n_0}\, \hat{g} < m_{n_0}(Z)$, then by Lemma 6.1 there is a Laurent polynomial $h$ so that $\hat{\mathrm{wt}}_{n_0}(\hat{g} - hf) > \hat{\mathrm{wt}}_{n_0}\, \hat{g}$. By our choice of $n_0$ and Lemma 6.2, $h$ is a polynomial, and so we can replace $\hat{g}$ with $\hat{g} - hf \in \mathcal{O}_{\mathbb{C}^3,0}$. After repeating this argument finitely many times, we can assume that $\hat{\mathrm{wt}}_{n_0}\, \hat{g} \geq m_{n_0}(Z)$.

Next, assume that $A \subset \mathcal{N}$ satisfies our inductive hypothesis, and that $n \in \mathcal{N}$ is a minimal element of $\mathcal{N} \setminus A$. It suffices to find a polynomial $h$ such that $\hat{g} - hf$ satisfies $P(A)$, as well as $\hat{\mathrm{wt}}_n(\hat{g} - hf) > \hat{\mathrm{wt}}_n(\hat{g})$.

By Lemma 6.1 there does exist a Laurent polynomial $h$ so that $\hat{\mathrm{wt}}_n(\hat{g} - hf) > \hat{\mathrm{wt}}_n\, \hat{g}$. Indeed, set $h = \hat{g}_n/f_n$. We can assume that $F_n(f)$ intersects the $x_1x_3$ and $x_2x_3$ coordinate hyperplanes. By Lemma 6.2 we have $\hat{\mathrm{wt}}_1\, h \geq 0$ and $\hat{\mathrm{wt}}_2\, h \geq 0$. In order to finish the proof, it therefore suffices to show $\hat{\mathrm{wt}}_3(h) \geq 0$ and $\mathrm{wt}_a(hf) \geq m_n(Z)$.

We construct a cycle $Z'$ as follows. Let $a$ be the unique vertex in $A$ adjacent to $n$ and $p$ the unique point on the $x_3$ axis satisfying $\ell_a(p) = m_a(Z)$. Set $m_k(Z') = m_k(Z)$ for all $k$ in the connected component of $G \setminus n$ containing $A$, otherwise set $m_k(Z') = \ell_k(p)$. As a result, the Newton polyhedron $\Gamma_+(Z')$ of $Z'$ is the convex closure of $\Gamma(Z)$ and the point $p$. In particular, if $k$ is in the connected component of $G \setminus n$ containing $A$, then either $F_k(Z') = F_k(Z)$, or $k = a$ and $F_a(Z') \subset F_a(Z)$. For any other vertex $k$, we have $F_k(Z') = \{p\}$. It follows from this that $Z' \in C$.

In fact, we find that

$$x \in \mathbb{R}^3_{\geq 0}, \quad \ell_a(x) = m_a(Z), \quad \ell_n(x) \leq m_n(Z) \quad \Rightarrow \quad x \in F_a(Z').$$

Now let $u \in \mathrm{supp}(h)$ and $w \in \mathrm{supp}(f_n)$. We then have $\ell_a(u + w) \geq m_a(Z)$ and $\ell_n(u + w) < m_n(Z)$. Since $\ell_a(0, 0, 1) > 0$, there is a $t > 0$ so that $\ell_a(u + w - (0, 0, t)) = m_n(Z)$, and we also have $\ell_n(u + w - (0, 0, t)) < m_n(Z)$. We have thus proved that $F_n(f) + u - (0, 0, t) \subset F_n(Z')$. Lemma 5.5 now gives the middle containment in

$$\Gamma_+(f) + u \subset \Gamma_+(f) + u - (0, 0, t) \subset \Gamma_+(Z') \subset \mathbb{R}^3_{\geq 0},$$

which implies, on one hand, that $\hat{\mathrm{wt}}_k(hf) \geq m_k(Z') = m_k(Z)$ for all $k \in A$, and on the other hand, $\hat{\mathrm{wt}}_3(h) = \hat{\mathrm{wt}}_3(hf) \geq 0$, finishing the proof. □

## 7 Suspension Singularities

In this section we consider suspension singularities. In this case, a stronger statement than Theorem 1.1 holds, namely, the three filtrations all coincide. Most of the work in this section, however, goes into proving the *reduced identity* for nodes for suspension singularities, see [8] Definition 6.1.5. This means that the Poincaré series associated with the filtration $\mathcal{F}$ (or $\mathcal{G}$ or $\mathcal{I}$, as they coincide in this case) is identified by a topological invariant, the *zeta function* associated with the link of the singularity.

In this section we assume that $(X, 0)$ is a suspension singularity, that is, there is an $f_0 \in \mathcal{O}_{\mathbb{C}^2,0}$ and an $N \in \mathbb{Z}_{>1}$ so that $(X, 0)$ is given by an equation $f = 0$, where $f(x, y, z) = f_0(x, y) + z^N$. Newton nondegeneracy for $f$ means that $f_0$ is Newton nondegenerate. For convenience, we will also assume that the diagram of $f$ is convenient. This is equivalent to $f_0$ not vanishing along the $x$ or $y$ axis.

***Proof of Theorem 1.2*** If $f$ is the $N$-th suspension of an equation of a plane curve given by $f_0 = 0$, so that $f(x, y, z) = f_0(x, y) + z^N$, then every compact facet of $\Gamma_+(f)$ is the convex hull of a compact facet of the Newton polyhedron of $f_0$ and the point $(0, 0, N)$. In particular, $\Gamma_+(f)$ is *bi-stellar*, and so by Proposition 4 of [5], we have $\mathcal{I} = \mathcal{G}$.

Now, let $n \in \mathcal{N}$ correspond to the facet $F_n \subset \Gamma_+(f)$. By the description above, $F_n$ intersects all coordinate hyperplanes. If $\hat{g} \in \mathcal{O}_{\mathbb{C}^3,0}$ and $\hat{\mathrm{wt}}_n\, \hat{g} < \mathrm{div}_n\, g|_X$, then by Lemmas 6.1 and 6.2, there is a polynomial $h$ so that $\hat{\mathrm{wt}}_n\, \hat{g} - fh > \hat{\mathrm{wt}}\, \hat{g}$. As a result, we find $\mathrm{wt}_n\, g = \mathrm{div}_n\, g$ for $g = \hat{g}|_X$, that is, $\mathcal{F} = \mathcal{G}$.

The formula for the Poincaré series is shown in Sect. 2 to follow from [5]. The formula for the zeta function is Theorem 7.9. □

Using a smooth subdivision of the normal fan to $\Gamma_+(f)$, we obtain an embedded resolution of $(X, 0)$, whose resolution graph we denote by $G$. This graph is obtained as follows. From $\bar{G}^*$, construct $G^*$ by replacing edges between $n, n' \in \mathcal{N}$ with a string, and an edge between $n \in \mathcal{N}$ and $n' \in \mathcal{N}^* \setminus \mathcal{N}$ with $t_{n,n'}$ bamboos. The graph $G$ is obtained from $G^*$ by removing the vertices in $\mathcal{N}^* \setminus \mathcal{N}$, see [11] for details. We denote by $\mathcal{V}$ the vertex set of $G$, and we have a natural inclusion $\mathcal{N} \subset \mathcal{V}$, where if $v \in \mathcal{V}$, then $v \in \mathcal{N}$ if and only if $v$ has degree $> 2$. We denote by $\mathcal{E}$ the set of vertices in $G$ with degree 1. Note that if $e \in \mathcal{E}$, then there are unique $n \in \mathcal{N}$ and $n' \in \mathcal{N}^* \setminus \mathcal{N}$ so that $e$ lies on a bamboo connecting $n$ and $n'$. We set $\alpha_e = \alpha_{n,n'}$ in this case, recall Definition 4.1. For a given $n$, we denote the set of such $e \in \mathcal{E}$ by $\mathcal{E}_n$. Thus, the family $(\mathcal{E}_n)_{n\in\mathcal{N}}$ is a partitioning of $\mathcal{E}$. A vertex $v \in \mathcal{V}$ corresponds to an irreducible component of the exceptional divisor $E_v$.

The associated intersection lattice is negative definite, in particular, the intersection matrix is invertible. Thus, for $v \in \mathcal{V}$, we have a well defined cycle $E_v^*$, that is, divisor supported on the exceptional divisor of the resolution, satisfying $(E_w, E_v^*) = 0$ if $w \neq 0$, but $(E_v, E_v^*) = -1$. We denote the lattice generated by $E_v$ by $L$, and the lattice generated by $E_v^*$ by $L'$. We then have $L = H_2(\tilde{X}, \mathbb{Z})$ and $L' = H_2(\tilde{X}, \partial\tilde{X}, \mathbb{Z}) = \mathrm{Hom}(L, \mathbb{Z})$.

Write $\Gamma_+(f_0) = \cup_{i=1}^r \Gamma_0^i$, where $\Gamma_0^i = [(a_{i-1}, b_{i-1}), (a_i, b_i)]$ are the facets of the Newton polyhedron of $f_0$, so that $0 = a_0 < \ldots < a_r$ and $b_r = 0$. Let $s_i$ be the length of the $i$-th segment, that is, the content of the vector $(a_i - a_{i-1}, b_i - b_{i-1})$. Let $F_{n_i}$ be the facet of $\Gamma_+(f)$ containing the segment $[(a_{i-1}, b_{i-1}), (a_i, b_i)]$. Furthermore, let $s_x = \gcd(N, b_0)$ and $s_y = \gcd(N, a_r)$. Then, in fact, if $n_x, n_y$ are the vertices in $\mathcal{N}^*$ corresponding to the $yz$ and $xz$ coordinate hyperplanes, respectively, then $s_x = t_{n_1,n_x}$ and $s_y = t_{n_r,n_y}$.

It can happen that the diagram $\Gamma(f)$ is not minimal in the sense of [1]. This is the case if $s_x = N$, $s_y = N$, $a_1 = 1$ or $b_{r-1} = 1$. If this is the case, we blow up the appropriate points to produce redundant legs consisting of a single $-1$ curve to make sure that nodes, that is, vertices of degree $> 2$ in $G$ correspond to facets in $\Gamma(f)$ and their legs correspond to primitive segments on the boundary of $\Gamma(f)$. In particular, we assume that $\mathrm{wt}\, xyz = \sum_{e\in\mathcal{E}} E_e^*$.

The sets $\mathcal{E}_{n_1}$ and $\mathcal{E}_{n_r}$ have special elements $e_j^x$, $1 \leq j \leq s_x$, and $e_j^y$, $1 \leq j \leq s_y$, corresponding to the segments $[(0, b_0, 0), (0, 0, N)]$ and $[(a_r, 0, 0), (0, 0, N)]$, respectively. Set $\mathcal{E}_1^x = \{e_i^x | 1 \leq i \leq s_x\}$ and $\mathcal{E}_i^x = \emptyset$ for $i > 1$. Similarly, set $\mathcal{E}_r^y = \{e_i^y | 1 \leq i \leq s_y\}$ and $\mathcal{E}_i^y = \emptyset$ for $i < r$. Further, let $\mathcal{E}_i^z = \mathcal{E}_{n_i} \setminus (\mathcal{E}_i^x \cup \mathcal{E}_i^y)$. Set also $\mathcal{E}^t = \cup_i \mathcal{E}_i^t$ for $t = x, y, z$. Note that we get $|\mathcal{E}_i^z| = s_i$. Define $s_z = \sum_i s_i$.

Write $\mathcal{E}_i^z = \{e_1^{z,i}, \ldots, e_{s_i}^{z,i}\}$. Note that the number $\alpha_e$ is constant for $e \in \mathcal{E}^x$ (in fact, we have $\alpha_e = a_1/s_1$). We denote this by $\alpha_x$. Define $\alpha_y$ similarly.

If $1 < i < r$ we have $\alpha_e = N$ for $e \in \mathcal{E}_i^z$. We have $\alpha_e = N/s_x$ for $e \in \mathcal{E}_1^z$ and $\alpha_e = N/s_y$ for $e \in \mathcal{E}_r^z$.

**Lemma 7.1** *Let $n \in \mathcal{N}$ and $e \in \mathcal{E}_n$. Then $\alpha_e E_e^* - E_n^* \in L$. Furthermore, $\alpha_e E_e^* - E_n^*$ is supported on the leg containing $e$, that is, the connected component of $G \setminus n$ containing $e$.*

***Proof*** This follows from Lemma 20.2 of [4]. □

**Definition 7.2** Let $H$ be the first homology group of the link of $(X, 0)$. Thus, $H = L'/L$, where $L \subset L'$ via the intersection product. If $l \in L'$, we denote its class in $H$ by $[l]$.

**Lemma 7.3** *The order of $H$ is $N^{s_z-1}\alpha_x^{s_x-1}\alpha_y^{s_y-1}$.*

***Proof*** From the proof of Theorem 8.5 of [6], we see that in fact, $|H| = \Delta(1)$, where $\Delta$ is the characteristic polynomial of the monodromy action on the second homology of the Milnor fiber. We leave to the reader to verify, using [12], that the characteristic polynomial is, in our case, given by the formula

$$\begin{aligned}\Delta(t) = &\left[\left(\prod_{i=1}^{r}(t^{m_i}-1)^{s_i}\right)\left(t^{m_1}-1\right)^{s_x-1}\left(t^{m_r}-1\right)^{s_y-1}\right]\\ &\left[\left(\prod_{i=1}^{r}(t^{\frac{m_i}{\alpha_i}}-1)^{s_i}\right)\left(t^{\frac{m_1}{\alpha_x}}-1\right)^{s_x}\left(t^{\frac{m_r}{\alpha_y}}-1\right)^{s_y}\right]^{-1}\\ &\left[\left(t^{\frac{m_1}{\alpha_1\alpha_x}}-1\right)\left(t^{\frac{m_r}{\alpha_r\alpha_y}}-1\right)\left(t^N-1\right)\right]\\ &(t-1)^{-1},\end{aligned}$$

where for $i = 1, \ldots, r$, we take $m_i \in \mathbb{Z}$ so that the facet $F_{n_i}$ of $\Gamma_+(f)$ containing $[(a_{i-1}, b_{i-1}), (a_i, b_i)]$ is contained in the hyperplane $\ell_{n_i} \equiv m_i$. This implies

$$\begin{aligned}\Delta(1) = &\frac{\left[\prod_{i=1} m_i^{s_i}\right] m_1^{s_x-1} m_r^{s_y-1}\left(\frac{m_i}{\alpha_1\alpha_x}\right)\left(\frac{m_r}{\alpha_r\alpha_y}\right)N}{\left[\prod_{i=1}\left(\frac{m_i}{\alpha_i}\right)^{s_i}\right]\left(\frac{m_i}{\alpha_1}\right)^{s_x}\left(\frac{m_r}{\alpha_r}\right)^{s_y}}\\ = &\left[\prod_{i=1}^{r}\alpha_i^{s_i}\right]\alpha_1^{-1}\alpha_r^{-1}\alpha_x^{s_x-1}\alpha_y^{s_y-1}N\end{aligned}$$

Now, for $1 < i < r$, we have $\alpha_i = N$. Furthermore, if $s_1 \neq 1$, then $s_x = 1$ and $\alpha_1 = N$. Similarly, if $s_r \neq 1$, then $s_y = 1$ and $\alpha_r = N$. As a result, the above product equals $N^{s_z-1}\alpha_x^{s_x-1}\alpha_y^{s_y-1}$. □

**Lemma 7.4** *For $1 \leq i \leq r$, let $g_i$ be a generic sum of $x^{\frac{a_{i+1}-a_i}{s_i}}$ and $y^{\frac{b_i-b_{i+1}}{s_i}}$. Then, for $1 < i < r$ we have* $\operatorname{div} g_i = E^*_{n_i}$. *In particular,* $[E^*_{n_i}] = 0 \in H$.

*Furthermore, we have* $\operatorname{div} g_1 = s_x E^*_{n_1}$ *and* $\operatorname{div} g_r = s_y E_{n_r}$. *In particular,* $s_x[E^*_{n_1}] = s_y[E^*_{n_r}] = 0 \in H$.

***Proof*** The curve $(C, 0) \subset (\mathbb{C}^2, 0)$ defined by $f_0$ splits into branches $C = \cup_{i,j} C_{i,j}$ where $C_{i,1} \cup \ldots \cup C_{i,s_i}$ correspond to the segment $[(a_{i-1}, b_{i-1}), (a_i, b_i)]$. Let $G_0$ be the graph associated with the minimal resolution $V \to \mathbb{C}^2$ of $f_0$. There are vertices $\bar{n}_i$ in $G_0$ so that the strict transforms $\tilde{C}_{i,j}$ intersect the component $E_{\bar{n}_i}$ transversely in one point each. The curve defined by $g_i$ is a curvette to $n_i$, that is, if we define $D_i = \{g_i = 0\} \subset \mathbb{C}^2$, then the strict transform $\tilde{D}_i$ in the resolution of $C$ is smooth and intersects $E_{\bar{n}_i}$ in one point, and is disjoint from the $\tilde{C}_i$.

The resolution of $(X, 0)$ is obtained by suspending the pull-back of $f_0$ to $V$, resolving some cyclic quotient singularities, and then blowing down some $(-1)$-curves, see e.g. Appendix 1 in [7]. In particular, we have a morphism $\tilde{X} \to V$, mapping $E_{n_i}$ to $E_{\bar{n}_i}$. The condition that $(X, 0)$ has a rational homology sphere link implies that this map is branched of order $N$ along this divisor. As a result, it restricts to an isomorphism $E_{n_i} \to E_{\bar{n}_i}$, and the preimage $D_i$ of $\tilde{C}_i$ intersects $E_{n_i}$ transversally in one point. Clearly, $D_i$ is the strict transform of the vanishing set of $g_i$ seen as a function on $X$. It follows that $\operatorname{div}_v g_i = E^*_{n_i}$.

Similarly, one verifies that we have maps $E_{n_1} \to E_{\bar{n}_i}$, which are branched covering maps of order $s_x$. Thus, the strict transform of the vanishing set of $g_1$ in $X$ consists of $s_x$ branches, each intersecting $E_{n_1}$ in one point. Thus, $\operatorname{div}_v g_1 = s_x E^*_{n_1}$. A similar argument holds for $g_r$. □

**Definition 7.5** Let $V'_{\mathcal{E}} = \mathbb{Z}\langle E^*_e | e \in \mathcal{E}\rangle$ and $V_{\mathcal{E}} = V'_{\mathcal{E}} \cap L$.

The group $H = L'/L$ is generated by residue classes of ends $[E^*_e]$, $e \in \mathcal{E}$. This is proved in Proposition 5.1 of [10]. In particular, the natural morphism $V'_{\mathcal{E}}/V_{\mathcal{E}} \to H$ is an isomorphism.

**Lemma 7.6** *The lattice $V_{\mathcal{E}}$ is generated by the following elements*

$$N E^*_e,\ e \in \mathcal{E}^z, \quad \alpha_x s_x E^*_e,\ e \in \mathcal{E}^x, \quad \alpha_y s_y E^*_e,\ e \in \mathcal{E}^x,$$

$$\alpha_x (E^*_{e^x_i} - E^*_{e^x_{i+1}}),\ 1 \leq i < s_i, \quad \alpha_y (E^*_{e^y_i} - E^*_{e^y_{i+1}}),\ 1 \leq i < s_y,$$

$$\operatorname{div}(t) = \sum_{e \in \mathcal{E}^t} E^*_e, \quad t = x, y, z.$$

***Proof*** We start by noting that by Lemmas 7.4 and 7.1, if $1 < i < r$ and $e \in \mathcal{E}_{n_r}$, then

$$N E^*_e = \alpha_e E^*_e \equiv E^*_n \equiv 0 \pmod{L},$$

i.e. $NE_e^* \in V_{\mathcal{E}}$. Similarly, if $e \in \mathcal{E}_1^z$, then

$$NE_e^* = s_x\alpha_e E_e^* \equiv s_x E_{n_1}^* \equiv 0 \pmod{L},$$

and $NE_e^* \in V_{\mathcal{E}}$ for $e \in \mathcal{E}_r^z$ as well. A similar argument shows $\alpha_x s_x E_e^* \in V_{\mathcal{E}}$ for $e \in \mathcal{E}^x$ and $\alpha_y s_y E_e^* \in V_{\mathcal{E}}$ for $e \in \mathcal{E}^y$. Let $A$ be the sublattice of $V'_{\mathcal{E}}$ generated by these elements, that is, the top row in the statement of the lemma. We then have $A \subset V_{\mathcal{E}}$, and $[V'_{\mathcal{E}} : A] = (\alpha_x s_x)^{s_x}(\alpha_y s_y)^{s_y} N^{s_z}$. By Lemma 7.3, we get

$$[V_{\mathcal{E}} : A] = [V'_{\mathcal{E}} : V_{\mathcal{E}}]^{-1}[V'_{\mathcal{E}} : A] = \alpha_x s_x^{s_x} \alpha_y s_y^{s_y} N. \tag{4}$$

The elements in the second row are also elements of $V_{\mathcal{E}}$, since, by Lemma 7.1 we have

$$\alpha_x \left(E^*_{e_i^x} - E^*_{e_{i+1}^x}\right) = \left(\alpha_x E^*_{e_i^x} - E^*_{n_1}\right) - \left(\alpha_x E^*_{e_{i+1}^x} - E^*_{n_1}\right) \in L,$$

and similarly for $\alpha_y \left(E^*_{e_i^y} - E^*_{e_{i+1}^y}\right)$. Let $A'$ be the subgroup of $V_{\mathcal{E}}$ generated by $A$ and these elements. Then $[A' : A] = s_x^{s_x-1} s_y^{s_y-1}$.

Finally, we have $\operatorname{div}(t) = \sum_{e\in\mathcal{E}^t} E_e^* \in L$ for $t = x, y, z$. Define $A''$ as the subgroup of $V_{\mathcal{E}}$ generated by $A'$ and $\operatorname{div}(t)$, $t = x, y, z$. Then $[A'' : A'] = (\alpha_x s_x) \cdot (\alpha_y s_y) \cdot N$, and so $[A'' : A] = \alpha_x s_x^{s_x} \alpha_y s_y^{s_y} N = [V_{\mathcal{E}} : A]$, which gives $A'' = V_{\mathcal{E}}$. □

**Lemma 7.7** *We have* $\hat{\mathrm{wt}}\, f|_{\mathcal{N}} = N\, \hat{\mathrm{wt}}\, z|_{\mathcal{N}}$.

***Proof*** Indeed, every compact facet of $\Gamma_+(f)$ contains $(0, 0, N)$. □

**Definition 7.8 ([8])** The *zeta function* associated with the graph $G$ is the expansion at the origin of the rational function $Z(t) = \prod_{v\in\mathcal{V}} \left(1 - [E_v^*]t^{E_v^*}\right)^{\delta_v - 2}$. Thus, if $G$ has more than one vertex, then we can write

$$Z(t) = \left[\prod_{n\in\mathcal{N}} \left(1 - [E_n^*]t^{E_n^*}\right)^{\delta_n - 2}\right]\left[\prod_{e\in\mathcal{E}}\sum_{k=0}^{\infty}\left([E_e^*]t^{E_e^*}\right)^k\right] \in \mathbb{Z}[H][[t^{L'}]],$$

whereas if $G$ has exactly one vertex, say $v$, then

$$Z(t) = (1 - [E_v^*]t^{E_v^*})^{-2} = \sum_{k=0}^{\infty}(k+1)\left([E_v^*]t^{E_v^*}\right)^k.$$

This latter case does not appear in our study of suspension singularities. Here, $t$ denotes variables indexed by $\mathcal{V}$, and so if $l = \sum_{v\in\mathcal{V}} l_v E_v \in L'$ with $l_v \in \mathbb{Q}$, then we write $t^l = \prod_{v\in\mathcal{V}} t_v^{l_v}$.

We have $Z(t) \in \mathbb{Z}[H][[t^{L'}]] \cong \mathbb{Z}[[t^{L'}]][H]$, and the coefficient in front of $t^l$ is in $[l] \cdot \mathbb{Z} \subset \mathbb{Z}[H]$. Therefore, we have a decomposition $Z(t) = \sum_{h \in H} h \cdot Z_h(t)$ with $Z_h(t) \in \mathbb{Z}[[t^{L'}]]$ for each $h \in H$. In particular, $Z_0(t) \in \mathbb{Z}[[t^L]]$.

The *reduced zeta function* $Z^{\mathcal{N}}(t)$ with respect to $\mathcal{N}$ is obtained from $Z(t)$ by restricting $t_v = 1$ for $v \notin \mathcal{N}$. By restricting $Z_0(t)$ similarly, we obtain $Z_0^{\mathcal{N}}(t) \in \mathbb{Z}[[t^{\bar{L}}]]$.

In general, if $A(t) = \sum_{l \in L'} a_l t^l$ is a powerseries, then we discard terms corresponding to $l \notin L$ by setting $A_0(t) = \sum_{l \in L} a_l t^l$.

**Theorem 7.9** *Assume that G is the resolution of a Newton nondegenerate suspension singularity, with rational homology sphere link. Then*

$$Z_0^{\mathcal{N}}(t) = \frac{1 - t^{\hat{\mathrm{wt}} f}}{(1 - t^{\hat{\mathrm{wt}} x})(1 - t^{\hat{\mathrm{wt}} y})(1 - t^{\hat{\mathrm{wt}} z})},$$

*where, on the right hand side, we restrict to variables associated with nodes only, i.e. we set $t_v = 1$ if $v \notin \mathcal{N}$.*

***Proof*** We assume that $s_x > 1$ and $s_y = 1$. The other cases are obtained by a small variation of this proof. Note that in this case we have $s_1 = 1$.

In what follows, we always assume all divisors to be restricted to $\mathcal{N}$. In particular, in view of Lemma 7.1, we can make the identification $([E_e^*]t^{E_e})^{\alpha_e} = [E_n^*]t^{E_n}$ for any $n \in \mathcal{N}$ and $e \in \mathcal{E}_n$. Given our assumption, we have $E^*_{e_1^y} = \mathrm{wt}\, y \in L$. This means that if we write $Z'(t) = Z(t)(1 - t^{E^*_{e_1^y}})$ we have $Z_0(t) = Z'_0(t)/(1 - t^{\mathrm{wt}\, y})$. We can therefore focus on $Z'_0$ instead of $Z_0$. Write

$$\begin{aligned} Z'(t) =& \frac{\left(1 - [E^*_{n_1}]t^{E^*_{n_1}}\right)^{s_x}}{\prod_{i=1}^{s_x}\left(1 - [E^*_{e_i^x}]t^{E^*_{e_i^x}}\right)} \cdot \frac{1}{1 - [E^*_{e^z_{1,1}}]t^{E^*_{e_i^x}}} \cdot \prod_{i=2}^{r} \frac{\left(1 - [E^*_{n_i}]t^{E^*_{n_i}}\right)^{s_i}}{\prod_{k_i=1}^{s_i} 1 - [E^*_{e^z_{i,k_i}}]t^{E^*_{e^z_{i,k_i}}}} \\ =& \prod_{i=1}^{s_x}\sum_{j_i=0}^{\alpha_x - 1}\left([E^*_{e_i^x}]t^{E^*_{e_i^x}}\right)^{j_i} \cdot \sum_{l=0}^{\infty}\left([E^*_{e^z_{1,1}}]t^{E^*_{e_i^x}}\right)^{l} \cdot \prod_{i=2}^{r}\prod_{k_i=1}^{s_i}\sum_{l_{i,k_i}=0}^{N-1}\left([E^*_{e^z_{i,k_i}}]t^{E^*_{e^z_{i,k_i}}}\right)^{l_{i,k_i}} \end{aligned}$$

Considering the presentation for $H$ given in Lemma 7.6, one sees that if the coefficient

$$\prod_{i=1}^{s_x}[E^*_{e_i^x}]^{j_i} \cdot [E^*_{e^z_{1,1}}]^{l} \cdot \prod_{i=2}^{r}\prod_{k_i=1}^{s_i}[E^*_{e^z_{i,k_i}}]^{l_{k_i}} = \left[\sum_{i=1}^{s_x} j_i E^*_{e_i^x} + l E^*_{e^z_{1,1}} + \sum_{i=2}^{r}\sum_{k_i=1}^{s_i} l_{k_i} E^*_{e^z_{i,k_i}}\right]$$

is trivial and $0 \leq j_i < \alpha_x$, then in fact $j_i$ is constant and both $\prod_{i=1}^{s_x}[E^*_{e^x_i}]^{j_i}$ and $[E^*_{e^z_{1,1}}]^l \cdot \prod_{i=2}^{r}\prod_{k_i=1}^{s_i}[E^*_{e^z_{i,k_i}}]^{l_{k_i}}$ are trivial. Therefore we get

$$Z_0'(t) = \left(\prod_{i=1}^{s_x}\sum_{j_i=0}^{\alpha_x-1}\left([E^*_{e^x_i}]t^{E^*_{e^x_i}}\right)^{j_i}\right)_0 \cdot \left(\frac{1}{1-[E^*_{e^z_{1,1}}]t^{E^*_{e^x_i}}}\cdot\prod_{i=2}^{r}\prod_{k_i=1}^{s_i}\sum_{l_{i,k_i}=0}^{N-1}\left([E^*_{e^z_{i,k_i}}]t^{E^*_{e^z_{i,k_i}}}\right)^{k_i}\right)_0$$

and

$$\left(\prod_{i=1}^{s_x}\sum_{j_i=0}^{\alpha_x-1}\left([E^*_{e^x_i}]t^{E^*_{e^x_i}}\right)^{j_i}\right)_0 = \sum_{j=0}^{\alpha_x-1} t^{j\left(E^*_{e^x_1}+\cdots+E^*_{e^x_{s_x}}\right)} = \frac{1-t^{\alpha_x \operatorname{wt} x}}{1-t^{\operatorname{wt} x}}$$

We have $t^{\alpha_x \operatorname{wt} x} = t^{s_x E^*_{n1}} = ([E^*_{e^z_{1,1}}]t^{E^*_{e^z_{1,1}}})^N$ by Lemma 7.1. Thus, we may continue

$$\begin{aligned}
Z_0'(t) =& \frac{1}{1-t^{\operatorname{wt} x}} \cdot \left(\frac{1-\left([E^*_{e^z_{1,1}}]t^{E^*_{e^z_{1,1}}}\right)^N}{1-[E^*_{e^z_{1,1}}]t^{E^*_{e^x_i}}}\prod_{i=2}^{r}\prod_{k_i=1}^{s_i}\sum_{l_{i,k_i}=0}^{N-1}\left([E^*_{e^z_{i,k_i}}]t^{E^*_{e^z_{i,k_i}}}\right)^{k_i}\right)_0 \\
=& \frac{1}{1-t^{\operatorname{wt} x}} \cdot \left(\prod_{i=1}^{r}\prod_{k_i=1}^{s_i}\sum_{l_{i,k_i}=0}^{N-1}\left([E^*_{e^z_{i,k_i}}]t^{E^*_{e^z_{i,k_i}}}\right)^{l_{i,k_i}}\right)_0 .
\end{aligned}$$

From Lemma 7.6 one sees that $\prod_{i=1}^{r}\prod_{k_i=1}^{s_i}[E^*_{e^z_{i,k_i}}]^{l_{i,k_i}}$ is trivial (assuming $0 \leq l_{i,k_i} < N$) if and only if $l_{i,k_i}$ is constant. Thus,

$$\left(\prod_{i=1}^{r}\prod_{k_i=1}^{s_i}\sum_{l_{i,k_i}=0}^{N-1}\left([E^*_{e^z_{i,k_i}}]t^{E^*_{e^z_{i,k_i}}}\right)^{l_{i,k_i}}\right)_0 = \sum_{l=0}^{N-1}\left(\prod_{i=1}^{r}\prod_{k_i=1}^{s_i}t^{E^*_{e^z_{i,k_i}}}\right)^l = \frac{1-t^{N\,\widehat{\operatorname{wt}}\,z}}{1-t^{\widehat{\operatorname{wt}}\,z}}.$$

We therefore get, using Lemma 7.7,

$$Z_0 = \frac{1}{1-t^{\widehat{\operatorname{wt}}\,y}}\cdot\frac{1}{1-t^{\widehat{\operatorname{wt}}\,x}}\cdot\frac{1-t^{\widehat{\operatorname{wt}}\,f}}{1-t^{\widehat{\operatorname{wt}}\,z}}$$

which finishes the proof. □

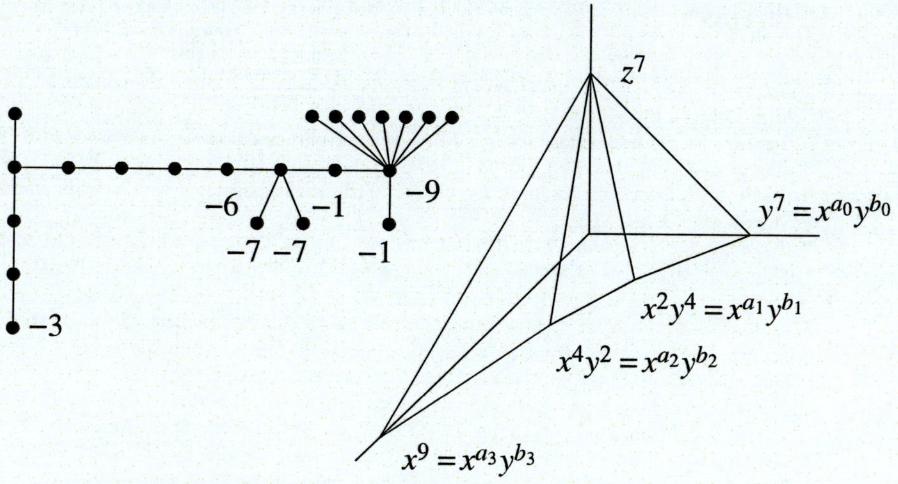


**Fig. 2** Unmarked vertices have Euler number −2

# 8 An Example

Let

$$f(x, y, z) = x^9 + x^4y^2 + x^2y^4 + y^7 + z^7.$$

In this case we have $N = 7$ and by Theorem 1.2 (Fig. 2)

$$s_x = 7, \quad \alpha_x = 2, \quad s_1 = 1, \quad s_y = 1, \quad \alpha_y = 2, \quad s_3 = 7,$$

$$s_z = s_1 + s_2 + s_3 = 1 + 2 + 1 = 4.$$

By Lemma 7.3, we have $|H| = 7^3 2^6 = 21{,}952$, and

$$P^{\mathcal{F}}(t) = Z_0^{\mathcal{N}}(t) = \frac{1 - t_1^{14}t_2^{42}t_3^{126}}{(1 - t_1^3t_2^7t_3^{14})(1 - t_1^2t_2^7t_3^{35})(1 - t_1^2t_2^6t_3^{18})}.$$

**Acknowledgments** I discovered the theorems proved in this article during the PhD program at Central European University under the supervision of András Némethi. I would like to thank András for the many fruitful discussions we have had, and for suggesting to me many interesting problems related to singularity theory.

## References

1. Braun, G., Némethi, A.: Invariants of Newton non-degenerate surface singularities. Compos. Math. **143**(4), 1003–1036 (2007)
2. Ebeling, W., Gusein-Zade, S.M.: Multi-variable Poincaré series associated with Newton diagrams. J. Singularities **1**, 60–68 (2010)
3. Ebeling, W., Gusein-Zade, S.M.: On divisorial filtrations associated with Newton diagrams. J. Singularities **3**, 1–7 (2011)
4. Eisenbud, D., Neumann, W.D.: Three-Dimensional Link Theory and Invariants of Plane Curve Singularities. Annals of Mathematics Studies, vol. 110. Princeton University Press, Princeton (1985)
5. Lemahieu, A.: Poincaré series of embedded filtrations. Math. Res. Lett. **18**(5), 815–825 (2011)
6. Milnor, J.: Singular Points of Complex Hypersurfaces. Ann. of Math. Stud., vol. 61. Princeton University Press, Princeton; University of Tokyo Press, Tokyo (1968)
7. Némethi, A.: Five lectures on normal surface singularities. In: Low Dimensional Topology. Bolyai Soc. Math. Stud., vol. 8, pp. 269–351. János Bolyai Math. Soc., Budapest (1999)
8. Némethi, A.: Poincaré series associated with surface singularities. In: Singularities I. Contemp. Math., vol. 474, pp. 271–297. Amer. Math. Soc., Providence (2008)
9. Némethi, A.: The cohomology of line bundles of splice-quotient singularities. Adv. Math. **229**(4), 2503–2524 (2012)
10. Neumann, W.D., Wahl, J.: Complete intersection singularities of splice type as universal Abelian covers. Geom. Topol. **9**, 699–755 (2005)
11. Oka, M.: On the resolution of the hypersurface singularities. In: Complex Analytic Singularities. Adv. Stud. Pure Math., vol. 8, pp. 405–436. North-Holland, Amsterdam (1987)
12. Varchenko, A.N.: Zeta-function of monodromy and Newton's diagram. Invent. Math. **37**, 253–262 (1976)

# Young Walls and Equivariant Hilbert Schemes of Points in Type *D*

**Ádám Gyenge**

**Abstract** We give a combinatorial proof for a multivariable formula of the generating series of type D Young walls. Based on this we give a motivic refinement of a formula for the generating series of Euler characteristics of Hilbert schemes of points on the orbifold surface of type D.

**Keywords** Hilbert scheme of points · Young walls · Generating function

**Subject Classifications** Primary 14C05; Secondary 05E10

## 1 Introduction

In this paper we survey and refine some existing formulas expressing a connection between affine Lie algebras, Young diagram combinatorics and singularity theory as investigated in [9, 10]. This connection is in the context of Hilbert schemes of points on orbifold surface singularities.

Let $G \subset SL(2, \mathbb{C})$ be a finite subgroup. The equivariant Hilbert scheme $\mathrm{Hilb}([\mathbb{C}^2/G])$ is the moduli space of $G$-invariant finite colength subschemes of $\mathbb{C}^2$, the invariant part of $\mathrm{Hilb}(\mathbb{C}^2)$ under the lifted action of $G$. This space decomposes as

$$\mathrm{Hilb}([\mathbb{C}^2/G]) = \bigsqcup_{\rho \in \mathrm{Rep}(G)} \mathrm{Hilb}^{\rho}([\mathbb{C}^2/G])$$

where

$$\mathrm{Hilb}^{\rho}([\mathbb{C}^2/G]) = \{I \in \mathrm{Hilb}(\mathbb{C}^2)^G \colon H^0(\mathcal{O}_{\mathbb{C}^2}/I) \simeq_G \rho\}$$

Á. Gyenge (✉)
Mathematical Institute, University of Oxford, Oxford, UK
e-mail: Adam.Gyenge@maths.ox.ac.uk



J. Fernández de Bobadilla et al. (eds.), *Singularities and Their Interaction with Geometry and Low Dimensional Topology*, Trends in Mathematics,
https://doi.org/10.1007/978-3-030-61958-9_3

for any finite-dimensional representation $\rho \in \mathrm{Rep}(G)$ of $G$; here $\mathrm{Hilb}(\mathbb{C}^2)^G$ is the set of $G$-invariant ideals of $\mathbb{C}[x, y]$, and $\simeq_G$ means $G$-equivariant isomorphism. Being components of fixed point sets of a finite group acting on smooth quasiprojective varieties, the orbifold Hilbert schemes themselves are smooth and quasiprojective [3].

The topological Euler characteristics of the equivariant Hilbert scheme can be collected into a generating function. Let $\rho_0, \ldots, \rho_n \in \mathrm{Rep}(G)$ denote the (isomorphism classes of) irreducible representations of $G$, with $\rho_0$ the trivial representation. The *orbifold generating series* of the orbifold $[\mathbb{C}^2/G]$ is

$$Z_{[\mathbb{C}^2/G]}(q_0, \ldots, q_n) = \sum_{m_0,\ldots,m_n=0}^{\infty} \chi\left(\mathrm{Hilb}^{m_0\rho_0+\ldots+m_n\rho_n}([\mathbb{C}^2/G])\right) q_0^{m_0} \cdots q_n^{m_n}$$

where $\rho_0, \ldots, \rho_n$ are the irreducible representations of $G_\Delta$, and $q_0, \ldots, q_n$ are formal variables.

Recall that finite subgroups of $SL(2, \mathbb{C})$ are the binary polyhedral groups. These are classified into three families: type $A_n$ for $n \geq 1$ (binary cyclic group of an $(n+1)$-gon), type $D_n$ for $n \geq 4$ (binary dihedral group of an $n$-gon) and type $E_n$ for $n = 6, 7, 8$ (binary tetrahedral, binary octahedral and binary icosahedral groups respectively) [11]. To each such type there also corresponds a simply laced finite type root system. For such a root system $\Delta$ we will denote by $G_\Delta$ the corresponding finite subgroup of $SL(2, \mathbb{C})$ and by $\mathfrak{g}_\Delta$ the corresponding Lie algebra. Moreover, to each such finite type root system there also corresponds an affine Lie algebra $\tilde{\mathfrak{g}}_\Delta$ obtained as a central extension of the loop algebra of $\mathfrak{g}_\Delta$. The corresponding affine root system is denoted as $\widetilde{\Delta}$.

Let $\mathfrak{heis}$ be the infinite Heisenberg algebra, and let $\tilde{\mathfrak{g}}_\Delta \oplus_{\mathfrak{z}} \mathfrak{heis}$ be the Lie algebra that is obtained from the direct sum of $\tilde{\mathfrak{g}}_\Delta$ and $\mathfrak{heis}$ by identifying the centers of the two components. Let $V_0$ be the basic representation of $\tilde{\mathfrak{g}}_\Delta$. Let furthermore $\mathcal{F}$ be the standard Fock space representation of $\mathfrak{heis}$, having central charge 1. Then $V = V_0 \otimes \mathcal{F}$ is a representation of $\tilde{\mathfrak{g}}_\Delta \oplus_{\mathfrak{z}} \mathfrak{heis}$ that is called the *extended basic representation*. A distinguished basis of this representation was introduced by Kashiwara in the context of the associated quantum groups; this is known as the "crystal basis".

It is known that the equivariant Hilbert schemes $\mathrm{Hilb}^\rho([\mathbb{C}^2/\Gamma])$ for all finite dimensional representations $\rho$ of $G$ are Nakajima quiver varieties [16] associated with $\widetilde{\Delta}$, with dimension vector determined by $\rho$, and a specific stability condition (see [4, 15] for more details for type $A$). The results of [16] on the relation between the cohomology of quiver varieties and affine Lie algebras, specialized to this case, imply that the direct sum of all cohomology groups $H^*(\mathrm{Hilb}^\rho([\mathbb{C}^2/G]))$ is graded isomorphic to the extended basic representation $V$ of the corresponding extended affine Lie algebra $\tilde{\mathfrak{g}}_\Delta \oplus_{\mathfrak{z}} \mathfrak{heis}$. This result combined with the Weyl-Kac character formula for the extended basic representation gives the following formula (see [10,

Appendix A]):

$$Z_{[\mathbb{C}^2/G_\Delta]}(q_0,\ldots,q_n) = \left(\prod_{m=1}^{\infty}(1-q^m)^{-1}\right)^{n+1} \cdot \sum_{\mathbf{m}=(m_1,\ldots,m_n)\in\mathbb{Z}^n} q_1^{m_1}\cdots q_n^{m_n}(q^{1/2})^{\mathbf{m}^\top\cdot C_\Delta\cdot\mathbf{m}} \tag{1.1}$$

where $q = \prod_{i=0}^n q_i^{d_i}$ with $d_i = \dim\rho_i$, and $C_\Delta$ is the finite type Cartan matrix corresponding to $\Delta$.

At least in types A and D an even stronger statement can be obtained. In these cases the elements of the crystal basis are in bijection with certain combinatorial objects called Young walls of type $\Delta$. The set of Young walls of type $\Delta$ will be denoted as $\mathcal{W}_\Delta$; these are endowed with an $n+1$ dimensional multi-weight: $\mathbf{wt}(\lambda) = (\mathrm{wt}_0(\lambda),\ldots,\mathrm{wt}_n(\lambda))$. The multi-variable generating series of objects in $\mathcal{W}_\Delta$ is

$$F_\Delta(q_0,\ldots,q_n) = \sum_{\lambda\in\mathcal{W}_\Delta}\mathbf{q}^{\mathbf{wt}(\lambda)}$$

where we used the multi-index notation

$$\mathbf{q}^{\mathbf{wt}(\lambda)} = \prod_{i=0}^{n} q_i^{\mathrm{wt}_i(\lambda)}.$$

Let $\Delta$ be of type A. It was observed first in [12] that there is a bijection

$$\mathcal{W}_\Delta \longleftrightarrow \mathcal{P}^{n+1}\times\mathbb{Z}^n \tag{1.2}$$

where $\mathcal{P}$ is the set of ordinary partitions and $n$ is the rank of the root system. This serves as the starting point of the following enhancement of expression (1.1).

**Theorem 1.1 ([4])** *Let $\Delta$ be of type A. Then*

1.

$$F_\Delta(q_0,\ldots,q_n) = \left(\prod_{m=1}^{\infty}(1-q^m)^{-1}\right)^{n+1} \cdot \sum_{\mathbf{m}=(m_1,\ldots,m_n)\in\mathbb{Z}^n} q_1^{m_1}\cdots q_n^{m_n}(q^{1/2})^{\mathbf{m}^\top\cdot C_\Delta\cdot\mathbf{m}}.$$

2. *There exist a locally closed decomposition of* $\mathrm{Hilb}([\mathbb{C}^2/G_\Delta])$ *into strata indexed by the elements of* $\mathcal{W}_\Delta$. *Each stratum is isomorphic to an affine space.*
3. *In particular,*

$$Z_{[\mathbb{C}^2/G_\Delta]}(q_0,\ldots,q_n) = F_\Delta(q_0,\ldots,q_n).$$

This not just gives a new proof of (1.1) in type A, but it also shows that the combinatorics of Young walls is directly related to an explicit stratification of $\mathrm{Hilb}([\mathbb{C}^2/G_\Delta])$. This relation is beneficial for example in motivic calculations (see e.g. Corollary 1.4 below). Although one can conclude Theorem 1.1 (3) from just formally comparing Eq. (1.1) with Theorem 1.1 (1), the above mentioned relation gives explanation for this coincidence on the level of the cells instead of just the homologies/Euler characteristics.

The main result of the paper is a complete combinatorial proof of the following statement.

**Theorem 1.2** *Let $\Delta$ be of type D. Then*

$$F_\Delta(q_0, \ldots, q_n) =$$

$$\left(\prod_{m=1}^{\infty}(1-q^m)^{-1}\right)^{n+1} \cdot \sum_{\mathbf{m}=(m_1,\ldots,m_n)\in\mathbb{Z}^n} q_1^{m_1}\cdots q_n^{m_n}(q^{1/2})^{\mathbf{m}^\top\cdot C_\Delta\cdot\mathbf{m}}.$$

This result was already announced in [10], and the proof was sketched in [7]; we flesh out the details in Sects. 3–4 below. Again, the starting point will be a decomposition as in (1.2) for the type D case (see Proposition 3.2 below). Combining Theorem 1.2 with the next result one gets a complete analogue of Theorem 1.1 for type D, and hence also an alternative proof of (1.1).

**Proposition 1.3 ([10, Theorem 4.1])** *Let $\Delta$ be of type D.*

1. *There exist a locally closed decomposition of* $\mathrm{Hilb}([\mathbb{C}^2/G_\Delta])$ *into strata indexed by the elements of $\mathcal{W}_\Delta$. Each stratum is isomorphic to an affine space.*
2. *Moreover,*

$$Z_{[\mathbb{C}^2/G_\Delta]}(q_0, \ldots, q_n) = F_\Delta(q_0, \ldots, q_n).$$

One can also consider a motivic enhancement of the series introduced above. Let $K_0(\mathrm{Var})$ be the Grothendieck ring of quasi-projective varieties over the complex numbers. The *motivic Hilbert zeta function* of the orbifold $[\mathbb{C}^2/G_\Delta]$ is

$$\mathcal{Z}_{[\mathbb{C}^2/G_\Delta]}(q_0, \ldots, q_n) = \sum_{m_0,\ldots,m_n=0}^{\infty} [\mathrm{Hilb}^{m_0\rho_0+\ldots+m_n\rho_n}([\mathbb{C}^2/G_\Delta])] q_0^{m_0}\cdots q_n^{m_n}.$$

Here $[X]$ denotes the class of $X$ in $K_0(\mathrm{Var})$, and it is not to be confused with orbifold quotients. The series $\mathcal{Z}_{[\mathbb{C}^2/G_\Delta]}(q_0, \ldots, q_n)$ is an element in $K_0(Var)[[q_0, \ldots, q_n]]$.

The combination of [2, Corollary 1.11] with Theorems 1.1 and 1.2 gives an explicit representation for the motivic Hilbert zeta function.

**Corollary 1.4** *Let $\Delta$ be of type A or D.*

$$\mathcal{Z}_{[\mathbb{C}^2/G_\Delta]}(q_0, \dots, q_n) = \left(\prod_{m=1}^{\infty}(1-\mathbb{L}^{m+1}q^m)^{-1}(1-\mathbb{L}^m q^m)^{-n}\right) \cdot \sum_{\mathbf{m}=(m_1,\dots,m_n)\in\mathbb{Z}^n} q_1^{m_1}\cdots q_n^{m_n}(q^{1/2})^{\mathbf{m}^\top\cdot C_\Delta\cdot\mathbf{m}}$$

*where* $\mathbb{L} = [\mathbb{A}^1] \in K_0(\mathrm{Var})$.

Once again, in type A this statement was proved in [4]. The above series has further specializations giving formulas for the Hodge polynomials and Poincaré polynomials of the equivariant Hilbert schemes.

Let $Y \subset \mathbb{C}^2$ be a closed subvariety invariant under the action of $G_\Delta$. One can consider the moduli space $\mathrm{Hilb}([\mathbb{C}^2/G_\Delta], Y) \subset \mathrm{Hilb}([\mathbb{C}^2/G_\Delta])$ of points supported on $Y$. The corresponding motivic generating series is

$$\mathcal{Z}_{([\mathbb{C}^2/G_\Delta],Y)}(q_0, \dots, q_n) = \sum_{m_0,\dots,m_n=0}^{\infty} [\mathrm{Hilb}^{m_0\rho_0+\dots+m_n\rho_n}([\mathbb{C}^2/G_\Delta], Y)] q_0^{m_0}\cdots q_n^{m_n}.$$

The techniques of [6] imply that

$$\mathcal{Z}_{[\mathbb{C}^2/G_\Delta]}(q_0, \dots, q_n) = \mathcal{Z}_{([\mathbb{C}^2/G_\Delta],Y)}(q_0, \dots, q_n) \cdot \mathcal{Z}_{[(\mathbb{C}^2\setminus Y)/G_\Delta]}(q_0, \dots, q_n).$$

This allows one to obtain further formulas from Corollary 1.4. For example,

$$\mathcal{Z}_{([\mathbb{C}^2/G_\Delta],0)}(q_0, \dots, q_n) = \frac{\mathcal{Z}_{[\mathbb{C}^2/G_\Delta]}(q_0, \dots, q_n)}{\mathcal{Z}_{[(\mathbb{C}^2\setminus 0)/G_\Delta]}(q_0, \dots, q_n)}$$

$$= \frac{\mathcal{Z}_{[\mathbb{C}^2/G_\Delta]}(q_0, \dots, q_n)}{\mathcal{Z}_{(\mathbb{C}^2\setminus 0)/G_\Delta}(q)} = \frac{\mathcal{Z}_{[\mathbb{C}^2/G_\Delta]}(q_0, \dots, q_n)}{\prod_{m=1}^{\infty}(1-\mathbb{L}^{m+1}q^m)^{-1}(1-\mathbb{L}^{m-1}q^m)} =$$

$$\left(\prod_{m=1}^{\infty}(1-\mathbb{L}^{m-1}q^m)^{-1}(1-\mathbb{L}^m q^m)^{-n}\right) \cdot \sum_{\mathbf{m}=(m_1,\dots,m_n)\in\mathbb{Z}^n} q_1^{m_1}\cdots q_n^{m_n}(q^{1/2})^{\mathbf{m}^\top\cdot C_\Delta\cdot\mathbf{m}},$$

where at the second equality we have used that $G_\Delta$ acts freely away from the origin, and at the third equality we have used the main result of [5] and that $[(\mathbb{C}^2\setminus 0)/G_\Delta] = [\mathbb{L}^2] - [pt]$ in $K_0(\mathrm{Var})$.

Suppose that $\Delta$ is of type D. Let $E \subset \mathbb{C}^2$ be the divisor defined by the ideal $(xy)$. This is invariant under the action of $G_\Delta$. Then

$$\begin{aligned}
\mathcal{Z}_{([\mathbb{C}^2/G_\Delta],E)}(q_0,\dots,q_n) &= \mathcal{Z}_{([\mathbb{C}^2/G_\Delta],0)}(q_0,\dots,q_n)\cdot \mathcal{Z}_{([\mathbb{C}^2/G_\Delta],E\setminus 0)}(q_0,\dots,q_n)\\
&= \mathcal{Z}_{([\mathbb{C}^2/G_\Delta],0)}(q_0,\dots,q_n)\cdot \mathcal{Z}_{(\mathbb{C}^2/G_\Delta,(E\setminus 0)/G_\Delta)}(q)\\
&= \mathcal{Z}_{([\mathbb{C}^2/G_\Delta],0)}(q_0,\dots,q_n)\cdot \prod_{m=1}^{\infty}(1-\mathbb{L}^m q^m)^{-1}(1-\mathbb{L}^{m-1}q^m)\\
&= \left(\prod_{m=1}^{\infty}(1-\mathbb{L}^m q^m)^{-n-1}\right)\cdot \sum_{\mathbf{m}=(m_1,\dots,m_n)\in\mathbb{Z}^n} q_1^{m_1}\cdots q_n^{m_n}(q^{1/2})^{\mathbf{m}^\top\cdot C_\Delta\cdot \mathbf{m}},
\end{aligned}$$

where again at the second equality we have used that $G_\Delta$ acts freely away from the origin, and at the third equality we have used $[(E\setminus 0)/G_\Delta]=[\mathbb{L}]-[pt]$ in $K_0(\mathrm{Var})$.

**Corollary 1.5**

1. *There exist a locally closed decomposition of* $\mathrm{Hilb}([\mathbb{C}^2/G_\Delta],E)$ *into strata indexed by the elements of* $\mathcal{W}_\Delta$. *Each stratum is isomorphic to an affine space.*
2. *The class in* $K_0(\mathrm{Var})$ *of the stratum* $\mathrm{Hilb}([\mathbb{C}^2/G_\Delta],E)_Y$ *corresponding to a Young wall* $Y=(\lambda_1,\dots,\lambda_{n+1},\mathbf{m})\in\mathcal{W}_\Delta\cong\mathcal{P}^{n+1}\times\mathbb{Z}^n$ *is*

$$[\mathrm{Hilb}([\mathbb{C}^2/G_\Delta],E)_Y]=[\mathbb{L}]^{\sum_{i=1}^{n+1}|\lambda_i|},$$

*where* $|\lambda_i|=\sum_j \lambda_i^j$.

***Proof*** The proof of Part (1) is very similar to that of [10, Theorem 4.1]. The divisor $E$ is preserved by the diagonal torus action on $\mathbb{C}^2$ used in [10] for the stratification of $\mathrm{Hilb}([\mathbb{C}^2/G_\Delta])$. It follows that the torus action on $\mathrm{Hilb}([\mathbb{C}^2/G_\Delta],E)$ has the same fixed points as the torus action on $\mathrm{Hilb}([\mathbb{C}^2/G_\Delta])$. By [10, Theorem 4.3],

$$\mathrm{Hilb}([\mathbb{C}^2/G_\Delta])^{\mathbb{C}^*}=\bigsqcup_{Y\in\mathcal{W}_\Delta} S_Y$$

where each $S_Y$ is an affine space. Let $\mathrm{Hilb}([\mathbb{C}^2/G_\Delta],E)_Y\subset \mathrm{Hilb}([\mathbb{C}^2/G_\Delta],E)$ denote the locus of ideals which flow to $S_Y$ under the torus action. Since $(E\setminus 0)/G_\Delta\cong\mathbb{C}^*$, the Zariski locally trivial fibration $\mathrm{Hilb}([\mathbb{C}^2/G_\Delta])_Y\to S_Y$ explored in [10, Theorem 4.1] restricts to a Zariski locally trivial fibration $\mathrm{Hilb}([\mathbb{C}^2/G_\Delta],E)_Y\to S_Y$ with affine space fibers, and a compatible torus action on the fibers. By [1, Sect.3, Remarks] this fibration is an algebraic vector bundle over $S_Y$, and hence trivial (Serre–Quillen–Suslin).

Part (2) follows from Part (1) and the formula for $\mathcal{Z}_{([\mathbb{C}^2/G_\Delta],E)}(q_0,\dots,q_n)$ above.

The aim of the current paper is twofold. First, we give an exposition about the combinatorics of the Young walls in type D. Second, we give a complete combinatorial proof of Theorem 1.2 in the type D case.

The structure of the rest of the paper is as follows. In Sect. 2 we review the combinatorics of the Young walls in type D. In Sect. 3 we introduce an associated combinatorial tool called the abacus. Using this we will calculate the generating series $F_\Delta(q_0, \ldots, q_n)$ of Young walls of type D and prove Theorem 1.2 in Sect. 4.

## 2 Young Walls of Type $D_n$

It is known that when $\Delta = A_n$, $n \geq 1$, the set of Young walls $\mathcal{W}_\Delta = \mathcal{P}$, the set of all Young diagrams/partitions equipped with the diagonal coloring (see [8]). We describe here the type $D$ analogue of the set of diagonally colored partitions used in type $A$, following [13, 14].

First we define the *Young wall pattern of type* $D_n$. This is the following infinite pattern, consisting of two types of blocks: half-blocks carrying possible labels $j \in \{0, 1, n-1, n\}$, and full blocks carrying possible labels $1 < j < n-1$:

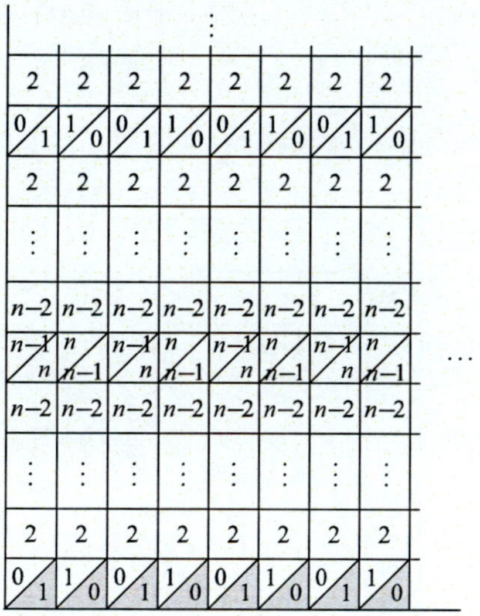


A *Young wall*[1] *of type* $D_n$ is a subset $Y$ of the infinite Young wall pattern of type $D_n$, satisfying the following rules.

[1]In [13, 14], these arrangements are called *proper Young walls*. Since we will not meet any other Young wall, we will drop the adjective *proper* for brevity.

(YW1) $Y$ contains all grey half-blocks, and a finite number of the white blocks and half-blocks.

(YW2) $Y$ consists of continuous columns of blocks, with no block placed on top of a missing block or half-block.

(YW3) Except for the leftmost column, there are no free positions to the left of any block or half-block. Here the rows of half-blocks are thought of as two parallel rows; only half-blocks of the same orientation have to be present.

(YW4) A full column is a column with a full block or both half-blocks present at its top; then no two full columns have the same height.[2]

Let $\mathcal{W}_\Delta$ denote the set of all Young walls of type $D_n$. For any $Y \in \mathcal{W}_\Delta$ and label $j \in \{0, \ldots, n\}$ let $wt_j(Y)$ be the number of white half-blocks, respectively blocks, of label $j$. These are collected into the multi-weight vector $\mathbf{wt}(Y) = (wt_0(Y), \ldots, wt_n(Y))$. The total weight of $Y$ is the sum

$$|Y| = \sum_{j=0}^{n} wt_j(Y),$$

and for the formal variables $q_0, \ldots, q_n$,

$$\mathbf{q}^{\mathbf{wt}(Y)} = \prod_{j=0}^{n} q_j^{wt_j(Y)}.$$

*Example 2.1* The following is an example of a Young wall for $\Delta = D_4$:

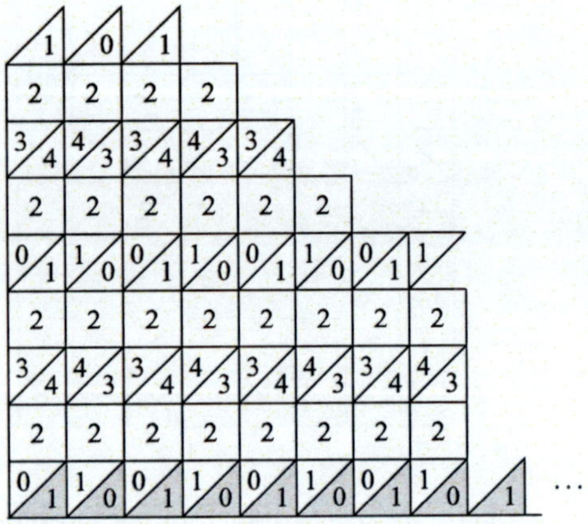

[2]This is the properness condition of [13].

## 3 Abacus Combinatorics

Recalling the Young wall rules (YW1)–(YW4), it is clear that every $Y \in \mathcal{W}_\Delta$ can be decomposed as $Y = Y_1 \sqcup Y_2$, where $Y_1 \in \mathcal{W}_\Delta$ has full columns only, and $Y_2 \in \mathcal{W}_\Delta$ has all its columns ending in a half-block. These conditions define two subsets $\mathcal{Z}_\Delta^f, \mathcal{Z}_\Delta^h \subset \mathcal{W}_\Delta$. Because of the Young wall rules, the pair $(Y_1, Y_2)$ uniquely reconstructs $Y$, so we get a bijection

$$\mathcal{W}_\Delta \longleftrightarrow \mathcal{Z}_\Delta^f \times \mathcal{Z}_\Delta^h. \tag{3.1}$$

Given a Young wall $Y \in \mathcal{W}_\Delta$ of type $D_n$, let $\lambda_k$ denote the number of blocks (full or half blocks both contributing 1) in the $k$-th vertical column. By the rules of Young walls, the resulting positive integers $\{\lambda_1, \dots, \lambda_r\}$ form a partition $\lambda = \lambda(Y)$ of weight equal to the total weight $|Y|$, with the additional property that its parts $\lambda_k$ are distinct except when $\lambda_k \equiv 0 \mod (n-1)$. Corresponding to the decomposition (3.1), we get a decomposition $\lambda(Y) = \mu(Y) \sqcup \nu(Y)$. In $\mu(Y)$, no part is congruent to 0 modulo $(n-1)$, and there are no repetitions; all parts in $\nu(Y)$ are congruent to 0 modulo $(n-1)$ and repetitions are allowed. Note that the pair $(\mu(Y), \nu(Y))$ does almost, but not quite, encode $Y$, because of the ambiguity in the labels of half-blocks on top of non-complete columns.

We now introduce another combinatorial object, *the abacus of type $D_n$* [13, 14]. This is the arrangement of positive integers, called positions, in the following pattern:

$$\begin{array}{cccccc} 1 & \dots\ n-2 & n-1 & n & \dots\ 2n-3 & 2n-2 \\ 2n-1 & \dots\ 3n-4 & 3n-3 & 3n-2 & \dots\ 4n-5 & 4n-4 \\ \vdots & \vdots & \vdots & \vdots & \vdots & \vdots \end{array}$$

For any integer $1 \le k \le 2n-2$, the set of positions in the $k$-th column of the abacus is the *k-th ruler*, denoted $R_k$. Several *beads* are placed on these rulers. For $k \not\equiv 0 \mod (n-1)$, the rulers $R_k$ can only contain normal (uncolored) beads, with each position occupied by at most one bead. On the rulers $R_{n-1}$ and $R_{2n-2}$, the beads are colored white and black. An arbitrary number of white or black beads can be put on each such position, but each position can only contain beads of the same color.

Given a type $D_n$ Young wall $Y \in \mathcal{W}_\Delta$, let $\lambda = \mu \sqcup \nu$ be the corresponding partition with its decomposition. For each nonzero part $\nu_k$ of $\nu$, set

$$n_k = \#\{1 \le j \le l(\mu) \mid \mu_j < \nu_k\}$$

to be the number of full columns shorter than a given non-full column. The abacus configuration of the Young wall $Y$ is defined to be the set of beads placed at positions $\lambda_1, \dots, \lambda_r$. The beads at positions $\lambda_k = \mu_j$ are uncolored; the color of the bead at position $\lambda_k = \nu_l$ corresponding to a column $C$ of $Y$ is

$$\begin{cases} \text{white,} & \text{if the block at the top of } C \text{ is ◿ and } n_l \text{ is even;} \\ \text{white,} & \text{if the block at the top of } C \text{ is ◸ and } n_l \text{ is odd;} \\ \text{black,} & \text{if the block at the top of } C \text{ is ◸ and } n_l \text{ is even;} \\ \text{black,} & \text{if the block at the top of } C \text{ is ◿ and } n_l \text{ is odd.} \end{cases}$$

One can check that the abacus rules are satisfied, that all abacus configurations satisfying the above rules, with finitely many uncolored, black and white beads, can arise, and that the Young wall $Y$ is uniquely determined by its abacus configuration.

*Example 3.1* The abacus configuration associated with the Young wall of Example 2.1 is

| $R_1$ | $R_2$ | $R_3$ | $R_4$ | $R_5$ | $R_6$ |
|---|---|---|---|---|---|
| 1 | 2 | 3 | 4 | 5 | ❻ |
| ⑦ | ⑧ | 9 | ⑩ | ⑪ | ⑫$^3$ |
| 13 | 14 | 15 | 16 | 17 | 18 |
| ⋮ | | | | ⋮ | |

The superscript at 12 indicates that there are 3 white beads at that position.

We now introduce certain distinguished Young walls of type $D_n$, and a method to obtain them with moving the beads on the abacus. On the Young wall side, define a *bar* to be a connected set of blocks and half-blocks, with each half-block occurring once and each block occurring twice. A Young wall $Y \in \mathcal{W}_\Delta$ will be called a *core* Young wall, if no bar can be removed from it without violating the Young wall rules. For an example of bar removal, see [13, Example 5.1(2)]. Let $\mathcal{C}_\Delta \subset \mathcal{W}_\Delta$ denote the set of all core Young walls.

Based on the calculations of [13, 14] the following result was obtained in [10, Proposition 7.2]. For completeness we include also its proof.

**Proposition 3.2** *Given a Young wall $Y \in \mathcal{W}_\Delta$, any complete sequence of bar removals through Young walls results in the same core* core$(Y) \in \mathcal{C}_\Delta$*, defining a map of sets*

$$\text{core}\colon \mathcal{W}_\Delta \to \mathcal{C}_\Delta.$$

*The process can be described on the abacus, respects the decomposition* (3.1)*, and results in a bijection*

$$\mathcal{W}_\Delta \longleftrightarrow \mathcal{P}^{n+1} \times \mathcal{C}_\Delta \tag{3.2}$$

*where $\mathcal{P}$ is the set of ordinary partitions. Finally, there is also a bijection*

$$\mathcal{C}_\Delta \longleftrightarrow \mathbb{Z}^n. \tag{3.3}$$

***Proof*** Decompose $Y$ into a pair of Young walls $(Y_1, Y_2)$ as above. Let us first consider $Y_1$. On the corresponding rulers $R_k$, $k \not\equiv 0 \mod (n-1)$, the following steps correspond to bar removals [13, Lemma 5.2].

(B1) If $b$ is a bead at position $s > 2n-2$, and there is no bead at position $(s-2n+2)$, then move $b$ one position up and switch the color of the beads at positions $k$ with $k \equiv 0 \mod (n-1)$, $s-2n+2 < k < s$.

(B2) If $b$ and $b'$ are beads at position $s$ and $2n-2-s$ $(1 \leq s \leq n-2)$ respectively, then remove $b$ and $b'$ and switch the color of the beads at positions $k \equiv 0 \mod (n-1)$, $s < k < 2n-2-s$.

Performing these steps as long as possible results in a configuration of beads on the rulers $R_k$ with $k \not\equiv 0 \mod (n-1)$ with no gaps from above, and for $1 \leq s \leq n-2$, beads on only one of $R_s$, $R_{2n-2-s}$. This final configuration can be uniquely described by an ordered set of integers $\{z_1, \dots, z_{n-2}\}$, $z_s$ being the number of beads on $R_s$ minus the number of beads on $R_{2n-2-s}$ [14, Remark 3.10(2)]. In the correspondence (3.3) this gives $\mathbb{Z}^{n-2}$. It turns out that the reduction steps in this part of the algorithm can be encoded by an $(n-2)$-tuple of ordinary partitions, with the summed weight of these partitions equal to the number of bars removed [13, Theorem 5.11(2)].

Let us turn to $Y_2$, represented on the rulers $R_k$, $k \equiv 0 \mod (n-1)$. On these rulers the following steps correspond to bar removals [14, Sections 3.2 and 3.3].

(B3) Let $b$ be a bead at position $s \geq 2n-2$. If there is no bead at position $(s-n+1)$, and the beads at position $(s-2n+2)$ are of the same color as $b$, then shift $b$ up to position $(s-2n+2)$.

(B4) If $b$ and $b'$ are beads at position $s \geq n-1$, then move them up to position $(s-n+1)$. If $s-n+1 > 0$ and this position already contains beads, then $b$ and $b'$ take that same color.

During these steps, there is a boundary condition: there is an imaginary position 0 in the rightmost column, which is considered to contain invisible white beads; placing a bead there means that this bead disappears from the abacus. It turns out that the reduction steps in this part of the algorithm can be described by a triple of ordinary partitions, again with the summed weight of these partitions equal to the number of bars removed [14, Proposition 3.6]. On the other hand, the final result can be encoded by a pair of ordinary partitions, or Young diagrams, which have the additional property of being a pyramid.

The different bar removal steps (B1)–(B4) construct the map $c$ algorithmically and uniquely. The stated facts about parameterizing the steps prove the existence of the bijection (3.2). To complete the proof of (3.3), we only need to remark further that the set of ordinary Young diagrams having the shape of a pyramid is in bijection with the set of integers (see [14, Remark 3.10(2)]). This gives the remaining $\mathbb{Z}^2$ factor in the bijection (3.3).

*Example 3.3* A possible sequence of bar removals on the abacus and Young wall of Examples 2.1 and 3.1 is as follows. Perform step (B1) on the beads at positions 7, 8, 10, 11. Perform step (B2) on the pairs of beads at positions (1,5) and (2,4). Perform step (B4) four times on two beads at position 12 by moving them consecutively to positions 9, 6 (where they take the color black), 3 and then 0 (which means removing them from the abacus). The resulting abacus configuration is then

| $R_1$ | $R_2$ | $R_3$ | $R_4$ | $R_5$ | $R_6$ |
|---|---|---|---|---|---|
| 1 | 2 | 3 | 4 | 5 | (6) |
| 7 | 8 | 9 | 10 | 11 | (12) |
|  | ⋮ |  |  | ⋮ |  |

This configuration describes the following core Young wall:

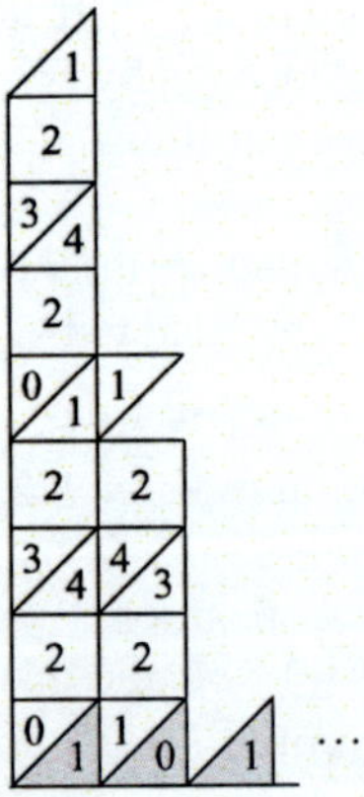


## 4 Enumeration of Young Walls

We next determine the multi-weight of a Young wall $Y$ in terms of the bijections (3.2)–(3.3). The quotient part is easy: the multi-weight of each bar is $(1, 1, 2, \ldots, 2, 1, 1)$, so the $(n + 1)$-tuple of partitions contributes a factor of

$$\left(\prod_{m=1}^{\infty}(1-q^m)^{-1}\right)^{n+1}.$$

Turning to cores, under the bijection $\mathcal{C}_\Delta \leftrightarrow \mathbb{Z}^n$, the total weight of a core Young wall $Y \in \mathcal{C}_\Delta$ corresponding to $(z_1, \ldots, z_n) \in \mathbb{Z}^n$ is calculated in [14, Remark 3.10]:

$$|Y| = \frac{1}{2}\sum_{i=1}^{n-2}\left((2n-2)z_i^2 - (2n-2i-2)z_i\right) + (n-1)\sum_{i=n-1}^{n}\left(2z_i^2 + z_i\right). \quad (4.1)$$

The next result gives a refinement of this formula for the multi-weight of $Y$.

**Theorem 4.1** *Let $q = q_0 q_1 q_2^2 \ldots q_{n-2}^2 q_{n-1} q_n$, corresponding to a single bar.*

(1) *Composing the bijection (3.3) with an appropriate $\mathbb{Z}$-change of coordinates in the lattice $\mathbb{Z}^n$, the multi-weight of a core Young wall $Y \in \mathcal{C}_\Delta$ corresponding to an element $\mathbf{m} = (m_1, \ldots, m_n) \in \mathbb{Z}^n$ can be expressed as*

$$q_1^{m_1} \cdot \ldots \cdot q_n^{m_n} (q^{1/2})^{\mathbf{m}^\top \cdot C \cdot \mathbf{m}},$$

*where $C$ is the Cartan matrix of type $D_n$.*

(2) *The multi-weight generating series*

$$F_\Delta(q_0, \ldots, q_n) = \sum_{Y \in \mathcal{W}_\Delta} \mathbf{q}^{\mathrm{wt}(Y)}$$

*of Young walls for $\Delta$ of type $D_n$ can be written as*

$$F_\Delta(q_0, \ldots, q_n) = \frac{\sum\limits_{\mathbf{m}=(m_1,\ldots,m_n)\in\mathbb{Z}^n}^{\infty} q_1^{m_1} \cdot \ldots \cdot q_n^{m_n} (q^{1/2})^{\mathbf{m}^\top \cdot C \cdot \mathbf{m}}}{\prod\limits_{m=1}^{\infty} (1-q^m)^{n+1}}.$$

(3) *The following identity is satisfied between the coordinates $(m_1, \ldots, m_n)$ and $(z_1, \ldots, z_n)$ on $\mathbb{Z}^n$:*

$$\sum_{i=1}^{n} m_i = -\sum_{i=1}^{n-2}(n-1-i)z_i - (n-1)c(z_{n-1}+z_n) - (n-1)b.$$

*Here $z_1 + \cdots + z_{n-2} = 2a - b$ for integers $a \in \mathbb{Z}$, $b \in \{0, 1\}$, and $c = 2b - 1 \in \{-1, 1\}$.*

Statement (2) clearly follows from (1) and the discussion preceding Theorem 4.1. Statement (3) is used to achieve additional results in [10].

Let us write $z_I = \sum_{i \in I} z_i$ for $I \subseteq \{1, \ldots, n-2\}$. Each such number decomposes uniquely as $z_I = 2a_I - b_I$, where $a_I \in \mathbb{Z}$ and $b_I \in \{0, 1\}$. Let us introduce also

$c_I = 2b_I - 1 \in \{-1, 1\}$. We will make use of the relations

$$a_I = \sum_{i \in I} a_i - \sum_{i_1 \in I, i_2 \in I \setminus \{i_1\}} b_{i_1} b_{i_2} + \sum_{i_1 \in I, i_2 \in I \setminus \{i_1\}, i_3 \in I \setminus \{i_1, i_2\}} 2 b_{i_1} b_{i_2} b_{i_3} - \dots ,$$

$$b_I = \sum_{i \in I} b_i - \sum_{i_1 \in I, i_2 \in I \setminus \{i_1\}} 2 b_{i_1} b_{i_2} + \sum_{i_1 \in I, i_2 \in I \setminus \{i_1\}, i_3 \in I \setminus \{i_1, i_2\}} 4 b_{i_1} b_{i_2} b_{i_3} - \dots .$$

To simplify notations let us introduce

$$r_I := a_I - \sum_{i \in I} a_i = - \sum_{i_1 \in I, i_2 \in I \setminus \{i_1\}} b_{i_1} b_{i_2} + \sum_{i_1 \in I, i_2 \in I \setminus \{i_1\}, i_3 \in I \setminus \{i_1, i_2\}} 2 b_{i_1} b_{i_2} b_{i_3} - \dots .$$

Using these notations the colored refinement of the weight formula (4.1) is the following.

**Lemma 4.2** *Given a core Young wall $Y \in \mathcal{C}_\Delta$ corresponding to $(z_i) \in \mathbb{Z}^n$ in the bijection of* (3.3)*, its content is given by the formula*

$$\mathbf{q}^{\mathbf{wt}(Y)} =$$

$$q_1^{-\sum_{i=1}^{n-2} b_i} q_2^{-2a_1 - \sum_{i=2}^{n-2} b_i} \dots q_{n-2}^{-\sum_{i=1}^{n-3} 2a_i - b_{n-2}} (q_0 q_1^{-1} q_{n-1} q_n)^{-\sum_{i=1}^{n-2} a_i} (q_0 q_1^{-1})^{a_{1\dots n-2}}$$

$$\cdot q^{\frac{1}{2} \sum_{i=1}^{n-2} (z_i^2 + b_i) + z_{n-1}^2 + z_n^2}$$

$$\cdot (q^{b_{1\dots n-2}} (q_1^{-1} \dots q_{n-2}^{-1} q_{n-1}^{-1})^{c_{1\dots n-2}})^{z_{n-1}} (q^{b_{1\dots n-2}} (q_1^{-1} \dots q_{n-2}^{-1} q_n^{-1})^{c_{1\dots n-2}})^{z_n} .$$

When forgetting the coloring a straightforward check shows that Lemma 4.2 gives back (4.1). Notice also that $z_i^2 + b_i = 4a_i^2 - 4a_i b_i + 2b_i$ is always an even number, so the exponents are always integers.

***Proof of Lemma* 4.2** Suppose that we restrict our attention to blocks of color $i$ by substituting $q_j = 1$ for $j \neq i$. Clearly,

$$wt_i(Y) \leq \sum_j wt_j(Y) = |Y|$$

where $|Y|$ is the total weight of $Y$. This inequality is true for each $0 \leq i \leq n$, and $|Y|$ is a linear combination of the parameters $\{z_i\}_{1 \leq i \leq n}$, their squares and a constant. It follows from the definition of the $\{z_i\}_{1 \leq i \leq n}$ that each $wt_i$ is a convex, increasing function of them. These imply that, when considered over the reals, each $wt_i$ are at most quadratically growing, convex analytic functions of $\{z_i\}_{1 \leq i \leq n}$. As a consequence, each $wt_i$ is again a linear combination of constants, the parameters $\{z_i\}_{1 \leq i \leq n}$ and their products. Hence, it is enough to check that the claimed formula is correct in two cases:

1. when any of the $z_i$'s is set to a given number and the others are fixed to 0; and
2. when all of the parameters are fixed to 0 except for an arbitrary pair $z_i$ and $z_j$, $i \neq j$.

First, consider that $z_i \neq 0$ for a fixed $i$, and $z_j = 0$ in case $j \neq i$.

(a) When $1 \leq i \leq n-2$, then the colored weight of the corresponding core Young wall is

$$(q_1 \dots q_i)^{-b_i} (q_{i+1}^2 \dots q_{n-2}^2 q_{n-1} q_n)^{-a_i} q^{2a_i^2 - 2a_i b_i + b_i} .$$

(b) When $i \in \{n-1, n\}$, then the associated core Young wall has colored weight

$$q^{z_i^2} (q_1 q_2 \dots q_{n-2} q_i)^{z_i} .$$

Both of these follow from (4.1) and its proof in [14] by taking into account the colors of the blocks in the pattern.

Second, assume that $z_i$ and $z_j$ are nonzero, but everything else is zero. Then the total weight is not the product of the two individual weights, but some correction term has to be introduced. The particular cases are:

(a) $1 \leq i, j \leq n-2$. There can only be a difference in the numbers of $q_0$'s and $q_1$'s which comes from the fact that in the first row there are only half blocks with 0's in the odd columns and 1's in the even columns. Exactly $-r_{ij}$ blocks change color from 0 to 1 when both $z_i$ and $z_j$ are nonzero compared to when one of them is zero. In general, this gives the correction term $(q_0 q_1^{-1})^{r_{1\dots n-2}} = (q_0 q_1^{-1})^{a_{1\dots n-2} - \sum_{i=1}^{n-2} a_i}$.

(b) $1 \leq i \leq n-2$, $j \in \{n-1, n\}$. For the same reason as in the previous case the parity of $z_i$ modifies the colored weight of the contribution of $z_j$, but not the total weight of it. If $z_i$ is even, then the linear term of the contribution of $z_j$ is $q_1 q_2 \dots q_{n-2} q_j$. In the odd case it is $q_0 q_2 \dots q_{n-2} q_{\kappa(j)}$. This is encoded in the correction term $(q^{b_{1\dots n-2}} (q_1^{-1} \dots q_{n-2}^{-1} q_j^{-1})^{c_{1\dots n-2}})^{z_j}$ where $j \to \kappa(j)$ swaps $n-1$ and $n$.

(c) $i = n-1$, $j = n$. $z_{n-1}$ and $z_n$ count into the total colored weight completely independently, so no correction term is needed.

Putting everything together gives the claimed formula for the colored weight of an arbitrary core Young wall.

Now we turn to the proof of Theorem 4.1. After recollecting the terms in the formula of Lemma 4.2 it becomes

$$q_1^{-b_{1\dots n-2} - c_{1\dots n-2}(z_{n-1}+z_n)} \prod_{i=2}^{n-2} q_i^{-2a_{1\dots i-1} + c_{1\dots i-1} b_{i\dots n-2} - c_{1\dots n-2}(z_{n-1}+z_n)}$$

$$\cdot q_{n-1}^{-a_{1\dots n-2} - c_{1\dots n-2} z_{n-1}} q_n^{-a_{1\dots n-2} - c_{1\dots n-2} z_n}$$

$$\cdot q^{\sum_{i=1}^{n-2}(2a_i^2 - 2a_i b_i + b_i) + b_{1\dots n-2} z_{n-1} + z_{n-1}^2 + b_{1\dots n-2} z_n + z_n^2 + r_{1\dots n-2}}$$

Let us define the following series of integers:

$$\begin{aligned}
m_1 &= -b_{1\ldots n-2} - c_{1\ldots n-2}(z_{n-1} + z_n)\ , \\
m_2 &= -2a_1 + c_1 b_{2\ldots n-2} - c_{1\ldots n-2}(z_{n-1} + z_n)\ , \\
&\vdots \\
m_{n-2} &= -2a_{1\ldots n-3} + c_{1\ldots n-3} b_{n-2} - c_{1\ldots n-2}(z_{n-1} + z_n)\ , \\
m_{n-1} &= -a_{1\ldots n-2} - c_{1\ldots n-2} z_{n-1}\ , \\
m_n &= -a_{1\ldots n-2} - c_{1\ldots n-2} z_n\ .
\end{aligned}$$

It is an easy and enlightening task to verify that the map

$$\mathbb{Z}^n \to \mathbb{Z}^n, \quad (z_1, \ldots, z_n) \mapsto (m_1, \ldots, m_n)$$

is a bijection, which is left to the reader.

***Proof of Theorem 4.1*** (1) One has to check that

$$\begin{aligned}
&\sum_{i=1}^{n} m_i^2 - m_1 m_2 - m_2 m_3 - \cdots - m_{n-2}(m_{n-1} + m_n) = \\
&= \sum_{i=1}^{n-2} (2a_i^2 - 2a_i b_i + b_i) + b_{1\ldots n-2} z_{n-1} + z_{n-1}^2 + b_{1\ldots n-2} z_n + z_n^2 + r_{1\ldots n-2}\ .
\end{aligned}$$

The terms containing $z_{n-1}$ or $z_n$ on the left hand side are

$$\begin{aligned}
&(n-2)(z_{n-1} + z_n)^2 + z_{n-1}^2 + z_n^2 - (n-3)(z_{n-1} + z_n)^2 - z_{n-1}^2 - z_n^2 - 2z_{n-1} z_n \\
&+ \left( 2b_{1\ldots n-2} + \sum_{i=1}^{n-3} 2(2a_{1\ldots i} - c_{1\ldots i} b_{i+1\ldots n-2}) + 2a_{1\ldots n-2} \right) c_{1\ldots n-2}(z_{n-1} + z_n) \\
&- \left( b_{1\ldots n-2} + \sum_{i=1}^{n-3} 2(2a_{1\ldots i} - c_{1\ldots i} b_{i+1\ldots n-2}) + 2a_{1\ldots n-2} \right) c_{1\ldots n-2}(z_{n-1} + z_n) \\
&\qquad = b_{1\ldots n-2} z_{n-1} + z_{n-1}^2 + b_{1\ldots n-2} z_n + z_n^2\ ,
\end{aligned}$$

since $b_{1\ldots n-2} c_{1\ldots n-2} = b_{1\ldots n-2}$.

The terms containing neither $z_{n-1}$ nor $z_n$ on the left hand side are

$$b_{1\ldots n-2} + \sum_{i=1}^{n-3}(2a_{1\ldots i} - c_{1\ldots i}b_{i+1\ldots n-2})^2 + 2a_{1\ldots n-2}^2 - b_{1\ldots n-2}(2a_1 - c_1 b_{2\ldots n-2})$$
$$- \sum_{i=1}^{n-4}(2a_{1\ldots i} - c_{1\ldots i}b_{i+1\ldots n-2})(2a_{1\ldots i+1} - c_{1\ldots i+1}b_{i+2\ldots n-2})$$
$$-2(2a_{1\ldots n-3} - c_{1\ldots n-3}b_{n-2})a_{1\ldots n-2}\,.$$

**Lemma 4.3**

$$2a_{1\ldots i} - c_{1\ldots i}b_{i+1\ldots n-2} = \sum_{j=1}^{i}(2a_j - b_j) + b_{1\ldots n-2}\,,$$

***Proof***

$$\begin{aligned}
&2a_{1\ldots i} - c_{1\ldots i}b_{i+1\ldots n-2}\\
&= 2a_{1\ldots i-1} + 2a_i - 2b_{1\ldots i-1}b_i + c_{1\ldots i-1}c_i b_{i+1\ldots n-2}\\
&= 2a_{1\ldots i-1} + 2a_i - 2b_{1\ldots i-1}b_i + 2c_{1\ldots i-1}b_i b_{i+1\ldots n-2} - c_{1\ldots i-1}b_{i+1\ldots n-2}\\
&= 2a_{1\ldots i-1} + 2a_i - b_i - c_{1\ldots i-1}(b_{i+1\ldots n-2} + b_i - 2b_i b_{i+1\ldots n-2})\\
&= 2a_{1\ldots i-1} - c_{1\ldots i-1}b_{i\ldots n-2} + 2a_i - b_i\,,
\end{aligned}$$

and then use induction.

Applying Lemma 4.3 and the last intermediate expression in its proof to the terms considered above, they simplify to

$$b_{1\ldots n-2} + \sum_{i=1}^{n-3}(2a_{1\ldots i} - c_{1\ldots i}b_{i+1\ldots n-2})^2 + 2a_{1\ldots n-2}^2 - b_{1\ldots n-2}(2a_1 - c_1 b_{2\ldots n-2})$$
$$- \sum_{i=1}^{n-4}(2a_{1\ldots i} - c_{1\ldots i}b_{i+1\ldots n-2})(2a_{1\ldots i} - c_{1\ldots i}b_{i+1\ldots n-2} + 2a_{i+1} - b_{i+1})$$
$$-2(2a_{1\ldots n-3} - c_{1\ldots n-3}b_{n-2})a_{1\ldots n-2}$$
$$= b_{1\ldots n-2} + (2a_{1\ldots n-3} - c_{1\ldots n-3}b_{n-2})^2 + 2a_{1\ldots n-2}^2 - b_{1\ldots n-2}(2a_1 - c_1 b_{2\ldots n-2})$$
$$- \sum_{i=1}^{n-4}(2a_{1\ldots i} - c_{1\ldots i}b_{i+1\ldots n-2})(2a_{i+1} - b_{i+1}) - 2(2a_{1\ldots n-3} - c_{1\ldots n-3}b_{n-2})a_{1\ldots n-2}$$
$$= b_{1\ldots n-2} + (2a_{1\ldots n-3} - c_{1\ldots n-3}b_{n-2})(2a_{1\ldots n-3} - c_{1\ldots n-3}b_{n-2} - 2a_{1\ldots n-2})$$

$$+2a_{1\ldots n-2}^2 - b_{1\ldots n-2}(2a_1 - c_1 b_{2\ldots n-2}) - \sum_{i=1}^{n-4}(2a_{1\ldots i} - c_{1\ldots i}b_{i+1\ldots n-2})(2a_{i+1} - b_{i+1})$$

$$= b_{1\ldots n-2} + 2a_{1\ldots n-2}^2 - b_{1\ldots n-2}(2a_1 - c_1 b_{2\ldots n-2})$$

$$- \sum_{i=1}^{n-3}(2a_{1\ldots i} - c_{1\ldots i}b_{i+1\ldots n-2})(2a_{i+1} - b_{i+1})$$

$$= 2a_{1\ldots n-2}^2 - b_{1\ldots n-2}(2a_1 - b_1) - \sum_{i=1}^{n-3}\left(\sum_{j=1}^{i}(2a_j - b_j) + b_{1\ldots n-2}\right)(2a_{i+1} - b_{i+1}) .$$

Let us denote this expression temporarily as $s_{n-2}$. Taking into account that

$$a_{1\ldots n-2} = a_{1\ldots n-3} + a_{n-2} - b_{1\ldots n-3}b_{n-2} ,$$

$$b_{1\ldots n-2} = b_{1\ldots n-3} + b_{n-2} - 2b_{1\ldots n-3}b_{n-2} ,$$

$s_{n-2}$ can be rewritten as

$$2a_{1\ldots n-3}^2 + 2a_{n-2}^2 + 2b_{1\ldots n-3}b_{n-2} + 4a_{1\ldots n-3}a_{n-2}$$
$$-4a_{1\ldots n-3}b_{1\ldots n-3}b_{n-2} - 4a_{n-2}b_{1\ldots n-3}b_{n-2}$$
$$-(b_{1\ldots n-3} + b_{n-2} - 2b_{1\ldots n-3}b_{n-2})(2a_1 - b_1)$$
$$-\sum_{i=1}^{n-4}\left(\sum_{j=1}^{i}(2a_j - b_j) + b_{1\ldots n-3}\right)(2a_{i+1} - b_{i+1})$$
$$-\sum_{i=1}^{n-3}(b_{n-2} - 2b_{1\ldots n-3}b_{n-2})(2a_{i+1} - b_{i+1})$$
$$-\left(\sum_{j=1}^{n-3}(2a_j - b_j) + b_{1\ldots n-3}\right)(2a_{n-2} - b_{n-2})$$
$$= s_{n-3} + 2a_{n-2}^2 + 2b_{1\ldots n-3}b_{n-2} + 4a_{1\ldots n-3}a_{n-2}$$
$$-4a_{1\ldots n-3}b_{1\ldots n-3}b_{n-2} - 4a_{n-2}b_{1\ldots n-3}b_{n-2}$$
$$-(b_{n-2} - 2b_{1\ldots n-3}b_{n-2})\left(\sum_{i=1}^{n-2}2a_i - b_i\right)$$
$$-\left(\sum_{j=1}^{n-3}(2a_j - b_j) + b_{1\ldots n-3}\right)(2a_{n-2} - b_{n-2})$$

$$= s_{n-3} + 2a_{n-2}^2 - 2a_{n-2}b_{n-2} + b_{n-2} + 2a_{n-2}\left(2a_{1\ldots n-3} - \sum_{j=1}^{n-3}(2a_j - b_j)\right)$$
$$+b_{1\ldots n-3}b_{n-2} - 2a_{n-2}b_{1\ldots n-3}$$
$$= s_{n-3} + 2a_{n-2}^2 - 2a_{n-2}b_{n-2} + b_{n-2} - b_{1\ldots n-3}b_{n-2}\,,$$

where at the last equality the identity

$$\sum_{j=1}^{n-3}(2a_j - b_j) = z_{1\ldots n-3} = 2a_{1\ldots n-3} - b_{1\ldots n-3}$$

was used.

It can be checked that $s_1 = 2a_1^2 - 2a_1b_1 + b_1$, so induction shows that

$$s_{n-2} = \sum_{i=1}^{n-2}(2a_i^2 - 2a_ib_i + b_i + b_{1\ldots i-1}b_i)\,.$$

It remains to show that

$$\sum_{i=2}^{n-2} b_{1\ldots i-1}b_i = r_{1\ldots n-2}\,,$$

which requires another induction argument, and is left to the reader.

(3) Apply Lemma 4.3 on

$$\sum_{i=1}^{n} m_i = -b_{1\ldots n-2}$$
$$-\sum_{i=1}^{n-2}(2a_{1\ldots i-1} - c_{1\ldots i-1}b_{i\ldots n-2}) - 2a_{1\ldots n-2} - (n-1)c_{1\ldots n-2}(z_{n-1} + z_n).$$

□

**Acknowledgments** The author is thankful to Jim Bryan, András Némethi and Balázs Szendrői for fruitful conversations about the topic. The author was supported by the EPSRC grant EP/R045038/1 while completing this work.

## References

1. Białynicki-Birula, A.: Some theorems on actions of algebraic groups. Ann. Math. **98**, 480–497 (1973)

2. Bryan, J., Gyenge, Á.: $G$-fixed Hilbert schemes on K3 surfaces, modular forms, and eta products (2019). Preprint, arXiv:1907.01535
3. Cartan, H.: Quotient d'un espace analytique par un groupe d'automorphismes. In: Algebraic Geometry and Topology: A Symposium in Honor of S. Lefschetz, pp. 90–102. Princeton University Press, Princeton (1957)
4. Fujii, Sh., Minabe, S.: A combinatorial study on quiver varieties. Symm. Integr. Geom. Methods Appl. **13**, 052 (2017)
5. Göttsche, L.: The Betti numbers of the Hilbert scheme of points on a smooth projective surface. Math. Ann. **286**(1), 193–207 (1990)
6. Gusein-Zade, S.M., Luengo, I., Melle-Hernández, A.: A power structure over the Grothendieck ring of varieties. Math. Res. Lett. **11**(1), 49–57 (2004)
7. Gyenge, Á.: Hilbert schemes of points on some classes of surface singularities. Ph.D. thesis, Eötvös Lóránd University, Budapest (2016)
8. Gyenge, Á.: Enumeration of diagonally colored young diagrams. Monatsh. Math. **183**(1), 143–157 (2017)
9. Gyenge, Á., Némethi, A., Szendrői, B.: Euler characteristics of Hilbert schemes of points on surfaces with simple singularities. Int. Math. Res. Not. **2017**(13), 4152–4159 (2016)
10. Gyenge, Á., Némethi, A., Szendrői, B.: Euler characteristics of Hilbert schemes of points on simple surface singularities. Eur. J. Math. **4**(2), 439–524 (2018)
11. Ito, Y., Nakamura, I.: Hilbert schemes and simple singularities. In: New Trends in Algebraic Geometry. London Mathematical Society Lecture Note Series, 151–234. Cambridge University Press, Cambridge (1999)
12. James, G., Kerber, A.: The Representation Theory of the Symmetric Group. Encyclopedia Math. Appl. Addison-Wesley, Reading (1981)
13. Kang, S.-J., Kwon, J.-H.: Crystal bases of the Fock space representations and string functions. J. Algebra **280**(1), 313–349 (2004)
14. Kwon, J.-H.: Affine crystal graphs and two-colored partitions. Lett. Math. Phys. **75**(2), 171–186 (2006)
15. Nagao, K.: Quiver varieties and Frenkel–Kac construction. J. Algebra **321**(12), 3764–3789 (2009)
16. Nakajima, H.: Geometric construction of representations of affine algebras. In: Proceedings of the International Congress of Mathematicians (Beijing, 2002), vol. 1, IMU, pp. 423–438. Higher Education Press, Beijing (2002)

# Real Seifert Forms, Hodge Numbers and Blanchfield Pairings

**Maciej Borodzik and Jakub Zarzycki**

**Abstract** In this survey article we present connections between Picard–Lefschetz invariants of isolated hypersurface singularities and Blanchfield forms for links. We emphasize the unifying role of Hermitian Variation Structures introduced by Némethi.

**Keywords** Seifert forms · Hodge numbers · Milnor fibration · Linking pairings · Blanchfield pairings

## 1 Introduction

Understanding a mathematical object via decomposing it into simple pieces is a very general procedure in mathematics, which can be seen in various branches and various fields. These procedures, often very different from each other, sometimes share common properties. In some cases, one mathematical object is defined in several fields and one procedure of decomposing is known under different names in different areas of mathematics.

The aim of this article is to show a bridge between real Blanchfield forms in knot theory and real Hermitian Variation Structures in singularity theory. In fact, we want to explain that these two apparently distant objects describe (almost) the same concept. Moreover, the methods for studying these two objects are very close. To be more specific, classification of simple Hermitian Variation Structures is an instance of a procedure known in algebraic geometry and algebraic topology as *dévissage*, which—in a vague sense—can be seen as a refinement of a primary decomposition of a torsion module over a PID. Dévissage is an important method of studying abstract linking forms, in particular, Blanchfield forms.

M. Borodzik (✉) · J. Zarzycki
Institute of Mathematics, University of Warsaw, Warsaw, Poland
e-mail: mcboro@mimuw.edu.pl; jz371722@students.mimuw.edu.pl



J. Fernández de Bobadilla et al. (eds.), *Singularities and Their Interaction with Geometry and Low Dimensional Topology*, Trends in Mathematics,
https://doi.org/10.1007/978-3-030-61958-9_4

These two points of view: the Hodge-theoretical one and the algebraic one, give possibility to apply methods of one field to answer questions that arise in another field. In this way, the first author and Némethi gave a proof of semicontinuity of a spectrum of a plane curve singularity [5] using Murasugi inequality of signatures. Conversely, the Hodge theoretic approach to Blanchfield forms, allows us to define Hodge-type invariants for links in $S^3$. Using these invariants we can quickly compute knot invariants based on a small piece of data: an exemplary calculation is shown in Example 4.1.

Another feature of Hodge-theoretical perspective is the formula for the Tristram–Levine signature, which we state in Proposition 4.7. This formula allows us to define the analog of the Tristram–Levine signature for twisted Blanchfield pairings, compare Definition 6.2. Many existing constructions of similar objects involve a choice of a matrix *representing* a pairing, see [7, Section 3.4]. However, finding a matrix representing given pairing, even for pairings over $\mathbb{C}[t, t^{-1}]$ is not a completely trivial task, see e.g. [7, Proposition 3.12]. The approach through Hodge numbers allows us to bypass this difficulty.

The structure of the paper is the following. In Sect. 2 we recall the basics of Picard–Lefschetz theory. This section serves as a motivation for introducing abstract Hermitian Variation Structures in Sect. 3. Section 4 recalls the construction of a Hermitian Variation Structure for general links in $S^3$. We also clarify several results of Keef, which were not completely correctly referred to in [6]. In Sect. 5 we give a definition of Blanchfield forms. We show that there is a correspondence between real Blanchfield forms and real Hermitian Variation Structures associated with the link. Moreover, the classification of the two objects is very similar.

In the last Sect. 6 we sketch the construction of twisted Blanchfield pairings and introduce Hodge numbers for such structures. We show how to recover the signature function from such a pairing. An example is given by Casson–Gordon signatures.

## 2 Milnor Fibration and Picard-Lefschetz Theory

Let $f : (\mathbb{C}^{n+1}, 0) \to (\mathbb{C}, 0)$ be a polynomial map with $0 \in \mathbb{C}^{n+1}$ an isolated critical point.

**Theorem 2.1 (Milnor's Fibration Theorem, See [23])** *For $\varepsilon > 0$ sufficiently small, the map $\Psi : S^{2n+1}_\varepsilon \setminus f^{-1}(0) \to S^1$ given by $\Psi(z) = \frac{f(z)}{\|f(z)\|}$ is a locally trivial fibration. The fiber $\Psi^{-1}(1)$ has the homotopy type of a wedge sum of some finite number of spheres $S^n$.*

The map $\Psi$ can be explicitly described near the set $f^{-1}(0)$. Namely, choose a sufficiently small regular neighborhood $N$ of $f^{-1}(0)$. It has a structure of a trivial $D^2$ bundle over $f^{-1}(0)$. Choose polar coordinates $(r, \phi)$ on $D^2 \setminus \{0\}$. For some choice of trivialization $N \cong f^{-1}(0) \times D^2$, the map $\Psi$ is given by $\Psi(x, r, \phi) = \frac{1}{2\pi}\phi$ (here $x \in f^{-1}(0)$, $r, \phi$ are coordinates on the disk). See [35, Section 2.4.13] for more details.

Let $F_t$ be the fiber $\Psi^{-1}(t)$. The *geometric monodromy* $h_t$ is a diffeomorphism $h_t : F_1 \to F_t$, smoothly depending on $t$, which corresponds to the trivialization of the Milnor fibration on the arc of $S^1$ from 1 to $t$. Note that $h_t$ is well-defined only up to homotopy.

**Definition 2.1** The *homological monodromy* is the map $h : H_n(F_1;\mathbb{Z}) \to H_n(F_1;\mathbb{Z})$ induced by the monodromy.

*Remark 2.1* The monodromy map in $N \setminus f^{-1}(0)$ can be defined to be $h_t(x, r, 0) = (x, r, 2\pi t)$. In particular, $h_1$ is the identity on $F_1 \cap (N \setminus f^{-1}(0))$.

The homological monodromy is not the only invariant that can be associated with the Milnor fibration. Let $\widetilde{F}_1 = \overline{F_1 \setminus N}$. Then $\widetilde{F}_1$ is a manifold with boundary homotopy equivalent to $F_1$. The monodromy map $h_1$ takes $\widetilde{F}_1$ to $\widetilde{F}_1$ and it is the identity on $\partial\widetilde{F}_1$.

Consider now a relative cycle $\alpha \in H_n(\widetilde{F}_1, \partial\widetilde{F}_1;\mathbb{Z})$. The cycle $h_1(\alpha) - \alpha$ has no boundary, hence it is an absolute cycle.

**Definition 2.2** The *variation map* var: $H_n(\widetilde{F}_1, \partial\widetilde{F}_1;\mathbb{Z}) \to H_n(\widetilde{F}_1;\mathbb{Z})$ is the map defined as $\operatorname{var}\alpha = h_1(\alpha) - \alpha$.

*Remark 2.2* Poincaré–Lefschetz duality for $F_1$ implies that $H_n(\widetilde{F}_1, \partial\widetilde{F}_1;\mathbb{Z}) \cong \operatorname{Hom}(H_n(\widetilde{F}_1;\mathbb{Z}),\mathbb{Z})$. Therefore, the variation map can be regarded as a map from $H_n(F_1;\mathbb{Z})^*$ to $H_n(F_1;\mathbb{Z})$.

We can also define a bilinear form based on linking numbers of $n$-cycles in $S^{2n+1}$.

**Definition 2.3** The *Seifert form* is the map $L : H_n(F_1,\mathbb{Z}) \times H_n(F_1,\mathbb{Z}) \to \mathbb{Z}$ given by $L(\alpha,\beta) = lk(\alpha, h_{\frac{1}{2}}\beta)$.

Here $\operatorname{lk}(A, B)$ is the generalized linking pairing of two disjoint $n$-cycles in $S^{2n+1}$. A classical definition deals first with the case when $H_n(B;\mathbb{Z}) \cong \mathbb{Z}$, e.g. $B$ is a closed connected orientable $n$-dimensional manifold. In this case, the choice of isomorphism $H_n(B;\mathbb{Z}) \cong \mathbb{Z}$ (eg. given by choosing an orientation of $B$) gives, via Alexander duality, an isomorphism $H_n(S^{2n+1} \setminus B;\mathbb{Z}) \cong \mathbb{Z}$. Then, we define $\operatorname{lk}(A, B)$ as the class of $A$ in $H_n(S^{2n+1} \setminus B;\mathbb{Z})$. The definition is later extended to the case when $B$ is a sum of cycles with $H_n(B;\mathbb{Z}) \cong \mathbb{Z}$.

There are relations between the variation map, the Seifert form and the monodromy. These are usually called Picard–Lefschetz formulae. References include [35, Lemma 4.20] and [1].

**Theorem 2.2** *The Seifert form, the variation map, the monodromy and the intersection form on $H_n(F_1;\mathbb{Z})$ are related by the following formulae:*

$$L(\operatorname{var} a, b) = \langle a, b\rangle$$

$$\langle a, b\rangle = -L(a, b) + (-1)^{n+1}L(b, a)$$

$$h = (-1)^{n+1}\operatorname{var}(\operatorname{var}^{-1})^*.$$

*Here* $\langle\cdot,\cdot\rangle$ *denotes the intersection form on* $H_n(F_1;\mathbb{Z})$.

Theorem 2.2 is a motivation to introduce Hermitian Variation Structures, which are the subject of the next section.

# 3 Hermitian Variation Structures and Their Classification

## 3.1 Abstract Definition

Let $\mathbb{F}$ be a field of characteristic zero. By $\bar{\cdot}$ we denote the involution of $\mathbb{F}$: if $\mathbb{F}=\mathbb{C}$, then it is a complex conjugation, if $\mathbb{F}=\mathbb{R},\mathbb{Q}$, then the involution is the identity. Set $\zeta=\pm 1$.

**Definition 3.1** A *$\zeta$-Hermitian variation structure* over $\mathbb{F}$ is a quadruple $(U;b,h,V)$ where

(HVS1) $U$ is a finite dimensional vector space over $\mathbb{F}$;
(HVS2) $b:U\to U^*$ is a $\mathbb{F}$-linear endomorphism with $\overline{b^*\circ\theta}=\zeta b$, where $\theta: U\to U^{**}$ is the natural isomorphism;
(HVS3) $h:U\to U$ is $b$-orthogonal, that is $\overline{h}^*\circ b\circ h=b$;
(HVS4) $V:U^*\to U$ is a $\mathbb{F}$-linear endomorphism with $\overline{\theta^{-1}\circ V^*}=-\zeta V\circ\overline{h^*}$ and $V\circ b=h-I$.

The motivation is clearly Picard–Lefschetz theory. Suppose $f\colon(\mathbb{C}^{n+1},0)\to(\mathbb{C},0)$ is a polynomial map as in Sect. 2. The following result is a direct consequence of Theorem 2.2.

**Proposition 3.1** *Consider the quadruple* $(U,b,h,V)$, *where* $U=H_n(F_1;\mathbb{C})$, $b\colon H_n(F_1;\mathbb{C})\to H_n(F_1,\partial F_1;\mathbb{C})=\operatorname{Hom}_{\mathbb{C}}(H_n(F_1;\mathbb{C});\mathbb{C})$ *is the Poincaré–Lefschetz duality,* $h\colon U\to U$ *is the homological monodromy and* $V$ *is the variation map. Then* $(U,b,h,V)$ *is a Hermitian Variation Structure over* $\mathbb{C}$ *with* $\zeta=(-1)^n$.

Relations (HVS3) and (HVS4) suggest that having two of the three operators $b,h$ and $V$ we can recover the third one. This is true under some conditions, which we are now going to spell out.

**Lemma 3.1**

*(a) If* $b$ *is an isomorphism then* $V=(h-I)b^{-1}$. *The HVS is determined by the triple* $(U;h,b)$
*(b) If* $V$ *is an isomorphism then* $h=-\zeta V\overline{(\theta^{-1}\circ V^*)^{-1}}$ *and* $b=-V^{-1}-\zeta\overline{(\theta^{-1}\circ V^*)^{-1}}$. *So* $V$ *determines the HVS.*

**Definition 3.2** The HVS such that $b$ is an isomorphism is called *nondegenerate*. If $V$ is an isomorphism, we say that the HVS is *simple*.

### 3.2 Classification of HVS Over $\mathbb{C}$

In [25] Némethi provides a classification of simple HVS over $\mathbb{F} = \mathbb{C}$. This classification is based on a Jordan block decomposition of the operator $h$. Note that we do not usually assume that all the eigenvalues of the monodromy operator are roots of unity, as is the case of the HVS associated with isolated hypersurface singularities.

Following [25] we first list examples of HVS. Then we state the classification result. In the following we let $J_k$ denote the $k$-dimensional matrix $\{c_{ij}\}$, with $c_{ij} = 1$ for $j = i, i+1$ and $c_{ij} = 0$ otherwise, that is, $J_k$ is the single Jordan block of size $k$.

*Example 3.1* Let $\nu \in \mathbb{C}^* \setminus S^1$ and $\ell \geq 1$. Define

$$\mathcal{V}_\nu^{2\ell} = \left(\mathbb{C}^{2\ell}; \begin{pmatrix} 0 & I \\ \zeta I & 0 \end{pmatrix}, \begin{pmatrix} \nu J_\ell & 0 \\ 0 & \frac{1}{\overline{\nu}} J_\ell^{*-1} \end{pmatrix}, \begin{pmatrix} 0 & \zeta(\nu J_\ell - I) \\ \frac{1}{\overline{\nu}} J_\ell^{*-1} - I & 0 \end{pmatrix}\right).$$

Then $\mathcal{V}_\nu^{2\ell}$ is a HVS. Furthermore, $\mathcal{V}_\nu^{2\ell}$ and $\mathcal{V}_{1/\overline{\nu}}^{2\ell}$ are isomorphic.

Before we state the next example, we need a simple lemma.

**Lemma 3.2** *Let $k \geq 1$ and $\zeta = \pm 1$. Up to a real positive scaling, there are precisely two non-degenerate matrices $b_\pm^k$ such that*

$$\overline{b_\pm^k}^* = \zeta b \text{ and } J_k^* b_\pm^k J_k = b_\pm^k.$$

The entries of $b_\pm^k$ satisfy $(b_\pm^k)_{i,j} = 0$ for $i + j \leq k$ and $b_{i,k+1-i} = (-1)^{i+1} b_{1,k}$. Moreover, $(b_\pm^k)_{1,k}$ is a power of $i$.

**Convention 3.1** *By convention, we choose signs in such a way that $(b_\pm^k)_{1,k} = \pm i^{-n^2-k+1}$, where $n$ is such that $\zeta = (-1)^n$.*

Using $b_\pm^k$ we can give an example of a HVS corresponding to the case $\mu \in S^1$.

**Lemma 3.3** *Let $\mu \in S^1$ and $k \geq 1$ be an integer. Up to isomorphism, there are two non-degenerate HVS such that $h = \mu J_k$. These structures have $b = b_+^k$ and $b = b_-^k$, respectively.*

For these two structures we use the notation:

$$\mathcal{V}_\mu^k(\pm 1) = \left(\mathbb{C}^k; b_\pm^k, \mu J_k, (\mu J_k - I)(b_\pm^k)^{-1}\right).$$

These two structures are simple unless $\mu = 1$. For $\mu = 1$ we need another construction of a simple HVS.

**Lemma 3.4** *Suppose $k \geq 2$. There are two degenerate HVS with $h = J_k$. These are:*

$$\widetilde{\mathcal{V}}_1^k(\pm 1) = \left(\mathbb{C}^k; \widetilde{b}_{\pm}, J_k, \widetilde{V}_{\pm}^k\right),$$

*where*

$$\widetilde{b}_{\pm}^k = \begin{pmatrix} 0 & 0 \\ 0 & b_{\pm}^{k-1} \end{pmatrix}$$

*and $\widetilde{V}_{\pm}^k$ is uniquely determined by b and h. Moreover, $\widetilde{\mathcal{V}}_1^k(\pm 1)$ is simple.*

While Lemma 3.4 deals with the case $k \geq 2$, there remains the case $k = 1$. Then, with $\mu = 1$, that is, $h = 1$, all possible structures can be enumerated explicitly. These are the following.

$$\mathcal{V}_1^1(\pm 1) = (\mathbb{C}, \pm i^{-n^2}, I, 0)$$

$$\widetilde{\mathcal{V}}_1^1(\pm 1) = (\mathbb{C}, 0, I, \pm i^{n^2+1})$$

$$\mathcal{T} = (\mathbb{C}, 0, I, 0).$$

In the above list, the structures $\mathcal{V}_1^1(\pm 1)$ and $\mathcal{T}$ are non-simple, and $\widetilde{\mathcal{V}}_1^1(\pm 1)$ are simple.

Concluding, for any $\mu \in S^1$ and in each dimension $k$, there are precisely two non-equivalent simple variation structures with $h = \mu J_k$. We use the following uniform notation for them:

$$\mathcal{W}_\mu^k(\pm 1) = \begin{cases} \mathcal{V}_\mu^k(\pm 1) & \text{if } \mu \neq 1 \\ \widetilde{\mathcal{V}}_1^k(\pm 1) & \text{if } \mu = 1. \end{cases} \tag{3.1}$$

The following result is one of the main results of [25].

**Theorem 3.2** *A simple HVS is uniquely expressible as a sum of indecomposable ones up to ordering of summands and up to an isomorphism. The indecomposable pieces are*

$$\mathcal{W}_\mu^k(\pm 1) \quad \textit{for } k \geq 1, \mu \in S^1$$

$$\mathcal{V}_\nu^{2\ell} \quad \textit{for } \ell \geq 1, 0 < |\nu| < 1.$$

**Definition 3.3** Let $\mathcal{M}$ be a simple HVS. The *Hodge number* $p_\mu^k(\pm 1)$ for $\mu \in S^1$ is the number of times the structure $\mathcal{W}_\mu^k(\pm 1)$ enters $\mathcal{M}$ as a summand. The Hodge number $q_\nu^\ell$ for $|\nu| \in (0, 1)$ is the number of times the structure $\mathcal{V}_\nu^{2\ell}$ enters $\mathcal{M}$ as a summand.

For an isolated hypersurface singularity, the whole 'Picard–Lefschetz package', that is, the monodromy, the variation map, the intersection form and the Seifert form, are defined over the integers. Passing to $\mathbb{C}$ in the definition of a Hermitian Variation Structure means that some information is lost. While we do not know how to recover the part coming from integer coefficients, the part of data coming from real coefficients is easy to see.

Suppose $\mathcal{M} = (U, b, h, V)$ is a HVS over $\mathbb{R}$. We construct a complexification of $\mathcal{M}$ by considering $\mathcal{M}_{\mathbb{C}} = (U \otimes \mathbb{C}, b \otimes \mathbb{C}, h \otimes \mathbb{C}, V \otimes \mathbb{C})$. Using Definition 3.3 we can associate Hodge numbers with $\mathcal{M}_{\mathbb{C}}$. The following result is implicit in [25], see also [6, Lemma 2.14].

**Lemma 3.5** *The Hodge numbers of $\mathcal{M}$ satisfy*

$$p^k_{\overline{\mu}}(u) = p^k_{\mu}((-1)^{k+1+s}\zeta u) \text{ and } q^{\ell}_{\overline{\nu}} = q^{\ell}_{\nu}.$$

*Here $s = 1$ if $\mu = 1$, otherwise $s = 0$.*

The definition of a HVS is a generalization of the definition of Milnor's isometric structure [24]; compare also [26]. Lemma 3.1 implies that if the intersection form is an isomorphism, then the HVS is determined by the underlying isometric structure. Classification Theorem 3.2 shows, that the only simple degenerate HVS correspond to the eigenvalue $\mu = 1$. This is the main feature of the concept of a HVS: it allows us to deal with the case $\mu = 1$.

## *3.3 The Mod 2 Spectrum*

The spectrum of an isolated hypersurface singularity was introduced by Steenbrink in [31]. It is an unordered $s$-tuple of rational numbers $a_1, \dots, a_s \in (0, n+1]$, where $n$ is the dimension of the hypersurface and $s$ is the Milnor number. The spectrum is one of the deepest invariants of hypersurface singularities. The definition of the spectrum involves the study of mixed Hodge structures associated with a singular point. We now show, following Némethi, that the mod 2 reduction (the tuple $a_1 \bmod 2, \dots, a_s \bmod 2$) of the spectrum can be recovered from Hodge numbers. In particular, for plane curve singularities, the whole spectrum is determined by the Hodge numbers.

**Theorem 3.3 (See [25, Theorem 6.5])** *Let $p^k_{\mu}(u)$ be the Hodge numbers of an isolated hypersurface singularity in $\mathbb{C}^{n+1}$. For any $\alpha \in (0, 2) \setminus \{1\}$, the multiplicity of $\alpha$ in the mod 2 spectrum is equal to*

$$\sum_{k=1}^{\infty} \sum_{\epsilon=\pm 1} k p^{2k}_{\mu}(\epsilon) + \sum_{k=1}^{\infty} \sum_{\epsilon=\pm 1} (k + 1 - \epsilon \lfloor \alpha \rfloor) p^{2k+1}_{\mu}(\epsilon),$$

*where $\mu = e^{2\pi i \alpha}$.*

The integer part of the spectrum, i.e. the case $\alpha \in \{1, 2\}$ can be treated in a similar manner.

## 4 HVS for Knots and Links

From now on we assume that $\zeta = -1$, so we consider only $(-1)$-variation structures.

### 4.1 Three Results of Keef

The monodromy, the variation and the intersection form for an isolated hypersurface singularity are defined homologically. The construction does not involve any analytic structure, that is, we need only existence of a topological fibration of the complement of the link of singularity over $S^1$. Therefore, if we have any fibered link $L \subset S^3$, we can use the same approach as above to define a HVS for such link. With a choice of a basis of $H_1(F)$, where $F$ is the fiber, the variation map is the inverse of the Seifert matrix.

The construction can be extended further: take a link with Seifert matrix $S$ and associate to it a simple HVS with variation map $S^{-1}$. Now the Seifert matrix is defined only up to $S$-equivalence (see [16, Section 5.2]) and need not be invertible in general. We shall use results of Keef to show that every Seifert matrix is $S$-equivalent to a block sum of an invertible matrix and a zero matrix. This invertible matrix is well-defined up to rational congruence (for an analogous result for knots refer to [16, Theorem 12.2.9]). Therefore, a HVS for any link in $S^3$ is defined.

Hereafter, where we mean S-equivalence, we mean S-equivalence with rational coefficients. As shown in [33], not all the results carry over to the case of $\mathbb{Z}$.

**Proposition 4.1 (See [19, Proposition 3.1])** *Any Seifert matrix $S$ for a link $L$ is S-equivalent over $\mathbb{Q}$ to a matrix $S'$ which is a block sum of a zero matrix and an invertible matrix $S_{in}$.*

**Proposition 4.2 (See [19, Theorem 3.5])** *Suppose $S = S_0 + S_{in}$ and $T = T_0 + T_{in}$ be two matrices over $\mathbb{Q}$, presented as block sums of a zero matrix (that is, $S_0$ and $T_0$) and an invertible matrix (that is, $S_{in}$ and $T_{in}$). The matrices $S$ and $T$ are S-equivalent if and only if they are congruent over $\mathbb{Q}$. Furthermore, if $S$ and $T$ have the same size, then congruence of $S$ and $T$ is equivalent to congruence of $S_{in}$ and $T_{in}$.*

**Proposition 4.3 (See [19, Theorem 3.6])** *Two matrices $S$ and $T$ are S-equivalent over $\mathbb{Q}$ if and only if their Seifert systems are isomorphic.*

Here, a *Seifert system* relative to a square matrix $S$ consists of the module $A_S = \mathbb{Q}[t, t^{-1}]/(tS - S^T)$ and a pairing on the torsion part of $A_S$ as defined in [19, Section 2].

From these three results we deduce the following fact. This result was often used in [6], but actually its proof was never written down in detail.

**Proposition 4.4** *Suppose $S$ is S-equivalent to matrices $S'$ and $S''$, which are both block sums of zero matrices $S'_0$ and $S''_0$ and $S'_{in}$, $S''_{in}$, such that $S'_{in}$, $S''_{in}$ are non-degenerate. Then $S'_{in}$ and $S''_{in}$ are congruent over $\mathbb{Q}$.*

***Proof*** As $\mathbb{Q}[t, t^{-1}]$ is a PID, the module $A_{S'} = A_{S''}$ decomposes as a direct sum of the free part and the torsion part. The sizes of $S'_0$ and $S''_0$ are equal to the rank over $\mathbb{Q}[t, t^{-1}]$ of the free part of the module.

Let $TA$ denote the torsion-part of $A_{S'} = A_{S''}$. The order of $TA$ is the degree of the polynomial $\det(tS'_{in} - {S'_{in}}^T) = \det(tS''_{in} - {S''_{in}}^T)$. As $S'_{in}$ and $S''_{in}$ are invertible, the degree of $\det(tS'_{in} - {S'_{in}}^T)$ is equal to the size of $S'_{in}$. Therefore, the sizes of $S'_{in}$ and $S''_{in}$ are equal. By Proposition 4.2, this shows that $S'_{in}$ and $S''_{in}$ are congruent over $\mathbb{Q}$.

*Remark 4.1* One would be tempted to guess that given a matrix $S$, the size of $S_0$ is $\dim(\ker S \cap \ker S^T)$. Such remark was made in [5, Section 2.2] but it was nowhere used. In fact, it is false. For a counterexample, take

$$S = \begin{pmatrix} 0 & 0 & 1 \\ 1 & 0 & 0 \\ 0 & 0 & 0 \end{pmatrix}.$$

One readily checks that $\ker S \cap \ker S^T = 0$ but $S$ is S-equivalent to the matrix $(0)$. So $\dim S_0 = 1$.

**Definition 4.1** Let $L \subset S^3$ be a link with Seifert matrix $S$. Suppose $S$ is S-equivalent to $S'$, which is a block sum of a zero matrix and an invertible matrix $S_{in}$. The Hermitian Variation Structure for $L$ is the Hermitian Variation Structure $\mathcal{M}(L)$ for which the variation operator is the inverse of $S_{in}$.

>From Proposition 4.4 we deduce the following result.

**Corollary 4.1** *The Hermitian Variation Structure $\mathcal{M}(L)$ is independent on the S-equivalence class of the matrix $S$, i.e. it is an invariant of $L$.*

## 4.2 HVS for Links and Classical Invariants

Given the link $L \subset S^3$ and the HVS $\mathcal{M}(L)$ we define Hodge numbers for $L$. Denote them $p^k_\mu(\pm 1)$ and $q^\ell_\nu$. The Hodge numbers determine the one-variable Alexander polynomial of $L$ over $\mathbb{R}$ and the signature function. To describe the relation in more detail, we introduce a family of polynomials.

$$
\begin{aligned}
&B_1(t) = (t-1), \quad B_{-1}(t) = (t+1) \\
&B_\mu(t) = (t-\mu)(1-\overline{\mu}t^{-1}) && \mu \in S^1,\ \operatorname{im}\mu > 0 \\
&B_\mu(t) = (t-\mu)(1-\mu^{-1}t^{-1}) && \mu \in \mathbb{R},\ 0<|\mu|<1 \\
&B_\mu(t) = (t-\mu)(t-\overline{\mu})(1-\mu^{-1}t^{-1})(1-\overline{\mu}^{-1}t^{-1}) && \mu \notin S^1 \cup \mathbb{R},\ 0<|\mu|<1.
\end{aligned}
\tag{4.1}
$$

The (Laurent) polynomials $B_\mu$ for $\mu \notin \{1,-1\}$ are characterized by the property that they have real coefficients, they are symmetric ($B_\mu(t) = B_\mu(t^{-1})$) and they cannot be presented as products of real symmetric polynomials. Moreover, these are (up to multiplication by $t$) the characteristic polynomials of the monodromy operators associated with HVS $\mathcal{W}^k_\mu$. With notation (4.1) we obtain (see [7, Section 4.1]):

**Proposition 4.5** *Let L be a knot. Then—up to multiplication by a unit in* $\mathbb{R}[t,t^{-1}]$*— the Alexander polynomial of L is equal to*

$$
\Delta_L(t) = \prod_{\substack{\mu\in S^1 \\ \operatorname{im}\mu\geq 0}} \prod_{\substack{k\geq 1 \\ u=\pm 1}} B_\mu(t)^{p^k_\mu(u)} \cdot \prod_{\substack{0<|\nu|<1 \\ \operatorname{im}\nu\geq 0}} \prod_{\ell\geq 1} B_\nu(t)^{q^\ell_\nu}. \tag{4.2}
$$

Another result gives the minimal number of generators of the Alexander module of a knot $L$ over $\mathbb{R}[t,t^{-1}]$; see [6, Section 4.3].

**Proposition 4.6** *Suppose* $\Delta_L$ *is not identically zero. The minimal number of generators of the Alexander module over* $\mathbb{R}[t,t^{-1}]$ *is equal to*

$$
\max\left(\max_{\mu\in S^1} \sum_{k,u} p^k_\mu(u),\ \max_{0<|\nu|<1} \sum_{\ell} q^\ell_\nu\right).
$$

The Hodge numbers of a link determine its Tristram–Levine signature. Recall that for a link $L$, the Tristram–Levine signature $\sigma_L(z)$ is the signature of the Hermitian matrix $(1-z)S + (1-\overline{z})S^T$, where $S$ is s Seifert matrix for $L$; see the recent survey of Conway [11] for the definition, properties and recent applications of signatures.

We will now show that the Hodge numbers determine the Tristram–Levine signature. The following two results can be deduced from [6, Proposition 4.14], see also [7, Section 5].

**Proposition 4.7** *Let L be a link and* $z_0 = e^{ix} \in S^1$ $(x \in (0,\pi))$ *be such that* $z_0$ *is not a zero of the Alexander polynomial of L. Then*

$$
\sigma_L(z_0) = -\sum_{y\in[0,x)} \sum_{\substack{u\in\{-1,1\} \\ k \text{ odd}}} u p^k_{e^{iy}}(u) + \sum_{y\in(x,\pi)} \sum_{\substack{u\in\{-1,1\} \\ k \text{ odd}}} u p^k_{e^{iy}}(u). \tag{4.3}
$$

**Proposition 4.8** *With notation of Proposition 4.7 if $z_0$ is a root of the Alexander polynomial, then:*

$$\sigma_L(z_0) - \frac{1}{2}\left(\lim_{t\to 0^+} \sigma_L(e^{it}z) + \sigma_L(e^{-it}z)\right) = \sum_{\substack{u\in\{-1,1\}\\ k \text{ even}}} u p^k_{z_0}(u). \tag{4.4}$$

The bottomline of Propositions 4.7 and 4.8 is that the Hodge numbers $p^k_\mu(u)$ with $k$ odd determine the values of signature functions outside of roots of the Alexander polynomial, while the Hodge numbers with $k$ even determine the jumps at the roots. Note that the jumps at the roots of the signature function (i.e. the left-hand side of (4.4)) are not concordance invariants.

Propositions 4.5, 4.7, 4.8 and 4.6 can be used to determine the Hodge numbers directly, without referring to explicit study of the Jordan block decomposition.

*Example 4.1* Let $K = 8_{20}$. It is known that $K$ is slice. From [10] we read off that $\Delta_K = (t-\mu)^2(t-\overline{\mu})^2$ for $\mu = \frac{1}{2}(1+i\sqrt{3})$. Moreover, the Nakanishi index (the minimal number of generators of the Alexander module of $K$) is 1. This data alone will determine all the Hodge numbers and—up to sign—also the signature function.

Namely, by Proposition 4.5 we know that for all $\lambda \neq \mu$, $k \geq 1$ and $u \in \{-1, 1\}$, we must have $p^k_\lambda(u) = 0$. Otherwise the Alexander polynomial for $K$ has a root a $\lambda$.

As the Alexander polynomial of $K$ has a double root at $\mu$, we infer that $\sum_{k,u} p^k_\mu(u) = 2$. As $p^k_\mu(u) \geq 0$ for all $k, u$, we have essentially two possibilities.

- $p^1_\mu(+1) + p^1_\mu(-1) = 2$;
- for some $u \in \{-1, 1\}$, $p^2_\mu(u) = 0$ and for $(k, w) \neq (2, u)$ we have $p^k_\mu(w) = 0$.

Now the Nakanishi index of $K$ is 1; by Proposition 4.6 the first case does not occur. Therefore $p^2_\mu(u) = 1$ and without extra data we cannot determine the sign $u$.

We conclude from Propositions 4.7 and 4.8 that the Tristram–Levine signature of $K$ is zero except for $\mu$ and $\overline{\mu}$, where it attains the value $u$.

Note that the maximum absolute value of the signature function is a lower bound for the unknotting number; see [11, Theorem 2.6] or [4, Theorem 4.1]. In particular, the $n$-fold connected sum of $K$, $nK$ has unknotting number at least $n/2$.

*Remark 4.2* Finding bounds for the unknotting number of smoothly slice knots is a notoriously difficult problem, because most known invariants that bound the unknotting number, are actually bounds for the 4-genus.

## *4.3 Signatures, HVS and Semicontinuity of the Spectrum*

Hodge numbers can be used to provide the relation between the signature of the link of singularity and the mod 2 spectrum. For simplicity, we state the result for curve singularities in $\mathbb{C}^2$.

**Theorem 4.1 (See [6, Corollary 4.15])** *Let $f\colon \mathbb{C}^2 \to \mathbb{C}$ define an isolated singularity with link $L$ and spectrum $Sp$. Suppose $x \in (0,1)$ is such that neither $x$ nor $x+1$ belong to the spectrum. Then*

$$\sigma_L(e^{2\pi i x}) = -\#Sp \cap (x, x+1) + \#Sp \setminus [x, x+1].$$

Theorem 4.1 can be regarded as a generalization of Litherland's formula expressing the signature of a torus knot in terms of the number elements in $Sp_{p,q} \cap (x, x+1)$, where $Sp_{p,q} = \{\frac{i}{p} + \frac{j}{q},\ 1 \le i < p,\ 1 \le j < q\}$ is the spectrum of singularity $x^p - y^q = 0$; see [21].

Spectrum of singularity is semicontinuous under deformation of singularities. While stating the result of Steenbrink and Varchenko [32, 34] is beyond the scope of this survey, we note that in [5], Murasugi inequality for signatures of links was used to obtain semicontinuity results.

# 5 Blanchfield Forms

We now pass to defining Blanchfield forms. In some sense, Blanchfield forms generalize Hermitian Variation Structures, although the connection might be hard to observe at first. We restrict to the case of knots, referring to [15] for the case of links. First, we need to set up some conventions. Suppose $R$ be a ring with involution (usually we consider $R = \mathbb{Z}, \mathbb{Q}, \mathbb{R}$ with trivial involution or $R = \mathbb{C}$ with complex conjugation). The ring $R[t, t^{-1}]$ has an involution given by $\overline{\sum a_j t^j} = \sum \overline{a_j} t^{-j}$.

## 5.1 Definitions

Let $K \subset S^3$ be a knot. Let $X = S^3 \setminus K$. By Alexander duality $H_1(X; \mathbb{Z}) = \mathbb{Z}$. Hurewicz theorem implies the existence of a surjetion $\pi_1(X) \to \mathbb{Z}$. We call the cover of $X$ corresponding to this surjection the *universal abelian cover* of $X$. We denote it by $\widetilde{X}$. The first homology group $H_1(\widetilde{X}; \mathbb{Z})$ has a structure of $\mathbb{Z}[t, t^{-1}]$-module, with multiplication by $t$ being induced by the action of the deck transformation on $\widetilde{X}$. This module is called the *Alexander module* of $K$. Usually it is denoted by $H_1(X; \mathbb{Z}[t, t^{-1}])$; in Sect. 5 we will denote it by $H$.

Blanchfield [2] defined a bilinear pairing $H \times H \to \mathbb{Q}(t)/\mathbb{Z}[t, t^{-1}]$. He also proved that it is Hermitian and non-degenerate. The pairing is nowadays called the *Blanchfield pairing* of $K$. Instead of going through the definition of the form, we will show how the pairing is computed.

**Theorem 5.1** *Let $K$ be a knot and let $S$ be a Seifert matrix for $K$, assume the size of $S$ is $n$. Denote $\Lambda = \mathbb{Z}[t, t^{-1}]$. Then $H = \Lambda^n/(tS - S^T)\Lambda^n$ and with this identification the Blanchfield pairing is $(x, y) \mapsto x^T(t-1)(S - tS^T)^{-1}\overline{y} \in \mathbb{Q}(t)/\Lambda$.*

*Remark 5.1* There is some confusion in the literature about the correct statement of Theorem 5.1. We refer the reader to [13], where various possibilities are discussed and some commonly appearing mistakes are corrected.

Theorem 5.1 shows that a Seifert matrix of $K$ determines the Blanchfield pairing. The reverse implication is also true; see e.g. [30, 33].

**Theorem 5.2** *The S-equivalence class of a Seifert matrix of a knot K is determined by the Blanchfield form.*

The importance of a Blanchfield form in knot theory justifies the following abstract definition.

**Definition 5.1** Let $R$ be an integral domain with (possibly trivial) involution. Let $\Omega$ be the field of fractions of $R$.

A *linking form* over $R$ is a pair $(M, \lambda)$, where $M$ is a torsion $R$-module and $\lambda\colon M \times M \to \Omega/R$ is a non-singular sesquilinear pairing. Here 'non-singular' means that the map $M \to \overline{\mathrm{Hom}_R(M, \Omega/R)}$ induced by $\lambda$ is an isomorphism.

We refer to Ranicki's books [28] and [29] for a detailed study of abstract linking forms and their properties.

## 5.2 Blanchfield Pairing Over $\mathbb{R}[t, t^{-1}]$

We will now study classification of Blanchfield pairings over $\mathbb{R}[t, t^{-1}]$. As in Sect. 3.2 we will first give some examples and then, based on these examples, we state the classification result. First we deal with the case $\mu \in S^1$.

**Definition 5.2** Let $\mu \in S^1$, $\operatorname{im} \mu > 0$. Let $k > 0$, $\epsilon \in \{-1, 1\}$. The hermitian form $\mathfrak{e}(\mu, k, \epsilon)$ is the pair $(M, \lambda)$, where

$$M = \mathbb{R}[t, t^{-1}]/B_\lambda(t)^k$$

$$\lambda(x, y) = \frac{\epsilon x \overline{y}}{B_\mu(t)^k}.$$

The second definition is for $\mu \notin S^1$.

**Definition 5.3** Suppose $\nu \in \mathbb{C}$, $\operatorname{im} \nu \geq 0$ and $0 < |\nu| < 1$. For $\ell > 0$ we define the hermitian form $\mathfrak{f}(\nu, \ell)$ as the pair $(M, \lambda)$, where

$$M = \mathbb{R}[t, t^{-1}]/B_\lambda(t)^\ell$$

$$\lambda(x, y) = \frac{x \overline{y}}{B_\nu(t)^\ell}.$$

Note that Definitions 5.2 and 5.3 do not cover the case $\mu = \pm 1$. These two cases are special, because $B_{\pm 1}(t)$ is not symmetric, but they do not occur in knot case, because $\pm 1$ is never a root of the Alexander polynomial of a knot.

The following result goes back at least to Milnor, see [24, Theorem 3.3]. We present the statement from [3], see also [7].

**Theorem 5.3** *Suppose $(M, \lambda)$ is a non-degenerate linking form over $\mathbb{R}[t, t^{-1}]$ such that the multiplication by $(t \pm 1)$ is an isomorphism of $M$. Then $(M, \lambda)$ decomposes into a finite sum:*

$$(M, \lambda) = \bigoplus_{i \in I} \mathfrak{e}(\mu_i, k_i, \epsilon_i) \oplus \bigoplus_{j \in J} \mathfrak{f}(\nu_j, \ell_j), \tag{5.1}$$

*where $\mu_i \in S^1$, $0 < |\nu_j| < 1$, and $\operatorname{im} \mu_i > 0$, $\operatorname{im} \nu_j \geq 0$ Such a decomposition is unique up to permuting summands.*

Theorem 5.3 motivates the following definition.

**Definition 5.4** Let $(M, \lambda)$ be as in the statement of Theorem 5.3. The number $e^k_\mu(\epsilon)$ (respectively $f^\ell_\nu$) is the number of times the form $\mathfrak{e}(\mu, k, \epsilon)$ (respectively $\mathfrak{f}(\nu, \ell)$) enters $(M, \lambda)$ as a direct summand.

## 5.3 Variation Operators and Linking Forms

Let $\mathcal{M}$ be a simple HVS over $\mathbb{R}$ with variation operator $V$ with $\zeta = -1$. Let $S = V^{-1}$. Motivated by Theorem 5.1 define the pairing $(M, \lambda)$ by

$$M = \mathbb{R}[t, t^{-1}]^n / (tS - S^T)\mathbb{R}[t, t^{-1}]^n, \;\; \lambda(x, y) = x^T (t-1)(S - tS^T)^{-1} \overline{y}. \tag{5.2}$$

We call this form the *linking form associated to* $\mathcal{M}$. We have the following result.

**Proposition 5.1** *Let $\mu \in S^1$, $\operatorname{im} \mu > 0$. Suppose $\mathcal{M} = \mathcal{V}^k_\mu(\epsilon) \oplus \mathcal{V}^k_{\overline{\mu}}((-1)^k \epsilon)$. Then, the linking form associated with $\mathcal{M}$ is equal to $\mathfrak{e}(\mu, k, \epsilon)$.*

***Proof*** The statement is well-known to the experts. The underlying $\mathbb{R}[t, t^{-1}]$-modules are clearly isomorphic and the sign $\epsilon$ is determined by comparing appropriate signatures, see [17, 18] and also Conway's survey [11, Section 4.2].

It is instructive to give an elementary proof of Proposition 5.1 in case $k = 1$. The method of computing the sign of a non-degenerate pairing over $\mathbb{R}[t, t^{-1}]/B_\mu(t)^k$ is as follows. Take an element $v \in \mathbb{R}[t, t^{-1}]/B_\mu(t)^k$ and compute $\lambda(v, v) = q/B_\mu(t)^k$. If $q$ is coprime with $B_\mu$, then the sign of $q(\mu)$ (this is clearly a real number) is precisely the sign of $\mathfrak{e}(\mu, k, \epsilon)$. A proof of the last statement follows quickly from the proof of [3, Proposition 4.2].

We will first compute the Seifert matrix $S$ and then $\lambda(v, v)$ via (5.2). From Lemma 3.2 we have $b^1_\epsilon = -\epsilon i$. Therefore, the variation operator associated with $\mathcal{V}^1_\mu(\epsilon)$ is $\epsilon i(\mu - 1)$. The variation operator corresponding to $\mathcal{V}^1_\mu(\epsilon) \oplus \mathcal{V}^1_{\overline{\mu}}(-\epsilon)$ is thus equal to

$$V = \epsilon \begin{pmatrix} i(\mu - 1) & 0 \\ 0 & -i(\overline{\mu} - 1) \end{pmatrix}.$$

Hence

$$S = V^{-1} = \frac{-i\epsilon}{|\mu - 1|^2} \begin{pmatrix} \overline{\xi} & 0 \\ 0 & \xi \end{pmatrix},$$

where $\xi = i(\mu - 1)$. Write in polar coordinates $\xi = r \cos\phi + ir \sin\phi$. Then, $S$ is congruent to the matrix

$$S = \frac{\epsilon}{r} \begin{pmatrix} \cos\phi & \sin\phi \\ -\sin\phi & \cos\phi \end{pmatrix},$$

The module $\mathbb{R}[t, t^{-1}]/B_\mu(t)$ is isomorphic to the module $\mathbb{R}[t, t^{-1}]^2/(tS - S^T)\mathbb{R}[t, t^{-1}]^2$.

Since $\det(S - tS^T) = tB_\mu(t)$, we have for any $v \in \mathbb{R}[t, t^{-1}]^2$:

$$\lambda(v, v) = v^T(t-1)(S - tS^T)^{-1}v = v^T \frac{(t-1)\epsilon r}{tB_\mu(t)} \begin{pmatrix} (1-t)\cos\phi & -(1+t)\sin\phi \\ (1+t)\sin\phi & (1-t)\cos\phi \end{pmatrix} v$$

Now take the vector $v = (1, 0)$ and consider its class in $\mathbb{R}[t, t^{-1}]^2/(tS - S^T)$, which we denote by the same letter. We obtain

$$\lambda(v, v) = \frac{\epsilon(t - 2 + t^{-1})r\cos\phi}{B_\mu(t)}.$$

Now the sign of $2 - \mu - \overline{\mu}$ is positive. To see the sign of $\cos\phi$ we note that $\operatorname{im}\mu > 0$, hence $\mu - 1$ is in the second quadrant, so $i(\mu - 1)$ is in the third one, thus $\cos\phi$ is negative.

*Remark 5.2* An analog of Proposition 5.1 for $\mu \notin S^1$ is trivial, because the pairing is determined by the underlying module structure.

The following result is an easy consequence of Proposition 5.1.

**Theorem 5.4** *There is an equality* $p^k_\mu(\epsilon) = e^k_\mu(\epsilon)$, $q^\ell_\nu = f^\ell_\nu$.

# 6 Twisted Blanchfield Forms and Applications

One of the features of the Hodge-theoretic point of view on Blanchfield pairings is that we can define signature-type invariants of pairings on torsion $\mathbb{R}[t, t^{-1}]$-modules, which do not necessarily come from Seifert matrices. In particular, we can easily define signature-type invariants for twisted Blanchfield pairings. This includes for instance so-called Casson-Gordon signatures.

## *6.1 Construction of Twisted Pairings*

We begin with a general construction. For a 3-manifold $X$ we consider its universal cover $\widetilde{X}$. This space is acted upon by $\pi_1(X)$. With $C_*(\widetilde{X})$ denoting the singular chain complex of $\widetilde{X}$, we can regard $C_*(\widetilde{X})$ as a left module over $\mathbb{Z}[\pi_1(X)]$. Suppose that $M$ is a $(R, \mathbb{Z}[\pi_1(X)])$-module for some ring $R$ (by this we mean a left $R$-module and a right $\mathbb{Z}[\pi_1(X)$-module). We define $C_*(X; M) = M \otimes_{\mathbb{Z}[\pi_1(X)]} C_*(\widetilde{X})$. This chain complex of left $R$-modules is called the *twisted chain complex* of $X$. Its homology is called the *twisted homology* of $X$; see [7, Section 6.1], [20].

A special instance of this operation is when we consider a representation $\beta\colon \pi_1(X) \to GL_d(R)$ for some ring $R$ with involution and some integer $d > 0$. The space $R^d$ has a structure of right $\mathbb{Z}[\pi_1(X)]$-module: an action of $\gamma \in \pi_1(X)$ is the multiplication the vector in $R^d$ by $\beta(\gamma)$ from the right. Taking $M = R^d$ and repeating the construction from the paragraph above, we obtain the twisted chain complex $C_*(X; R^d_\beta)$ (we write the subscript $\beta$) to stress that this is a twisted chain complex).

Let us specify our situation more. Restrict to the case $X$ is a closed 3-manifold (the case of manifolds with boundary has also been studied, but there are more technical details). Suppose $R = \mathbb{F}[t, t^{-1}]$ for some field $\mathbb{F}$ and $\beta\colon \pi_1(X) \to GL_d(R)$ is a *unitary* representation.

If the twisted homology group $H_1(X; \mathbb{F}[t, t^{-1}]^d)$ is $\mathbb{F}[t, t^{-1}]$-torsion, then one can define a hermitian non-singular pairing

$$H_1(X; \mathbb{F}[t, t^{-1}]^d_\beta) \times H_1(X; \mathbb{F}[t, t^{-1}]^d_\beta) \to \mathbb{F}(t)/\mathbb{F}[t, t^{-1}];$$

see [7, 22, 27]. This pairing is usually called the *twisted Blanchfield pairing*.

## *6.2 Twisted Hodge Numbers and Twisted Signatures*

We specify now to the situation, when $\mathbb{F} = \mathbb{R}$ and $X = M(K)$, the zero-framed surgery on a knot $K$. Let $\beta\colon \pi_1(X) \to GL_d(\mathbb{R}[t, t^{-1}])$ be a unitary representation such that $H_1(X; \mathbb{R}[t, t^{-1}]^d_\beta)$ is $\mathbb{R}[t, t^{-1}]$-torsion. Assume furthermore that

$H_1(X; \mathbb{R}[t, t^{-1}]^d_\beta)$ has no $(t \pm 1)$-torsion. Then the twisted Blanchfield pairing is defined and by Theorem 5.3 above, it decomposes as a sum of $\mathfrak{e}(\mu, k, \epsilon)$ and $\mathfrak{f}(\nu, \ell)$.

**Definition 6.1** The *twisted Hodge number* $p^k_\mu(\epsilon)_\beta$ and $f^\ell_{\nu,\beta}$ is the number of times the summand $\mathfrak{e}(\mu, k, \epsilon)$, respectively $\mathfrak{f}(\nu, \ell)$ enters the decomposition (5.1).

Having defined twisted Hodge numbers, we can define twisted signatures via an analog of (4.3).

**Definition 6.2** Suppose $\mu = e^{2\pi i x}$, $x \in (0, 1/2)$. The function

$$\mu \mapsto \sigma_\beta(\mu) = \sum_{\substack{k \text{ odd} \\ \epsilon=\pm 1}} \left( p^k_\mu(\epsilon)_\beta + 2 \sum_{y \in (0,x)} p^k_{e^{2\pi i y}}(\epsilon)_\beta \right)$$

is called the *twisted signature function*. The function is extended via $\sigma_\beta(\overline{\mu}) = \sigma_\beta(\mu)$.

There is a subtle difference between Definition 6.2 and Proposition 4.7. The classical result, Proposition 4.7, sums contributions of the Hodge numbers in a range including 0. Therefore it is perfectly possible that the signature function is equal to 1 for all values close to 1. This is the case for example for the Hopf link.

Definition 6.2 sums over $y$ in an open interval $(0, x)$, so the previous behavior is impossible. This is not merely a technical issue: it seems difficult to extend the definition of twisted signature to get a meaningful contribution of $\mu = 1$.

## 6.3 A Few Words on Case $\mathbb{F} = \mathbb{C}$

The construction of Hodge numbers via classification of linking pairings can be done over $\mathbb{C}[t, t^{-1}]$. We can define $\mathfrak{e}(\mu, k, \epsilon)$ for $\mu \in S^1$, and $\mathfrak{f}(\mu, k)$ for $0 < |\mu| < 1$. The underlying module structure is $\mathbb{C}[t, t^{-1}]/(t - \mu)^k$. However, the specific construction seems to be harder than in case over $\mathbb{R}$; see [7, Section 2]. Once this technical difficulty is overcome, we can define twisted Hodge numbers and twisted signatures essentially via Definitions 6.1 and 6.2.

An important instance of twisted signatures over $\mathbb{C}[t, t^{-1}]$ are signatures defined from Casson–Gordon invariants introduced by Casson and Gordon, see [8, 9]. In short, let $K$ be a knot and let $n$ be an integer. Consider the $n$-fold cyclic branched cover $L_n(K)$. Let $m$ be a prime power coprime with $n$. For any non-trivial homomorphism $\chi : H_1(L_n(K); \mathbb{Z}) \to \mathbb{Z}_m$ we can construct a unitary representation $\pi_1(M(K)) \to GL_n(\mathbb{C}[t, t^{-1}])$. The concrete formula for the representation is beyond the scope of this article, we refer to [7, Section 8.1] The signature associated to this representation via Definition 6.2 is called a *Casson-Gordon signature* $\sigma_{\chi,m} : S^1 \to \mathbb{Z}$. Casson–Gordon sliceness obstruction can be translated into vanishing of some Casson–Gordon signatures. The following result is stated in [7, Theorem 8.8, Corollary 8.16] as a corollary of a result of Miller and Powell [22].

**Theorem 6.1** *Let $K$ be a topologically slice knot. Then for any prime power $n$, there exists a metabolizer $P$ of the linking form on $H_1(L_n(K);\mathbb{Z})$ such that for any prime power $q^a$ and any non-trivial homomorphism $\chi\colon H_1(L_n(K);\mathbb{Z}_{q^a})\to\mathbb{Z}$ vanishing on $P$, there is $b\geq a$ such that $\sigma_{\chi,q^b}$ is zero almost everywhere on $S^1$.*

The main feature of Theorem 6.1 is computability. Miller and Powell [22] gave an algorithm to compute the twisted Blanchfield pairing using Fox differential calculus. The methods of [7], which we presented in this article, allow us to compute the Casson-Gordon signatures. As an application [7] and later [12] could prove non-sliceness of some linear combinations of iterated torus knots, generalizing previous results of Hedden et al. [14].

### *6.4 A Closing Remark*

The two decomposition results: the classification of HVS of Theorem 3.2 and the classification of real Blanchfield forms in Theorem 5.3 share many properties. There are some differences, which we now want to summarize.

The classification of HVS deals much more efficiently with the case $\mu = 1$, because of the special definition of a simple HVS for $\mu = 1$. The presence of $(t-1)$-torsion modules in the theory of linking forms is a source o notorious technical difficulties.

The classification of Blanchfield forms is more general and is usually much easier to generalize. The construction of twisted Blanchfield pairings is a straightforward generalization of the construction of the classical pairing. Also, in many classification results, it is more convenient to have a single object (a pairing), than a quadruple of objects.

**Acknowledgments** The first authors is greatly indepted to András Némethi for years of fruitful collaboration and for his guidance since they first met in 2009. The authors would like to thank to Anthony Conway and Wojciech Politarczyk for their comments on the draft version of the paper. They are also grateful to the referees for their remarks. The first author is supported by the National Science Center grant 2016/22/E/ST1/00040.

## References

1. Arnold, V., Gusein-Zade, S., Varchenko, A.: Singularities of Differentiable Maps. Volume 2. Modern Birkhäuser Classics. Birkhäuser/Springer, New York (2012)
2. Blanchfield, R.: Intersection theory of manifolds with operators with applications to knot theory. Ann. Math. **65**(2), 340–356 (1957)
3. Borodzik, M., Friedl, S.: The unknotting number and classical invariants II. Glasg. Math. J. **56**(3), 657–680 (2014)
4. Borodzik, M., Friedl, S.: The unknotting number and classical invariants, I. Algebr. Geom. Topol. **15**(1), 85–135 (2015)

5. Borodzik, M., Némethi, A.: Spectrum of plane curves via knot theory. J. Lond. Math. Soc. **86**(1), 87–110 (2012)
6. Borodzik, M., Némethi, A.: Hodge-type structures as link invariants. Ann. Inst. Fourier (Grenoble) **63**(1), 269–301 (2013)
7. Borodzik, M., Conway, A., Politarczyk, W.: Twisted Blanchfield pairings, twisted signatures and Casson-Gordon invariants (2018). Preprint. arXiv:1809.08791
8. Casson, A., Gordon, C.: On slice knots in dimension three. In: Algebraic and geometric Topology (Proceedings of Symposia in Pure Mathematics, Stanford University, Stanford, California, 1976), Part 2, Proceedings of Symposia in Pure Mathematics, XXXII, pp. 39–53. American Mathematical Society, Providence (1978)
9. Casson, A., Gordon, C.: Cobordism of classical knots. In: À la recherche de la topologie perdue. Progress in Mathematic, vol. 62, pp. 181–199. Birkhäuser Boston, Boston (1986). With an appendix by P. M. Gilmer
10. Cha, J., Livingston, C.: Knotinfo: Table of knot invariants (2019). http://www.indiana.edu/~knotinfo. Accessed 12 July 2019
11. Conway, A.: The Levine-Tristram signature: a survey (2019). arXiv:1903.04477
12. Conway, A., Kim, M., Politarczyk, W.: Non-slice linear combinations of iterated torus knots (2019). Preprint, arXiv:1910.01368
13. Friedl, S., Powell, M.: A calculation of Blanchfield pairings of 3-manifolds and knots. Mosc. Math. J. **17**(1), 59–77 (2017)
14. Hedden, M., Kirk, P., Livingston, C.: Non-slice linear combinations of algebraic knots. J. Eur. Math. Soc. (JEMS) **14**(4), 1181–1208 (2012)
15. Hillman, J.: Algebraic Invariants of Links: Series on Knots and Everything, vol. 52, 2nd edn. World Scientific Publishing, Hackensack (2012)
16. Kawauchi, A.: A Survey of Knot Theory. Birkhäuser Verlag, Basel (1996). Translated and revised from the 1990 Japanese original by the author
17. Kearton, C.: Signatures of knots and the free differential calculus. Quart. J. Math. Oxford Ser. **30**(118), 157–182 (1979)
18. Kearton, C.: Quadratic forms in knot theory. In: Quadratic Forms and Their Applications (Dublin, 1999). Contemporary Mathematics, vol. 272, pp. 135–154. American Mathematical Society, Providence (2000)
19. Keef, P.: On the S-equivalence of some general sets of matrices. Rocky Mountain J. Math. **13**(3), 541–551 (1983)
20. Kirk, P., Livingston, C.: Twisted Alexander invariants, Reidemeister torsion, and Casson-Gordon invariants. Topology **38**(3), 635–661 (1999)
21. Litherland, R.: Signatures of iterated torus knots. In: Topology of Low-Dimensional Manifolds (Proceedings of the Second Sussex Conference, Chelwood Gate, 1977). Lecture Notes in Mathematics, vol. 722, pp. 71–84. Springer, Berlin (1979)
22. Miller, A., Powell, M.: Symmetric chain complexes, twisted Blanchfield pairings and knot concordance. Algebr. Geom. Topol. **18**(6), 3425–3476 (2018)
23. Milnor, J.: Singular Points of Complex Hypersurfaces. Annals of Mathematics Studies, No. 61. Princeton University Press/University of Tokyo Press, Princeton/Tokyo (1968)
24. Milnor, J.: On isometries of inner product spaces. Invent. Math. **8**, 83–97 (1969)
25. Némethi, A.: The real Seifert form and the spectral pairs of isolated hypersurface singularities. Compositio Math. **98**(1), 23–41 (1995). http://www.numdam.org/item?id=CM_1995_98_1_23_0
26. Neumann, W.: Invariants of plane curve singularities. In: Knots, Braids and Singularities (Plans-sur-Bex, 1982). Monographs of L'Enseignement Mathematique, vol. 31, pp. 223–232. Enseignement Mathematique, Geneva (1983)
27. Powell, M.: Twisted Blanchfield pairings and symmetric chain complexes. Q. J. Math. **67**(4), 715–742 (2016)
28. Ranicki, A.: Exact Sequences in the Algebraic Theory of Surgery. Mathematical Notes, vol. 26. Princeton University Press/University of Tokyo Press, Princeton/Tokyo (1981)

29. Ranicki, A.: Algebraic L-Theory and Topological Manifolds. Cambridge Tracts in Mathematics, vol. 102. Cambridge University Press, Cambridge (1992)
30. Ranicki, A.: Blanchfield and Seifert algebra in high-dimensional knot theory. Mosc. Math. J. **3**(4), 1333–1367 (2003)
31. Steenbrink, J.: Mixed Hodge structure on the vanishing cohomology. In: Real and Complex Singularities (Proceedings of the Nordic Summer School/NAVF Symposium in Mathematics, Oslo, 1976), pp. 525–563 (1977)
32. Steenbrink, J.: Semicontinuity of the singularity spectrum. Invent. Math. **79**(3), 557–565 (1985)
33. Trotter, H.: On $S$-equivalence of Seifert matrices. Invent. Math. **20**, 173–207 (1973)
34. Varchenko, A.: Semicontinuity of the spectrum and an upper bound for the number of singular points of the projective hypersurface. Dokl. Akad. Nauk SSSR **270**(6), 1294–1297 (1983)
35. Żołądek, H.: The Monodromy Group. Instytut Matematyczny Polskiej Akademii Nauk. Monografie Matematyczne (New Series) [Mathematics Institute of the Polish Academy of Sciences. Mathematical Monographs (New Series)], vol. 67. Birkhäuser Verlag, Basel (2006)

# $\hbar$-Deformed Schubert Calculus in Equivariant Cohomology, K-Theory, and Elliptic Cohomology

**Richárd Rimányi**

*Dedicated to András Némethi, on the occasion of his 60th birthday.*

**Abstract** In this survey paper we review recent advances in the calculus of Chern-Schwartz-MacPherson, motivic Chern, and elliptic classes of classical Schubert varieties. These three theories are one-parameter ($\hbar$) deformations of the notion of fundamental class in their respective extraordinary cohomology theories. Examining these three classes in conjunction is justified by their relation to Okounkov's stable envelope notion. We review formulas for the $\hbar$-deformed classes originating from Tarasov–Varchenko weight functions, as well as their orthogonality relations. As a consequence, explicit formulas are obtained for the Littlewood-Richardson type structure constants.

**Keywords** Schubert calculus · Elliptic cohomology

**Subject Classifications** 14N15, 55N34

## 1 Introduction

A basic structure of traditional Schubert calculus is the cohomology ring of a homogeneous space $X$, together with a distinguished basis. The elements of the distinguished basis are associated with the geometric subvarieties (called Schubert varieties) of $X$. The first objects to study are the structure constants of the ring with respect to the distinguished basis. These structure constants satisfy various positivity, stabilization, saturation, and other properties, and can be related with

---

R. Rimányi (✉)
Department of Mathematics, University of North Carolina at Chapel Hill, Chapel Hill, NC, USA
e-mail: rimanyi@email.unc.edu



J. Fernández de Bobadilla et al. (eds.), *Singularities and Their Interaction with Geometry and Low Dimensional Topology*, Trends in Mathematics,
https://doi.org/10.1007/978-3-030-61958-9_5

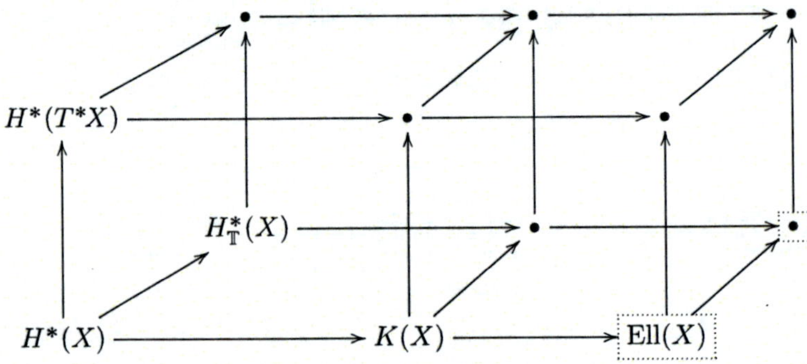


**Fig. 1** Three "orthogonal" directions to generalize classical Schubert calculus

other mathematical fields, such as combinatorics, representation theory, integrable systems.

In this paper we survey some generalizations of this traditional setup, organized as vertices of the diagram in Fig. 1. The traditional setup, mentioned in the first paragraph, is the bottom left corner, the cohomology ring of the homogeneous space $X$. Going from the front face to the back face of the diagram represents the change from ordinary "cohomologies" to equivariant ones. Equivariant cohomologies of $X$ take into account the geometry of $X$ together with the natural group (say torus) action on it. Due to some techniques that only exist in equivariant theories it can be an *easier* theory to work with. Nevertheless, formulas in non-equivariant theories can be recovered from equivariant ones by plugging in 0 (or 1, depending on conventions) in the equivariant variables.

Stepping one step to the right on the diagram from cohomology we arrive at K theory. K theoretic Schubert calculus, ordinary or equivariant, has been studied extensively, see [32] and references therein. Stepping one further step to the right we get to elliptic cohomology, ordinary or equivariant: the two vertices in the diagram that are in dotted frames. We framed these vertices in the diagram because these settings lack the notion of a well-defined distinguished basis, which was present in $H^*$, $H^*_{\mathbb{T}}$, $K$, and $K_{\mathbb{T}}$. Namely, it turns out that in elliptic cohomology the notion of fundamental class depends on choices [13]. There are important results in these settings (e.g. [24, 36, 37] and references therein) that follow from making certain choices (of a resolution, or a basis in a Hecke algebra).

*Remark 1.1* There are other extraordinary cohomology theories, for example the (universal) complex cobordism theory; their position would be further to the right on the diagram. Yet, we restrict our attention to the ones depicted in Fig. 1, as their associated formal group law is an algebraic group.

The focus of this paper is the rest of the diagram, namely the top face. Besides pioneering works, e.g. [4, 5, 46], this direction of generalization is very recent.

There are two ways of introducing this direction of generalization. One way is that we study Schubert calculus not on the homogeneous space $X$ but on its cotangent bundle $T^*X$, using its extra holomorphic symplectic structure. Although $X$ and $T^*X$ are homotopically equivalent so their cohomology rings are isomorphic, their geometric subvarieties are different (in $T^*X$ one studies the conical Lagrangian cycles). We will not touch upon this interpretation, except in notation: In the diagram above we indicate this generalization by replacing $X$ with $T^*X$.

The other way of explaining the step from the bottom face to the top face is that we change the notion of the characteristic class that we associate with Schubert cells. While traditionally (on the bottom face) we associate the *fundamental class* to the Schubert variety, in this generalization we associate a particular one-parameter deformation of the fundamental class. The parameter will be denoted by $\hbar$, and the class will be called the $\hbar$-deformed Schubert class. In cohomology this $\hbar$-deformed class was conjectured/discovered by Grothendieck, Deligne, Schwartz, MacPherson [39, 56], and is called the *Chern-Schwartz-MacPherson class.*[1] Its equivariant theory is worked out by Ohmoto [43, 44, 60]. In K theory the $\hbar$-deformed class[2] was defined by Brasselet-Schürmann-Yokura [12] under the name of *motivic Chern class*. The equivariant version is defined in [7, 20].[3]

The most recent discovery is the definition of the (ordinary or equivariant) $\hbar$-deformed elliptic class associated with a Schubert cell [31, 49], that is, Schubert calculus in the rightmost two vertices on the top face of the diagram. A pleasant surprise of such Schubert calculus is that the $\hbar$-deformed elliptic class does not depend on choices—the corresponding vertices in the diagram are not framed. While it is a general fact that one can recover the non-$\hbar$-deformed theory from the $\hbar$-deformed theory by plugging in an obvious value ($0, 1, \infty$, depending on setup) for $\hbar$, it turns out that at such specialization the $\hbar$-deformed elliptic class has a singularity. This fact is another incarnation of the phenomenon mentioned above that the non-$\hbar$-deformed elliptic Schubert calculus depends on choices.

*Remark 1.2* Let us comment on a principle that unifies the three $\hbar$-deformations in $H^*$, $K$, Ell, which is actually the reason for the attention $\hbar$-deformed Schubert calculus is getting recently. In works of Okounkov and his co-authors Maulik, Aganagic [1, 40, 45] (see also [26, 50–52, 54]), a remarkable bridge is built between quantum integrable systems and geometry. Via this bridge the (extraordinary) cohomology of a geometrically relevant space is identified with the Bethe algebra of a quantum integrable system.

---

[1] In the classical CSM literature the parameter $\hbar$ is not indicated, because it can be recovered from the grading in $H^*$.

[2] In most of the literature the letter $y$ is used for $\hbar$ in K theory, to match the classical notion of $\chi_y$-genus.

[3] The CSM, motivic Chern, and elliptic classes were *not* discovered as $\hbar$-deformations of the notion of the fundamental class, but as generalizations of the notion of total Chern class for singular varieties with covariant functoriality; their interpretation as $\hbar$-deformations of the fundamental class suggested in this paper is post-factum. Thus, the present paper is a re-interpretation of the story of characteristic classes of singular varieties from the mid-70s to the present.

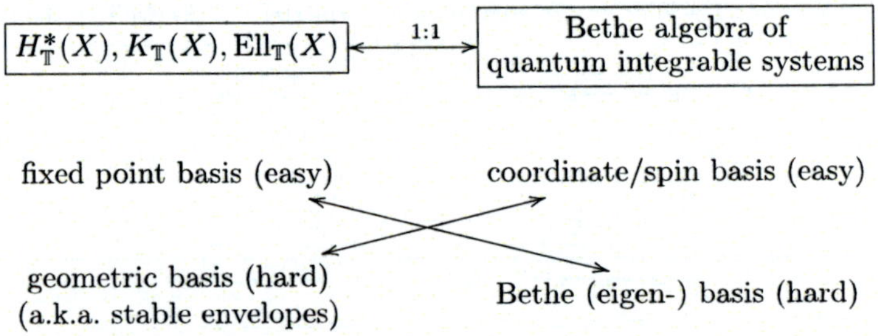


On both sides of the identification we have an "easy" and a "hard" basis, and the identification matches the easy basis of one side with the hard basis of the other side. The geometric basis that matches the spin basis of the Bethe algebra side is named the cohomological, K theoretic, and elliptic *stable envelope*. It is now proved that in type A Schubert calculus settings the three flavors of stable envelopes coincide (through some identifications, and convention matching) with the three $\hbar$-deformed Schubert classes: the CSM class, the motivic Chern class, and the elliptic class. This relation with quantum integrable system is the reason we denote the deformation parameter by $\hbar$.

*Remark 1.3* A fourth direction to generalize classical Schubert calculus is *quantum Schubert calculus*. While quantum cohomology and K theory (and possibly elliptic cohomology) are related with their $\hbar$-deformations see e.g. [40], [51, App.3], we will not study them in this paper.

The topic of $\hbar$-deformed Schubert calculus is rather fresh and the available literature is rather technical. Moreover, as explained in Remark 1.2 above, some of the existing literature is hidden in quantum integrable system papers. The goal of the present survey is twofold.

On the one hand we want to give an accessible, motivated, and technicality-free presentation of the main achievement, what we call Main Theorem, see Sect. 3. The content of this theorem is the existence of localization type formulas for $\hbar$-deformed Schubert classes (in $H^*_{\mathbb{T}}$, $K_{\mathbb{T}}$, $\mathrm{Ell}_{\mathbb{T}}$), their orthogonality relations, and their direct consequence, explicit formulas for $\hbar$-deformed Schubert constants.

On the other hand we give precise formulas of the key ingredients (weight functions, inner products, orthogonality statements) consistent with usual Schubert calculus usage. While these formulas exist in some conventions in the literature, the conventions used there are optimized for some other purposes. Also, we tried to separate the complicated formulas (they are exiled to the penultimate section) from the main part of the paper where the idea is presented.

# 2 Ordinary and Equivariant Cohomological Schubert Calculus

## 2.1 *Schubert Classes and Structure Coefficients*

Consider the compact smooth variety $\mathrm{Gr}(m, n)$, the Grassmannian of $m$-planes in $\mathbb{C}^n$. For an $m$-element subset $I$ of $[n] \stackrel{\text{def}}{=} \{1, \ldots, n\}$ one defines the Schubert cell

$$\Omega_I = \{V^m \subseteq \mathbb{C}^n : \dim(V^m \cap \mathbb{C}^q) = |\{i \in I : i \leq q\}| \; \forall q\},$$

where $\mathbb{C}^1 \subset \mathbb{C}^2 \subset \ldots \subset \mathbb{C}^{n-1} \subset \mathbb{C}^n$ is the standard full flag. The collection of cohomological fundamental classes $[\overline{\Omega}_I]$ forms a basis in the cohomology ring of the Grassmannian, hence via

$$[\overline{\Omega}_I] \cdot [\overline{\Omega}_J] = \sum_K \mathrm{c}_{I,J}^K \cdot [\overline{\Omega}_K]$$

the structure coefficients (a.k.a. Littlewood Richardson coefficients) $\mathrm{c}_{I,J}^K \in \mathbb{Z}$ are defined.

To name an example, let us "encode" the subset $I = \{i_1 < \ldots < i_m\}$ with the partition $(\lambda_1 \geq \lambda_2 \geq \ldots \geq \lambda_m)$, by $\lambda_j = n - m - (i_j - j)$. With a slight abuse of notation let the Young diagram of $\lambda$ mean the corresponding fundamental class $[\overline{\Omega}_I]$. Then in $H^*(\mathrm{Gr}(3, 6))$ we have

$$\ydiagram{2,1} \cdot \ydiagram{2,1} = \ydiagram{3,3} + 2\,\ydiagram{3,2,1} + \ydiagram{2,2,2}, \tag{2.1}$$

that is, e.g. $\mathrm{c}_{\{2,4,6\},\{2,4,6\}}^{\{1,3,5\}} = 2$.

The natural action of the torus $\mathbb{T} = (\mathbb{C}^*)^n$ on $\mathbb{C}^n$ induces an action of $\mathbb{T}$ on $\mathrm{Gr}(m, n)$. The Schubert cells are invariant, and hence their closures carry a fundamental class in $\mathbb{T}$ equivariant cohomology as well. These classes form a basis of $H_{\mathbb{T}}^*(\mathrm{Gr}(m, n))$ over the ring $H_{\mathbb{T}}^*(\mathrm{pt}) = \mathbb{Z}[z_1, \ldots, z_n]$, where $z_i$ is the first Chern class of the tautological line bundle corresponding to the $i$'th factor of $\mathbb{T}$. Hence the structure constants are polynomials in $z_i$'s. For example, the $\mathbb{T}$ equivariant version of (2.1) now reads

$$\begin{aligned} \ydiagram{2,1} \cdot \ydiagram{2,1} = {} & \ydiagram{3,3} + 2\,\ydiagram{3,2,1} + \ydiagram{2,2,2} + (2z_5 - z_1 - z_2)\,\ydiagram{3,2} \\ & + (z_3 + z_5 + z_6 - z_1 - z_2 - z_4)\,\ydiagram{3,1,1} + (z_5 - z_6 - 2z_2)\,\ydiagram{2,2,1} \\ & + (z_5 - z_4)(z_3 + z_5 - z_1 - z_2)\,\ydiagram{3,1} + (z_3 - z_2)(z_5 + z_6 - z_2 - z_4)\,\ydiagram{2,1,1} \\ & + (z_5 - z_2)^2\,\ydiagram{2,2} + (z_5 - z_4)(z_5 - z_2)(z_3 - z_2)\,\ydiagram{2,1}. \end{aligned} \tag{2.2}$$

To recover the non-equivariant version (2.1) from the equivariant (2.2), one needs to substitute all $z_i = 0$.

There are multiple ways of calculating the structure constants presented above, see e.g. [30] for an effective algorithm tailored to this situation. Now we show an approach which generalizes to the more general settings in Fig. 1. This method has two ingredients:

**(i) Formulas representing fundamental classes** Consider the $t = t_1, \dots, t_m$ and $z = z_1, \dots, z_n$ variables. Define the rational functions

$$U_I = \prod_{a=1}^{m} \prod_{b=i_a+1}^{n} (z_b - t_a) \prod_{1 \le a < b \le m} \frac{1}{t_b - t_a}, \qquad W_I = \mathrm{Sym}_{t_1,\dots,t_m}(U_I),$$

and, for a permutation $\sigma \in S_n$ define

$$W_{\sigma,I} = W_{\sigma^{-1}(I)}(t_1, \dots, t_m, z_{\sigma(1)}, \dots, z_{\sigma(n)}).$$

Despite its appearance as a rational function, $W_{\sigma,I}$ is in fact a polynomial. Interpreting

- $t$ as the Chern roots of the tautological rank $m$ subbundle over $\mathrm{Gr}(m, n)$, and
- $z$ as the tautological Chern roots of the torus $\mathbb{T}$ (cf. Sect. 2.1),

the function $W_{\mathrm{id},I}$ represents the fundamental class $[\overline{\Omega}_I]$.
This is a classical result in equivariant Schubert calculus which is stated more-or-less explicitly already in [33, 34], for more recent literature see [17, 42]. It also follows from our Main Theorem below by specialization.

**(ii) Orthogonality** Define

$$\langle f(t,z), g(t,z)\rangle = \sum_{K \subseteq [n]} \frac{f(z_K, z) g(z_K, z)}{R_K}, \qquad R_K = \prod_{i \in K} \prod_{j \in [n]-K} (z_j - z_i)$$

where $|K| = m$, $z_K$ is the collection of $z$ variables with index from $K$. Let $s_0$ be the longest permutation of $n$. Then

$$\langle W_{\mathrm{id},I}, W_{s_0,J}\rangle = \delta_{I,J}.$$

This statement is the algebraic way of phrasing the geometric Duality Theorem [23, 9.4] on the intersection of Schubert varieties corresponding to opposite reference full flags.

A direct consequence of the statements in (i) and (ii) is an explicit expression for the structure constants of $H^*_{\mathbb{T}}(\mathrm{Gr}(m,n))$ with respect to the fundamental classes of Schubert varieties.

**Corollary 2.1** *We have*

$$c_{I,J}^{K} = \langle W_{\mathrm{id},I} W_{\mathrm{id},J}, W_{s_0,K} \rangle. \tag{2.3}$$

The explicit expression in Corollary 2.1 can be coded to a computer, and it can produce expressions like the ones presented in (2.2). It has, however, disadvantages. One of them is the denominators: due to the nature $\langle\, , \rangle$ is defined we obtain the structure constants as a (large) sum of rational functions. Part of the claim is that this rational function in fact simplifies to a polynomial. Yet, such a simplification is usually rather time- and memory-consuming for computers.[4] Even if we are only interested in the non-equivariant structure constants, i.e. the substitution $z_i = 0$ in (2.3), we must carry out the simplification from rational function to polynomial first, because the denominators of the rational functions are products of $(z_i - z_j)$ factors. Another disadvantage of the formula for $c_{I,J}^{K}$ in Corollary 2.1 is that it does not display known positivity properties of the structure constants.

## 3 The Main Theorem

The advantage of Corollary 2.1 is that it generalizes to the other vertices in Fig. 1. This feature is the content of the recent development of ℏ-deformed Schubert calculus in cohomology, K theory, and elliptic cohomology.

**Theorem 3.1 (Main Theorem)** *Let $\mathcal{F}_\lambda$ be a partial flag variety of type A.*

**Formulas for Schubert classes.** *There are explicit formulas $W^{\mathbb{H}}_{\mathrm{id},I}$, $W^{\mathbb{K}}_{\mathrm{id},I}$, $W^{\mathbb{E}}_{\mathrm{id},I}$ for the ℏ-deformed Schubert classes in*

$$H^*_{\mathbb{T}}(\mathcal{F}_\lambda), \qquad K_{\mathbb{T}}(\mathcal{F}_\lambda), \qquad \mathrm{Ell}_{\mathbb{T}}(\mathcal{F}_\lambda).$$

*These are functions in terms of equivariant variables $z_i$, Chern roots of tautological bundles over $\mathcal{F}_\lambda$ called $t_j^{(i)}$, and ℏ (as well as other parameters in case of* Ell*).*

**Orthogonality.** *The given formulas satisfy orthogonality relations for appropriate inner products $\langle\, , \rangle_{\mathbb{H}}$, $\langle\, , \rangle_{\mathbb{K}}$, $\langle\, , \rangle_{\mathbb{E}}$.*

**Structure constant formulas.** *Hence, we have the explicit formulas*

$$\begin{aligned}
c_{I,J}^{K} &= \langle W^{\mathbb{H}}_{\mathrm{id},I} W^{\mathbb{H}}_{\mathrm{id},J}, W^{\mathbb{H}}_{s_0,K} \rangle_{\mathbb{H}}, \\
c_{I,J}^{K} &= \langle W^{\mathbb{K}}_{\mathrm{id},I} W^{\mathbb{K}}_{\mathrm{id},J}, (-\hbar)^{-\dim_K} \iota[W^{\mathbb{K}}_{s_0,K}] \rangle_{\mathbb{K}}, \\
c_{I,J}^{K} &= \langle W^{\mathbb{E}}_{\mathrm{id},I} W^{\mathbb{E}}_{\mathrm{id},J}, (\vartheta(\hbar)/\vartheta'(1))^{\dim_\lambda} \tau[W^{\mathbb{E}}_{s_0,K}] \rangle_{\mathbb{E}}.
\end{aligned}$$

*for the ℏ-deformed Schubert structure constants in $H^*_{\mathbb{T}}$, $K_{\mathbb{T}}$, and* $\mathrm{Ell}_{\mathbb{T}}$.

[4]Reconstructing the polynomial from its values at a large set of well-chosen points is faster than symbolic simplification, but it is still time-consuming.

The statement of this theorem is deliberately vague, as the details of the theorem are rather technical. The rigorous mathematical meaning of this theorem follows from the explanation of all of its terms throughout the rest of the paper.

Notations about the partial flag variety $\mathcal{F}_\lambda$ and various structures on it (such as bundles, torus action, Schubert cells and varieties) are set up in Sect. 4.

After introducing elliptic functions and their trisecant identity in Sect. 5, we present a down-to-earth introduction to the equivariant elliptic cohomology of flag varieties in Sect. 6.

The $\hbar$-deformed Schubert classes—namely the Chern-Schwartz-MacPherson class, the motivic Chern class, and the elliptic class, in $H^*_{\mathbb{T}}$, $K_{\mathbb{T}}$, and $\mathrm{Ell}_{\mathbb{T}}$—are introduced in Sect. 7.

The formulas $W^{\mathbb{H}}_{\mathrm{id},I}$, $W^{\mathbb{K}}_{\mathrm{id},I}$, $W^{\mathbb{E}}_{\mathrm{id},I}$ as well as their orthogonality relations are given in Sect. 8.

Some examples for structure constant obtained from the Main Theorem are shown in Sect. 9.

## 4 The Partial Flag Variety

Let $N$ be a non-negative integer, $\lambda = (\lambda_1, \ldots, \lambda_N) \in \mathbb{Z}^N_{\geq 0}$, and define

$$\lambda^{(j)} \stackrel{\text{def}}{=} \sum_{i=1}^{j} \lambda_i, \qquad n \stackrel{\text{def}}{=} \lambda^{(N)} = \sum_{i=1}^{N} \lambda_i.$$

The partial flag variety $\mathcal{F}_\lambda$ parametrizes nested subspaces

$$\{0\} = V_0 \subseteq V_1 \subseteq \ldots \subseteq V_{N-1} \subseteq V_N = \mathbb{C}^n \tag{4.1}$$

with $\dim V_j = \lambda^{(j)}$. It is a smooth variety of dimension $\dim_\lambda \stackrel{\text{def}}{=} \sum_{1\leq i<j\leq N} \lambda_i\lambda_j$. Let us recall the usual structures on $\mathcal{F}_\lambda$.

Bundles The tautological rank $\lambda^{(j)}$ bundle, whose fiber over the point (4.1) is $V_j$ will be called $\mathcal{V}_j$.

Torus action The standard action of the torus $\mathbb{T} \stackrel{\text{def}}{=} (\mathbb{C}^*)^n$ on $\mathbb{C}^n$ induces its action on $\mathcal{F}_\lambda$.

Combinatorial gadgets Consider tuples $I = (I_1, \ldots, I_N)$ where $I_j \subseteq [n]$, satisfying $|I_j| = \lambda_j$, $I_i \cap I_j = \emptyset$. Their collection will be denoted by $\mathcal{I}_\lambda$. For example $\mathcal{I}_{(1,2)} = \{(\{1\}, \{2, 3\}), (\{2\}, \{1, 3\}), (\{3\}, \{1, 2\})\}$. For $I \in \mathcal{I}_\lambda$ we will use the notation $I^{(k)} = \bigcup_{s=1}^{k} I_s = \{i_1^{(k)} < i_2^{(k)} < \ldots < i_{\lambda^{(k)}}^{(k)}\}$.

Torus fixed points The fixed points $x_I$ of the $\mathbb{T}$ action on $\mathcal{F}_\lambda$ are parametrized by $\mathcal{I}_\lambda$:

$$x_I = \left(\text{span}\{\epsilon_i\}_{i\in I_1} \subseteq \text{span}\{\epsilon_i\}_{i\in I_1\cup I_2} \subseteq \ldots\right) \in \mathcal{F}_\lambda,$$

where $\epsilon_1, \ldots, \epsilon_n$ is the standard basis of $\mathbb{C}^n$.

Schubert cells Define the Schubert cell

$$\Omega_I = \{(V_\bullet) \in \mathcal{F}_\lambda : \dim(V_p \cap \mathbb{C}^q) = |\{i \in I_1 \cup \ldots \cup I_p : i \le q\}| \; \forall p, q\},$$

where $\mathbb{C}^k = \text{span}\{\epsilon_1, \ldots, \epsilon_k\}$. We have $x_I \in \Omega_I$ and $\Omega_I$ has dimension

$$\dim_I \overset{\text{def}}{=} |\cup_{j<k} \{(a,b) \in I_j \times I_k : a > b\}|.$$

# 5 Elliptic Functions

## 5.1 *Theta Functions*

We will use the following version of theta-functions:

$$\vartheta(x) = (x^{1/2} - x^{-1/2}) \prod_{s=1}^{\infty} (1 - q^s x)(1 - q^s/x).$$

We treat $q \in \mathbb{C}$, $|q| < 1$ as a fixed parameter, and will not indicate dependence on it. The function $\vartheta$ is defined on a double cover of $\mathbb{C}$. Theta functions will often appear through

$$\delta(x, y) \overset{\text{def}}{=} \frac{\vartheta(xy)\vartheta'(1)}{\vartheta(x)\vartheta(y)},$$

which is meromorphic on $\mathbb{C}^* \times \mathbb{C}^*$.

*Remark 5.1* The $q \to 0$ limit we call *trigonometric limit* because at $q = 0$ the function $\vartheta(x)$ is $\sin(y)$ (up to a constant) in the new variable $x^{1/2} = e^{iy}$. By disregarding the constant factor and denoting the new variable by the same letter as the old one, we say $\vartheta(x) \to \sin(x)$ is our trigonometric limit. The further approximation of $\sin(x)$ with $x$ will be called the *rational limit*. The three levels $\vartheta(x) \to \sin(x) \to x$ correspond to the Euler class formulas of line bundles in the three cohomology theories Ell, $K$, $H^*$.[5] Equivalently, the formal group laws of the

[5] It is more customary to regard $1 - x$ as the K theoretic Euler class of a line bundle, but again, up to a unit ($x$ is invertible in K theory!) this is $\sin(y)$ in a new variable.

three theories are (up to constants and change of variables)

$$(\vartheta(x), \vartheta(y)) \mapsto \vartheta(xy), \qquad (\sin(x), \sin(y)) \mapsto \sin(x+y), \qquad (x, y) \mapsto x+y.$$

In the three versions the $\delta$-functions (up to constant) are

$$\frac{\vartheta(xy)}{\vartheta(x)\vartheta(y)}, \qquad \frac{\sin(x+y)}{\sin(x)\sin(y)} = \cot(x) + \cot(y), \qquad \frac{x+y}{xy} = \frac{1}{x} + \frac{1}{y}.$$

Observe that the $\delta$-function separates to a sum of two terms, one depending on $x$ the other on $y$, in the trigonometric and rational limits. However, such separation does not hold in the elliptic case.

## 5.2 *Fay's Trisecant Identity*

**Proposition 5.2 ([15] and [21, Thm. 7.3])** *For variables satisfying* $x_1x_2x_3 = 1$ *and* $y_1y_2y_3 = 1$ *we have*

$$\delta(x_1, y_2)\delta(x_2, 1/y_1) + \delta(x_2, y_3)\delta(x_3, 1/y_2) + \delta(x_3, y_1)\delta(x_1, 1/y_3) = 0. \tag{5.1}$$

Note that in the trigonometric limit, that is, substituting $\delta(x, y) = \sin(x + y)/(\sin(x)\sin(y))$, identity (5.1) takes the form

$$\begin{aligned} x_1 + x_2 + x_3 = 0, y_1 + y_2 + y_3 = 0 \Rightarrow \\ \cot(x_1)\cot(x_2) + \cot(x_2)\cot(x_3) + \cot(x_3)\cot(x_1) = \\ \cot(y_1)\cot(y_2) + \cot(y_2)\cot(y_3) + \cot(y_3)\cot(y_1). \end{aligned} \tag{5.2}$$

In the rational limit, that is, substituting $\delta(x, y) = (x+y)/(xy)$, identity (5.1) takes the form

$$\begin{aligned} x_1 + x_2 + x_3 = 0, y_1 + y_2 + y_3 = 0 \Rightarrow \\ \frac{1}{x_1x_2} + \frac{1}{x_2x_3} + \frac{1}{x_3x_1} = \frac{1}{y_1y_2} + \frac{1}{y_2y_3} + \frac{1}{y_3y_1}. \end{aligned} \tag{5.3}$$

However, in these two limits, more is true. Namely, not only the two sides of (5.2) are equal to each other, but *both sides* of (5.2) are 1 (the reader is invited to verify this statement using high school memories about trigonometric identities). Similarly, the two sides of (5.3) are both 0. In the elliptic version (5.1) no such "separation of $x$ and $y$ variables" holds.

It is worth recording (5.2) in "exponential variables" (that is, let $x$ denote what was $e^{ix}$ in (5.2)):

$$x_1x_2x_3 = y_1y_2y_3 = 1 \Rightarrow$$
$$\frac{1+x_1}{1-x_1}\frac{1+x_2}{1-x_2} + \frac{1+x_2}{1-x_2}\frac{1+x_3}{1-x_3} + \frac{1+x_3}{1-x_3}\frac{1+x_1}{1-x_1} =$$
$$\frac{1+y_1}{1-y_1}\frac{1+y_2}{1-y_2} + \frac{1+y_2}{1-y_2}\frac{1+y_3}{1-y_3} + \frac{1+y_3}{1-y_3}\frac{1+y_1}{1-y_1},$$

which holds because both sides are equal to $-1$.

There are various other identities involving theta function, e.g. the ones in [54, Sect. 2.1] or [41, Sect. 4.1] are direct generalizations of (5.1).

## 6 Equivariant Elliptic Cohomology of $\mathcal{F}_\lambda$

This section is an informal introduction to equivariant elliptic cohomology of type A partial flag manifolds, or more generally, of Goresky-Kottwitz-MacPherson (GKM)-spaces [27]. Our general references are [25], [1, Section 2], [22, Section 4], [54, Section 7].

Before explaining what we mean by equivariant *elliptic* cohomology of $\mathcal{F}_\lambda$ let us revisit its equivariant cohomology and K theory. According to equivariant localization, the restriction maps to the (finitely many) fixed points induce injections of algebras:

$$H^*_{\mathbb{T}}(\mathcal{F}_\lambda) \hookrightarrow \bigoplus_{x\in\mathcal{F}_\lambda^{\mathbb{T}}} H^*_{\mathbb{T}}(x) = \bigoplus_{x\in\mathcal{F}_\lambda^{\mathbb{T}}} \mathbb{Z}[z_1,\dots,z_n],$$

$$K_{\mathbb{T}}(\mathcal{F}_\lambda) \hookrightarrow \bigoplus_{x\in\mathcal{F}_\lambda^{\mathbb{T}}} K_{\mathbb{T}}(x) = \bigoplus_{x\in\mathcal{F}_\lambda^{\mathbb{T}}} \mathbb{Z}[z_1^{\pm1},\dots,z_n^{\pm1}].$$

Recall that the $\mathbb{T}$ fixed points $x_I$ of $\mathcal{F}_\lambda$ are parametrized by $\mathcal{I}_\lambda$; the map is $f \mapsto (f|_{x_I})_{I\in\mathcal{I}_\lambda}$.

Moreover, the image of these injections have the following (so-called GKM-) descriptions. The tuple $(f_I)_{I\in\mathcal{F}_\lambda^{\mathbb{T}}}$ belongs to the image, if and only if, for "$(i,j)$-neighboring" fixed points $x_I$ and $x_J$ the difference of components $f_I - f_J$ is divisible by $z_i - z_j$. Here "$(i,j)$-neighboring" means that $J$ is obtained from $I$ by replacing the numbers $i$ and $j$. Divisibility is meant in the ring of polynomials and in the ring of Laurent polynomials, respectively. It is convenient to rephrase this divisibility condition to

$$f_I|_{z_i=z_j} = f_J|_{z_i=z_j} \qquad \text{for } (i,j)\text{-neighboring fixed points } I \text{ and } J. \tag{6.1}$$

Further encoding our descriptions we can say that

$$H^*_{\mathbb{T}}(\mathcal{F}_\lambda) \hookrightarrow \bigoplus_{I\in\mathcal{I}_\lambda} \text{"natural functions" on } \mathbb{C}^n,$$

$$K_{\mathbb{T}}(\mathcal{F}_\lambda) \hookrightarrow \bigoplus_{I\in\mathcal{I}_\lambda} \text{"natural functions" on } (\mathbb{C}^*)^n,$$

such that the image is characterized by (6.1). Polynomials and Laurent polynomials are indeed the "natural functions" on $\mathbb{C}^n$ and $(\mathbb{C}^*)^n$. This description has a built-in flexibility needed in several applications: in certain studies of $\mathcal{F}_\lambda$ one replaces the coefficient ring $\mathbb{Z}$ with other rings—and this can be achieved by just redefining what "natural functions" mean. In some other studies we want to permit some denominators, then again we just need to redefine the notion of "natural functions", and still the same description holds.

In this contexts $\mathbb{T}$ equivariant elliptic cohomology of $\mathcal{F}_\lambda$ would be natural to define by blindly replacing $\mathbb{C}$ or $\mathbb{C}^*$ above with the third kind of 1-dimensional algebraic group, the torus $E = \mathbb{C}^*/(q^{\mathbb{Z}})$ for a fixed $|q| < 1$. The choice of $q$ is immaterial for us, we will treat it as a parameter of the theory. However, $E$ being compact, there are no functions on $E$ or $E^n$. But there are sections of line bundles. Hence, $\mathbb{T}$ equivariant elliptic cohomology of $\mathcal{F}_\lambda$ is defined by

$$\mathrm{Ell}_{\mathbb{T}}(\mathcal{F}_\lambda) \hookrightarrow \bigoplus_{I\in\mathcal{I}_\lambda} \text{sections of line bundles on } E^n,$$

with image characterized by (6.1). Like above, we use $z_i$ as coordinates, now on $E^n$. Of course one may ask which line bundles are we allowing, and sections of what properties (meromorphic, etc.). In our point of view, the answer to those questions reflect different flavors of equivariant elliptic cohomology, similarly to how the exact choice of "natural functions" on $\mathbb{C}^n$ or $(\mathbb{C}^*)^n$ resulted different flavors of cohomology and K theory.

*Example 6.1* Let $\mathcal{F}_\lambda = \mathbb{P}^1$ and let us permit extra parameters $\hbar$ and $\mu_1, \mu_2$. Then the ordered pairs

$$(\vartheta(z_2/z_1), 0), \qquad \left(\vartheta'(1)\frac{\vartheta(z_2\mu_2/(z_1\mu_1))}{\vartheta(\mu_2/\mu_1)}, \vartheta'(1)\frac{\vartheta(z_1\hbar/z_2)}{\vartheta(\hbar)}\right) \tag{6.2}$$

are both equivariant elliptic cohomology classes on $\mathbb{P}^1$ (verify the property (6.1) for both). In fact these two classes will be the $\hbar$-deformed Schubert classes associated with the Schubert cells $\{(1:0)\}$ and $\mathbb{P}^1 - \{(1:0)\}$.

*Remark 6.1* An easy way to guarantee for a tuple to satisfy condition (6.1) is to describe the components of the tuple as suitable substitutions of the same function depending on suitable new variables. For example, consider the functions

$$\frac{\vartheta(z_1\hbar\mu_2/(t\mu_1))\vartheta(z_2/t)}{\vartheta(\hbar\mu_2/\mu_1)}, \qquad \vartheta'(1)\frac{\vartheta(z_1\hbar/t)\vartheta(z_2\mu_2/(t\mu_1))}{\vartheta(\hbar)\vartheta(\mu_2/\mu_1)},$$

and for each one consider the ordered pair of its $t = z_1$ and $t = z_2$ substitutions. We obtain exactly the tuples in the Example above. The very fact that they are $t = z_1$ and $t = z_2$ substitutions of the same function guarantees condition (6.1).

*Remark 6.2* In some circumstances specifying the permitted line bundles and the permitted sections is important. For example, if a uniqueness theorem claims that an equivariant elliptic cohomology class is determined by a list of axioms, then one must precisely define which line bundles and what kind of sections are permitted, see [1, Section 3.5], [22, Appendix A] and [54, Section 7.8]. However, if we have some concrete tuple of theta-functions on $E^n$ then we can state that this tuple is an equivariant elliptic cohomology class for the line bundle determined by the transformation properties of the theta functions, as long as the tuple satisfies (6.1).

# 7 ℏ-Deformed Schubert Classes in $H^*$, $K$, and Ell

## 7.1 ℏ-Deformed Schubert Class in Cohomology: CSM Class

Here we sketch the definition of Chern-Schwartz-MacPherson classes, following [2, 3, 43, 44], see also [4, 18, 39].

Let $F^{\mathbb{T}}(-)$ be the covariant functor of $\mathbb{T}$ invariant constructible functions (on complex algebraic varieties, with an appropriately defined push-forward map using the notion of Euler characteristic). Let $H_*^{\mathbb{T}}(-)$ be the functor of $\mathbb{T}$ equivariant homology as in [14]. The $\mathbb{T}$ equivariant MacPherson transformation is the unique natural transformation

$$C_*^{\mathbb{T}} : F^{\mathbb{T}}(-) \to H_*^{\mathbb{T}}(-)$$

satisfying the normalization $C_*^{\mathbb{T}}(\mathbb{1}_X) = c(TX) \cap \mu_X^{\mathbb{T}}$ for smooth projective $X$. Here $\mathbb{1}_X$ is the constant 1 function on $X$, $c(TX)$ the equivariant total Chern class, and $\mu_X^{\mathbb{T}}$ the equivariant fundamental homology class. If $i : Y \subseteq X$ is a subvariety of a smooth ambient space $X$ with $\mathbb{T}$ equivariant Poincaré duality $\mathcal{P}$, then we define the (cohomological) Chern-Schwartz-MacPherson (CSM) class $\mathrm{c}^{\mathrm{sm}}(Y) = \mathrm{c}^{\mathrm{sm}}(Y \subseteq X) \overset{\mathrm{def}}{=} \mathcal{P}(i_*(C_*^{\mathbb{T}}(\mathbb{1}_Y))) \in H_{\mathbb{T}}^*(X)$.[6]

[6]Although traditionally the class living in *homology* is called Chern-Schwartz-MacPherson class—transforming it to the *cohomology* of an ambient smooth space is convenient for the purposes of this paper, just like in [18, 43, 44].

The CSM class of $Y \subseteq X$ is an inhomogeneous cohomology class in $H^*_{\mathbb{T}}(X)$. Its lowest degree component is the fundamental cohomology class $[\overline{Y}] \in H^*_{\mathbb{T}}(X)$.

It is customary to homogenize it with an extra variable $\hbar$, making it of homogeneous degree $\dim X$. This version contains the same information as the original $\hbar = 1$ substitution of it, but in some other setups this $\hbar$-version is more natural (and has its own name "characteristic cycle class"). For the purpose of this paper we use the $\hbar$-homogenized one, that is, from now on $\mathrm{c}^{\mathrm{sm}}(Y) = \mathrm{c}^{\mathrm{sm}}(Y \subseteq X) \in H^*(X)[\hbar]$. In this version it is the coefficient of the highest power of $\hbar$ in $\mathrm{c}^{\mathrm{sm}}(Y)$ which equals $[\overline{Y}]$. This justifies our vocabulary of calling the CSM class the *$\hbar$-deformed Schubert class.*

Let us comment on how one deals with $\mathrm{c}^{\mathrm{sm}}$ classes in practice. There are three standard approaches.

The first approach is based directly on the fact that $C_*$ is a natural transformation of functors, and compares the CSM class of $Y$ with the CSM class of its closure and some geometry of the resolution of the closure—taking into account Euler characteristics of fibers. Typically we arrive at an inclusion-exclusion (sieve) type formula for $\mathrm{c}^{\mathrm{sm}}(Y)$. This approach can be modified by arranging the inclusion-exclusion argument 'upstairs', in the resolution itself.

The second approach is based on the fact that $\mathrm{c}^{\mathrm{sm}}$, besides the defining axioms, satisfies another strong rigidity property. Its "Segre version" $\mathrm{s}^{\mathrm{sm}}(Y) = \mathrm{c}^{\mathrm{sm}}(Y)/c(TX)$ is consistent with pull-back: $\mathrm{s}^{\mathrm{sm}}(f^{-1}(Y)) = f^*\mathrm{s}^{\mathrm{sm}}(Y)$, for closed $Y$ and sufficiently transversal $f$ to $Y$.

The third approach is that in certain situations the $\mathrm{c}^{\mathrm{sm}}$ classes satisfy a collection of interpolation constraints, and those constraints uniquely determine them. This approach was triggered by Maulik-Okounkov's notion of *cohomological stable envelope*, and was proved in [18, 48].

## 7.2 *$\hbar$-Deformed Schubert Class in K Theory: Motivic Chern Class*

The non-$\hbar$-deformed (equivariant) Schubert calculus has a large literature, going back to [35, 38], see references in the more recent [28]. In this section we follow [7, 12, 20, 61] and sketch the definition of the $\hbar$-deformed Schubert class in K theory: the equivariant motivic Chern class.

Let $X$ be a quasi-projective complex algebraic $\mathbb{T}$-variety. Let $G_0^{\mathbb{T}}(Var/X)$ denote the Grothendieck group of equivariant varieties and morphisms over $X$, modulo the scissors relation. There is a unique natural transformation

$$\mathrm{mC} : G_0^{\mathbb{T}}(Var/X) \to K_{\mathbb{T}}(X)[\hbar]$$

satisfying[7]

- functoriality: $\mathrm{mC}[g \circ f] = f_* \,\mathrm{mC}[g]$, and
- normalization: $\mathrm{mC}[\mathrm{id}_X] = \lambda_\hbar(T^*X) \overset{\text{def}}{=} \sum \hbar^i [\Lambda^i T^*X]$ for smooth $X$.

For $i : Y \subseteq X$ we write $\mathrm{mC}[Y] = \mathrm{mC}[i]$. The class $\mathrm{mC}[Y] \in K_{\mathbb{T}}(X)[\hbar]$ is called the *motivic Chern class* of $Y$ in $X$.

*Remark 7.1* For subvarieties $Y$ with mild (so-called Du Bois) singularities, the $\hbar = 1$ substitution recovers the K theoretic fundamental class of $Y$ in $X$. This justifies the name $\hbar$-deformed class. For $Y$ with non-Du Bois singularities the notion of *K theoretic fundamental class* is in fact ambiguous [47, Section 5], [16], and one may argue that the "right" choice for that notion is $\mathrm{mC}[Y]_{\hbar=1}$.

Let us comment on how one deals with mC classes in practice. Similarly to CSM theory there are three different approaches.

The first one is based directly on the functoriality property: We find a resolution $\tilde{Y} \to \overline{Y}$, and calculate the mC class of the composition $\tilde{Y} \to X$ using the normalization property. This will not equal $\mathrm{mC}[Y]$, but the difference is supported on the singular locus of $\tilde{Y} \to \overline{Y}$. To find that difference we resolve the singular locus, then the singular locus of that, etc. Finally we arrive at an inclusion-exclusion (sieve) type formula for $\mathrm{mC}[Y]$. This approach can be modified by arranging the inclusion-exclusion argument upstairs, in the resolution itself.

The second approach is based on the fact that mC, besides the defining axioms, satisfies another strong rigidity property. Its "Segre version" $\mathrm{mS}[Y] = \mathrm{mC}[Y]/\lambda_\hbar(T^*X)$ is consistent with pull-back: $\mathrm{mS}[f^{-1}(Y)] = f^*\,\mathrm{mS}[Y]$, for closed $Y$ and sufficiently transversal $f$ to $Y$.

The third approach is that in certain situations the mC classes satisfy a collection of interpolation constraints, and those constraints uniquely determine them. This approach was triggered by Okounkov's notion of *K theoretic stable envelope*, and is proved in [19, 20].

## 7.3 ħ-Deformed Schubert Class in Elliptic Cohomology: The Elliptic Class

The elliptic characteristic class $E(\Omega_I)$ associated with a Schubert cell (in arbitrary $G/P$ type) was defined in [31, 49]. This class necessarily depends on the $\hbar$-variable, as well as a new set of variables $\mu_i$, which are called Kähler-, or dynamical variables. Namely,

$$E(\Omega_I) \overset{\text{def}}{=} \tilde{\mathcal{E}\ell\ell}(\overline{\Omega}_I, D_I)$$

[7] It is more customary to denote the auxiliary parameter $\hbar$ by $y$, in accordance with the fact that the integral of the class $\mathrm{mC}[\mathrm{id}_X]$ is the $\chi_y$ genus of $X$ with $\hbar = y$. Yet, we keep the $\hbar$ notation to have consistent notation throughout $H^*$, $K$, Ell.

where $\mathcal{E}\ell\ell$ is an equivariant and elliptic version of the Borisov-Libgober class [9–11, 58, 59], and $D_I$ is an appropriate divisor on $\overline{\Omega}_I$ (for details see [31, 49]). By setup the Borisov-Libgober class depends on $\hbar$, and the divisor $D_I$ depends on a character of the corresponding parabolic subgroup $P$. Some of the properties of $E(\Omega_I)$ include the following.

- $E(\Omega_I)$ specializes to $\mathrm{mC}(\Omega_I)$ and further to $c^{\mathrm{sm}}(\Omega_I)$ in the trigonometric, and rational limit of elliptic cohomology.
- $E(\Omega_I)$ is computable from a resolution of $\overline{\Omega}_I$ through a process similar to the process computing $\mathrm{mC}(\Omega_I)$, $c^{\mathrm{sm}}(\Omega_I)$.
- $E(\Omega_I)$ satisfies and is determined by a small set of axioms which are essentially of interpolation flavor (cf. the interpolation characterization of $\mathrm{mC}(\Omega_I)$, $c^{\mathrm{sm}}(\Omega_I)$).
- In type A the class $E(\Omega_I)$ coincides with the notion of *elliptic stable envelope* of [1].
- The switch "equivariant parameters $\leftrightarrow$ dynamic parameters" is an incarnation of $d = 3, \mathcal{N} = 4$ mirror symmetry [53, 55].

## 8 Weight Functions and Their Orthogonality Relations

Weight functions, in three flavors—rational, trigonometric, and elliptic—were defined and studied by Tarasov–Varchenko and others in relation with hypergeometric solutions to KZ equations, [22, 29, 50–52, 54, 57]. Here we define weight functions adjusted to our geometric needs, and in Theorem 8.4 we show that they represent $\hbar$-deformed Schubert classes.

Let $N \in \mathbb{Z}_{\geq 0}$ and $\lambda \in \mathbb{Z}^N_{\geq 0}$. Define the initial sums $\lambda^{(k)} = \sum_{i=1}^k \lambda_i$, and set $n = \lambda^{(N)}$. We will consider functions $W(\boldsymbol{t}, \boldsymbol{z}, \hbar)$ and $W(\boldsymbol{t}, \boldsymbol{z}, \boldsymbol{\mu}, \hbar)$ in the variables

$$\boldsymbol{t} = (t_1^{(1)}, \dots, t_{\lambda^{(1)}}^{(1)}, \quad t_1^{(2)}, \dots, t_{\lambda^{(2)}}^{(2)}, \quad \dots \quad t_1^{(N-1)}, \dots, t_{\lambda^{(N-1)}}^{(N-1)}),$$

$$\boldsymbol{z} = (z_1, \dots, z_n), \qquad \boldsymbol{\mu} = (\mu_1, \dots, \mu_N), \qquad \text{and} \qquad \hbar.$$

When $t_a^{(N)}$ appears in the formulas, it is interpreted as $z_a$. For a function in these variables we define

$$\mathrm{Sym}_\lambda(f) = \mathrm{Sym}_{t^{(1)}} \dots \mathrm{Sym}_{t^{(N-1)}}(f)$$

where $\mathrm{Sym}_{t^{(k)}}(g)$ denotes the symmetrization in the $t_*^{(k)}$ variables, i.e.

$$\sum_{\sigma \in S_{\lambda^{(k)}}} g\left(t_a^{(k)} \mapsto t_{\sigma(a)}^{(k)}\right).$$

## 8.1 Rational Weight Functions

$$\psi^{\mathbb{H}}_{I,k,a,b}(x) = \begin{cases} x+\hbar & \text{if } i_b^{(k+1)} < i_a^{(k)} \\ \hbar & \text{if } i_b^{(k+1)} = i_a^{(k)} \\ x & \text{if } i_b^{(k+1)} > i_a^{(k)}, \end{cases}$$

$$U_I^{\mathbb{H}} = \prod_{k=1}^{N-1} \left( \prod_{a=1}^{\lambda^{(k)}} \prod_{b=1}^{\lambda^{(k+1)}} \psi^{\mathbb{H}}_{I,k,a,b}(t_b^{(k+1)} - t_a^{(k)}) \cdot \prod_{a<b\le\lambda^{(k)}} \frac{1}{t_b^{(k)} - t_a^{(k)}} \prod_{b\le a\le\lambda^{(k)}} \frac{1}{t_b^{(k)} - t_a^{(k)} + \hbar} \right),$$

$$W_I^{\mathbb{H}} = \mathrm{Sym}_\lambda(U_I^{\mathbb{H}}), \qquad W^{\mathbb{H}}_{\sigma,I} = W^{\mathbb{H}}_{\sigma^{-1}(I)}(\boldsymbol{t}, z_{\sigma(1)}, \dots, z_{\sigma(n)}, \hbar) \qquad (\sigma \in S_n).$$

## 8.2 Trigonometric Weight Functions

$$\psi^{\mathbb{K}}_{I,k,a,b}(x) = \begin{cases} 1+\hbar x & \text{if } i_b^{(k+1)} < i_a^{(k)} \\ (1+\hbar)x & \text{if } i_b^{(k+1)} = i_a^{(k)} \\ 1-x & \text{if } i_b^{(k+1)} > i_a^{(k)}, \end{cases}$$

$$U_I^{\mathbb{K}} = \prod_{k=1}^{N-1} \left( \prod_{a=1}^{\lambda^{(k)}} \prod_{b=1}^{\lambda^{(k+1)}} \psi^{\mathbb{K}}_{I,k,a,b}(t_a^{(k)}/t_b^{(k+1)}) \cdot \prod_{a<b\le\lambda^{(k)}} \frac{1}{1 - t_a^{(k)}/t_b^{(k)}} \prod_{b\le a\le\lambda^{(k)}} \frac{1}{1 + \hbar t_a^{(k)}/t_b^{(k)}} \right),$$

$$W_I^{\mathbb{K}} = \mathrm{Sym}_\lambda(U_I^{\mathbb{K}}), \qquad W^{\mathbb{K}}_{\sigma,I} = W^{\mathbb{K}}_{\sigma^{-1}(I)}(\boldsymbol{t}, z_{\sigma(1)}, \dots, z_{\sigma(n)}, \hbar) \qquad (\sigma \in S_n).$$

## 8.3 Elliptic Weight Functions

Define the integer invariants

- $p(I, j, i) = |I_j \cap \{1, \dots, i-1\}|$;
- $j(I, k, a)$ is defined by $i_a^{(k)} \in I_{j(I,k,a)}$,

and the functions

$$\psi^{\mathbb{E}}_{I,k,a,b}(x) = \begin{cases} \vartheta(x\hbar)/\vartheta(\hbar) & \text{if } i_b^{(k+1)} < i_a^{(k)} \\ \dfrac{\vartheta(x\frac{\mu_{k+1}}{\mu_{j(I,k,a)}}\hbar^{1+p(I,j(I,k,a),i_a^{(k)})-p(I,k+1,i_a^{(k)})})}{\vartheta(\frac{\mu_{k+1}}{\mu_{j(I,k,a)}}\hbar^{1+p(I,j(I,k,a),i_a^{(k)})-p(I,k+1,i_a^{(k)})})} & \text{if } i_b^{(k+1)} = i_a^{(k)} \\ \vartheta(x) & \text{if } i_b^{(k+1)} > i_a^{(k)}, \end{cases}$$

$$U_I^{\mathbb{E}} = \vartheta'(1)^{\dim_I} \prod_{k=1}^{N-1} \left( \prod_{a=1}^{\lambda^{(k)}} \prod_{b=1}^{\lambda^{(k+1)}} \psi^{\mathbb{E}}_{I,k,a,b}(t_b^{(k+1)}/t_a^{(k)}) \cdot \prod_{a<b\le\lambda^{(k)}} \frac{1}{\vartheta(t_b^{(k)}/t_a^{(k)})} \prod_{b<a\le\lambda^{(k)}} \frac{\vartheta(\hbar)}{\vartheta(\hbar t_b^{(k)}/t_a^{(k)})} \right),$$

$$W_I^{\mathbb{E}} = \mathrm{Sym}_\lambda(U_I^{\mathbb{E}}), \qquad W_{\sigma,I}^{\mathbb{E}} = W_{\sigma^{-1}(I)}^{\mathbb{E}}(\boldsymbol{t}, z_{\sigma(1)}, \ldots, z_{\sigma(n)}, \hbar, \boldsymbol{\mu}) \qquad (\sigma \in S_n).$$

## 8.4 Orthogonality

For $I \in \mathcal{I}_\lambda$ let

$$R_I^{\mathbb{H}} = \prod_{k<l} \prod_{a\in I_k} \prod_{b\in I_l} (z_b - z_a), \qquad Q_I^{\mathbb{H}} = \prod_{k<l} \prod_{a\in I_k} \prod_{b\in I_l} (z_b - z_a + \hbar),$$

$$R_I^{\mathbb{K}} = \prod_{k<l} \prod_{a\in I_k} \prod_{b\in I_l} (1 - z_a/z_b), \qquad Q_I^{\mathbb{K}} = \prod_{k<l} \prod_{a\in I_k} \prod_{b\in I_l} (1 + z_b/(z_a\hbar)),$$

$$R_I^{\mathbb{E}} = \prod_{k<l} \prod_{a\in I_k} \prod_{b\in I_l} \vartheta(z_b/z_a), \qquad Q_I^{\mathbb{E}} = \prod_{k<l} \prod_{a\in I_k} \prod_{b\in I_l} \vartheta(\hbar z_b/z_a).$$

Given $\lambda$ and $I \in \mathcal{I}_\lambda$, for a function $f(\boldsymbol{t})$ and $I \in \mathcal{I}_\lambda$ let $f(\boldsymbol{z}_I)$ denote the result of substituting $t_a^{(k)} = z_{i_a^{(k)}}$, for all $k = 1, \ldots, N-1$, $a = 1, \ldots, \lambda^{(k)}$. Define the inner products $\langle\ ,\ \rangle_{\mathbb{H}}$, $\langle\ ,\ \rangle_{\mathbb{K}}$, $\langle\ ,\ \rangle_{\mathbb{E}}$

$$\langle f(\boldsymbol{t}), g(\boldsymbol{t})\rangle = \sum_{I\in\mathcal{I}_\lambda} \frac{f(\boldsymbol{z}_I)g(\boldsymbol{z}_I)}{R_I Q_I}$$

by using the relevant versions of $R_I$ and $Q_I$ in the denominator. In practice the functions $f, g$ will also depend on other variabes $z, \hbar$ (and $\boldsymbol{\mu}$ in case of $\mathbb{E}$), but the substitution does not affect those.

**Theorem 8.1 (Rational Orthogonality)** *Let $s_0$ be the longest permutation in $S_n$. We have*

$$\langle W^{\mathbb{H}}_{\mathrm{id},I}, W^{\mathbb{H}}_{s_0,J}\rangle_{\mathbb{H}} = \delta_{I,J}.$$

**Theorem 8.2 (Trigonometric Orthogonality)** *Let $\iota[f(z,\hbar)]$ be obtained from the function $f(z,\hbar)$ by substituting $1/t_a^{(k)}$ for $t_a^{(k)}$, $1/z_i$ for $z_i$ (for all possible indexes) and $1/\hbar$ for $\hbar$. We have*

$$\langle W^{\mathbb{K}}_{\mathrm{id},I}, (-\hbar)^{-\dim_J} \iota[W^{\mathbb{K}}_{s_0,J}]\rangle_{\mathbb{K}} = \delta_{I,J}.$$

**Theorem 8.3 (Elliptic Orthogonality)** *Let $\tau[f(z,\hbar,\boldsymbol{\mu})]$ be obtained from the function $f(z,\hbar,\boldsymbol{\mu})$ by substituting $\hbar^{\lambda_i}/\mu_i$ for $\mu_i$ (for all $i$). We have*

$$\langle W^{\mathbb{E}}_{\mathrm{id},I}, (\vartheta(\hbar)/\vartheta'(1))^{\dim_\lambda} \tau[W^{\mathbb{E}}_{s_0,J}]\rangle_{\mathbb{E}} = \delta_{I,J}.$$

To illustrate the non-triviality of the elliptic orthogonality relations, let us mention that the special case of elliptic orthogonality

$$\langle W^{\mathbb{E}}_{\mathrm{id},(\{3\},\{1,2\})}, (\vartheta(\hbar)/\vartheta'(1))^2\tau[W^{\mathbb{E}}_{s_0,(\{1\},\{2,3\})}]\rangle_{\mathbb{E}} = 0$$

is equivalent to the trisecant identity (5.1) with the variables

$$\begin{array}{ll} x_1 = z_2/z_1 & y_1 = \mu_2/(\mu_1\hbar) \\ x_2 = z_1/z_3 & y_2 = \hbar \\ x_3 = z_3/z_2 & y_3 = \mu_1/\mu_2. \end{array}$$

## 8.5 Weight Functions Represent ħ-Deformed Schubert Classes

**Theorem 8.4** *Interpreting the $t_j^{(k)}$ variables as Chern roots of the tautological bundles $\mathcal{V}_k$ of rank $\lambda^{(k)}$ and the $z_i$ variables as equivariant variables, the weight functions express the ħ-deformed Schubert classes:*

$$\begin{aligned} \mathrm{c}^{\mathrm{sm}}(\Omega_I) &= \mathrm{W}^{\mathbb{H}}_{\mathrm{id},I} \quad [18, 48], \\ \mathrm{mC}(\Omega_I) &= \mathrm{W}^{\mathbb{K}}_{\mathrm{id},I} \quad [20], \\ E(\Omega_I) &= \mathrm{W}^{\mathbb{E}}_{\mathrm{id},I} \quad [31, 49]. \end{aligned}$$

*The $\mu_i$ variables in $\mathrm{W}^{\mathbb{E}}_{\mathrm{id},I}$ express the dependence of $E(\Omega_I)$ on the character of the parabolic subgroup.*

*Remark 8.5* The elliptic weight function formulas have singularities at $\hbar = 1$, see the general formulas above, or Remark 6.1 and Example 6.1. As mentioned, this is another incarnation of the fact that defining non-ħ-deformed elliptic classes of Schubert cells in $\mathrm{Ell}(\mathcal{F}_\lambda)$, $\mathrm{Ell}_{\mathbb{T}}(\mathcal{F}_\lambda)$ is problematic.

# 9 Sample Schubert Structure Constants

Having our Main Theorem (with all its ingredients), it is now only a question of computer power to find Schubert structure constants at each vertex of Fig. 1.[8] In the three subsections below we show some sample calculations in

$$H^*(T^*\mathrm{Gr}(3,6)) \text{ and } H^*_{\mathbb{T}}(T^*\mathrm{Gr}(3,6)), \qquad \mathrm{Ell}_{\mathbb{T}}(T^*\mathbb{P}^n), \qquad \mathrm{Ell}(T^*\mathbb{P}^1),$$

then, in Sect. 9.4 we discuss questions about these structure constants.

[8] In fact, just 10 of the 12 vertices, see Remark 8.5.

## 9.1 Cohomology

In $H^*(\mathrm{T^*Gr}(3,6))$

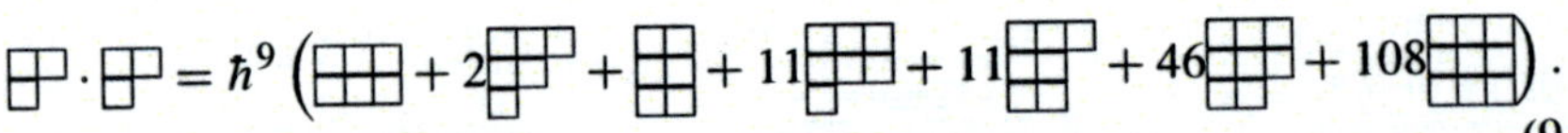

(9.1)

Observe that the two extensions (2.2) and (9.1) of (2.1) go the opposite directions: in one of them the non-zero coefficients extend to "smaller" partitions, in the other one to "larger" partitions. A combination of the two extensions is, of course, $H^*_{\mathbb{T}}(T^*\mathrm{Gr}(3,6))$. Those formulas involve $z_i$ and $\hbar$, and tend to get very large, yet, for example for $I = (\{2,4,6\},\{1,3,5\})$ "= ", in $H^*_{\mathbb{T}}(T^*\mathrm{Gr}(3,6))$ we have

$$\mathrm{c}^I_{I,I} = (z_5 - z_4)(z_5 - z_2)(z_3 - z_2)\times$$
$$(z_1 - z_2 + \hbar)(z_1 - z_4 + \hbar)(z_3 - z_4 + \hbar)(z_1 - z_6 + \hbar)(z_3 - z_6 + \hbar)(z_5 - z_6 + \hbar).$$

This coefficient is 0 after substituting $z_i = 0$, so the corresponding term is not visible in (9.1). The coefficient of $\hbar^{\dim_I + \dim_I - \dim_I} = \hbar^6$ is $(z_5 - z_4)(z_5 - z_2)(z_3 - z_2)$, which turns up in (2.2) as $c^I_{I,I}$.

## 9.2 Equivariant Elliptic Cohomology

Consider the elliptic classes for $\mathcal{F}_\lambda = \mathbb{P}^n$. For $k \in [n]$ denote $I_k := (\{k\}, [n] - \{k\})$, and let $\mathrm{c}^m_{k,l} \overset{\text{def}}{=} \mathrm{c}^{I_m}_{I_k, I_l}$.

**Theorem 9.1** *Let $k \leq l$. For $m > k$ we have $\mathrm{c}^m_{k,l} = 0$ and*

$$\mathrm{c}^k_{k,l} = \vartheta'(1)^{l-1} \frac{\vartheta(z_l/z_k \cdot \mu_2/\mu_1 \cdot \hbar^{2-l})}{\vartheta(\mu_2/\mu_1 \cdot \hbar^{2-l})} \prod_{i=1}^{l-1} \frac{\vartheta(z_i\hbar/z_k)}{\vartheta(\hbar)} \prod_{i=l+1}^{n} \vartheta(z_i/z_k).$$

*In particular,*

$$\mathrm{c}^k_{k,k} = \vartheta'(1)^{k-1} \prod_{i=1}^{k-1} \frac{\vartheta(z_i\hbar/z_k)}{\vartheta(\hbar)} \prod_{i=k+1}^{n} \vartheta(z_i/z_k).$$

## 9.3 Non-equivariant Elliptic Cohomology

Plugging in $z_i = 1$ for all $i$ in equivariant elliptic cohomology formulas yields non-equivariant elliptic cohomology formulas. The actual analysis of occurring functions is intriguing. For example, in $\mathrm{Ell}(T^*\mathbb{P}^1)$ we obtain

$$c^{(1,2)}_{(2,1),(2,1)} = \vartheta'(1)^2 \lim_{z_2/z_1 \to 1} \left( \frac{\frac{\vartheta(z_2/z_1 \cdot \mu_2/\mu_1)}{\vartheta(\mu_2/\mu_1)} - \frac{\vartheta(z_2/z_1 \cdot \hbar)}{\vartheta(\hbar)}}{\vartheta(z_2/z_1)} \right).$$

Observe that the numerator vanishes at $z_2/z_1 = 1$, and the denominator has a simple 0 there. Hence the ratio has a removable singularity at $z_2/z_1 = 1$, and the limit is the value when that singularity is removed.

The two terms in the numerator of the limit above have different transformation properties (a.k.a. factors of automorphy) with respect to $z_2 \to z_2 q$. Hence those terms are sections of different line bundles; therefore, they should not be added unless we choose our vector space to be the direct sum of the vector spaces of sections of different bundles. This is a questionable property of elliptic structure constants which deserves future study.

## 9.4 Positivity?

A fundamental feature of both characteristic class formulas and structure constant formulas in various situations is positivity. For example, the integer structure coefficients in ordinary cohomology (the classical Littlewood-Richardson coefficients) are known to be non-negative. The $z$-polynomial structure coefficients in equivariant cohomology are known to be polynomials of $z_{\text{large}} - z_{\text{small}}$ with non-negative coefficients (see the example in (2.2)). Analogous results hold in K theory and equivariant K theory. In the $\hbar$-deformed worlds, for CSM classes and for motivic Chern classes, some positivity as well as log-concavity results and conjectures can be found in [6, 8, 19].

It is reasonable to expect positivity in the $\hbar$-deformed equivariant elliptic cohomology—generalizing the analogous properties in K theory and cohomology. The challenge is to figure out what positivity actually means for an elliptic function. We plan to study these expected elliptic positivity properties in the future.

*Remark 9.2* Besides positivity, other features of characteristic classes and structure constants in various versions of Schubert calculus include stabilization and saturation properties. While hints of stabilization appear in $\hbar$-deformed cohomology and K theory, nothing is known in elliptic cohomology so far.

**Acknowledgments** The author is grateful to A. Weber for several helpful comments on this paper, and to L. M. Fehér and A. Varchenko for many useful conversations on this topic. The author is supported by a Simons foundation grant.

## References


1. Aganagic, M., Okounkov, A.: Elliptic stable envelopes (2016). Preprint. arXiv:1604.00423
2. Aluffi, P.: Classes de Chern des variétés singulières, revisitées. C. R. Math. Acad. Sci. Paris **342**(6), 405–410 (2006)
3. Aluffi, P., Brasselet, J.-P.: Une nouvelle preuve de la concordance des classes définies par M.-H. Schwartz et par R. MacPherson. Bull. de la Soc. Math. de France **136**(2), 159–166 (2008)
4. Aluffi, P., Mihalcea, L.C.: Chern classes of Schubert cells and varieties. J. Algebraic Geom. **18**(1), 63–100 (2009)
5. Aluffi, P., Mihalcea, L.C.: Chern-Schwartz-MacPherson classes for Schubert cells in flag manifolds. Comp. Math. **152**(12), 2603–2625 (2016)
6. Aluffi, P., Mihalcea, L.C., Schürmann, J., Su, C.: Shadows of characteristic cycles, Verma modules, and positivity of Chern-Schwartz-MacPherson classes of Schubert cells (2017). arXiv:1709.08697
7. Aluffi, P., Mihalcea, L.C., Schürmann, J., Su, C.: Motivic Chern classes of Schubert cells, Hecke algebras, and applications to Casselman's problem (2020). arXiv:1902.10101
8. Aluffi, P., Mihalcea, L.C., Schürmann, J., Su, C.: Positivity of Segre-MacPherson classes (2020). arXiv:1902.00762
9. Borisov, L., Libgober, A.: Elliptic genera of toric varieties and applications to mirror symmetry. Inv. Math. 140, 453 (2000)
10. Borisov, L., Libgober, A.: Elliptic genera of singular varieties. Duke Math. J. 116(2), 319–351 (2003)
11. Borisov, L., Libgober, A.: McKay correspondence for elliptic genera. Ann. Math. **161**(3), 1521–1569 (2005)
12. Brasselet, J.-P., Schürmann, J., Yokura, S.: Hirzebruch classes and motivic Chern classes for singular spaces. J. Topol. Anal. **2**(1), 1–55 (2010)
13. Bressler, P., Evens, S.: The Schubert calculus, braid relations, and generalized cohomology. Trans. Amer. Math. Soc. **317**(2), 799–811 (1990)
14. Edidin, D., Graham, W.: Equivariant intersection theory. Invent. Math. **131**, 595–634 (1998)
15. Fay, J.D.: Theta Functions on Riemann Surfaces. Springer Lecture Notes in Mathematics, vol. 352. Springer, Berlin (1973)
16. Fehér, L.M.: Motivic Chern classes of cones. In: Fernández de Bobadilla, J., László, T., Stipsicz, A. (eds.) Singularities and Their Interaction with Geometry and Low Dimensional Topology, Trends in Mathematics. Springer, Cham (2021). https://doi.org/10.1007/978-3-030-61958-9_9
17. Fehér, L., Rimányi, R.: Schur and Schubert polynomials as Thom polynomials—cohomology of moduli spaces. Cent. Eur. J. Math. **4**, 418–434 (2003)
18. Fehér, L., Rimányi, R.: Chern-Schwartz-MacPherson classes of degeneracy loci. Geom. Top. **22**, 3575–3622 (2018)
19. Fehér, L.M., Rimányi, R., Weber, A.: Characteristic classes of orbit stratifications, the axiomatic approach. In: Hu, J., Li, C., Mihalcea, L.C. (eds.) Schubert Calculus and Its Applications in Combinatorics and Representation Theory, Guangzhou, China, November 2017. Springer Proceedings in Mathematics & Statistics, vol. 332, pp. 223–249. Springer, Singapore (2020)
20. Fehér, L.M., Rimányi, R., Weber, A.: Motivic Chern classes and K-theoretic stable envelopes. Proc. London Math. Soc. **122**(1), 153–189 (2021)

21. Felder, G., Rimányi, R., Varchenko, A.: Poincare-Birkhoff-Witt expansions of the canonical elliptic differential form. In: Etingof, P., Gelaki, S., Shnider, S. (eds.) Quantum Groups. Contemporary Mathematics, vol. 433, pp. 191–208. American Mathematical Society, Providence (2007)
22. Felder, G., Rimányi, R., Varchenko, A.: Elliptic dynamical quantum groups and equivariant elliptic cohomology. SIGMA **14**, 132, 41 (2018)
23. Fulton, W.: Young Tableau. Cambridge University Press, Cambridge (1996)
24. Ganter, N., Ram, A.: Generalized Schubert calculus. J. Ramanujan Math. Soc. **28A** (Special Issue-2013), 1–42 (2012)
25. Ginzburg, V., Kapranov, M., Vasserot, E.: Elliptic algebras and equivariant elliptic cohomology I, unpublished. arXiv:q-alg/9505012
26. Gorbounov, V., Rimányi, R., Tarasov, V., Varchenko, A.: Cohomology of the cotangent bundle of a flag variety as a Yangian Bethe algebra, J. Geom. Phys. **74**, 56–86 (2013)
27. Goresky, M., Kottwitz, R., MacPherson, R.: Equivariant cohomology, Koszul duality, and the localization theorem. Invent. Math. **131**, 25–83 (1998)
28. Graham, W., Kreiman, V.: Excited Young diagrams, equivariant K-theory, and Schubert varieties. Trans. AMS **367**, 6597–6645 (2015)
29. Konno, H.: Elliptic Stable Envelopes and Finite-dimensional Representations of Elliptic Quantum Group (2018). Preprint. arXiv:1803.01540
30. Knutson, A., Tao, T.: Puzzles and (equivariant) cohomology of Grassmannians. Duke Math. J. **119**(2), 221–260 (2003)
31. Kumar, S., Rimányi, R., Weber, A.: Elliptic classes of Schubert varieties. Math. Ann. **378**(1), 703–728, (2020). DOI 10.1007/s00208-020-02043-z
32. Lam, T., Schilling, A., Shimozono, M.: K-theory Schubert calculus of the affine Grassmannian. Comp. Math. **146**(4), 811–852 (2019)
33. Lascoux, A.: Puissances extérieures, déterminants et cycles de Schubert. Bull Soc. Math. France **102**, 161–179 (1974)
34. Lascoux, A.: Interpolation. Lectures at Tianjin University. American Mathematical Society, Providence (1996)
35. Lascoux, A., Schützenberger, M.-P.: Structure de Hopf de l'anneau de cohomologieet de l'anneau de Grothendieck d'une variété de drapeaux. C. R. Acad. Sci. Paris Ser. I Math. **295**(11), 629–633 (1982)
36. Lenart, C., Zainoulline, K.: A Schubert basis in equivariant elliptic cohomology. New York J. Math. **23**, 711–737 (2015)
37. Lenart, C., Zainoulline, K., Zhong, C.: Parabolic Kazhdan-Lusztig basis, Schubert classes and equivariant oriented cohomology. J. Inst. Math. Jussieu **19**, 1889–1929 (2019)
38. Lusztig, G.: Equivariant K-theory and representations of Hecke algebras. Proc. AMS **94**(2), 337–342 (1985)
39. MacPherson, R.: Chern classes for singular algebraic varieties. Ann. Math. **100**, 421–432 (1974)
40. Maulik, D., Okounkov, A.: Quantum Groups and Quantum Cohomology. Astérisque No. 408 (2019)
41. Mikosz, M., Weber, A.: Elliptic classes, McKay correspondence and theta identities. J. Alg. Comb. (2019). arXiv:1909.07303
42. Molev, A.I., Sagan, B.E.: A Littlewood-Richardson rule for factorial Schur functions. Trans. AMS **351**(11), 4429–4443 (1999)
43. Ohmoto, T.: Equivariant Chern classes of singular algebraic varieties with group actions. Math. Proc. Cambridge Phil. Soc. **140**, 115–134 (2006)
44. Ohmoto, T.: Singularities of maps and characteristic classes. In: Advanced Studies in Pure Mathematics , School on Real and Complex Singularities in São Carlos 2012, Araújo dos Santos, R.N., Jorge Pérez, V.H., Nishimura, T., Saeki, O. (eds.), pp. 191–265. Mathematical Society of Japan, Tokyo (2016)

45. Okounkov, A.: Lectures on K-theoretic computations in enumerative geometry. In: Geometry of Moduli Spaces and Representation Theory. IAS/Park City Mathematics Series, vol. 24, pp. 251–380. American Mathematical Society, Providence (2017)
46. Parusiński, A., Pragacz, P.: Chern-Schwartz-MacPherson classes and the Euler characteristic of degeneracy loci and special divisors, J. Amer. Math. Soc. **8**(4), 793–817 (1995)
47. Rimányi, R., Szenes, A.: Residues, Grothendieck polynomials, and K-theoretic Thom polynomials (2018). Preprint. arXiv:1811.02055
48. Rimányi, R., Varchenko, A.: Equivariant Chern-Schwartz-MacPherson classes in partial flag varieties: interpolation and formulae. In: Schubert Varieties, Equivariant Cohomology and Characteristic Classes—IMPANGA 15, pp. 225–235. European Mathematical Society, Zürich (2018)
49. Rimányi, R., Weber, A.: Elliptic classes of Schubert varieties via Bott-Samelson resolution. J. of Topol. **13**(3), 1139–1182 (2020). DOI 10.1112/topo.12152
50. Rimányi, R., Tarasov, V., Varchenko, A.: Cohomology classes of conormal bundles of Schubert varieties and Yangian weight fucntions. Math. Z. **277**(3–4), 1085–1104 (2014)
51. Rimányi, R., Tarasov, V., Varchenko, A.: Trigonometric weight functions as K-theoretic stable envelope maps for the cotangent bundle of a flag variety. J. Geometry Phys. **94**, 81–119 (2015)
52. Rimányi, R., Tarasov, V., Varchenko, A.: Partial flag varieties, stable envelopes and weight functions. Quantum Topol. **6**(2), 333–364 (2015)
53. Rimányi, R., Smirnov, A., Varchenko, A., Zhou, Z.: Three dimensional mirror self-symmetry of the cotangent bundle of the full flag variety. SIGMA **15**, 093, 22 (2019)
54. Rimányi, R., Tarasov, V., Varchenko, A.: Elliptic and K-theoretic stable envelopes and Newton polytopes. Selecta Math. **25**, 16 (2019)
55. Rimányi, R., Smirnov, A., Varchenko, A., Zhou, Z.: 3d mirror symmetry and elliptic stable envelopes. Int. Math. Res. Not. (to appear, 2021)
56. Schwartz, M.-H.: Classes caractéristiques définies par une stratification d'une variété analytique complexe. I. C. R. Acad. Sci. Paris **260**, 3262–3264 (1965)
57. Varchenko, A., Tarasov, V.: Jackson integral representations for solutions of the Knizhnik-Zamolodchikov quantum equation. Leningrad Math. J. **6**, 275–313 (1994)
58. Waelder, R.: Equivariant elliptic genera. Pacific J. Math. **235**(2), 345–377 (2008)
59. Waelder, R.: Equivariant elliptic genera and local McKay correspondences. Asian J. Math. **12**(2), 251–284 (2008)
60. Weber, A.: Equivariant Chern classes and localization theorem. J. Singularities **5**, 153–176 (2012)
61. Weber, A.: Equivariant Hirzebruch class for singular varieties. Selecta Math. (N.S.) **22**(3), 1413–1454 (2016)

# Fundamental Groups and Path Lifting for Algebraic Varieties

János Kollár

**Abstract** We study 3 basic questions about fundamental groups of algebraic varieties. For a morphism, is being surjective on $\pi_1$ preserved by base change? What is the connection between openness in the Zariski and in the Euclidean topologies? Which morphisms have the path lifting property?

**Keywords** Algebraic varieties · Fundamental groups · Path lifting

## 1 Introduction

The aim of these notes is to study three questions involving maps between the fundamental groups of algebraic varieties.

- Let $X \to Y$ be a morphism of schemes that induces a surjection on the algebraic fundamental groups. Does the same hold after a base change $X \times_S Z \to Y \times_S Z$?
- Let $X \to Y$ be a morphism between $\mathbb{C}$-schemes. When can we lift every continuous path in $Y(\mathbb{C})$ to a path in $X(\mathbb{C})$?
- Let $X \to Y$ be a morphism between $\mathbb{C}$-schemes. What is the connection between openness in the Zariski and in the Euclidean topologies?

An answer to the first question was used in the study of Pell surfaces [5]. While this application involves only maps between algebraic curves, the curves in question are singular and non-proper, and it turns out to be not much harder to consider the general case. This is treated in Sect. 2.

The proof uses some basic properties of open and universally open morphisms, some of which I did not find in the literature. These are worked out in Sect. 3.

J. Kollár (✉)
Princeton University, Princeton, NJ, USA
e-mail: kollar@math.princeton.edu



J. Fernández de Bobadilla et al. (eds.), *Singularities and Their Interaction with Geometry and Low Dimensional Topology*, Trends in Mathematics,
https://doi.org/10.1007/978-3-030-61958-9_6

Various forms of path lifting are studied in Sect. 4. The answer is most complete for arc lifting (Definition 4.1) which is equivalent to openness of the morphism in the Euclidean topology and universal openness in the Zariski topology; see Theorem 4.2.

While the main applications are to schemes of finite type, the discussions in Sects. 2–3 are formulated for arbitrary Noetherian schemes.

## 2 Maps Between Fundamental Groups

**Definition 2.1** Let $X$ be a connected scheme and $x \to X$ a geometric point. The *fundamental group* of $X$ with base point $x$ is denoted by $\pi_1(X, x)$. Working with schemes, we use the algebraic fundamental group.

Let $f : X \to Y$ be a morphism of connected schemes. Fix a base point $x \to X$ and its $f$-image $y \to Y$. We get a natural group homomorphism $f_* : \pi_1(X, x) \to \pi_1(Y, y)$.

(1) We say that $f$ is $\pi_1$*-surjective* if $f_* : \pi_1(X, x) \to \pi_1(Y, y)$ is surjective.

Using the correspondence between quotients of the fundamental group and finite, étale covers we get the following equivalent form.

(2) $f$ is $\pi_1$-surjective iff for every connected, finite, étale cover $Y'$, the fiber product $X \times_Y Y'$ is also connected.

The latter formulation shows that the base point can be ignored in the first version of the definition. More generally, choosing a different $x' \to y$, the image of $\pi_1(X, x') \to \pi_1(Y, y)$ is a conjugate of the image of $\pi_1(X, x) \to \pi_1(Y, y)$.

One of the main aims is to show that connectedness of the fiber product $X \times_Y Y'$ also holds if $Y' \to Y$ is proper and universally open, see Theorem 2.2.5. We discuss open and universally open morphisms in Sect. 3.

**Theorem 2.2** *Let $X, S$ be connected schemes and $g : X \to S$ a proper and universally open morphism. The following are equivalent.*

(1) *$g$ is $\pi_1$-surjective.*
(2) *$X \times_S X$ is connected.*
(3) *$X \times_S \cdots \times_S X$ ($n$ copies of $X$) is connected for some $n \geq 2$.*
(4) *The number of connected components of $X \times_S \cdots \times_S X$ is bounded, independent of the number of factors.*
(5) *For every connected scheme $Y$ and proper, universally open morphism $h : Y \to S$, the fiber product $X \times_S Y$ is connected.*
(6) *For every connected scheme $Y$ and proper, universally open morphism $h : Y \to S$, the second projection $X \times_S Y \to Y$ is $\pi_1$-surjective.*

The following example shows that in (2.2.5–6) we need $Y \to S$ to be universally open. It would not be enough to assume only that $Y \to S$ is finite and open.

*Example 2.3* Let $C$ be a projective, nodal rational curve over an algebraically closed field with normalization $\pi : \mathbb{P}^1 \cong \bar{C} \to C$ and $c \in C$ a smooth point. Note that $\pi_1(C, c) \cong \mathbb{Z}$. Let $g_n : (c_n, C_n) \to (c, C)$ be its unique degree $n$, connected, étale cover. For $n, m \geq 1$ set $(x, X_{n,m}) := (c_n, C_n) \amalg_{c_n=c_m} (c_m, C_m)$. (We use $\amalg$ to denote disjoint union. The subscript $c_n = c_m$ means that we take the disjoint union and then identify the points $c_n$ and $c_m$.) The maps $g_n, g_m$ glue to $g_{n,m} : (x, X_{n,m}) \to (c, C)$. Then

(1) $g_{n,m}$ is $\pi_1$-surjective iff $(n, m) = 1$,
(2) $\bar{C} \to C$ is finite, open and surjective, yet
(3) $X_{n,m} \times_C \bar{C}$ is the disjoint union of $\bar{C} \amalg_{c=c} \bar{C}$ and of $n + m - 2$ copies of $\bar{C}$. Thus it is disconnected iff $n + m > 2$.

We start the proof of Theorem 2.2 with a series of remarks and lemmas that establish various special cases, and then use them to settle the general case.

*Remark 2.4 (Stein Factorization)* Let $h : X \to X'$ be a proper morphism such that $h_* \mathcal{O}_X = \mathcal{O}_{X'}$. Then $X$ is connected if $X'$ is connected. Since $h_*$ commutes with any flat base change, (2.1.2) shows that $h_* : \pi_1(X, x) \to \pi_1(X', x')$ is an isomorphism.

Applying this to the Stein factorization $g : X \to X' \to S$ shows that Theorem 2.2 holds for proper morphisms iff it holds for finite morphisms.

**Lemma 2.5 (2.2.2⇒ 2.2.5)** *Let $X, Y, S$ be connected schemes and $g : X \to S, h : Y \to S$ finite, universally open morphisms. Assume that $X \times_S X$ is connected. Then $X \times_S Y$ is also connected.*

***Proof*** Fix a geometric point $y \to Y$ and then choose $x_1, x_2 \to X$ such that $g(x_1) = g(x_2) = h(y)$. Let $W \subset X \times_S X \times_S Y$ be the connected component that contains $(x_1, x_2, y)$. Since $h$ is finite and universally open, so is the projection $\pi_{12} : X \times_S X \times_S Y \to X \times_S X$. Thus $\pi_{12} : W \to X \times_S X$ is surjective by (3.3.3). In particular, there is a point $(x_1, x_1, y') \in W$ for some $y' \to Y$. Consider now the 2 projections $\pi_i : W \to X \times_S Y$ for $i = 1, 2$. Note that

$$(x_1, y), (x_1, y') \in \pi_1(W) \quad \text{and} \quad (x_2, y), (x_1, y') \in \pi_2(W).$$

Since $W$ is connected, this shows that $(x_1, y), (x_2, y)$ are in the same connected component of $X \times_S Y$.

Let $V_j \subset X \times_S Y$ be the connected components. Projection to $Y$ is universally open, so the projections $V_j \to Y$ are surjective by (3.3.3). All preimages of $y \in Y$ are in the same connected component by the above argument, hence $X \times_S Y$ has only 1 connected component. □

**Lemma 2.6** *Let $g : X \to S$ be a quasi-finite, universally open morphism. Assume that the diagonal $\Delta_{X/S}$ is a connected component of $X \times_S X$. Then $g$ uniquely factors as $g : X \to S' \to S$ where $X \to S'$ is a universal homeomorphism and $S' \to S$ is étale.*

*Proof* Such a factorization is unique, so it is enough to construct it étale locally on $S$. By Corollary 3.8, after an étale base change we may assume that $g$ is of the form $g : \amalg_i (x_i, X_i) \to (s, S)$ where each restriction $g_i : (x_i, X_i) \to (s, S)$ is finite, local and $k(x_i)/k(s)$ is purely inseparable. Thus $X_i \times_S X_i$ is connected, since all of its irreducible components contain the 1-pointed scheme $x_i \times_s x_i$. Thus $\Delta_{X_i/S} = X_i \times_S X_i$, so $k(x_i')/k(g(x_i'))$ is also purely inseparable for every $x_i' \in X_i$. Hence $g_i$ is a universal homeomorphism by [2, I.3.7–8]. Thus the factorization is $g : \amalg_i (x_i, X_i) \to \amalg_i (s, S) \to (s, S)$. □

**Lemma 2.7** *Let $X$ be a connected scheme and $g : X \to S$ a finite, universally open morphism. Then $g$ uniquely factors as $g : X \to S' \to S$ where $S' \to S$ is finite, étale and $X \times_{S'} X$ is connected.*

*Proof* Let $\Delta^{\mathrm{conn}}_{X/S}$ denote the connected component of $\Delta_{X/S}$ in $X \times_S X$. It is a finite equivalence relation on $X$ and the geometric quotient $S_1 := X/\Delta^{\mathrm{conn}}_{X/S}$ exists by [4, Lem.17]. The natural map $X \times_{S_1} X \to X \times_S X$ is a universal homeomorphism onto $\Delta^{\mathrm{conn}}_{X/S}$, thus $X \times_{S_1} X$ is connected. We apply Lemma 2.6 to $S_1 \to S$ to get $S_1 \to S' \to S$ where $S_1 \to S'$ is a universal homeomorphism and $S' \to S$ is étale. □

**Corollary 2.8 (2.2.1 ⇔ 2.2.2 ⇔2.2.5)** *Let $X, S$ be connected schemes and $g : X \to S$ a finite, universally open morphism. Assume that $X \times_S X$ is connected. Then $\pi_1(X) \to \pi_1(S)$ is surjective.*

*Proof* (2.2.2) ⇒ (2.2.5) was proved in Lemma 2.5 and we noted in (2.1.2) that (2.2.5) ⇒ (2.2.1). Finally Lemma 2.7 reduced (2.2.2) to the étale case, hence (2.2.1) ⇒ (2.2.2). □

Combining Lemma 2.7 and Corollary 2.8 we get the finite case of the following. For proper morphisms we also use Remark 2.4.

**Corollary 2.9** *Let $g : X \to S$ be a proper, universally open morphism of connected schemes. Let $x \to X$ be a geometric point and $s \to S$ its image. Then* $\mathrm{im}[\pi_1(X, x) \to \pi_1(S, s)]$ *has finite index in* $\pi_1(S, s)$. □

The normalization of the nodal plane cubic shows that the conclusion does not hold if $g$ is only assumed finite and open.

**2.10 (Proof of Theorem 2.2)** *As we noted in Remark 2.4, it is enough to prove the special case when $X \to S$ and $Y \to S$ are both finite.*

*We already proved that 2.2.1 ⇔ 2.2.2 and that they imply 2.2.5. Setting $X = Y$ shows that 2.2.5 ⇒ 2.2.2 and we get 2.2.3 by induction on the number of factors.*

*We use the shorthand $X^n_S := X \times_S \cdots \times_S X$ for the fiber product with $n$ factors. The coordinate projection $\pi_{12} : X^n_S \to X \times_S X$ is surjective, thus 2.2.3 ⇒ 2.2.2.*

*Since every connected component of $X^n_S$ dominates $S$, if $X^n_S$ has at least 2 connected components then $X^{mn}_S$ has at least $2^m$ connected components. Thus (2.2.3) ⇔ (2.2.4) and taking $Y = S$ shows that (2.2.6) ⇒ (2.2.1).*

*Conversely, assume (2.2.1) and fix* $h : Y \to S$. *We already know that* $X \times_S Y$ *is connected. If* $X \times_S Y \to Y$ *is not* $\pi_1$*-surjective then by (2.1.2) there is a nontrivial, finite étale cover* $Y' \to Y$ *such that* $(X \times_S Y) \times_Y Y'$ *is not connected. Applying (2.2.5) to* $Y' \to S$ *we get that* $X \times_S Y' \cong X \times_S Y \times_Y Y'$ *is connected, a contradiction. Thus (2.2.1)* $\Rightarrow$ *(2.2.6).* □

Next we consider 3 variants of $\pi_1$-surjectivity.

**2.11 (Topological Fundamental Group)** *Let* $X$ *be a connected* $\mathbb{C}$*-scheme of finite type and* $x \in X$ *a point. We then have the topological fundamental group* $\pi^{\text{top}}(X(\mathbb{C}), x)$. *A morphism* $f : X \to Y$ *between connected* $\mathbb{C}$*-schemes of finite type is* $\pi^{\text{top}}$*-surjective if* $f_* : \pi^{\text{top}}(X) \to \pi^{\text{top}}(Y)$ *is surjective.*

*Although the natural map* $\pi^{\text{top}}(X) \to \pi_1(X)$ *can have infinite kernel, as a consequence of Lemma 4.10 we see that if* $f$ *is proper and universally open then the index of* $\operatorname{im}[\pi^{\text{top}}(X) \to \pi^{\text{top}}(Y)]$ *in* $\pi^{\text{top}}(Y)$ *equals the index of* $\operatorname{im}[\pi(X) \to \pi(Y)]$ *in* $\pi(Y)$. *In particular,* $f$ *is* $\pi^{\text{top}}$*-surjective iff it is* $\pi$*-surjective. Thus Theorem 2.2 holds for the topological fundamental group as well.*

**2.12 (First Homology Group)** *Let* $X$ *be a connected scheme and* $x \to X$ *a geometric point. The first homology group, denoted by* $H_1(X)$, *is defined as the abelianization of* $\pi_1(X, x)$; *it is independent of the base point. Since we start with the algebraic fundamental group,* $H_1(X)$ *is a* $\hat{\mathbb{Z}}$ *module, where* $\hat{\mathbb{Z}} \cong \oplus_p \mathbb{Z}_p$ *is the profinite completion of* $\mathbb{Z}$. *If* $k = \mathbb{C}$ *then* $H_1(X)$ *is the profinite completion of* $H_1(X(\mathbb{C}), \mathbb{Z})$.

*We say that* $f$ *is* $H_1$*-surjective if* $f_* : H_1(X) \to H_1(Y)$ *is surjective. Note that* $\pi_1$*-surjective implies* $H_1$*-surjective.*

*In [5] we needed to understand whether* $H_1$*-surjectivity is preserved by base change as in Theorem 2.2.6. The following example shows that it is not.*

*Let* $X$ *be a simply connected manifold (or variety over* $\mathbb{C}$*) on which* $A_n$ *acts freely. Assume that* $n \geq 6$ *is odd. Let* $A_{n-1} \subset A_n$ *be a point stabilizer and* $C_n \subset A_n$ *a subgroup generated by an* $n$*-cycle. We get a commutative diagram*

$$\begin{array}{ccc} X & \xrightarrow{g'} & X/C_n \\ \downarrow & & \downarrow \\ X/A_{n-1} & \xrightarrow{g} & X/A_n, \end{array} \tag{2.12.1}$$

*which is a fiber product square. Here* $g$ *is* $H_1$*-surjective but* $g'$ *is not.*

*This is the reason why, although in [5] the main interest is in* $H_1$*-surjectivity, we needed to understand the base change behavior of* $\pi_1$*-surjectivity.*

**2.13 (Tame Fundamental Group)** *Let* $X$ *be a* $k$*-scheme and* $\operatorname{char} k = p > 0$. *Let* $\pi_1^{(p)}(X, x)$ *denote the largest prime to* $p$ *quotient of* $\pi_1(X, x)$, *that is, the inverse limit of all quotients* $\pi_1(X, x) \twoheadrightarrow H$ *where* $p \nmid |H|$. *We say that* $f$ *is* $\pi_1$*-surjective modulo* $p$ *if* $f_* : \pi_1^{(p)}(X) \to \pi_1^{(p)}(Y)$ *is surjective.*

*The diagram (2.12.1) also shows that* $\pi_1$*-surjectivity modulo* $p$ *is not preserved by base change.*

## 3 Open and Universally Open Maps

We discuss properties of universally open maps that were used in the proofs of Theorems 2.2 and 4.2. We aim to treat these in their natural generality and also establish various results that are of independent interest.

**Definition 3.1** A morphism $f : X \to S$ is *open at* $x \in X$ if for every open $x \in U \subset X$, its image $f(U)$ contains an open neighborhood of $f(x)$. A morphism $f : X \to S$ is *open* (resp. *open along* $W \subset X$) if $f$ is open at every $x \in X$ (resp. $x \in W$).

A morphism $f : X \to S$ is *universally open at* $x \in X$ (resp. along $W \subset X$) if $f_T : X \times_S T \to T$ is open along $g_X^{-1}(x)$ (resp. $g_X^{-1}(W)$) for every $g : T \to S$ where $g_X : X \times_S T \to X$ is the first projection. We say that $f$ is *universally open* if it is universally open at every $x \in X$. (Note that the Zariski topology on the product of 2 varieties is not the product topology, this is why open does not imply universally open.)

The following examples are good to keep in mind.

(1) Let $g : X \to Y$ be a morphism of finite type $\mathbb{C}$-schemes. We see in Theorem 4.2 that $g$ is universally open iff $g(\mathbb{C}) : X(\mathbb{C}) \to Y(\mathbb{C})$ is open in the Euclidean topology. Thus universal openness should be viewed as the more geometric notion.
(2) If $g : X \to Y$ is open then every irreducible component of $X$ dominates an irreducible component of $Y$. See Lemma 3.15 for a partial converse statement.
(3) Let $g : C_1 \to C_2$ be a quasi-finite morphism of purely one-dimensional $k$-schemes. If $C_2$ is irreducible then $g$ is open since every dense, constructible subset of $C_2$ is open.
(4) Let $p : \bar{X} \to X$ be a finite, birational morphism. Then $p$ is universally open iff it is a bijection on geometric points.
(5) As a consequence we see that if $C$ is an irreducible curve with nodes then the normalization $\bar{C} \to C$ is open but not universally open.
(6) Let $p : \bar{X} \to X$ be a finite, birational morphism. Assume that every irreducible component of $X$ has dimension $\geq 2$. Then $p$ is open $\Leftrightarrow$ $p$ is universally open $\Leftrightarrow$ $p$ is a bijection on geometric points.

Thus the difference between open and universally open appears mostly for one-dimensional targets. Nonetheless, many proofs involve localization and induction on the dimension, so the difference between the 2 notions can be significant.

*Example 3.2 (Openness Is Not an Open Property)* The following examples show that openness of a morphism at a point is not an open property. I assume this is why this notion is not defined in [6, Tag 004E]. Nonetheless, I think that the notion is natural and has several useful properties.

(3.2.1) Set $X := (z = 0) \cup (y = z - x = 0) \subset \mathbb{C}^3$, $S := (z = 0)$ and $p : X \to S$ the coordinate projection. Then $p$ is universally open at all points of the plane $(z = 0)$ but not open at the points of the punctured line $(y = z - x =$

$0)\setminus\{(0,0,0)\}$. It turns out to be a general feature that openness along a fiber says a lot about the maximal dimensional irreducible components of $X$ but very little about the lower dimensional ones; see Theorem 3.13.

(3.2.2) Let $Y$ be the pinch point $(x^2 = y^2 z) \subset \mathbb{C}^3$ and $X \cong \mathbb{C}^2_{uv}$ its normalization. Here $p : X \to Y$ is given by $(u, v) \mapsto (uv, u, v^2)$. Note that $g$ is universally open at $(0, 0)$ (see (3.4.4) for a more general claim). However it is not open at any other point of the $v$-axis.

Furthermore, although all fibers of $p : X \to Y$ have dimension 0, it does not have pure relative dimension 0. Indeed, the fiber product $X \times_Y X$ has 2 irreducible components; one is $X \cong \mathbb{C}^2$ and the other is isomorphic to $\mathbb{C}^1$ lying over the $v$-axis.

See Example 4.11 for other properties of this surface.

**3.3 (Basic Properties)** *The following are some obvious properties.*

(1) *Let $f : X \to S$ be a morphism and $Z \subset X$ a locally closed subscheme. If $f|_Z$ is open at $x \in Z$ then $f$ is open at $x$.*

(2) *Let $g : Y \to X$ and $f : X \to S$ be morphisms, $y \in Y$.*

   a. *If $g$ is open at $y$ and $f$ is open at $g(y)$ then $f \circ g$ is open at $y$.*
   b. *If $f \circ g$ is open at $y$ then $f$ is open at $g(y)$.*

(3) *Let $g : X \to Y$ be proper and open. Let $X_i \subset X$ be a connected component. Then $g(X_i)$ is closed and open in $Y$, hence a connected component.*

(4) *If $f : X \to S$ is open at $x \in X$ then $f_T : f^{-1}(T) \to T$ is also open at $x$ for every subvariety $f(x) \in T \subset S$.*

(5) *Let $f : X \to S$ be morphism and $S_i \subset S$ the irreducible components. Set $X_i := f^{-1}(S_i)$. By restriction we get $f_i : X_i \to S_i$. Then $f$ is open (resp. universally open) at $x$ iff $f_i$ is open (resp. universally open) at $x$ whenever $x \in X_i$.*

(6) *A flat morphism of finite presentation is universally open [6, Tag 01UA]. This is probably easiest to see using (3.4.3).*

(7) *[6, Tag 04R3] Being universally open is étale local on source and target. That is, if we have a commutative diagram*

$$\begin{array}{ccc} (x', X') & \xrightarrow{h'} & (x, X) \\ g' \downarrow & & \downarrow g \\ (s', S') & \xrightarrow{h} & (s, S), \end{array}$$

*where $h, h'$ are étale, then $g$ is universally open at $x$ iff $g'$ is universally open at $x'$.*

(8) *Let $X \to S$ be a proper morphism with Stein factorization $X \to X' \to S$. By (2.b), if $X \to S$ is open then so is $X' \to S$. The following example shows that $X \to X'$ need not be open. Take 2 copies of $\mathbb{A}^1 \times \mathbb{P}^1$ and glue them along the points $(0_i, p) \in \mathbb{A}^1_i \times \mathbb{P}^1$ for some $p \in \mathbb{P}^1$ to get $X$. The Stein factorization of*

*the first coordinate projection is*

$$\pi : (\mathbb{A}^1_1 \times \mathbb{P}^1) \amalg_{(0_1,p)=(0_2,p)} (\mathbb{A}^1_2 \times \mathbb{P}^1) \xrightarrow{\rho} \mathbb{A}^1_1 \amalg_{0_1=0_2} \mathbb{A}^1_2 \xrightarrow{\sigma} \mathbb{A}^1.$$

*Note that $\pi$ and $\sigma$ are universally open but $\rho$ is not open.*

**3.4 (Valuative and Base Change Criteria)** *The simplest non-open morphism is the embedding of a closed point into an irreducible curve $\{p\} \hookrightarrow C$. It turns out that this example is quite typical. If there is an irreducible, positive dimensional subvariety $s \in T \subset S$ such that $f^{-1}(T) = f^{-1}(s)$ then, by (3.3.4), $f^{-1}(T) \to \{s\} \hookrightarrow T$ shows that $f$ is not open at $x$, giving a necessary openness criterion.*

We claim that if $f : X \to S$ is of finite type, then the criterion is also sufficient, after a small change.

(3.4.1) Let $f : X \to S$ be a morphism of finite type, $x \in X$ a point and $s := f(x)$. Then $f$ is not open at $x$ iff there is an open subset $x \in U \subset X$ and an irreducible subscheme $s \in C \subset S$ such that $s$ has codimension 1 in $C$ and the generic point of $C$ is not contained in $f(U)$. In particular,

$$f|_{U\cap f^{-1}(C)} : U \cap f^{-1}(C) \to \bar{s} \to C \tag{3.4.1.a}$$

shows that $f$ is not open at $x$.

***Proof*** Choose $U$ such that $f(U)$ does not contain any open neighborhood of $s$. Since $f(U)$ is constructible, $S \setminus f(U)$ is constructible and its closure contains $s$. There is thus an irreducible component $W \subset S \setminus f(U)$ whose closure contains $s$. Note that $f(U) \cap \bar{W}$ is a nowhere dense constructible subset. Take any irreducible subscheme $\{s\} \in C \subset S$ that is not contained in the closure of $f(U) \cap \bar{W}$. □

Since $\mathcal{O}_{s,C}$ is dominated by a valuation ring, we can restate (3.4.1) in the following variant forms.

(3.4.2) A finite type morphism $f : X \to S$ is open at $x \in X$ if $f_T : X \times_S T \to T$ is open at $(x, s)$ for every one-dimensional, irreducible subscheme $s \in T \subset \operatorname{Spec} \mathcal{O}_{s,S}$ where $s := f(x)$. □

(3.4.3) [6, Tag 01TZ]. A finite type morphism $f : X \to S$ is universally open at $x \in X$ if $f_T : X \times_S T \to T$ is open along $g_X^{-1}(x)$ for every $g : T \to S$ where $T$ is the spectrum of a valuation ring and $g$ maps its closed point to $s := f(x)$. □

(3.4.4) Let $g : X \to S$ be a morphism. Assume that there is a closed subscheme $x \in Z \subset X$ such that $h := g|_Z : Z \to S$ is finite, dominant, $h^{-1}(s) = \{x\}$ and $k(x)/k(s)$ is purely inseparable. Then $g$ is universally open at $x$.

***Proof*** Let $x \in U \subset X$ be open. Then $Z \setminus U$ is closed, hence so is $h(Z \setminus U)$, which does not contain $s$. So

$$s \in S \setminus h(Z \setminus U) \subset h(Z \cap U) \subset g(U)$$

shows that $g$ is open at $x$. The assumptions are preserved by base change, so $g$ is universally open at $x$. □

(3.4.5) Let $g : X \to (s, S)$ be a morphism of finite type. Assume that $X, S$ are integral and $g$ is open along $X_s$. Then $\dim X = \dim X_s + \dim S$.

*Proof* We need to prove that the generic fiber also has dimension $X_s$. This is clear after base change to $g : T \to S$ as in (3.4.3), where $g$ maps the closed point of $T$ to $s$ and the generic point of $T$ to the generic point of $S$. □

(3.4.6) We prove in Theorem 3.6 that a finite type morphism $f : X \to (s, S)$ is universally open along $X_s$ iff $f_{S'} : X \times_S S' \to (s', S')$ is open along $(X \times_S S')_{s'}$ for every étale $g : (s', S') \to (s, S)$.

**3.5 (Openness and Pure Dimensional Morphisms)** *Let $g : X \to S$ be a morphism of finite type.*

(3.5.1) Set $X^{(n)} := \{x \in X : \dim_x X_{g(x)} = n\}$ and $X^{(\le n)} := \{x \in X : \dim_x X_{g(x)} \le n\}$.

By the upper semicontinuity of the fiber dimension [6, Tag 02FZ], $X^{(\le n)}$ is open in $X$ and $X^{(n)} = X^{(\le n)} \setminus X^{(\le n-1)}$ is closed in $X^{(\le n)}$ and locally closed in $X$.

(3.5.2) Let $g : X \to (s, S)$ be a morphism of finite type. Then $g$ is open (resp. universally open) along $X_s$ iff $g^{(n)} : X^{(n)} \to S$ is open (resp. universally open) along $X_s^{(n)}$ for every $n$.

*Proof* As we noted, $X^{(\le n)}$ is open in $X$, thus we may as well assume that all fibers have dimension $\le n$. The formation of $X^{(n)}$ commutes with base change, and over a one-dimensional base the claims are clear. □

Note that the punctual version does not hold. As an example, set $S := (xy = 0) \subset \mathbb{A}^2$ and $X = (x = 0) \cup (y = z = 0) \subset \mathbb{A}^3$. The coordinate projection is open at the origin yet $X^{(1)} \to S$ is not open at the origin. A much worse example is given in Example 3.12.

(3.5.3) Let $g : X \to S$ be a morphism of finite type whose fibers have pure dimension $n$. Let $x \in Z \subset X$ be a relative complete intersection, that is, $Z$ is a complete intersections of Cartier divisors $D_1, \dots, D_r \subset X$ and the fibers of $g|_Z$ have pure dimension $n - r$. Then $g$ is open (resp. universally open) at $x$ iff $g|_Z$ is open (resp. universally open) at $x$.

*Proof* Assume that $g|_Z$ is open at $x$ and let $x \in U \subset X$ be open. Then $Z \cap U \subset Z$ is open and $g(x) \in g(Z \cap U) \subset g(U)$ shows that $g$ is open at $x$.

Conversely, assume that $g|_Z$ is not open at $x$. By (3.4.1–2) it is enough to check the claim when $S$ is a spectrum of a local ring of dimension 1. Thus, after replacing $X$ with an open neighborhood of $x$ we may assume that $Z_s = Z$; this follows from (3.4.1). Since $\dim_x Z = \dim_x X - r$ and $\dim_x Z_s = \dim_x X_s - r$, we conclude that $\dim_x X = \dim_x X_s$. Thus $X_s$ is an irreducible component of $X$, hence $g$ is not open. □

(3.5.4) Let $g : X \to S$ be a morphism of finite type. If $g$ has *pure relative dimension* $n$ (that is, $X \times_S T$ has pure dimension $\dim T + n$ for every local morphism $T \to S$ from an integral scheme to $S$) then $g$ is also universally open. This follows from (3.4.3). Note, however, that the local version of this is not true; see (3.2.2).

Next we show that it is enough to use étale base changes in the definition of universal openness.

**Theorem 3.6** *Let $g : X \to (s, S)$ be a morphism of finite type. Then $g$ is universally open along $X_s$ iff $g'$ is open along $X'_{s'}$ for every étale base change diagram as in (3.3.7).*

***Proof*** If $g$ is universally open at $x$ then it is open after every base change.

Conversely, pick $x \in X_s$ and assume that $g$ is open at $x$ after every étale base change. Set $n = \dim_x X_s$. By (3.5.2) $g^{(n)} : X^{(n)} \to S$ is also open at $x$ after every étale base change and by (3.5.3) the same holds for every relative complete intersection $x \in Z \subset X^{(n)}$ of codimension $n$. Then $g|_Z$ is quasi-finite. Take an étale base change $(s', S') \to (s, S)$ as in Proposition 3.7 to get

$$Z \times_S S' = W \coprod \amalg_i (z'_i, Z'_i).$$

By construction every $(z'_i, Z'_i) \to (s', S')$ is finite and $k(z'_i)/k(s')$ is purely inseparable. Furthermore, $(z'_i, Z'_i) \to (s', S')$ is open by assumption, hence dominant. Thus $(z'_i, Z'_i) \to (s', S')$ is universally open by (3.4.4) and so is $g|_Z : Z \to S$. Thus $g$ is universally open by (3.5.3) and (3.5.2). □

We used some results on étale localization of quasi-finite morphisms; see [6, Tag 04HF] for proofs and further generalizations.

**Proposition 3.7** *Let $g : X \to S$ be a quasi-finite morphism and $s \in S$. Then there is an étale morphism $(s', S') \to (s, S)$ such that*

$$X \times_S S' = W \coprod \amalg_i (x'_i, X'_i),$$

*where $W$ does not have any points lying over $s'$ and, for every $i$, the morphism $g'_i : (x'_i, X'_i) \to (s', S')$ is finite, $(g'_i)^{-1}(s') = \{x'_i\}$ and $k(x'_i)/k(s')$ is purely inseparable.* □

**Corollary 3.8** *Let $g : (x, X) \to (s, S)$ be a quasi-finite morphism. Then there is a commutative diagram*

$$\begin{array}{ccc} (x', X') & \xrightarrow{h'} & (x, X) \\ g' \downarrow & & \downarrow g \\ (s', S') & \xrightarrow{h} & (s, S), \end{array} \tag{3.8.1}$$

*where $h, h'$ are étale, $g'$ finite, $g'^{-1}(s') = \{x'\}$ and $k(x')/k(s')$ is purely inseparable.* □

We refer to (3.8.1) as an *étale base change diagram* of $g$.

**Corollary 3.9** *[6, Tag 02LO] Let $g : (x, X) \to (s, S)$ be a finite morphism. Then there is an étale morphism $(s', S') \to (s, S)$ such that*

$$X \times_S S' = \amalg_i (x'_i, X'_i),$$

*where, for every $i$, the morphism $g'_i : (x'_i, X'_i) \to (s', S')$ is finite, $(g'_i)^{-1}(s') = \{x'_i\}$ and $k(x'_i)/k(s')$ is purely inseparable.* □

**Definition 3.10** A local scheme $(x, X)$ is called *geometrically unibranch* if for every étale morphism $(x', X') \to (x, X)$, the local scheme $(x', X')$ is irreducible. See [6, Tag 06DT] for other definitions and basic properties.

Openness is very well behaved for geometrically unibranch targets; cf. [6, Tag 0F32]

**Proposition 3.11** *Let $g : (x, X) \to (s, S)$ be a finite type morphism. Assume that $X$ is irreducible and $S$ is geometrically unibranch at $s$. The following are equivalent.*

(1) $\dim X = \dim X_s + \dim S$.
(2) *$g$ is open along $X_s$.*
(3) *$g$ is universally open along $X_s$.*

***Proof*** Note that (3) implies (2) by definition, and (2) implies (1) by (3.4.5). It remains to show that (1) $\Rightarrow$ (3).

Note that $\dim_x X \le \dim_x X_s + \dim_s S$ for every $x \in X_s$, thus $X_s$ is pure dimensional.

Let $x \in Z \subset X$ be a relative complete intersection of codimension $\dim X_s$. By (3.5.3) it is enough to show that $g|_Z$ is universally open at $x$. Thus we may assume that $g$ is quasi-finite. Apply Proposition 3.7 to get an étale morphism $(s', S') \to (s, S)$ and a decomposition

$$X \times_S S' = W \coprod \amalg_i (x'_i, X'_i),$$

where the $k(x'_i) \supset k(s')$ are purely inseparable and the projections $g'_i : X'_i \to S'$ are finite. The $g'_i$ are also dominant by assumption, hence universally open at $x'_i$ by (3.4.4). Thus $X \times_S S'$ is universally open along $X'_{s'}$, and so $X \to S$ is universally open along $X_s$. □

*Example 3.12* It seems natural to hope that the equivalence (3.11.2) $\Leftrightarrow$ (3.11.3) holds point-wise. This is, however, not the case. To see this, we construct below a projective morphism of surfaces $g : (x, X) \to (s, S)$ such that $S$ is normal and $g$ is open at $x$, yet $g$ is not universally open at $x$.

Let $(s, S)$ be a normal surface singularity with a (non-minimal) resolution $\tau : Y \to S$ and exceptional curves $E_1, \dots, E_n$. Assume that

(1) for every $i = 1, \dots, n-1$ there is a morphism $(E_i, Y) \to (x_i, X_i)$ that contracts only the curve $E_i$,
(2) $X_i$ is normal and $\pi_i : X_i \to S$ is projective,
(3) if an algebraic curve $C \subset Y$ is disjoint from $E_1 \cup \cdots \cup E_{n-1}$ then it is also disjoint from $E_n$.

Let $(x, X)$ be obtained from the surfaces $X_1, \dots, X_{n-1}$ by identifying the points $x_1, \dots, x_{n-1}$. The morphisms $\pi_i$ glue to a projective morphism $\pi : (x, X) \to (s, S)$.

*Claim 3.12.4 $\pi$ is open but not universally open at $x$.*

***Proof*** Let $B \subset Y$ be an algebraic curve that meets $E_n$ transversally at a general point $b \in E_n$. Then $\operatorname{Spec} \mathcal{O}_{b,B} \to S$ shows that $\pi$ is not universally open at $x$.

Assume to the contrary that $\pi$ is not open. Then, by (3.4.1), there is a curve $s \in C_S \subset S$ such that the closure $C_X$ of the preimage of $C_S \setminus \{s\}$ does not pass through $x$. Note that $C_X$ is the union of the birational transforms $(\pi_i)_*^{-1} C_S$ and these in turn are the images of $C := \tau_*^{-1} C_S \subset Y$. Thus $C$ is disjoint from $E_1 \cup \cdots \cup E_{n-1}$. By (3) then $C$ is also disjoint from $E_n$, which contradicts $s \in C_S$. □

In order to construct such an $(s, S)$, we start with $\mathbb{P}^2$ and 3 general lines $L_1, L_2, L_3$. Let $B \subset \mathbb{P}^2$ be a general quartic curve. We obtain $Y'$ by blowing up the 12 intersection points of $B$ with the lines. The birational transforms of the $L_i$ become $E'_i \subset Y'$. Contracting them we get $(s, S)$. Pick next a very general point $p \in E'_3$ and let $Y = B_p Y'$ denote the blow-up. The birational transforms of the $L_i$ give $E_1$, $E_2$, $E_3$ and $E_4$ is the exceptional curve of $Y \to Y'$.

Each $E_i \subset Y$ is a rational curve with negative self-intersection, so it can be contracted projectively. (This is essentially due to Castelnuovo; the proof in [3, V.5.7] is easy to modify.) It remains to check assumption (3). Slightly stronger, we claim that an algebraic curve $C' \subset Y'$ can not intersect $E' := E'_1 \cup E'_2 \cup E'_3$ only at $p$. To see this note that $\operatorname{Pic}(Y')$ is finitely generated (in fact, isomorphic to $\mathbb{Z}^{13}$), hence the image of the restriction map

$$\operatorname{Pic}(Y') \to \operatorname{Pic}(E') \cong \mathbb{Z} + \mathbb{C}^*$$

is finitely generated. Thus, for a very general point $p \in E'$, the intersection of $\mathbb{Z}[p] \subset \operatorname{Pic}(E')$ with the image is the trivial element $[0] \in \operatorname{Pic}(E')$.

As a concrete example, we can take $(s, S)$ to be the projectivisation of the affine surface

$$(s, S^0) := \left(xyz + x^4 + y^4 + z^4 = 0\right) \subset \mathbb{A}^3. \qquad \square$$

If $X$ is reducible and $g : X \to (s, S)$ is universally open along $X_s$ then, as shown by Example 4.12.2, we can say very little about the lower dimensional irreducible components of $X$. The next result shows that the maximal dimensional irreducible components behave better.

**Theorem 3.13** *Let $g : X \to (s, S)$ a finite type morphism that is universally open along $X_s$. Set $n := \dim X_s$ and let $X^{\max} \subset X^{(n)}$ be the union of all those irreducible components of $X^{(n)}$ that dominate some irreducible component of $S$ and have nonempty intersection with $X_s$.*

*Then $X^{\max} \to S$ is universally open along $X_s \cap X^{\max}$.*

***Proof*** The conclusions can be checked after an étale base change, hence, using Lemma 3.14, we may assume that every irreducible component of $(s, S)$ is unibranch. Then, by (3.3.7), we may also assume that $(s, S)$ is irreducible, hence unibranch.

Let $X_i \subset X^{\max}$ be an irreducible component. By assumption $X_i$ dominates $S$ and the generic fiber of $X_i \to S$ has dimension $n$ since $X_i \subset X^{(n)}$. Thus $\dim X_i = n + \dim S$ and so $X_i \to S$ is universally open along $X_s \cap X_i$ by Proposition 3.11.

**Lemma 3.14 ([6, Tag 0CB4])** *Let $(s, S)$ be a local scheme. Then there is an étale morphism $(s', S') \to (s, S)$ such that every irreducible component $(s', S'_i) \subset (s', S')$ is geometrically unibranch.* □

The following is a partial converse to (3.1.2).

**Lemma 3.15** *Let $S$ be a connected scheme and $g : X \to S$ a dominant morphism of finite type whose fibers have pure dimension $n$. The following are equivalent.*

(1) *$g : X \to S$ is universally open,*
(2) *$g \times g : X \times_S X \to S$ is open,*
(3) *Every irreducible component of $X$ dominates an irreducible component of $S$ and every irreducible component of $X \times_S X$ dominates an irreducible component of $S$.*

***Proof*** If $g : X \to S$ is universally open then so is $X \times_S X \to X$, hence also their composite $X \times_S X \to S$. Thus (1) ⇒ (2) and (2) ⇒ (3) is clear.

It remains to show that if (3) holds then $g$ is universally open along $X_s$ for every $s \in S$. Using Lemma 3.14, after an étale base change we may assume that every irreducible component $(s, S_i) \subset (s, S)$ is geometrically unibranch.

Set $X_i := g^{-1}(S_i)$. By (3.3.7) it is enough to show that each $X_i \to S_i$ is universally open along $(X_i)_s$. If this does not hold, then, by Proposition 3.11, there is an irreducible component $Z_i \subset X_i$ that does not dominate $S_i$. By assumption there is an irreducible component $X_j \subset X$ that contains $Z_i$ and this $X_j$ dominates some irreducible component $S_j \subset S$. By assumption $S_i \neq S_j$.

Let $X_i^* \subset X$ be an irreducible component that does dominate $S_i$. Then $X_j \times_S X_i^*$ is an union of irreducible components of $X \times_S X$ that lie over $S_i \cap S_j$, hence they do not dominate any irreducible component of $S$. This is impossible by (3),

*Example 3.16* This example shows that the equidimensionality assumption in the previous Lemma 3.15 is necessary. Set

$$X := (x_0y_0 + x_1y_1 + x_2y_2 = 0) \subset \mathbb{A}^3_{\mathbf{x}} \times \mathbb{P}^2_{\mathbf{y}}$$

with projection $\pi : X \to S = \mathbb{A}^3_{\mathbf{x}}$. The central fiber has dimension 2, so $\pi$ is not open. The fiber product

$$X \times_S X \subset \mathbb{A}^3_{\mathbf{x}} \times \mathbb{P}^2_{\mathbf{y}} \times \mathbb{P}^2_{\mathbf{z}}$$

is given by 2 equations, hence its irreducible components have dimension $\geq 7-2 = 5$. The central fiber of $X \times_S X \to S$ has dimension 4, so it can not be an irreducible component. Thus $X \times_S X$ is irreducible.

Note also that if $X \to S$ is not pure dimensional then the fiber product $X \times_S \cdots \times_S X$ of more than $\dim S$ copies of $X$ always has a non-dominant irreducible component.

## 4 Path Lifting in the Euclidean Topology

Let $g : X \to Y$ be a morphism of $\mathbb{C}$-schemes of finite type. In this section we compare scheme-theoretic properties of $g$ with properties of $g(\mathbb{C}) : X(\mathbb{C}) \to Y(\mathbb{C})$ in the Euclidean topology.

It is easy to see that universal openness in the Zariski topology (Definition 3.1) is equivalent to openness in the Euclidean topology; see Lemma 4.3. Next we study 3 versions of the path lifting property. We see in Theorems 4.2 and 4.6 that 2 of them have very satisfactory scheme-theoretic descriptions.

**Definition 4.1 (Arc and Path Lifting)** Let $h : M \to N$ be a continuous map of topological spaces and $\gamma : [0, 1] \to N$ a path. A *lift* of $\gamma$ with starting point $m \in h^{-1}(\gamma(0))$ is a continuous map $\gamma' : [0, 1] \to M$ such that $\gamma'(0) = m$ and $\gamma = h \circ \gamma'$.

(1) We say that $h$ has the *path lifting* property if a lift $\gamma'$ exists for every $\gamma : [0, 1] \to N$ and $m \in h^{-1}(\gamma(0))$. We do not require $\gamma'$ to be unique. If this holds then every path $\gamma : [0, 1) \to N$ also lifts.
(2) We say that $h$ has the *arc lifting* (or *local path lifting*) property if the lift $\gamma'$ exists over some subinterval $[0, \epsilon]$, where $\epsilon > 0$ may depend on $m \in h^{-1}(\gamma(0))$.
(3) We say that $h$ has the *2-point path lifting* property if given $\gamma : [0, 1] \to N$ and $m_i \in h^{-1}(\gamma(i))$ for $i = 0, 1$, there is a lifting $\gamma'$ such that $\gamma'(0) = m_0$ and $\gamma'(1) = m_1$. The basic example is a fiber bundle $h : M \to N$ with path-connected fiber $F$.

Note also that if $N$ is path-connected, $M \neq \emptyset$ and $h : M \to N$ has the path lifting property then $h$ is surjective.

Lifting constant paths shows that if $h$ has the 2-point path lifting property then all fibers of $h$ are path-connected.

The concept of path lifting occurs most frequently in the topological literature, but the next result shows that, from the scheme-theoretic point of view, arc lifting is the most natural property.

**Theorem 4.2** *Let $g : X \to Y$ be a morphism of $\mathbb{C}$-schemes of finite type. The following are equivalent.*

(1) *$g : X \to Y$ is universally open (Definition 3.1).*
(2) *$g(\mathbb{C}) : X(\mathbb{C}) \to Y(\mathbb{C})$ is open in the Euclidean topology.*
(3) *$g(\mathbb{C}) : X(\mathbb{C}) \to Y(\mathbb{C})$ has the arc lifting property (4.1.2).*

***Proof*** The equivalence of (4.2.1) and (4.2.2) is established in Lemma 4.3.

Assume next that $g$ is not universally open at $x$. By Theorem 3.6, after a suitable étale base change, it is not open at $x$. An étale base change is a local homeomorphism in the Euclidean topology, thus it does not alter the liftability of arcs.

Thus, by (3.4.2) there is a curve $g(x) \in C \subset Y$ such that $f(U) \cap C = \{g(x)\}$. Choosing a path in this curve $C$ shows that (4.2.3) $\Rightarrow$ (4.2.2). Thus it remains to prove that (4.2.2) $\Rightarrow$ (4.2.3). This done in 3 steps.

If $g$ is finite, arc lifting is proved in Lemmas 4.4–4.5. If $X_{g(x)}$ has pure dimension $n$ at $x$, then we take a relative complete intersection $x \in Z \subset X$ of codimension $n$. Then $g|_Z : Z \to Y$ is quasi-finite at $x$ and open by (3.5.3). An étale base change as in Proposition 3.7 reduces this to the already discussed finite case. Thus an arc $\gamma : [0, \epsilon] \to Y$ has a lift $\gamma' : [0, \epsilon] \to Z$. This is also a lift to $X$.

Finally, in general set $n = \dim_x X_{g(x)}$. By (3.5.2) the restriction of $g$ to $X^{(n)}$ is universally open. The fibers of $X^{(n)} \to Y$ have pure dimension $n$, thus $X^{(n)} \to Y$ has the arc lifting property and so does $X$.

**Lemma 4.3** *$g : X \to Y$ be a morphism between $\mathbb{C}$-schemes of finite type. Then $g$ is universally open at $x$ in the Zariski topology iff $g(\mathbb{C}) : X(\mathbb{C}) \to Y(\mathbb{C})$ is open at $x$ in the Euclidean topology.*

***Proof*** Assume that $g(\mathbb{C})$ is open at $x$ in the Euclidean topology and let $x \in U \subset X$ be a Zariski open neighborhood. Then $U$ is also Euclidean open, hence $g(U)$ contains a Euclidean open neighborhood of $g(x)$. Since $g(U)$ is also constructible, it also contains a Zariski open neighborhood of $x$. Thus $g$ is open at $x$ in the Zariski topology. An étale base change is a local homeomorphism in the Euclidean topology, hence $g$ is also universally open at $x$ in the Zariski topology by Theorem 3.6.

Conversely, assume that $g$ is not universally open at $x$. By Theorem 3.6, after a suitable étale base change, it is not open at $x$. Thus, by (3.4.1) this means that there is a curve $g(x) \in C \subset Y$ such that $f(U) \cap C = \{g(x)\}$. This shows that $g(\mathbb{C})$ is not open at $x$ in the Euclidean topology either.

**Lemma 4.4** *Let $g : X \to Y$ be a finite morphisms of $\mathbb{C}$-schemes of finite type. Then one can choose triangulations on them such that $g(\mathbb{C}) : X(\mathbb{C}) \to Y(\mathbb{C})$ is simplicial.*

***Proof*** We may assume that $X, Y$ are reduced. Set $n := \dim Y$, $X_n := X$ and $Y_n := Y$. If $g_r : X_r \to Y_r$ is already defined then let $U_r \subset Y_r$ be the largest smooth, open subset of pure dimension $r$ such that $g_r$ is étale. Set $Y_{r-1} := Y_r \setminus U_r$ and $X_{r-1} := \operatorname{red} g^{-1}(Y_{r-1})$.

Triangulate $Y$ by starting with $Y_0$, extending it to a triangulation of $Y_1$ (possibly adding new vertices), then extending it to a triangulation of $Y_2$ (possibly after a refinement) and so on. At the end we get a triangulation of $Y$ such that the interior $\Delta^\circ$ of every simplex lies in some $U_r$. Thus the preimages of the simplices give a triangulation of $X$ and $g$ is simplicial.

**Lemma 4.5** *Let $M, N$ be connected simplicial complexes and $g : M \to N$ a proper, open, simplicial map with finite fibers. Then $g$ has the path lifting property.*

***Proof*** We are given a path $\gamma : [0, 1] \to N$ and a lifting $\gamma'(0)$. Choose a maximal lifting $\gamma'_c : [0, c) \to N$. (At the beginning we allow $c = 0$.) First we define $\gamma'(c)$. Set $n = \gamma(c)$ and let $m_i \in M$ be the preimages of $n$. Choose an open neighborhood $n \in U \subset N$ such that $g^{-1}(U) = \cup_i V_i$ is a disjoint union of connected neighborhoods of the $m_i$. Then $\gamma'(c-\epsilon, c)$ is contained in one of them, say $V_j$, and setting $\gamma'(c) = m_j$ is the unique continuous extension of $\gamma' : [0, c) \to M$ to $\gamma' : [0, c] \to M$.

In order to extend beyond $c$, we may as well assume that $n := \gamma(c)$ is a vertex. Then $m := \gamma'(c)$ is also a vertex. We can choose the neighborhood $n \in U \subset N$ to be a cone over the link $L_n$. Then $m \in V \subset M$ is also a cone over its link $L_m$ and $g|_{L_m} : L_m \to L_n$ is finite, open and simplicial. If $\gamma$ maps $[c, c+\epsilon]$ to $V$ then we define $\gamma'$ on $[c, c+\epsilon]$ as follows. If $\gamma(x) = n$ then set $\gamma'(x) = m$. Set

$$B := \{x \in [c, c+\epsilon] : \gamma(x) \neq n\} \subset [c, c+\epsilon].$$

It is a countable union of open intervals and $\gamma$ maps $B$ to

$$U \setminus \{n\} \sim (0, 1) \times L_n.$$

Correspondingly we can write $\gamma|_B = (\alpha_B, \gamma_B)$ where $\gamma_B$ maps $B$ to $L_n$. Since $g|_{L_m} : L_m \to L_n$ is simplicial, by induction on the dimension, $\gamma_B$ lifts to $\gamma'_B : B \to L_m$. (As we noted in (4.1.1), open paths also lift.) Then set $\gamma'|_B := (\alpha_B, \gamma'_B)$.

Next we consider the scheme-theoretic description of the 2-point path lifting property.

**Theorem 4.6** *Let $g : X \to Y$ be a morphism of $\mathbb{C}$-schemes of finite type. The following are equivalent.*

(1) *$g : X \to Y$ is universally open, surjective and has connected fibers.*
(2) *$g(\mathbb{C}) : X(\mathbb{C}) \to Y(\mathbb{C})$ is open in the Euclidean topology, surjective and has connected fibers.*

(3) $g(\mathbb{C}) : X(\mathbb{C}) \to Y(\mathbb{C})$ *is surjective, has connected fibers and satisfies the arc lifting property (4.1.2).*
(4) $g(\mathbb{C}) : X(\mathbb{C}) \to Y(\mathbb{C})$ *has the 2-point path lifting property (4.1.3).*

***Proof*** The equivalence of (1), (2), and (3) follows from Theorem 4.2, once we note that a $\mathbb{C}$-scheme of finite type is connected in the Zariski topology iff it is connected in the Euclidean topology. (This is usually proved using Chow's theorem, but one can use Bertini's hyperplane section theorem to reduce it to the one-dimensional case, which was known to Riemann.)

We noted at the end of Definition 4.1 that if (4) holds then $g$ is surjective and has connected fibers. Thus (4) implies (1–3) and it remains to show that (1–3) imply (4).

We have $\gamma : [0, 1] \to Y(\mathbb{C})$ and liftings $m_0$ of $\gamma(0)$ and $m_1$ of $\gamma(1)$. Assume that we already have a lifting $\gamma'$ defined on $[0, c]$. By (3) we have a lifting $\gamma_c'' : [c, c+\epsilon] \to X(\mathbb{C})$ such that $\gamma_c''(c) = \gamma'(c)$. The concatenation of $\gamma'$ on $[0, c]$ with $\gamma_c''$ defines $\gamma'$ on $[0, c+\epsilon]$.

Assume next that we have $\gamma'$ defined on $[0, c)$. We do not claim that it extends to $[0, c]$, but we show that the restriction of $\gamma'$ to $[0, c-\epsilon]$ extends to $\gamma'' : [0, c+\epsilon]$ for some $\epsilon > 0$. Moreover, if $c = 1$ then we can also arrange that $\gamma''(1) = m_1$. For other values of $c$ pick any lifting $m_c$ of $\gamma(c)$. Applying (3) gives a lifting

(5) $\gamma_c'' : [c-\epsilon, c+\epsilon] \to X$ such that $\gamma_c''(c) = m_c$.

Starting from $X \to Y$, we get a stratification $\{Y_i\}$ of $Y$ as in (4.7.1). Applying Lemma 4.5 to $\amalg_i Y_i \to Y$ we get a triangulation of $Y$ with $\gamma(c)$ as a vertex. Let $L$ be the smallest subcomplex that contains $\gamma\big((c-\delta, c)\big)$ for some $\delta > 0$ and let $\Delta \subset L$ be a maximal dimensional simplex with interior $\Delta^\circ$. Then $\gamma^{-1}(\Delta^\circ)$ is open in $(c-\delta, c)$ and its closure contains $c$. In particular, we see that

(a) there is a homeomorphism $\beta : g^{-1}(\Delta^\circ) \sim \Delta^\circ \times M$ that commutes with projection to $\Delta^\circ$ for some $M$ and
(b) there are $0 < \eta_1 < \eta_2 < \epsilon$ such that $\gamma$ maps $[c-\eta_2, c-\eta_1]$ to $\Delta^\circ$.

As we noted in (4.1.3), the restriction of $\gamma$ to $[c-\eta_2, c-\eta_1]$ has a lifting $\bar\gamma$ such that

$$\bar\gamma(c-\eta_2) = \gamma'(c-\eta_2) \quad \text{and} \quad \bar\gamma(c-\eta_1) = \gamma_c''(c-\eta_1).$$

Thus the concatenation of $\gamma'$ on $[0, c-\eta_2]$, $\bar\gamma$ on $[c-\eta_2, c-\eta_1]$ and $\gamma_c''$ on $[c-\eta_1, c+\epsilon]$ defines a lifting of $\gamma$ over $[0, c+\epsilon]$.

**4.7 (Stratification of Maps)** *Let $g : X \to Y$ be a morphism of varieties over $\mathbb{C}$. As in [1, Sec.I.1.7], $Y$ has a stratification by closed subvarieties*

$$Y = Y_n \supset Y_{n-1} \supset \cdots \supset Y_0 \tag{4.7.1}$$

*where $Y_i \setminus Y_{i-1}$ has pure dimension $i$ and each*

$$g^{-1}\big(Y_i \setminus Y_{i-1}\big) \to \big(Y_i \setminus Y_{i-1}\big)$$

*is a topologically locally trivial fiber bundle. (There may be different fibers over different connected components.) Thus path lifting could fail only at points where a path moves from one stratum to another.*

Let $g : X \to Y$ be a quasi-finite, universally open morphism. Using Lemma 4.5 we see that $g(\mathbb{C}) : X(\mathbb{C}) \to Y(\mathbb{C})$ has the path lifting property iff $g$ is proper. However, as shown by Example 4.12, the path lifting property does not seem to have an equivalent scheme-theoretic version for morphisms with positive dimensional fibers. Nonetheless, the following sufficient condition is quite natural and useful.

**Theorem 4.8** *Let $g : X \to Y$ be a proper, universally open, pure dimensional morphism of $\mathbb{C}$-schemes of finite type. Then $g(\mathbb{C}) : X(\mathbb{C}) \to Y(\mathbb{C})$ has the path lifting property.*

***Proof*** As (3.3.8) shows, we can not use Stein factorization to reduce Theorem 4.8 directly to Theorem 4.6, but a suitable modification of the proof will work.

We follow the proof of (4.6.3) ⇒ (4.6.4), but we need to make a different choice of $\gamma_c''$ in (4.6.5). Note that $\gamma'\big([c-\eta_1, c-\eta_2]\big)$ is contained in a connected component $\Delta^\circ \times M_1 \subset \Delta^\circ \times M$, and everything works as before if we can ensure that $\gamma_c''\big([c-\eta_1, c-\eta_2]\big)$ is also contained in $\Delta^\circ \times M_1$.

To do this, let $X \to Y_1 \to Y$ be the Stein factorization. The choice of the connected component $\Delta^\circ \times M_1 \subset \Delta^\circ \times M$ defines a section $\sigma : \Delta^\circ \to Y_1$. By Lemma 4.9, after an étale base change we may assume to have a relative complete intersection $Z \subset X$ such that the induced map $g_1 : Z \to Y_1$ is finite and surjective. In particular, there is an $m_z \in Z$ such that $g_1(m_z) = \sigma\big(\gamma(c - \eta_2)\big)$. Since $Z \to Y$ is finite and open, it satisfies the path lifting property by Lemma 4.5. Thus the restriction of $\gamma$ to $[c - \eta_2, c]$ has a lifting $\gamma_c''$ such that $\gamma_c''(c - \eta_2) = m_z$. By construction $\gamma_c''(c-\eta_2)$ and $\gamma'(c-\eta_2)$ are in the same connected component of the fiber and the rest of the proof now works as before.

**Lemma 4.9** *Let $g : X \to Y$ be a proper, universally open morphism of finite type and of relative dimension $n$. Let $X \to Y_1 \to Y$ denote its Stein factorization. Let $y \in Y$ be a point and $P \subset X_y$ a finite subset that has nonempty intersection with every irreducible component of $X_y$. Let $P \subset Z \subset X$ be a relative complete intersection of codimension $n$. Then there is an étale neighborhood $(y', Y') \to (y, Y)$ such that the induced morphism $Z' \to Y_1'$ is finite and surjective.*

***Proof*** By construction $Z \to Y$ is quasi-finite, hence there is an étale neighborhood $(y', Y') \to (y, Y)$ such that $Z' \to Y'$ is finite. Thus $Z' \to Y_1'$ is also finite. We need to prove that it is surjective.

Let $\tilde{Y}_1'$ be the disjoint union of the irreducible components of $Y_1'$. By base change we get $\tilde{X}_1' \to \tilde{Y}_1'$, $\tilde{P}_1' \subset (\tilde{X}_1')_{y'}$ and $\tilde{P}_1' \subset \tilde{Z}_1' \subset \tilde{X}_1'$. An irreducible component of $(\tilde{X}_1')_{y'}$ has pure dimension $n$, thus its image in $X_y$ is an irreducible component. Thus $\tilde{P}_1'$ has nonempty intersection with every irreducible component of $(\tilde{X}_1')_{y'}$. Thus $\tilde{Z}_1'$ dominates $\tilde{Y}_1'$ and hence $Z'$ dominates $Y_1'$. Since $Z' \to Y_1'$ is finite, this implies that it is surjective.

Next we establish the Euclidean version of Corollary 2.9.

**Corollary 4.10** *Let $g : (x, X) \to (y, Y)$ be a proper, universally open morphism of pointed, connected $\mathbb{C}$-schemes. Then $\operatorname{im}[\pi_1(X(\mathbb{C}), x) \to \pi_1(Y(\mathbb{C}), y)]$ has finite index in $\pi_1(Y(\mathbb{C}), y)$.*

***Proof*** Let $n$ denote the maximal fiber dimension. Then $g^{(n)} : X^{(n)} \to Y$ is also proper and universally open by (3.5.2). We may choose $x \in X^{(n)}$ and then $g_*^{(n)}$ factors as

$$\pi_1(X^{(n)}(\mathbb{C}), x) \to \pi_1(X(\mathbb{C}), x) \to \pi_1(Y(\mathbb{C}), y).$$

Thus it is enough to show that $\operatorname{im}[\pi_1(X^{(n)}(\mathbb{C}), x) \to \pi_1(Y(\mathbb{C}), y)]$ has finite index in $\pi_1(Y(\mathbb{C}), y)$. The advantage is that $g(\mathbb{C}) : X^{(n)}(\mathbb{C}) \to Y(\mathbb{C})$ has the path lifting property by Theorem 4.8. The rest follows from (4.10.1).

*Claim 4.10.1 Let $h : M \to N$ be a continuous map of path-connected topological spaces that has the path lifting property. Assume that $h^{-1}(n)$ has finitely many path-connected components for some $n \in N$. Then the image of $\pi_1(M, m) \to \pi_1(N, n)$ has finite index in $\pi_1(N, n)$.*

***Proof*** Every loop $\gamma$ starting and ending in $n$ lifts to a path that starts at $m$ and ends in $h^{-1}(n)$. If 2 loops $\gamma_1, \gamma_2$ end at the same path-connected component then $\gamma_1^{-1}\gamma_2$ lifts to a loop on $M$. This shows that the index of the image of $\pi_1(M, m) \to \pi_1(N, n)$ is bounded by the number of path-connected components of the fiber.

## 4.1 Examples

The first example shows that Theorem 4.6 does not have a point-wise version.

*Example 4.11* Let $Y := (x^2 = y^2 z)$ be the pinch point as in Example 3.2.1 with normalization $p : \bar{Y} \to Y$. We saw that $p$ is universally open at the origin. However, it does not have the arc lifting property at the origin.

To see this consider the real curves $\gamma^{\pm} : t \mapsto (t, \pm t^2 \sin(t^{-1})) \subset \mathbb{R}^2$. Note that $p \circ \gamma^+$ and $p \circ \gamma^-$ intersect at the points $t = (m\pi)^{-1}$. Thus the arc

$$\gamma(t) := \begin{cases} p \circ \gamma^+(t) & \text{if} \quad \gamma^+(t) \geq 0 \quad \text{and} \\ p \circ \gamma^-(-t) & \text{if} \quad \gamma^-(-t) \geq 0 \end{cases}$$

has no lifting.

*Example 4.12 (Path Lifting and Properness)* It is natural to hope that arc lifting plus properness should imply path lifting, but this is not the case.

(4.12.1) Let $X$ be obtained from $X_1 := B_{(0,0)}\mathbb{C}^2$ and of $X_2 := \mathbb{P}^1 \times \mathbb{C}^2$ by identifying the exceptional divisor of the blow-up with $\mathbb{P}^1 \times \{(0, 0)\}$. The projection $X \to \mathbb{C}^2$ is universally open, and so is $X_2 \to \mathbb{C}^2$, but $X_1 \to \mathbb{C}^2$ is not even open. The path

$$t \mapsto \left(1-t, (1-t)\sin\left(\tfrac{1}{1-t}\right)\right)$$

does not lift to $B_{(0,0)}\mathbb{C}^2$. Thus $X \to \mathbb{C}^2$ is universally open and proper, it has the arc lifting property but not the path lifting property.

In the next 2 variants of the above construction, gluing of 2 irreducible components has the opposite effect on path lifting.

(4.12.2) Set $Y_1 := (ux = vy) \subset \mathbb{P}^2_{xyz} \times \mathbb{C}^2_{uv}$. Note that $\Sigma := (x = y = 0) \subset Y_1$ is a section of $\pi_2 : Y_1 \to \mathbb{C}^2_{uv}$. Thus every path starting in $\mathbb{C}^2_{uv} \setminus \{(0,0)\}$ can be lifted to $Y_1$ but $Y_1 \to \mathbb{C}^2_{uv}$ does not have the path lifting property.

Let $Y$ be obtained from $Y_1$ and of $Y_2 := \mathbb{P}^2 \times \mathbb{C}^2_{uv}$ by identifying the fibers over the origin. Then $p : Y \to \mathbb{C}^2_{uv}$ has the path lifting property.

(4.12.3) Let $Z_1$ be obtained from $\mathbb{P}^1 \times \mathbb{C}^2_{uv}$ by blowing up $\mathbb{P}^1 \times \{(0,0)\}$. The projection to $\mathbb{C}^2_{uv}$ factors as $Z_1 \to B_{(0,0)}\mathbb{C}^2 \to \mathbb{C}^2$, thus not all paths starting in $\mathbb{C}^2_{uv} \setminus \{(0,0)\}$ can be lifted to $Z_1$. Let $Z$ be obtained from $Z_1$ and of $Z_2 := \mathbb{P}^1 \times \mathbb{P}^1 \times \mathbb{C}^2_{uv}$ by identifying the fibers over the origin. Thus $Z \to \mathbb{C}^2$ is universally open and proper, it has the arc lifting property but not the path lifting property.

**Acknowledgments** Partial financial support was provided by the NSF under grant numbers DMS-1362960 and DMS-1440140 while the author was in residence at MSRI during the Spring 2019 semester.

## References

1. Goresky, M., MacPherson, R.: Stratified Morse theory. In: Ergebnisse der Mathematik und ihrer Grenzgebiete (3), vol. 14, Springer, Berlin (1988). MR 932724 (90d:57039)
2. Grothendieck, A.: Éléments de géométrie algébrique. I., Springer, Heidelberg (1971)
3. Hartshorne, R.: Algebraic geometry. Graduate Texts in Mathematics, No. 52. Springer, New York (1977). MR 0463157 (57 #3116)
4. Kollár, J.: Quotients by finite equivalence relations. In: Current Developments in Algebraic Geometry. Math. Sci. Res. Inst. Publ., vol. 59, Cambridge University Press, Cambridge (2012). With an appendix by Claudiu Raicu, pp. 227–256. MR 2931872
5. Kollár, J.: Pell surfaces. Acta Math. Hungar. (2019)
6. The Stacks Project Authors: Stacks Project (2015). http://stacks.math.columbia.edu

# Cremona Transformations of Weighted Projective Planes, Zariski Pairs, and Rational Cuspidal Curves

**Enrique Artal Bartolo, José I. Cogolludo-Agustín, and Jorge Martín-Morales**

*To András Némethi, source of inspiration in singularity theory*

**Abstract** In this work, we study a family of Cremona transformations of weighted projective planes which generalize the standard Cremona transformation of the projective plane. Starting from special plane projective curves we construct families of curves in weighted projective planes with special properties. We explain how to compute the fundamental groups of their complements, using the blow-up-down decompositions of the Cremona transformations, we find examples of Zariski pairs in weighted projective planes (distinguished by the Alexander polynomial). As another application of this machinery we study a family of singularities called weighted Lê–Yomdin, which provide infinitely many examples of surface singularities with a rational homology sphere link. To end this paper we also study a family of surface singularities generalizing Brieskorn–Pham singularities in a different direction. This family contains infinitely many examples of integral homology sphere links, answering a question by Némethi.

**Keywords** Weighted projective planes · Homology spheres · Zariski pairs

**Subject Classifications** Primary 14H20, 14H30, 14F45, 14E07; Secondary 57M99, 57M12

---

E. A. Bartolo (✉) · J. I. Cogolludo-Agustín
Departamento de Matemáticas, IUMA, Universidad de Zaragoza, Zaragoza, Spain
e-mail: artal@unizar.es
http://riemann.unizar.es/~artal; jicogo@unizar.es
http://riemann.unizar.es/~jicogo

J. Martín-Morales
Centro Universitario de la Defensa, IUMA, Academia General Militar, Zaragoza, Spain
e-mail: jorge@unizar.es
http://cud.unizar.es/martin



J. Fernández de Bobadilla et al. (eds.), *Singularities and Their Interaction with Geometry and Low Dimensional Topology*, Trends in Mathematics,
https://doi.org/10.1007/978-3-030-61958-9_7

# 1 Introduction

This paper deals with curves in surfaces with normal singularities and the interplay between their topological and algebraic properties.

In this direction we provide a family of examples of curves in weighted projective planes using a generalization of the classical Cremona transformations. This allows us to construct infinitely many pairs of curves in weighted projective planes defining linearly equivalent divisors and the same local type of singularities, whose embeddings are not homeomorphic. Moreover, whose complements have non-isomorphic fundamental groups. This is known in the literature as Zariski pairs when referred to plane projective curves [2] since Zariski provided the first example of such a phenomenon in [36]. The curves are obtained from a smooth cubic and three tangent lines via a weighted Cremona transformation in Sect. 2.4. These groups are distinguished using two different techniques. In Sect. 3.1 a topological approach is given by obtaining presentations of the groups. These presentations, which in general are complicated to calculate, can be derived from those of the original curve after Cremona transformations in a very explicit geometric way. To complete this example, we also present a more algebraic approach via cyclic coverings as was originally used by Zariski and later developed by Steenbrink [32, Lemma 3.14], Libgober [21], Esnault-Viehweg [14], Vaquié [35], and the first author [2]. Our method uses a generalization of [14] given in [4], see Sect. 3.3. Section 4 is devoted to developing some methods to construct rational cuspidal curves in weighted projective planes which will be useful in the later sections.

The second part of the paper focuses on local properties of surface singularities. Our main goal is to provide examples of surface germs whose link is a rational (or even integral) homology sphere. A source of examples is given by superisolated singularities. In Sect. 5 we introduce the determinant of a surface singularity as the absolute value of the determinant of the intersection matrix of a resolution. This invariant of the surface singularity can also be calculated using a partial resolution, as shown in Sect. 5.1. Note that a surface singularity has a rational homology sphere link if and only if the dual graph of a (partial) resolution is a tree whose vertices are rational curves. Moreover, a rational homology sphere link is integral if the determinant of the singularity is one. We use this criterion to study weighted Lê–Yomdin singularities and to describe infinite families with rational and integral homology sphere links.

In particular, following the ideas in [1, 26], one can use the Zariski pairs obtained in Sect. 3 to construct weighted Lê–Yomdin singularities having the same Alexander polynomials, the same abstract topology, but different embedded topology. It would be hopeless to compute the Jordan form of the complex monodromy (the actual invariant that distinguishes the embedded topology) without the use of the techniques in this paper.

The last part is devoted to solving two problems on surface singularities with a rational sphere link. Namely, in Sect. 6.1 we study Brieskorn–Pham surface singularities $\{x^a + y^b + z^c = 0\} \subset \mathbb{C}^3$ as a special case of weighted Lê–Yomdin.

We illustrate how to recover classical results in a simple way, namely to characterize which ones have a rational sphere link and show that the only integral homology spheres occur in the classical case, that is, whenever $(a, b, c)$ are pairwise coprime. Besides Brieskorn–Pham singularities, more examples are provided in Sect. 6.2 using weighted Cremona transformations and Kummer covers.

András Némethi asked us if it was possible to find singularities with integral homology sphere links in the realm of weighted Lê–Yomdin singularities. The only ones we found are the already known Brieskorn–Pham singularities. As an alternative, in Sect. 6.3, a family of surface singularities is presented following [27], see also [29] for the *splice diagram* approach. We give conditions for this family to have a rational homology sphere link. Moreover, this family provides infinitely many examples of integral homology sphere links which may answer the question by András Némethi in the affirmative.

## 2 Quotient Singularities and Weighted Cremona Transformations

The main objects of this work will be weighted projective planes (and lines) and quotient singularities. A quotient singularity is a normal space which is locally isomorphic to $(X, 0)$ where $X$ is the quotient of $\mathbb{C}^n$ by the action of a cyclic group $\mu_m \subset \mathbb{C}^*$ given by

$$\zeta \cdot (x_1, \ldots, x_n) = (\zeta^{a_1} x_1, \ldots, \zeta^{a_n} x_n), \qquad \zeta^m = 1, (x_1, \ldots, x_n) \in \mathbb{C}^n.$$

If $\gcd(m, a_1, \ldots, a_n) = 1$, the action is faithful. We denote this singularity by $\frac{1}{m}(a_1, \ldots, a_n)$. There are some trivial equivalences of quotient singularities such as $\frac{1}{m}(a_1, \ldots, a_n) = \frac{1}{m}(da_1, \ldots, da_n)$ if $\gcd(m, d) = 1$. A less obvious one is given by

$$\frac{1}{m}(a_1, \ldots, a_n) \cong \frac{d}{m}\left(a_1, \frac{a_2}{d}, \ldots, \frac{a_n}{d}\right) \text{ if } d = \gcd(m, a_2, \ldots, a_n)$$

(see [12] as a general reference on the subject).

### 2.1 *Curves in Quotient Surface Singularities*

We introduce some notation for germs of curves in a quotient surface singularity $S := \frac{1}{d}(a, b)$ (with $a, b, d$ pairwise coprime and $d > 1$). Let $\pi : \mathbb{C}^2 \to S$ be the quotient map. Any germ of curve $C \subset S$ is defined as the zero locus of a non-constant equivariant germ $f \in \mathbb{C}\{x, y\}$, that is, a germ satisfying $f(\zeta \cdot (x, y)) = \zeta^k f(x, y)$ for some $k = 0, \ldots, d - 1$. For a fixed $k$, the collection of all such equivariant germs inherits an $\mathcal{O}_S$-module structure as a subset of $\mathbb{C}\{x, y\}$ and will be denoted by $\mathcal{O}_S(k)$. Note that an equivariant germ is a function on $S$ only when $k = 0$, that is, $\mathcal{O}_S = \mathcal{O}_S(0)$.

**Definition 2.1** A germ of curve $C$ is said to be *quasi-smooth* if $C$ is smooth as an abstract curve. If, in addition, a defining germ for $C$ can be found to have multiplicity one, then $C$ is said to be *extremely quasi-smooth*.

*Remark 2.2* There are simple characterizations of the above concepts in terms of a minimal resolution $\hat{S} \to S$; recall that its dual graph is a bamboo whose vertices represent smooth rational divisors. A curve is quasi-smooth if its strict transform in $\hat{S}$ is a curvette of an exceptional divisor, that is, smooth and transversal to it at a smooth point of the exceptional locus. Moreover, it is extremely quasi-smooth if this divisor is either end of the bamboo. In the particular case $\frac{1}{d}(1, 1)$, any quasi-smooth curve is extremely quasi-smooth, and any linear form can be the multiplicity-one component of $f$. Otherwise, in $\frac{1}{d}(a, b)$ with $(a, b) \neq (1, 1)$ the equivariant part of multiplicity 1 of an extremely quasi-smooth $f$ can only be given by the eigenspaces of the cyclic action, in our notation, either $x$ or $y$.

## 2.2 Weighted Projective Planes

In this section we briefly describe weighted projective planes in order to fix some notation. A *weight* is a triple $\omega := (e_1, e_2, e_3) \in \mathbb{Z}^3_{>0}$ such that $\gcd \omega = 1$. The *weighted projective plane* $\mathbb{P}^2_\omega$ is a normal surface obtained as the quotient of $\mathbb{C}^3 \setminus \{0\}$ by the action of $\mathbb{C}^*$ given by

$$t \cdot (x, y, z) = (t^{e_1} x, t^{e_2} y, t^{e_3} z), \qquad t \in \mathbb{C}^*, (x, y, z) \in \mathbb{C}^3 \setminus \{0\}.$$

Weighted projective lines are defined in a similar way. The symbol $[x : y : z]_\omega$ stands for points in $\mathbb{P}^2_\omega$, for orbits in $\mathbb{C}^3 \setminus \{0\}$ or their closure in $\mathbb{C}^3$. This variety is covered by three *quotient charts*. One of them is

$$\begin{array}{ccc} \frac{1}{e_3}(e_1, e_2) & \xrightarrow{\Psi_{\omega,3}} & \mathbb{P}^2_\omega \setminus \{z = 0\} \\ [(x, y)] & \longmapsto & [x : y : 1]_\omega. \end{array}$$

The other two quotient charts are defined accordingly.

Define $d_k := \gcd(e_i, e_j)$ and $\alpha_k := \frac{e_k}{d_i d_j}$, $\{i, j, k\} = \{1, 2, 3\}$. Note that $\eta := (\alpha_1, \alpha_2, \alpha_3)$ are pairwise coprime. According to the properties described above, the map

$$\begin{array}{ccc} \mathbb{P}^2_\omega & \xrightarrow{\pi_{\eta,\omega}} & \mathbb{P}^2_\eta \\ [x : y : z]_\omega & \longmapsto & [x^{d_1} : y^{d_2} : z^{d_3}]_\eta \end{array} \tag{2.1}$$

is well defined since

$$t\cdot[x:y:z]_\omega=[t^{e_1}x:t^{e_2}y:t^{e_3}z]_\omega\mapsto[t^{d_1e_1}x^{d_1}:t^{d_2e_2}y^{d_2}:t^{d_3e_3}z^{d_3}]_\eta=t^{d_1d_2d_3}\cdot[x^{d_1}:y^{d_2}:z^{d_3}]_\eta,$$

and $d_ie_i=\alpha_id_1d_2d_3$. Moreover, one can easily check that it is an isomorphism.

One may consider $\mathbb{P}^2_\omega$ and $\mathbb{P}^2_\eta$ in a slightly different way (see also [12]). The plane $\mathbb{P}^2_\eta$ has at most 3 singular points at $P_x:=[1:0:0]_\eta$ (if $\alpha_1>1$), $P_y:=[0:1:0]_\eta$ (if $\alpha_2>1$), and $P_z:=[0:0:1]_\eta$ (if $\alpha_3>1$). The plane $\mathbb{P}^2_\omega$ is an orbifold where the quotient charts $\Psi_{\omega,i}$ are not normalized; the associated analytic variety to $\mathbb{P}^2_\omega$ is $\mathbb{P}^2_\eta$ since the normalization of the source of $\Psi_{\omega,i}$ is precisely the source of $\Psi_{\eta,i}$.

## 2.3 Weighted Blow-ups

Let us consider now $\omega:=(e_1,e_2)\in\mathbb{Z}^2_{>0}$, $\gcd\omega=1$. The *ω-weighted blow-up* of $\mathbb{C}^2$ at the origin is the map $\pi_\omega:\widehat{\mathbb{C}}^2_\omega\to\mathbb{C}^2$ where

$$\widehat{\mathbb{C}}^2_\omega:=\{(\mathbf{x},\mathbf{u})\in\mathbb{C}^2\times\mathbb{P}^1_\omega\mid\mathbf{x}\in\mathbf{u}\}.$$

This normal variety is represented with two quotient charts. One of them is

$$\widehat{\Psi}_{\omega,2}:\frac{1}{e_2}(e_1,-1)\to\widehat{\mathbb{C}}^2_\omega,\qquad (x,y)\mapsto((xy^{e_1},y^{e_2}),[x:1]_\omega);$$

the other one is analogous and modeled on $\frac{1}{e_1}(-1,e_2)$. The exceptional divisor of $\pi_\omega$ is a weighted projective line which contains the singular points $(0,[1:0]_\omega)$ (if $e_1>1$) and $(0,[0:1]_\omega)$ (if $e_2>1$) of the surface $\widehat{\mathbb{C}}^2_\omega$. Note that the curvettes of this divisor are extremely quasi-smooth if either $e_1$ or $e_2$ equal 1.

Let us study now three-dimensional weighted blow-ups. We recover the notation introduced in Sect. 2.2 for a weight $\omega$ and its normalization $\eta$, both in $\mathbb{Z}^3_{>0}$. We consider $\Pi_\omega:\widehat{\mathbb{C}}^3_\omega\to\mathbb{C}^3$ where

$$\widehat{\mathbb{C}}^3_\omega:=\{(\mathbf{x},\mathbf{u})\in\mathbb{C}^3\times\mathbb{P}^2_\omega\mid\mathbf{x}\in\mathbf{u}\}.$$

The normal variety is now represented with three charts. One of them is

$$\widehat{\Psi}_{\omega,3}:\frac{1}{e_3}(e_1,e_2,-1)\to\widehat{\mathbb{C}}^3_\omega,\qquad (x,y,z)\mapsto((xz^{e_1},yz^{e_2},z^{e_3}),[x:y:1]_\omega);$$

the other two charts can analogously be defined and have as domains the quotients $\frac{1}{e_1}(-1,e_2,e_3)$ and $\frac{1}{e_2}(e_1,-1,e_3)$.

Let us study the local structure of $\widehat{\mathbb{C}}^3_\omega$ at $E_\omega:=\Pi_\omega^{-1}(0)$; since $\Pi_\omega$ is an isomorphism outside this exceptional divisor the points not in $E_\omega$ are smooth. Note that $E_\omega$ is naturally isomorphic to $\mathbb{P}^2_\omega$; in addition by (2.1), one has $\mathbb{P}^2_\omega \cong \mathbb{P}^2_\eta$. For the sake of simplicity we will denote the elements of $E_\omega$ only by their $\omega$-quasi-homogeneous coordinates. Let us denote:

$$P_x = [1:0:0]_\omega, \quad P_y = [0:1:0]_\omega, \quad P_z = [0:0:1]_\omega,$$

$$\check{X} = \{[0:y:z]_\omega \mid yz \neq 0\},\ \check{Y} = \{[x:0:z]_\omega \mid xz \neq 0\},\ \check{Z} = \{[x:y:0]_\omega \mid xy \neq 0\}.$$

In order to provide a stratification of $E_\omega$ in terms of the singular points of the ambient space we need a description of the singular locus.

**Proposition 2.3** *Let $P = [x_0 : y_0 : 1]_\omega \in E_\omega \cap \widehat{\Psi}_{\omega,3}(\mathbb{C}^3) \subset \widehat{\mathbb{C}}^3_\omega$. The following properties hold:*

(1) *If $x_0y_0 \neq 0$ then $(\widehat{\mathbb{C}}^3_\omega, P)$ is smooth.*
(2) *If $P \in \check{X}$, i.e. $y_0 \neq 0$ and $x_0 = 0$, then $(\widehat{\mathbb{C}}^3_\omega, P)$ is isomorphic to the germ at the origin of $\frac{1}{d_1}(e_1, 0, -1)$.*
(3) *If $P \in \check{Y}$, i.e. $x_0 \neq 0$ and $y_0 = 0$, then $(\widehat{\mathbb{C}}^3_\omega, P)$ is isomorphic to the germ at the origin of $\frac{1}{d_2}(0, e_2, -1)$.*
(4) *If $P = P_z$, i.e. $x_0 = y_0 = 0$, then $(\widehat{\mathbb{C}}^3_\omega, P_z)$ is isomorphic to the germ at the origin of $\frac{1}{e_3}(e_1, e_2, -1)$.*

***Proof*** It is only necessary to prove (2). Note that $P$ is obtained as the image by $\widehat{\Psi}_{\omega,3}$ of $(0, y_0, 0) \in \frac{1}{e_3}(e_1, e_2, -1)$. The isotropy subgroup of $(0, y_0, 0)$ by the action is the cyclic group of order $d_1 = \gcd(e_2, e_3)$. Hence at a neighborhood of $(0, y_0, 0)$ the space looks like $\frac{1}{d_1}(e_1, e_2, -1) = \frac{1}{d_1}(e_1, 0, -1)$.

*Remark 2.4* A similar statement holds for the other charts. Note that a point satisfying property (2) above, say $P = [0 : 1 : 1]_\omega$ belongs in the image of $\widehat{\Psi}_{\omega,3}$, $P = \widehat{\Psi}_{\omega,3}(0, 1, 0)$ as stated in Proposition 2.3, but also in the image of $\widehat{\Psi}_{\omega,2}$, $P = \widehat{\Psi}_{\omega,2}(0, 0, 1)$. Note that the notation for the quotient types given above do no match, that is, $\frac{1}{d_1}(e_1, 0, -1)$ if considered in $\widehat{\Psi}_{\omega,3}(\mathbb{C}^3)$ and $\frac{1}{d_1}(e_1, -1, 0)$ if considered in $\widehat{\Psi}_{\omega,3}(\mathbb{C}^3)$. To avoid this ambiguity we will simply say that given $P \in \check{X}$, then $(\widehat{\mathbb{C}}^3_\omega, P)$ is isomorphic to the product of $(\mathbb{C}, 0)$ and the germ at the origin of $\frac{1}{d_1}(e_1, -1)$. A similar property holds for $\check{Y}$, $\check{Z}$.

*Remark 2.5* If $d_1 = e_3$, i.e., if $e_3$ divides $e_2$, in the proof of Proposition 2.3 the condition $y_0 \neq 0$ is not needed and $P_z$ behaves as the points in $\check{X}$. A similar property holds for the other pairs of axes and vertices.

**Notation 2.6** *We fix the following notation for the strata of $E_\omega$.*

- *two-dimensional stratum. The stratum $\mathcal{T}$ is the intersection of $E_\omega$ with the smooth subvariety of $\widehat{\mathbb{C}}^3_\omega$; it contains $\{[x : y : z]_\omega \mid xyz \neq 0\}$. It contains also $\check{X}$ (resp. $\check{Y}$, resp. $\check{Z}$) if $d_1 = 1$ (resp. $d_2 = 1$, resp. $d_3 = 1$) and $P_x$ (resp. $P_y$, resp. $P_z$) if $e_1 = 1$ (resp. $e_2 = 1$, resp. $e_3 = 1$).*

- *one-dimensional strata. Following Proposition 2.3 and the above remarks we set:*

$$\mathcal{L}_x = \begin{cases} \emptyset & \text{if } d_1 = 1 \\ \check{X} \cup \{P_y\} & \text{if } 1 < d_1 = e_2 \neq e_3 \\ \check{X} \cup \{P_z\} & \text{if } 1 < d_1 = e_3 \neq e_2 \\ X & \text{if } 1 < d_1 = e_2 = e_3 \\ \check{X} & \text{otherwise.} \end{cases}$$

*The remaining strata $\mathcal{L}_y$ and $\mathcal{L}_z$ are defined accordingly.*

- *zero-dimensional strata.*

$$\mathcal{P}_x = \begin{cases} \emptyset & \text{if } e_1 \text{ divides either } e_2 \text{ or } e_3, \\ \{P_x\} & \text{otherwise.} \end{cases}$$

*The remaining strata $\mathcal{P}_y$ and $\mathcal{P}_z$ are defined accordingly.*

## 2.4 Weighted Cremona Transformations

The most well-known Cremona transformation of $\mathbb{P}^2$ corresponds to the birational map $[x : y : z] \mapsto [yz : xz : xy]$; geometrically, this map is the composition of the blow-ups at $[1 : 0 : 0]$, $[0 : 1 : 0]$, $[0 : 0 : 1]$ and the contractions of the strict transforms of the lines $x = 0$, $y = 0$, $z = 0$ which become pairwise disjoint $(-1)$-lines in the blown-up plane.

In this section we generalize this transformation to a birational map from a weighted projective plane to $\mathbb{P}^2$. Let us fix $\mathbb{P}^2_\omega$, $\omega := (e_1, e_2, e_3)$, where $e_1, e_2, e_3$ are pairwise coprime, i.e., $\omega = \eta$. In order to stress this property we will use the notation $e_i = \alpha_i$, $i = 1, 2, 3$. Consider two positive integers $\beta_1, \beta_2$ such that $\alpha_1\beta_1 + \alpha_2\beta_2 = \alpha_3 + \alpha_1\alpha_2$ (they exist from standard semigroup properties). These arithmetic data provide the following map

$$\begin{array}{ccc} \mathbb{P}^2_\omega & \overset{\Phi_{\omega,\beta_1,\beta_2}}{\dashrightarrow} & \mathbb{P}^2 \\ [x : y : z]_\omega & \longmapsto & [y^{\alpha_1} z : x^{\alpha_2} z : x^{\beta_1} y^{\beta_2}], \end{array}$$

which is a well-defined rational map (not a morphism) since the three coordinates have $\omega$-degree equal to $\alpha_1\alpha_2 + \alpha_3$. It is in fact a birational map whose inverse is given by

$$\begin{array}{ccc} \mathbb{P}^2 & \dashrightarrow & \mathbb{P}^2_\omega \\ [x : y : z] & \longmapsto & \left[ y^{\frac{1}{\alpha_2}} z^{\frac{\alpha_1}{\alpha_3}} : x^{\frac{1}{\alpha_1}} z^{\frac{\alpha_2}{\alpha_3}} : x^{\frac{\beta_2}{\alpha_1}} y^{\frac{\beta_1}{\alpha_2}} \right]_\omega . \end{array}$$

We will show that this map is well defined as long as the radicals $x^{\frac{1}{\alpha_1}}, y^{\frac{1}{\alpha_2}}, z^{\frac{1}{\alpha_3}}$ are chosen consistently throughout the formula. Assume $x_0$ (resp. $y_0$, $z_0$) is such that $x_0^{\alpha_1} = x$ (resp. $y_0^{\alpha_2} = y$, $z_0^{\alpha_3} = z$) and choose for instance $x_1 = \zeta_{\alpha_1} x_0$. Let $\hat{\alpha}_2 \in \mathbb{Z}$ be such that $\alpha_2\hat{\alpha}_2 \equiv 1 \bmod \alpha_1$. As a consequence, the following congruences hold: $\alpha_3\hat{\alpha}_2 \equiv (\alpha_3 + \alpha_1\alpha_2)\hat{\alpha}_2 \equiv (\alpha_1\beta_1 + \alpha_2\beta_2)\hat{\alpha}_2 \equiv \beta_2 \bmod \alpha_1$. Then

$$[y_0 z_0^{\alpha_1} : x_1 z_0^{\alpha_2} : x_1^{\beta_2} y_0^{\beta_1}]_\omega = [y_0 z_0^{\alpha_1} : \zeta_{\alpha_1} x_0 z_0^{\alpha_2} : \zeta_{\alpha_1}^{\beta_2} x_0^{\beta_2} y_0^{\beta_1}]_\omega =$$

$$[(\zeta_{\alpha_1}^{\hat{\alpha}_2})^{\alpha_1} y_0 z_0^{\alpha_1} : (\zeta_{\alpha_1}^{\hat{\alpha}_2})^{\alpha_2} x_0 z_0^{\alpha_2} : (\zeta_{\alpha_1}^{\hat{\alpha}_2})^{\alpha_3} x_0^{\beta_2} y_0^{\beta_1}]_\omega = [y_0 z_0^{\alpha_1} : x_0 z_0^{\alpha_2} : x_0^{\beta_2} y_0^{\beta_1}]_\omega.$$

A similar argument applies to other choices of roots of $y^{\frac{1}{\alpha_2}}$ and $z^{\frac{1}{\alpha_3}}$. These equations completely determine the birational map, but a more geometric description will be useful.

**Proposition 2.7** *The map $\Phi_{\omega,\beta_1,\beta_2}$ is the composition of the following blow-ups and downs:*

(1) *Three simultaneous blow-ups:*

(a) *Type $(\alpha_1, \alpha_2)$ at $[0:0:1]_\omega \cong \frac{1}{\alpha_3}(\alpha_1, \alpha_2)$.*

(b) *Type $(1, \beta_1)$ at $[0:1:0]_\omega$ isomorphic to*

$$\frac{1}{\alpha_2}(\alpha_1, \alpha_3) = \frac{1}{\alpha_2}(\alpha_1, \alpha_1\alpha_2 + \alpha_3) = \frac{1}{\alpha_2}(\alpha_1, \alpha_1\beta_1 + \alpha_2\beta_2) = \frac{1}{\alpha_2}(1, \beta_1).$$

(c) *Type $(1, \beta_2)$ at $[1:0:0]_\omega$ isomorphic to*

$$\frac{1}{\alpha_1}(\alpha_2, \alpha_3) = \frac{1}{\alpha_1}(\alpha_2, \alpha_1\alpha_2 + \alpha_3) = \frac{1}{\alpha_1}(\alpha_2, \alpha_1\beta_1 + \alpha_2\beta_2) = \frac{1}{\alpha_1}(1, \beta_2).$$

(2) *Three simultaneous blow-downs:*

(a) *Type $(1, 1)$ at $[0:0:1]$.*

(b) *Type $(\alpha_2, \beta_1)$ at $[1:0:0]$.*

(c) *Type $(\alpha_1, \beta_2)$ at $[0:1:0]$.*

***Proof*** Let us start with the three blow-ups in $\mathbb{P}^2_\omega$. We obtain a normal rational surface $S$. The preimage of the three axes appear in Fig. 1, containing the strict transforms $L_x, L_y, L_z$ of the lines and the exceptional components $E_x, E_y, E_z$. The self-intersections and the type of the singular points are computed using [8, Theorem 4.3].

The strict transforms of the lines coincide with the exceptional components of a $(\alpha_1, \beta_2)$-blowing-up ($L_y$), a $(\alpha_2, \beta_1)$-blowing-up ($L_x$) and a standard blowing-up ($L_z$). The result of the triple blowing-down is $\mathbb{P}^2$.

This geometric expression will be useful for the study of curves in $\mathbb{P}^2_\omega$ via their transforms in $\mathbb{P}^2$.

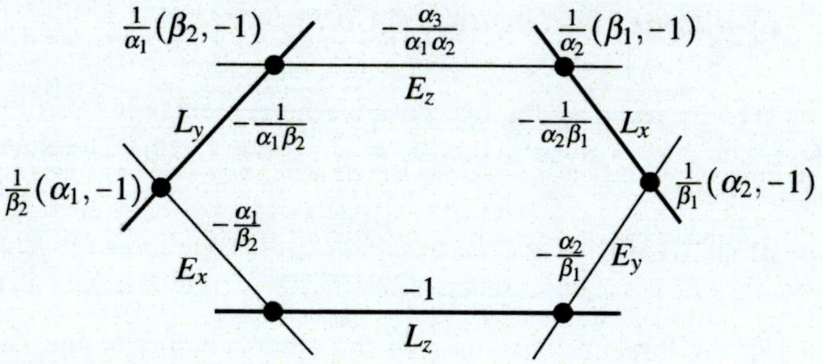


**Fig. 1** Weighted blow-ups of $\mathbb{P}^2$ in $S$

# 3 Zariski Pairs on Weighted Projective Planes

In this section, we are going to use the Cremona transformations in Sect. 2.4 to produce Zariski pairs in weighted projective planes. By a *Zariski pair* we mean two curves embedded in the same surface whose *combinatorics* are the same, but whose embeddings are non-homeomorphic. As in the classical case of curves in the projective plane, the combinatorics of a curve in a weighted projective plane is encoded by the degrees of its irreducible components and the dual graph of a minimal resolution of the curve (where the strict transforms of the irreducible components of the curve are marked).

In this section we will produce families of Zariski pairs of irreducible curves. Let us start with the combinatorics defined by a smooth projective cubic and three tangent lines at inflection points. Note that a generic choice of a smooth cubic can be made so that such lines are non-concurrent and hence the remaining singular points are three nodes. This combinatorics admits a Zariski pair of sextics, see [2], and their embeddings are distinguished by the algebraic property of whether or not the inflection points of the cubic, that is, the three non-nodal singular points of the sextic, which have type $\mathbb{A}_6$, are aligned. The image by a standard Cremona transformation of the smooth cubics (using the three tangent lines at the axes) produces a Zariski pair of irreducible sextics with three $\mathbb{E}_6$-points. In this case, the embeddings can be proven to be different showing that the fundamental group of their complements are not isomorphic.

Our strategy is to replace this Cremona transformation by the inverse of those described in Sect. 2.4.

### 3.1 Fundamental Groups of Complements

Let us start by recalling the two possible fundamental groups of the complements of the sextic curves given as the union of a smooth cubic and three tangent lines at inflection points.

**Proposition 3.1 ([3])** *Let $\mathcal{C}$ be a smooth cubic with three tangent lines $X, Y, Z$ at inflections which are* not *aligned. Then, $\pi_1(\mathbb{P}^2 \setminus (\mathcal{C} \cup X \cup Y \cup Z))$ is abelian.*

In [3], the fundamental group of the other member of the Zariski pair is also computed; since it is non-abelian, this invariant distinguishes the two members. For our purpose, we need a more geometrical presentation of the group involving meridians for all the irreducible components and such that the meridians close to the nodes are made explicit. Let us recall the concept of meridian in order to clarify what we mean by *meridians close to a singular point*.

**Definition 3.2** Let $Z$ be a connected quasi-projective manifold and let $H$ be a hypersurface of $Z$. Consider $P \in Z \setminus H$ and $K$ an irreducible component of $H$. A homotopy class $\gamma \in \pi_1(Z \setminus H; P)$ is called a *meridian about $K$ with respect to $H$* if $\gamma = [\delta]$ for some loop $\delta$ satisfying the following:

(1) there is a smooth complex analytic disk $\Delta \subset Z$ transverse to $H$ such that $\Delta \cap H = \{P'\} \subset K$ (transversality implies that $P'$ is a smooth point of $H$).
(2) there is a path $\alpha$ in $Z \setminus H$ starting at $P$ and ending at some point $P'' \in \partial\Delta$.
(3) $\delta = \alpha * \beta * \overline{\alpha}$, where the operation $*$ here means concatenation of paths from left to right, $\beta$ is the closed path obtained by traveling from $P''$ along $\partial\Delta$ in the positive direction and $\overline{\alpha}$ represents the path $\alpha$ traveled in the opposite direction, that is, $\overline{\alpha}(t) := \alpha(1-t)$.

It is well known that meridians with respect to the same irreducible component define a conjugacy class of members of the fundamental group.

*Example 3.3* Let $Z = \mathbb{C}^2$ and $H = \{xy = 0\}$ and let $P := (1, 1)$. The paths $\mu_x, \mu_y : [0, 1] \to Z \setminus H$ defined by

$$\mu_x(t) = (e^{2i\pi t}, 1), \quad \mu_y(t) = (1, e^{2i\pi t}),$$

define meridians with respect to the irreducible components of $H$ (for which the path $\alpha$ is trivial). They commute as elements in the fundamental group $\pi_1(Z \setminus H; P)$. If $Z$ is quasi-projective surface and $H$ is a curve containing a node, *two meridians are close to the node* if there is a common path $\alpha$ from the base point of $\pi_1(Z \setminus H; P)$ to a point *close* to the node such that the $\beta$-paths look like in this example.

**Proposition 3.4 ([6])** *Let $\mathcal{C}$ be a smooth cubic with three tangent lines $X, Y, Z$ at inflections which are aligned. Then, $\pi_1(\mathbb{P}^2 \setminus (\mathcal{C} \cup X \cup Y \cup Z))$ is*

$$\langle c, \ell_x, \ell_y, \ell_z \mid [\ell_x, \ell_y]=[\ell_y, \ell_z]=[\ell_z, \ell_x]=[c, \ell_x^{-1}\ell_z]=[c, \ell_y^{-1}\ell_z]=c\ell_x c\ell_y c\ell_z=1\rangle \tag{3.1}$$

*where c is a meridian of $\mathcal{C}$, and $\ell_x, \ell_y, \ell_z$ are meridians of $X, Y, Z$, respectively; moreover the meridians of the lines correspond to meridians close to the double points.*

Let us fix $\Phi := \Phi_{\omega,\beta_1,\beta_2}$ as in Sect. 2.4, and let us denote by $\tilde{\mathcal{C}} \subset \mathbb{P}^2_\omega$ the strict transform of the smooth cubic $\mathcal{C}$ by $\Phi$, where the lines $X, Y, Z$ have equations $x = 0, y = 0, z = 0$, respectively. Consider the following three homogeneous polynomials of degree 3

$$H_\lambda(x, y, z) := x^3 + y^3 + z^3 + 3xy(\lambda^{-1}x + \lambda y) + 3xz(x + z) + 3yz(\lambda^{-1}y + \lambda z),$$

where $\lambda^3 = 1$. The curve $\mathcal{C}_\lambda = \{H_\lambda = 0\}$ is a smooth cubic which is tangent to the line $L_x$ at the inflection point $[0 : 1 : -\lambda]$ and analogously for $Y$ at $[-1 : 0 : 1]$, and $Z$ at $[1 : -\lambda : 0]$. Note that for the cubic $\mathcal{C}_1$ the three inflection points are contained in the line $x+y+z = 0$. However, for the smooth cubic $\mathcal{C}_{\exp\frac{2i\pi}{3}}$ the three inflection points are not aligned.

**Corollary 3.5** *In the non-aligned case, $\pi_1(\mathbb{P}^2_\omega \setminus \tilde{\mathcal{C}})$ is isomorphic to $\mathbb{Z}/3(\alpha_1\alpha_2+\alpha_3)$.*

***Proof*** The space $\mathbb{P}^2_\omega \setminus \tilde{\mathcal{C}}$ is homeomorphic to $S \setminus (\widehat{\mathcal{C}} \cup L_x \cup L_y \cup L_z)$ (see Fig. 1) and the space $\mathbb{P}^2 \setminus (\mathcal{C} \cup X \cup Y \cup Z)$ is homeomorphic to $S \setminus (\widehat{\mathcal{C}} \cup L_x \cup L_y \cup L_z \cup E_x \cup E_y \cup E_z)$, where $\widehat{\mathcal{C}}$ denotes the strict transform of $\mathcal{C}$ in $S$. As a consequence of [17, Lemma 4.18] the kernel of the epimorphism

$$\pi_1(\mathbb{P}^2 \setminus (\mathcal{C} \cup X \cup Y \cup Z)) \twoheadrightarrow \pi_1(\mathbb{P}^2_\omega \setminus \tilde{\mathcal{C}}) \tag{3.2}$$

is the normal subgroup generated by the meridians of $E_x, E_y, E_z$ in $S$. Since the source is an abelian group by Proposition 3.1, the group $\pi_1(\mathbb{P}^2_\omega \setminus \tilde{\mathcal{C}})$ is abelian as well. Hence it coincides with $H_1(\mathbb{P}^2_\omega \setminus \tilde{\mathcal{C}}; \mathbb{Z}) \cong \mathbb{Z}/\deg(\tilde{\mathcal{C}})$, since $\tilde{\mathcal{C}}$ contains the vertices of $\mathbb{P}^2_\omega$.

In order to compute the other fundamental group we need a technical result.

**Lemma 3.6** *Let $\pi : \widehat{\mathbb{C}}^2_{(\alpha_1,\alpha_2)} \to \mathbb{C}^2$ be the $(\alpha_1, \alpha_2)$-blow-up of the origin in $\mathbb{C}^2$ and let $E$ denote its exceptional component. Let $X, Y \subset \mathbb{C}^2$ be the axes (curves of equations $x = 0$, $y = 0$, respectively), and let us keep this notation for their strict transforms. Let $U := \mathbb{C}^2 \setminus (X \cup Y) \equiv \widehat{\mathbb{C}}^2_{(\alpha_1,\alpha_2)} \setminus (E \cup X \cup Y)$.*

*If $\mu_X, \mu_Y, \mu_E$ denote meridians of the respective curves in $\pi_1(U) \cong \mathbb{Z}\mu_X \oplus \mathbb{Z}\mu_Y$, then (multiplicative notation) $\mu_E = \mu_X^{\alpha_1}\mu_Y^{\alpha_2}$.*

***Proof*** Consider $(1, 1)$ as the base point, then $\mu_X$ is the loop $t \mapsto (e^{2i\pi t}, 1)$, while $\mu_Y$ is the loop $t \mapsto (1, e^{2i\pi t})$. Let us pick a chart of $\widehat{\mathbb{C}}^2_{(\alpha_1,\alpha_2)}$, say

$$\begin{array}{ccc} \frac{1}{\alpha_1}(-1, \alpha_2) & \longrightarrow & \mathbb{C}^2 \\ [(x, y)] & \longmapsto & (x^{\alpha_1}, x^{\alpha_2} y). \end{array}$$

The base point in the chart is the class of $(1, 1)$; the equation of $E$ is $x = 0$ and hence $\mu_E$ is represented by $t \mapsto [(e^{2i\pi t}, 1)]$. Hence, in $\mathbb{C}^2$ is represented by $t \mapsto (e^{2i\alpha_1\pi t}, e^{2i\alpha_2\pi t})$ and the result follows.

**Proposition 3.7** *In the aligned case, $\pi_1(\mathbb{P}^2_\omega \setminus \tilde{\mathcal{C}})$ is isomorphic to $\mathbb{Z}/3(\alpha_1\alpha_2 + \alpha_3)$ if 2 divides $\alpha_1\alpha_2\alpha_3\beta_1\beta_2$ and to*

$$\langle \ell, u \mid \ell^{\alpha_1\alpha_2+\alpha_3} = 1, u^3 = \ell^2 \rangle \tag{3.3}$$

*otherwise. This group is a central extension of $\mathbb{Z}/2 * \mathbb{Z}/3$ by a cyclic group of order $\frac{\alpha_1\alpha_2+\alpha_3}{2}$.*

***Proof*** Following the proof of Corollary 3.5, the epimorphism described in (3.2) also holds in this case. Hence a presentation of $\pi_1(\mathbb{P}^2_\omega \setminus \tilde{\mathcal{C}})$ can be given once meridians of $E_x, E_y, E_z$ are written in terms of the generators provided in (3.1). Since the meridians $\ell_x, \ell_y, \ell_z$ of the lines in the presentation (3.1) are homotopic to meridians close to the double points, by Lemma 3.6 we have that $\ell_x\ell_y$ is a meridian of $E_z$, $\ell_x^{\alpha_1}\ell_z^{\beta_2}$ is a meridian of $E_x$, and $\ell_y^{\alpha_2}\ell_z^{\beta_1}$ is a meridian of $E_y$. Hence a presentation of $\pi_1(\mathbb{P}^2_\omega \setminus \tilde{\mathcal{C}})$ can be obtained by adding the relations

$$\ell_x\ell_y = \ell_x^{\alpha_1}\ell_z^{\beta_2} = \ell_y^{\alpha_2}\ell_z^{\beta_1} = 1 \tag{3.4}$$

to the presentation given in (3.1).

Finally, let us simplify this presentation. As a first step one can eliminate $\ell_x$, since $\ell_x = \ell_y^{-1}$. Also, choose $\hat{\alpha}_1, \hat{\alpha}_2 \in \mathbb{Z}$ such that $\alpha_2\hat{\alpha}_1 - \alpha_1\hat{\alpha}_2 = 1$. Note that $\ell_y, \ell_z$ commute; then the remaining two relations in (3.4) become

$$\ell_y^{-\alpha_1}\ell_z^{\beta_2} = \ell_y^{\alpha_2}\ell_z^{\beta_1} = 1 \Longrightarrow \begin{cases} 1 = \ell_z^{\alpha_1\beta_1+\alpha_2\beta_2} = \ell_z^{\alpha_1\alpha_2+\alpha_3}, \\ \ell_y = \ell_z^{-(\hat{\alpha}_1\beta_2+\hat{\alpha}_2\beta_1)}. \end{cases}$$

In fact, this is an equivalence. Let us denote $\ell := \ell_z$ and $u := c\ell$. Since $[c, \ell_y\ell] = [c, \ell_y^{-1}\ell] = 1$, one has

$$\begin{aligned} 1 =& c\ell_y^{-1}c\ell_y c\ell = c\ell_y^{-1}c\ell_y\ell\ell^{-1}c\ell = c\ell_y^{-1}(\ell_y\ell)c\ell^{-1}c\ell \Longleftrightarrow \\ 1 =& (c\ell)^2 c\ell^{-1} = (c\ell)^3\ell^{-2} \Longleftrightarrow \ell^2 = u^3. \end{aligned}$$

Hence $\pi_1(\mathbb{P}^2_\omega \setminus \tilde{\mathcal{C}})$ admits a presentation

$$\langle \ell, u \mid \ell^{\alpha_1\alpha_2+\alpha_3} = 1, [u, \ell^{\hat{\alpha}_1\beta_2+\hat{\alpha}_2\beta_1-1}] = 1, u^3 = \ell^2 \rangle. \tag{3.5}$$

Note that, using $\ell^2 = u^3$, the relation $[u, \ell^{\hat{\alpha}_1\beta_2+\hat{\alpha}_2\beta_1-1}] = 1$ can be either eliminated or replaced by $[u, \ell] = 1$ depending on the parity of $\hat{\alpha}_1\beta_2 + \hat{\alpha}_2\beta_1$. In addition, $\ell$ can also be eliminated using $\ell^{\alpha_1\alpha_2+\alpha_3} = 1$ and $u^3 = \ell^2$ in case $\alpha_1\alpha_2 + \alpha_3$ is odd. In particular, if $\hat{\alpha}_1\beta_2+\hat{\alpha}_2\beta_1$ is even or $\alpha_1\alpha_2+\alpha_3$ is odd, then (3.5) becomes an abelian group. Otherwise, one obtains the presentation (3.3).

It is immediate to verify that $\hat{\alpha}_1\beta_2+\hat{\alpha}_2\beta_1$ is odd and $\alpha_1\alpha_2+\alpha_3$ even if and only if $\alpha_1\alpha_2\alpha_3\beta_1\beta_2$ is odd, which ends the proof.

**Corollary 3.8** *The derived subgroup $F$ of $\pi_1(\mathbb{P}^2_\omega \setminus \tilde{C})$ (in the non-abelian case) is the direct product of $\mathbb{Z}/(\frac{\alpha_1\alpha_2+\alpha_3}{2})$ and a free group of rank 2. The characteristic polynomial of the action of the monodromy on $F/F' \otimes_{\mathbb{Z}} \mathbb{C}$ is $t^2-t+1$.*

### 3.2 A Family of Zariski Pairs of Irreducible Weighted Projective Curves

Summarizing the previous section, let $\omega = (\alpha_1,\alpha_2,\alpha_3)$ be pairwise coprime positive integers, and $\beta_1,\beta_2$ such that $\alpha_1\beta_1+\alpha_2\beta_2 = \alpha_1\alpha_2+\alpha_3$. Consider $\mathcal{C}$ a smooth projective cubic and $\Phi_1$ (resp. $\Phi_2$) the weighted Cremona transformation from $\mathbb{P}^2_\omega$ to $\mathbb{P}^2$ with respect to three tangent lines to $\mathcal{C}$ at aligned (resp. non-aligned) inflection points. Let us denote by $\tilde{\Phi}_i^*(\mathcal{C})$ the strict transform of $\mathcal{C}$ by the Cremona transformation $\Phi_i$.

**Theorem 3.9** *Under the conditions above, if $\alpha_1\alpha_2\alpha_3\beta_1\beta_2$ is odd then $(\tilde{\Phi}_1^*(\mathcal{C}), \tilde{\Phi}_2^*(\mathcal{C}))$ is a Zariski pair of irreducible weighted projective curves of degree $3(\alpha_1\alpha_2+\alpha_3)$ in $\mathbb{P}^2_\omega$.*

***Proof*** Since both $\Phi_i$, $i=1,2$ are birational and $\mathcal{C}$ is irreducible, then $\tilde{\Phi}_i^*(\mathcal{C})$, $i = 1,2$ are both irreducible as well. Also, the singularities of $\tilde{\Phi}_i^*(\mathcal{C})$ are determined locally by the singularities of the union of $\mathcal{C}$ and the lines used for the Cremona transformation $\Phi_i$. Hence, $\tilde{\Phi}_1^*(\mathcal{C})$ and $\tilde{\Phi}_2^*(\mathcal{C})$ have the same combinatorics. Finally, if $\alpha_1\alpha_2\alpha_3\beta_1\beta_2$ is odd, then by Proposition 3.7 and Corollary 3.5 the fundamental groups of their complements are not isomorphic. This ends the proof.

### 3.3 Cyclic Covers and Their Irregularity à la Esnault–Viehweg

The purpose of this section is to prove Theorem 3.9 via a generalization of the Alexander polynomial method, that is, the calculation of invariants associated with cyclic covers of the weighted projective plane ramified along the curves. In particular, we will calculate the dimension of the eigenspaces of the homology in degree 1 of the cover with respect to the action of the deck transformation. This approach was originally used by Zariski [36] for sextics with six cusps in the projective plane. Later on, Libgober [21] and Esnault [13] made significant progress in this direction for cyclic covers and projective plane. Also Esnault-Viehweg [14] gave the tools that allowed the first author in [2], Sabbah [31], and

Loeser-Vaquie [23] to find descriptions of the irregularity of cyclic covers. This approach was extended by Libgober [22] for abelian covers. The approach presented here is a generalization of Esnault-Viehweg's and was developed by the authors for cyclic covers of surfaces with abelian quotient singularities and $\mathbb{Q}$-resolutions (or partial resolutions) in [4].

Let $\rho : X \to \mathbb{P}^2_\omega$ be the cyclic cover of $\mathbb{P}^2_\omega$ ramified along a reduced curve $\mathcal{C}$ of degree $d$. Consider $\check{X} = \rho^{-1}(\mathbb{P}^2_\omega \setminus (\mathcal{C} \cup \operatorname{Sing} \mathbb{P}^2_\omega))$ the unramified part of the cover and let $\sigma : \check{X} \to \check{X}$ be a generator of the monodromy of the unramified cover.

Let $\pi : Y \to \mathbb{P}^2_\omega$ be a $\mathbb{Q}$-embedded resolution of $\mathcal{C}$. For $P \in \operatorname{Sing} \mathcal{C}$, let $\Gamma_P$ be the dual graph of the exceptional divisor of $\pi$ over $P$. For any $v$ vertex of $\Gamma_P$ we will denote by $E_v$ the associated exceptional divisor over $P$ and by $m_v$ (resp. $\nu_v - 1$) the coefficient of $E_v$ in the divisor $\pi^*\mathcal{C}$ (resp. in $K_\pi$, the relative canonical divisor).

The following result describes a method to recover the dimension of the different eigenspaces of $H^1(X, \mathbb{C})$ with respect to the monodromy action (or deck transformation of the cover). A more general result can be found in [4, Theorem 4.4] for non-reduced divisors, but we state it here for covers associated with reduced divisors.

**Theorem 3.10 ([4, Theorem 4.4])** *The dimension of the eigenspace of $\sigma^*$ acting on $H^1(X; \mathbb{C})$ for the eigenvalue $e^{\frac{2i\pi k}{d}}$, $0 < k < d$, equals* $\dim \operatorname{coker} \pi^{(k)} + \dim \operatorname{coker} \pi^{(d-k)}$ *where*

$$\pi^{(k)} : H^0\left(\mathbb{P}^2_\omega, \mathcal{O}_{\mathbb{P}^2_w}\left(kH + K_{\mathbb{P}^2_w}\right)\right) \longrightarrow \bigoplus_{P \in \operatorname{Sing} \mathcal{C}} \frac{\mathcal{O}_{\mathbb{P}^2_\omega, P}\left(kH + K_{\mathbb{P}^2_\omega}\right)}{\mathcal{M}^{(k)}_{\mathcal{C},P}},$$

*is naturally defined given H a divisor of degree* 1, *$K_{\mathbb{P}^2_\omega}$ denotes the canonical divisor, and $\mathcal{M}^{(k)}_{\mathcal{C},P}$ is the following $\mathcal{O}_{\mathbb{P}^2_\omega, P}$-module of quasi-adjunction*

$$\mathcal{M}^{(k)}_{\mathcal{C},P} := \left\{ g \in \mathcal{O}_{\mathbb{P}^2_\omega, P}\left(kH + K_{\mathbb{P}^2_\omega}\right) \;\middle|\; \operatorname{mult}_{E_v} \pi^* g > \frac{k m_v}{d} - \nu_v,\ \forall v \in \Gamma_P \right\}.$$

Note that the module of quasi-adjunction $\mathcal{M}^{(k)}_{\mathcal{C},P}$ is a submodule of the module of equivariant germs $\mathcal{O}_{\mathbb{P}^2_\omega, P}(\ell)$ for some $\ell = 0, \ldots, d-1$ as defined in Sect. 2.1, namely, $\ell$ is the local class of the divisor $kH + K_{\mathbb{P}^2_\omega}$ at $P$. Our purpose will be to calculate $\dim \operatorname{coker} \pi^{(k)} + \dim \operatorname{coker} \pi^{(d-k)}$ for certain $k$ and $d = 3(\alpha_1\alpha_2 + \alpha_3)$ for the $d$-cyclic cover of the curves in the family presented in Sect. 3.1.

Under the conditions of Theorem 3.9, that is, $\alpha_1\alpha_2\alpha_3\beta_1\beta_2$ odd, let us consider the curve $\tilde{\mathcal{C}}_\lambda := \tilde{\Phi}^*_{\omega,\beta_1,\beta_2} \mathcal{C}_\lambda$ as defined in Sect. 3.1. This curve has, in general, three singular points at the vertices $P_x, P_y, P_z$. Recall that for $\zeta := \exp \frac{2i\pi}{3}$ an easy computation given in Corollary 3.5 shows that the fundamental group of $\mathbb{P}^2_\omega \setminus \tilde{\mathcal{C}}_\zeta$ is abelian and hence the first cohomology group of any cyclic cover ramified along $\tilde{\mathcal{C}}_\zeta$ vanishes.

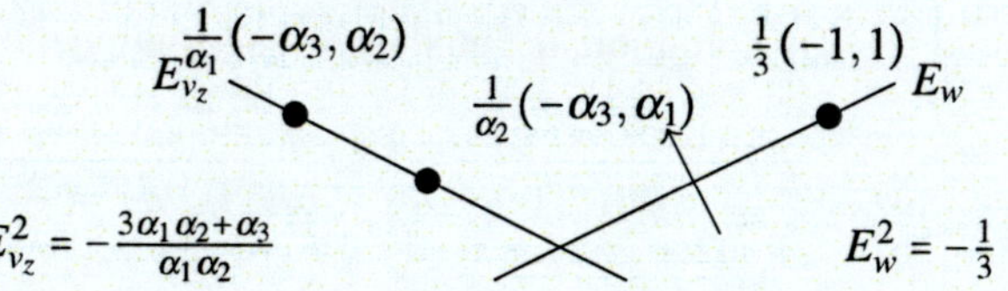


**Fig. 2** A $\mathbb{Q}$-resolution of $(\tilde{C}, P_z)$

In order to understand the maps $\pi^{(k)}$ and the corresponding modules of quasi-adjunction $\mathcal{M}^{(k)}_{\tilde{C},P}$ described in Theorem 3.10 one needs to study the singular points of $\tilde{C} := \tilde{C}_1$ in $\mathbb{P}^2_\omega$. Recall that $\operatorname{Sing}\tilde{C} \supseteq \{P_x, P_y, P_z\}$. More precisely, we will restrict our attention to the case $\frac{k}{d} = \frac{5}{6}$. Since $d = 3(\alpha_1\alpha_2 + \alpha_3)$, the degrees of the curves involved in $\pi^{(k)}$ is

$$d_k = \frac{5d}{6} - (\alpha_1 + \alpha_2 + \alpha_3) = \frac{5\alpha_1\alpha_2 + 3\alpha_3}{2} - (\alpha_1 + \alpha_2).$$

**Proposition 3.11** *A $\mathbb{Q}$-resolution of $(\tilde{C}, P_z)$ has a dual graph with two vertices and its exceptional set is shown in Fig. 2. Then $\mathcal{M}^{(k)}_{\tilde{C},P_z}$, $k = \frac{5d}{6}$, is defined by the following conditions on germs $g \in \mathcal{O}_{\mathbb{P}^2_\omega,P_z}(d_k)$:*

$$\operatorname{mult}_{E_{v_z}} \pi^* g \geq \frac{5\alpha_1\alpha_2 - 2(\alpha_1 + \alpha_2)}{2\alpha_3} + \frac{1}{2}, \quad \operatorname{mult}_{E_w} \pi^* g \geq \frac{15\alpha_1\alpha_2 - 6(\alpha_1 + \alpha_2)}{2\alpha_3} + 2.$$

***Proof*** The result is purely local, so one can assume $C$ is the cubic $zx^2 - (y+x)^3 = 0$ at the flex $[1 : -1 : 0]$. Then, the local equation of $\tilde{C}$ at $[0 : 0 : 1]_\omega$, regarded as $[(0, 0)] \in \frac{1}{\alpha_3}(\alpha_1, \alpha_2)$, is $x^{\beta_1}y^{\beta_2+2\alpha_1} - (x^{\alpha_2} + y^{\alpha_1})^3 = 0$. Note that $\alpha_1\beta_1 + \alpha_2(\beta_2 + 2\alpha_1) = 3\alpha_1\alpha_2 + \alpha_3 > 3\alpha_1\alpha_2$. Hence the Newton polygon of this equation is a segment of slope $-\frac{\alpha_1}{\alpha_2}$, and we perform an $(\alpha_1, \alpha_2)$-blowing-up. Since we start from a cyclic point one chart of this blow-up is given by

$$(x, y) \mapsto (x^{\frac{\alpha_1}{\alpha_3}}, x^{\frac{\alpha_2}{\alpha_3}} y),$$

i.e., the total transform is $x^{\frac{3\alpha_1\alpha_2}{\alpha_3}}(xy^{\beta_2+2\alpha_1} - (1 + y^{\alpha_1})^3) = 0$. We denote this exceptional divisor as $E_{v_z}$. Hence $m_{v_z} = \frac{3\alpha_1\alpha_2}{\alpha_3}$ and after a change of coordinates the strict transform (through a smooth ambient point) has equation $x - y^3 = 0$.

One can check that the multiplicity of the relative canonical divisor is $\nu_{v_z} = \frac{\alpha_1+\alpha_2}{\alpha_3}$. To complete the resolution, we perform a $(3, 1)$-blow up, producing a new component $E_w$ for which $m_w = 3\left(\frac{3\alpha_1\alpha_2}{\alpha_3} + 1\right)$ and $\nu_w = 3\frac{\alpha_1+\alpha_2}{\alpha_3} + 1$.

By definition, the module of quasi-adjunction $\mathcal{M}^{(k)}_{\tilde{C},P_z}$ is a submodule of

$$\mathcal{O}_z(d_k) := \mathcal{O}_{\mathbb{P}^2_\omega,P_z}(d_k), \quad d_k = \frac{5\alpha_1\alpha_2 + 3\alpha_3}{2} - (\alpha_1 + \alpha_2).$$

given by the germs $g \in \mathcal{O}_z(d_k)$ satisfying

$$\begin{aligned} \operatorname{mult}_{E_{v_z}} \pi^* g &> \frac{km_{v_z}}{d} - \nu_{v_z} = \frac{5\alpha_1\alpha_2 - 2(\alpha_1 + \alpha_2)}{2\alpha_3}, \\ \operatorname{mult}_{E_w} \pi^* g &> \frac{km_w}{d} - \nu_w = \frac{15\alpha_1\alpha_2 + 3\alpha_3 - 6(\alpha_1 + \alpha_2)}{2\alpha_3}. \end{aligned} \tag{3.6}$$

Finally, note that the class of $g$ imposes extra conditions, namely, if $H = V(h)$, $h \in \mathcal{O}_z(1)$, then $\operatorname{mult}_{E_v} \pi^*\left(\frac{g}{h^{d_k}}\right)$ must be an integer for $v \in \{v_z, w\}$. Using (3.6) we can write $\operatorname{mult}_{E_{v_z}} \pi^* g = \frac{5\alpha_1\alpha_2 - 2(\alpha_1+\alpha_2)}{2\alpha_3} + \varepsilon_{v_z}$, for some $\varepsilon_{v_z} \in \mathbb{Q}_{>0}$. Hence,

$$\operatorname{mult}_{E_{v_z}} \pi^*\left(\frac{g}{h^{d_k}}\right) = \frac{5\alpha_1\alpha_2 - 2(\alpha_1 + \alpha_2)}{2\alpha_3} + \varepsilon_{v_z} - \frac{d_k}{\alpha_3} = \varepsilon_{v_z} - \frac{3}{2}\alpha_3 \in \mathbb{Z}.$$

This implies $\varepsilon_{v_z} = \frac{1}{2} + n_{v_z}$, $n_{v_z} \in \mathbb{Z}_{\geq 0}$. Analogously for $v = w$ one obtains

$$\operatorname{mult}_{E_w} \pi^*\left(\frac{g}{h^{d_k}}\right) = \frac{15\alpha_1\alpha_2 + 3\alpha_3 - 6(\alpha_1 + \alpha_2)}{2\alpha_3} + \varepsilon_w - 3\frac{d_k}{\alpha_3} = \varepsilon_w + \frac{3}{2} \in \mathbb{Z},$$

which implies $\varepsilon_w = \frac{1}{2} + n_w$, $n_w \in \mathbb{Z}_{\geq 0}$ and this ends the proof.

**Proposition 3.12** *A $\mathbb{Q}$-resolution of $(\tilde{C}, P_x)$ is obtained with one weighted blow-up. Then $\mathcal{M}^{(k)}_{\tilde{C},P_x}$, $k = \frac{5d}{6}$, is defined by the following condition on germs $g \in \mathcal{O}_{\mathbb{P}^2_\omega,P_x}(d_k)$:*

$$\operatorname{mult}_{E_{v_x}} \pi^* g \geq \frac{3}{\gcd(3, \alpha_1)} \cdot \frac{\alpha_1 + 3\beta_2 - 2}{2\alpha_1} + 1$$

***Proof*** We follow the same ideas as in the proof of Proposition 3.11. Locally we work with the cubic $xy^2 - (y+z)^3 = 0$ (this cubic has a flex at $[0:1:-1]$). Then, the local equation of $\tilde{C}$ at $[1:0:0]_\omega$, regarded as $[(0,0)] \in \frac{1}{\alpha_1}(\alpha_2, \alpha_3) = \frac{1}{\alpha_1}(1, \beta_2)$, is $y^{\alpha_1}z^3 - (z + y^{\beta_2})^3 = 0$. We can change the coordinates (not affecting the action) where the equation becomes $y^{\alpha_1}(z - y^{\beta_2})^3 - z^3 = 0$. In these new coordinates the Newton polygon is non-degenerated and the singularity is resolved with a blowing-up with exceptional component $E_{v_x}$. Its weight is $(3, \alpha_1 + 3\beta_2)$ if $\gcd(3, \alpha_1) = 1$ and $\left(1, \frac{\alpha_1}{3} + \beta_2\right)$ otherwise.

The invariants are

$$m_{v_x} = 3\frac{\alpha_1 + 3\beta_2}{\alpha_1}, \quad \nu_{v_x} = \frac{\alpha_1 + 3\beta_2 + 3}{\alpha_1}.$$

Let us compute the quasi-adjunction module $\mathcal{M}^{(k)}_{\mathcal{D},P_x}$, as a submodule of $\mathcal{O}_x(\bar{d}_k) := \mathcal{O}_{\mathbb{P}^2_\omega, P_x}(\bar{d}_k)$, where $\bar{d}_k$ is such that $\alpha_2\bar{d}_k \equiv d_k \bmod \alpha_1$, which implies that $\bar{d}_k \equiv \frac{\alpha_1+3\beta_2-2}{2}$. The condition for a germ $g \in \mathcal{O}_x(\bar{d}_k)$ to be in $\mathcal{M}^{(k)}_{\mathcal{D},P_x}$ is:

$$\operatorname{mult}_{E_{v_x}} \pi^* g > 3\frac{\alpha_1 + 3\beta_2 - 2}{2\alpha_1}.$$

As above, the restriction given by $g \in \mathcal{O}_x(\bar{d}_k)$ leads to

$$\operatorname{mult}_{E_{v_x}} \left(\pi^* \frac{g}{h^{\bar{d}_k}}\right) = 3\frac{\alpha_1 + 3\beta_2 - 2}{2\alpha_1} + \varepsilon_{v_x} - 3\frac{\bar{d}_k}{\alpha_1} = \varepsilon_{v_x} \in \mathbb{Z}.$$

Hence, $\varepsilon_{v_x} \in \mathbb{Z}_{>0}$.

**Proposition 3.13** *Let $g(x, y, z)$ be a weighted homogeneous polynomial in* $\ker \pi^{(k)}$ *with* $\deg_\omega g = \frac{5(\alpha_1\alpha_2+\alpha_3)-2(\alpha_1+\alpha_2+\alpha_3)}{2}$.

*Then, there is a weighted homogeneous polynomial $f$, $\deg_\omega f = \alpha_1\alpha_2+\alpha_3$, such that $g(x, y, z) = x^{\frac{1}{2}(\alpha_2+\beta_1-2)} y^{\frac{1}{2}(\alpha_1+\beta_2-2)} f(x, y, z)$ and*

$$\operatorname{mult}_{E_{v_z}} \pi^* f(x, y, 1) \geq \frac{\alpha_1\alpha_2}{\alpha_3}, \qquad \operatorname{mult}_{E_w} \pi^* f(x, y, 1) \geq \frac{3\alpha_1\alpha_2}{\alpha_3} + \frac{1}{2},$$

$$\operatorname{mult}_{E_{v_x}} \pi^* f(1, y, z) \geq \frac{3\beta_2}{\gcd(3, \alpha_1)\alpha_1} + 1, \ \operatorname{mult}_{E_{v_y}} \pi^* f(x, 1, z) \geq \frac{3\beta_1}{\gcd(3, \alpha_2)\alpha_2} + 1.$$

***Proof*** The exponent of $x^n$ as a factor of $g$ is given by the maximal value $n \in \mathbb{Z}_{>0}$, such that the divisor $\mathrm{V}(g) - nY$ is effective. Using the generalization of Noether's multiplicity Theorem in this context, see [8, Theorem 4.3(4)], and Proposition 3.11 one obtains

$$\begin{aligned}(\mathrm{V}(g) \cdot Y)_{P_z} &\geq \frac{(\operatorname{mult}_{E_{v_z}} \pi^* g) \cdot (\operatorname{mult}_{E_{v_z}} \pi^* y)\alpha_3}{\alpha_1\alpha_2} \\ &\geq \frac{(5\alpha_1\alpha_2 - 2(\alpha_1 + \alpha_2))}{2\alpha_1\alpha_3} + \frac{1}{2\alpha_1} = \frac{1}{2\alpha_1\alpha_3}(5\alpha_1\alpha_2 - 2\alpha_1 - 2\alpha_2 + \alpha_3).\end{aligned}$$

Hence,

$$((\mathrm{V}(g) - nY) \cdot Y)_{P_z} \geq \frac{1}{2\alpha_1\alpha_3}(5\alpha_1\alpha_2 - 2\alpha_1 - 2\alpha_2(1 + n) + \alpha_3).$$

Analogously, at $P_x$ one can use Proposition 3.12 to obtain

$$(\mathrm{V}(g)\cdot Y)_{P_x} \geq \left(3\frac{\alpha_1+3\beta_2-2}{2\alpha_1}+1\right)\frac{1}{\alpha_1+3\beta_2}=\frac{5\alpha_1+9\beta_2-6}{2\alpha_1(\alpha_1+3\beta_2)},$$

regardless of the value of $\gcd(3,\alpha_1)$. Hence,

$$((\mathrm{V}(g)-nY)\cdot Y)_{P_x} \geq \frac{5\alpha_1+9\beta_2-6-6n}{2\alpha_1(\alpha_1+3\beta_2)}.$$

Then a global computation of the intersection multiplicity can be bounded by

$$\begin{aligned}
((\mathrm{V}(g)-nY)\cdot Y)_{\mathbb{P}^2_\omega} &\geq ((\mathrm{V}(g)-nY)\cdot Y)_{P_z}+((\mathrm{V}(g)-nY)\cdot Y)_{P_x}\\
&\geq \frac{1}{2\alpha_1\alpha_3}(5\alpha_1\alpha_2-2\alpha_1-2\alpha_2(1+n)+\alpha_3)+\frac{5\alpha_1+9\beta_2-6-6n}{2\alpha_1(\alpha_1+3\beta_2)}\\
=\frac{1}{2\alpha_1\alpha_3}(5(\alpha_1\alpha_2+\alpha_3)&-2(\alpha_1+\alpha_2+\alpha_3)-2n\alpha_2)-\frac{1}{\alpha_1}+\frac{5\alpha_1+9\beta_2-6-6n}{2\alpha_1(\alpha_1+3\beta_2)}\\
&=\frac{\deg_\omega(\mathrm{V}(g)-nY)\cdot\deg_\omega(Y)}{\alpha_1\alpha_2\alpha_3}+3\frac{\alpha_1+\beta_2-2-2n}{2\alpha_1(\alpha_1+3\beta_2)}.
\end{aligned}$$

By Bézout's Theorem for weighted projective planes, $n=\frac{1}{2}(\alpha_1+\beta_2-2)$. The same calculation applies to the divisor $X$. This shows that $g=x^m y^n f(x,y,z)$, $m=\frac{1}{2}(\alpha_2+\beta_1-2)$ where

$$\begin{aligned}
\deg(f) &= \frac{1}{2}(5(\alpha_1\alpha_2+\alpha_3)-2(\alpha_1+\alpha_2+\alpha_3)-2n\alpha_2-2m\alpha_1)\\
&= \frac{1}{2}(5(\alpha_1\alpha_2+\alpha_3)-2(\alpha_1+\alpha_2+\alpha_3)-(\alpha_1+\beta_2-2)\alpha_2-(\alpha_2+\beta_1-2)\alpha_1)\\
&= \alpha_1\alpha_2+\alpha_3.
\end{aligned}$$

The last equality follows from $\alpha_1\beta_1+\alpha_2\beta_2=\alpha_1\alpha_2+\alpha_3$.

The last part follows immediately from Propositions 3.11 and 3.12 and the additivity properties of the multiplicity.

The local algebraic information obtained in this section will help us effectively study the morphism $\pi^{(k)}$ described in Theorem 3.10. Let us use the notation introduced before Theorem 3.9 and at the beginning of this section, let us also denote by $X_1$ (resp. $X_2$) the cyclic cover of $\mathbb{P}^2_\omega$ of order $d=3(\alpha_1\alpha_2+\alpha_3)$ ramified along $\tilde{\Phi}_1^*(C)$ (resp. $\tilde{\Phi}_2^*(C)$). Finally, denote by $L_i^{(k)}$ the invariant part of $H^1(X_i,\mathcal{O}_{X_i})$ with respect to the action of the monodromy by multiplication by $\exp\frac{2\pi ik}{d}$. Likewise, we denote by $\pi_i^{(k)}$ the map described in Theorem 3.10 for the curve $\tilde{\Phi}_i^*(C)$. The discussion above shows the following.

**Proposition 3.14** *If the product* $\alpha_1\alpha_2\alpha_3\beta_1\beta_2$ *is odd and* $\frac{k}{d} = \frac{5}{6}$, *then* $\dim\ker\pi_1^{(k)} = 0$ *and* $\dim\ker\pi_2^{(k)} = 1$.

***Proof*** By Proposition 3.13, the image by $\Phi_i$ of $V(f)$ is a line passing through the three flexes. The existence of this line for $\tilde{\Phi}_2^*(\mathcal{C})$ but not for $\tilde{\Phi}_1^*(\mathcal{C})$ ends the proof.

The machinery developed in this section allows one to give an alternative proof of Theorem 3.9 which is independent of fundamental group calculations.

***Proof of Theorem 3.9*** Since the curves $\tilde{\Phi}_1^*(\mathcal{C})$ and $\tilde{\Phi}_2^*(\mathcal{C})$ have the same combinatorics and the same local type of singularities, the target space for $\pi_1^{(k)}$ and $\pi_2^{(k)}$ are the same. Therefore Proposition 3.14 implies $\dim\operatorname{coker}\pi_1^{(k)} = 1 + \dim\operatorname{coker}\pi_2^{(k)}$ for $\frac{k}{d} = \frac{5}{6}$. By Theorem 3.10, $\dim\operatorname{coker}\pi_i^{(k)} = \dim L_i^{(k)}$ is a birational invariant of $X_i$ and thus $X_1 \not\cong X_2$, which implies that the fundamental groups of $\mathbb{P}^2_\omega \setminus \tilde{\Phi}_1^*(\mathcal{C})$ and $\mathbb{P}^2_\omega \setminus \tilde{\Phi}_2^*(\mathcal{C})$ are not isomorphic and thus $(\tilde{\Phi}_1^*(\mathcal{C}), \tilde{\Phi}_2^*(\mathcal{C}))$ forms a Zariski pair.

## 4 Some Rational Cuspidal Curves on Weighted Projective Planes

The study of rational cuspidal curves in $\mathbb{P}^2$ is a classical subject. There is an extensive literature about them, and we recommend the beautiful paper [15] reviewing this topic, the most relevant conjectures, and bibliography. Two outstanding conjectures have been solved recently by Koras and Palka: the Nagata-Coolidge conjecture [19], that is, any rational cuspidal curve can be transported to a line via a Cremona transformation and such curves can have at most four singular points [20]. There is a strong knowledge of such curves in $\mathbb{P}^2$ which have helped for the solution of these conjectures and other important problems, like the semigroup conjecture in [15], which was proven in [10].

Only one rational cuspidal curve in $\mathbb{P}^2$ possesses four cusps: a quintic curve with singular locus $\mathbb{A}_6 + 3\mathbb{A}_2$. There are many of them with three singular points, see [16] for an infinite family. The simplest one is the cuspidal quartic with three ordinary cusps. The standard Cremona transformation is a way to produce this curve, namely, the standard Cremona transformation of a smooth conic with respect to three of its tangent lines produces a tricuspidal quartic. Note that the blowing-up at the vertices does not affect the curve, and the blowing-downs produce the three cusps.

## 4.1 Rational Cuspidal Curves via Weighted Cremona Transformations

In this section, we will study the strict transforms of the above tritangent conic using the inverse of the weighted Cremona transformations introduced in Sect. 2.4. As a first stage, let us compute their fundamental groups. As in Sect. 3, let us start with the arrangement of a smooth conic and three lines, giving a presentation which contains suitable meridians for all the components.

Let $\mathcal{C}$ be a smooth conic and let $X, Y, Z$ be three distinct tangent lines to $\mathcal{C}$. If the equations of the lines are $x = 0$, $y = 0$, $z = 0$, respectively, then the equation of $\mathcal{C}$ (up to a suitable change of coordinates) is

$$x^2 + y^2 + z^2 - 2(yz + xz + xy) = 0.$$

The fundamental group of the complement of the smooth conic and three tangent lines is the Artin group of the triangle $T(4, 4, 2)$ (i.e. [9]). However, for our purposes, it is more suitable to use a presentation with a more geometrical interpretation. We present it here for completeness, but its proof is immediate using the classical Zariski-van Kampen method (as in [11]).

**Proposition 4.1** *The fundamental group of* $\mathbb{P}^2 \setminus (\mathcal{C} \cup X \cup Y \cup Z)$ *is isomorphic to*

$$\langle c, \ell_x, \ell_y, \ell_z \mid [\ell_x, \ell_y] = [\ell_x, \ell_z] = [\ell_y^c, \ell_z] = \ell_y c \ell_x c \ell_z = 1\rangle. \tag{4.1}$$

*The element* $c$ *is a meridian of* $\mathcal{C}$, *and* $\ell_x, \ell_y, \ell_z$ *are meridians of* $X, Y, Z$, *respectively. Moreover,* $(\ell_x, \ell_y)$ *are meridians close to* $[0 : 0 : 1]$, $(\ell_x, \ell_z)$ *are meridians close to* $[0 : 1 : 0]$, *and* $(\ell_y^c = c^{-1}\ell_y c, \ell_z)$ *are meridians close to* $[0 : 0 : 1]$.

Considering $u := c\ell_z$, the above presentation of $\pi_1(\mathbb{P}^2 \setminus (\mathcal{C} \cup X \cup Y \cup Z))$ can be alternatively written as:

$$\langle u, \ell_x, \ell_y, \ell_z \mid [\ell_x, \ell_y] = [\ell_x, \ell_z] = [\ell_y^u, \ell_z] = u\ell_y u\ell_x \ell_z^{-1} = 1\rangle. \tag{4.2}$$

As in Sect. 3, fixing $\omega, \beta_1, \beta_2$, we consider the birational map $\Phi$ and we denote by $\tilde{\mathcal{C}}$ the strict transform of $\mathcal{C}$ by $\Phi$.

**Proposition 4.2** *Let* $d := \gcd(\alpha_1+2\beta_2, \alpha_2+2\beta_1)$. *Then* $\pi_1(\mathbb{P}^2_\omega \setminus \tilde{\mathcal{C}})$ *is the semidirect product* $(\mathbb{Z}/d)A \ltimes (\mathbb{Z}/2(\alpha_1\alpha_2+\alpha_3))B$ *where* $BAB^{-1} = A^{-1}$. *Hence the group has size* $2d(\alpha_1\alpha_2 + \alpha_3)$ *and it is abelian if and only if* $d = 1$.

*Remark 4.3* Note that $d$ is odd, since $\alpha_1, \alpha_2$ cannot be simultaneously even. Moreover,

$$\begin{aligned}\gcd(d, \alpha_1) &= \gcd(\alpha_1, \beta_2, \alpha_2 + 2\beta_1) = \gcd(\alpha_1, \alpha_2\beta_2, \alpha_2 + 2\beta_1)\\ &= \gcd(\alpha_1, \alpha_3, \alpha_2 + 2\beta_1) = 1\end{aligned}$$

and analogously $\gcd(d, \alpha_2) = 1$. Note also

$$\gcd(d, \beta_1) = \gcd(d, \beta_1, \alpha_1 + 2\beta_2, \alpha_2) = 1,$$

and hence $\gcd(d, \beta_2) = 1$. The following congruences can easily be checked:

$$\alpha_1\beta_1 \equiv -2\beta_1\beta_2 \equiv \alpha_2\beta_2 \mod d.$$

Moreover,

$$\begin{aligned}\gcd(d, \alpha_1\alpha_2 + \alpha_3) &= \gcd(d, \alpha_1\alpha_2 + \alpha_3, \alpha_2(\alpha_1 + 2\beta_2)) \\ &= \gcd(d, \alpha_1\alpha_2 + \alpha_3, \alpha_1(\alpha_2 - 2\beta_1)) \\ &= \gcd(d, \alpha_1\alpha_2 + \alpha_3, 2\alpha_1\beta_1) = 1.\end{aligned}$$

***Proof of Proposition* 4.2** The presentation of $\pi_1(\mathbb{P}^2_\omega \setminus \tilde{C})$ is obtained from (4.2) by adding the relations which *kill* the meridians of the exceptional divisors $E_x$, $E_y$, $E_z$:

$$\ell_x\ell_y = \ell_x^{\alpha_1}\ell_z^{\beta_2} = u^{-1}\ell_y^{\alpha_2}u\ell_z^{\beta_1} = 1. \tag{4.3}$$

Let us first check that the abelianization of this quotient is $\mathbb{Z}/2(\alpha_1\alpha_2 + \alpha_3)$. We will denote by $[\bullet]$ the class of $\bullet$ in the abelianization. Note that $[\ell_z] = [u]^2$; moreover, using Bézout's identity and the equations in (4.3), both $[\ell_x]$ and $[\ell_y]$ can be expressed in terms of $[\ell_z]$. Hence, the abelianization is cyclic. A presentation matrix in terms of the generators $[\ell_y]$, $[u]$ is given by

$$\begin{pmatrix} \alpha_2 & 2\beta_1 \\ -\alpha_1 & 2\beta_2 \end{pmatrix}$$

whose determinant, $2(\alpha_1\alpha_2 + \alpha_3)$, is the size of the abelianization.

Let us study now the group itself. Note first that $\ell_y$ can be eliminated from (4.3) as $\ell_y = \ell_x^{-1}$. Let us check that $u^2$ is central. The last relation in (4.2) can be written as

$$u^2 = \ell_z\ell_x^{-1}(u^{-1}\ell_x u).$$

We deduce that $u^2$ commutes with $\ell_z$, since it commutes with each factor; hence $u^2$ also commutes with $u\ell_z u^{-1}$. Also note that

$$\ell_x^{\alpha_1} = \ell_z^{-\beta_2}, \quad \ell_x^{\alpha_2} = u\ell_z^{\beta_1}u^{-1} \Longrightarrow \ell_x = \ell_z^{-\hat{\alpha}_1\beta_2}u\ell_z^{-\hat{\alpha}_2\beta_1}u^{-1}.$$

Then $u^2$ commutes with $\ell_x$ and it is central.

Using the last relation in (4.2) $\ell_z$ can also be eliminated as $\ell_z = u\ell_x^{-1}u\ell_x$. The presentation of the group becomes:

$$\langle u, \ell_x \mid [\ell_x, u\ell_x^{-1}u] = [\ell_x^u, u\ell_x^{-1}u\ell_x] = \ell_x^{\alpha_1}(u\ell_x^{-1}u\ell_x)^{\beta_2} = u^{-1}\ell_x^{-q}u(u\ell_x^{-1}u\ell_x)^{\beta_1} = 1\rangle$$

which can be further simplified using the centrality of $u^2$:

$$\langle u, \ell_x \| [\ell_x, u^2] = [\ell_x, u\ell_x u^{-1}] = u^{2\beta_2}\ell_x^{\alpha_1+\beta_2}(u\ell_x u^{-1})^{-\beta_2} = u^{2\beta_1}(u\ell_x u^{-1})^{-(\alpha_2+\beta_1)}\ell_x^{\beta_1} = 1\rangle$$

The map $\pi_1(\mathbb{P}^2_\omega \setminus \tilde{C}) \to \mathbb{Z}/2$ given by $u \mapsto 1$ and $\ell_x, \ell_z \mapsto 0$ is well defined. A presentation of its kernel $K$ is obtained using Reidemeister-Schreier method. The generators are $X_0 := \ell_x$, $X_1 := u\ell_x u^{-1}$, and $U := u^2$. The first two relations imply that the group is abelian; the other relations yield:

$$u^{2\beta_2}\ell_x^{\alpha_1+\beta_2}(u\ell_x u^{-1})^{-\beta_2} = 1 \Longrightarrow U^{\beta_2}X_1^{\alpha_1+\beta_2}X_0^{-\beta_2} = U^{\beta_2}X_0^{\alpha_1+\beta_2}X_1^{-\beta_2} = 1,$$

$$u^{2\beta_1}(u\ell_x u^{-1})^{-(\alpha_2+\beta_1)}\ell_x^{\beta_1} = 1 \Longrightarrow U^{\beta_1}X_1^{-(\alpha_2+\beta_1)}X_0^{\beta_1} = U^{\beta_1}X_0^{-(\alpha_2+\beta_1)}X_1^{\beta_1} = 1.$$

Let us express these relations in a matrix (the rows represent the relations and the columns stand for the generators). These relations become (recall $d = \gcd(\alpha_1 + 2\beta_2, \alpha_2 + 2\beta_1)$):

$$\left(\begin{array}{ccc} d & -d & 0 \\ -\beta_2 & \alpha_1+\beta_2 & \beta_2 \\ \beta_1 & -(\alpha_2+\beta_1) & \beta_1 \\ \hline X_0 & X_1 & U \end{array}\right) \sim \left(\begin{array}{ccc} d & 0 & 0 \\ -\beta_2 & \alpha_1 & \beta_2 \\ \beta_1 & -\alpha_2 & \beta_1 \\ \hline X_0X_1^{-1} & X_1 & U \end{array}\right).$$

The determinant of this matrix is $d(\alpha_1\alpha_2 + \alpha_3)$ which is the size of $K$; the greatest common divisor of the 2-minors divides $d$ and $2\alpha_1\beta_1$, i.e., it is 1. Hence, the group $K$ is cyclic of order $d(\alpha_1\alpha_2 + \alpha_3)$. One also has that $D := X_0X_1^{-1}$ is of order $d$. Considering the product of the second relation to the power $\alpha_2$ and the third relation to the power $\alpha_1$ one obtains

$$1 = D^{\alpha_1\beta_1-\alpha_2\beta_2}U^{\alpha_1\alpha_2+\alpha_3} = U^{\alpha_1\alpha_2+\alpha_3}.$$

>From the order in the abelianization, we deduce that $U$ is of order $\alpha_1\alpha_2 + \alpha_3$. Hence $K$ is the direct product of the cyclic group of order $d$ generated by $D$ and the cyclic group of order $\alpha_1\alpha_2 + \alpha_3$ generated by $U$. The conjugation by $u$ satisfies $uUu^{-1} = U$ and $uDu^{-1} = D^{-1}$. The result follows.

Embedded $\mathbb{Q}$-resolutions of the singularities of these curves can be computed as in Propositions 3.11 and 3.12, see Fig. 3.

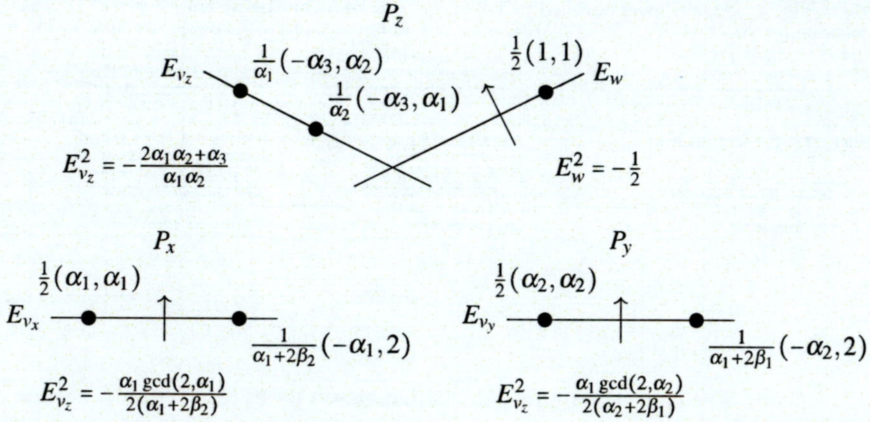


**Fig. 3** Singularities of the rational cuspidal curves

## 4.2 Rational Cuspidal Curves via Weighted Kummer Covers

There is another simple way to produce rational cuspidal curves in weighted projective planes from this arrangement of curves. It is quite simple but it will be shown to be useful in the upcoming sections. Let $d_1, d_2, d_3$ be pairwise coprime integers and let $\omega := (e_1, e_2, e_3)$, where $e_i := d_j d_k$, $\{i, j, k\} = \{1, 2, 3\}$. Following Sect. 2.2 note that $\eta = (1, 1, 1)$ and thus there is an isomorphism $\mathbb{P}^2_\omega \to \mathbb{P}^2$ given by $[x : y : z]_\omega \mapsto [x^{d_1} : y^{d_2} : z^{d_3}]$. This map gives a geometrical interpretation to the group

$$G_{d_1,d_2,d_3} := \pi_1(\mathbb{P}^2 \setminus (\mathcal{C} \cup X \cup Y \cup Z))/\langle \ell_x^{d_1} = \ell_y^{d_2} = \ell_z^{d_3} = 1 \rangle.$$

as the orbifold group of $\mathbb{P}^2_\omega$ with respect to the curve $\mathcal{C} \cup X \cup Y \cup Z$ and index $e(\mathcal{C}) = 0, n(X) = d_1, n(Y) = d_2, n(Z) = d_3$ as defined in [5]. Briefly, if $X$ is a smooth projective surface, $D = D_1 \cup \ldots \cup D_s$ is a normal crossing union of smooth hypersurfaces, and $n_i := n(D_i) \in \mathbb{Z}_{\geq 0}$, then one can define the *orbifold fundamental group* $\pi_1^{\text{orb}}(X)$ of $X$ with respect to $D$ with indices $n_i$ as the quotient of the group $\pi_1(X \setminus D)$ by the normal subgroup generated by $\gamma_i^{n_i}$, where $\gamma_i$ is a meridian of $D_i$. If $D$ do not have normal crossings, then one resolves to a normal crossing divisor by blowing up points and defines the index at an exceptional divisor as the least common multiple of the indices of the components passing through the point (in this context we set $\text{lcm}(0, n) = 0$).

**Proposition 4.4** *The abelianization of the group* $G_{d_1,d_2,d_3}$ *is* $\mathbb{Z}/2d_1d_2d_3$. *The group is abelian if and only if at least one of* $d_1, d_2, d_3$ *equals* 1.

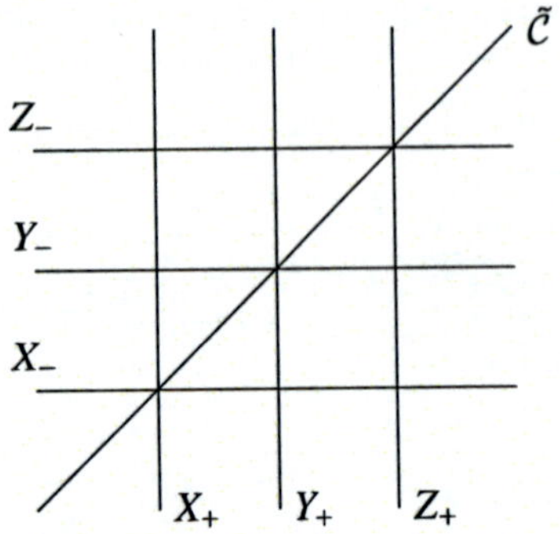


**Fig. 4** Double cover ramified along $C$

***Proof*** The computation for the abelianization is straightforward. Assume $d_3 = 1$, then

$$\begin{aligned} G_{d_1,d_2,1} &= \langle \ell_x, u \mid [\ell_x, \underbrace{u\ell_x\, u}_{v^{-1}}] = 1, \ell_x^{d_1} = (u\ell_x u)^{d_2} = 1\rangle \\ &= \langle \ell_x, v \mid [\ell_x, v^2] = 1, \ell_x^{d_1} = \ell_x^{d_2} v^{2d_2} = 1\rangle. \end{aligned}$$

Using Bézout's identity we can express $\ell_x$ in terms of $v$ and the result follows.

For the case $(d_1, d_2, d_3) \neq (1, 1, 1)$, let us consider the double cover of $\mathbb{P}^2$ ramified along $C$. It is $\mathbb{P}^1 \times \mathbb{P}^1$ and the preimage $\tilde{C}$ of $C$ is the diagonal; the three lines are transformed in pairs of vertical-horizontal lines $X_\pm, Y_\pm, Z_\pm$ intersecting $\tilde{C}$, see Fig. 4.

Note that the fundamental group of $\mathbb{P}^1 \times \mathbb{P}^1 \setminus (\tilde{C} \cup X_+ \cup X_- \cup Y_+ \cup Y_- \cup Z_+ \cup Z_-)$ is isomorphic to the fundamental group of the projective complement of Ceva's arrangement. The kernel $K_{d_1,d_2,d_3}$ of the map $G_{d_1,d_2,d_3} \to \mathbb{Z}/2$, $u \mapsto 1$ and the images of the other generators vanish, it is an index 2 subgroup of $G_{d_1,d_2,d_3}$ and it is an orbifold fundamental group for the above configuration in $\mathbb{P}^1 \times \mathbb{P}^1$; the projection on each component induces orbifold morphisms onto $\mathbb{P}^1_{d_1,d_2,d_3}$, where $\mathbb{P}^1_{d_1,d_2,d_3}$ is an orbifold modeled on $\mathbb{P}^1$, with three quotient points of order $d_1, d_2, d_3$, respectively, and its orbifold fundamental group is isomorphic to $\langle \mu_1, \mu_2, \mu_3 \mid \mu_1^{d_1} = \mu_2^{d_2} = \mu_3^{d_3} = \mu_3\mu_2\mu_1 = 1\rangle$, a triangle group. The combination of the two projections induces an epimorphism of $K_{d_1,d_2,d_3}$ onto $\pi_1^{\text{orb}}(\mathbb{P}^1_{d_1,d_2,d_3})$. The triangle is hyperbolic if $\{d_1, d_2, d_3\} \neq \{2, 3, 5\}$ and hence its group is infinite. If $\{d_1, d_2, d_3\} = \{2, 3, 5\}$, the triangle group is the alternating group $A_5$, with cardinal 60. Using `GAP` [34] we can compute the intersection of the kernels of the two projections (which is a subgroup of index 3600); its abelianization is $\mathbb{Z}^{59}$ and the result follows. The computation can be checked in https://github.com/enriqueartal/AnOrbifoldFundamentalGroup using `Sagemath` [33] and `Binder` [30].

### 4.3 A rational Cuspidal Curve with Four Cusps

We present in this section a nice example in $\mathbb{P}^2_{(1,1,2)}$, which is a rational curve of degree 6 with 4 ordinary cusps – incidentally, note that 4 is the maximal number of singular points a rational cuspidal curve can have in $\mathbb{P}^2$. Let us explain how to construct it via Cremona transformations. Let us start with a tricuspidal quartic $\mathcal{C}_0$; this curve, dual of the nodal cubic, has a bitangent line $L$. Let $P_0 \in \mathcal{C}_0 \cap L$; its blown-up produces a ruled surface $\Sigma_1$, where the negative section $E$ is the exceptional component and $\mathcal{C}_1$, the strict transform of $\mathcal{C}_0$ has three cusps and one tangent fiber. Let us consider the Nagata transformation at $\mathcal{C}_1 \cap E$; the result is $\Sigma_2$ and the blow-down of the negative section produces $\mathbb{P}^2_{(1,1,2)}$. The strict transform $\mathcal{C}$ of $\mathcal{C}_1$ is the desired curve.

**Proposition 4.5** *The fundamental group* $\pi_1(\mathbb{P}^2_{(1,1,2)} \setminus (\mathcal{C} \cup \{P_z\}))$ *has a presentation*

$$\langle s, t, u \mid sts = tst, sus = usu, tut = utu, (stu)^2 = 1 \rangle.$$

***Proof*** Following the construction it is the fundamental group of $\Sigma_2 \setminus (\mathcal{C}_2 \cup E_2)$ where $\mathcal{C}_2$ is the strict transform of $\mathcal{C}$ and $E_2$ is the negative section. The Zariski-van Kampen method applied to the ruling yields the result.

### 4.4 Milnor Fibers

We have presented in Sect. 3 and in Sect. 4 several examples of irreducible quasi-projective curves such that their (maybe orbifold) fundamental groups are non-abelian. As a consequence their cones are quasi-homogeneous non-isolated surface singularities in $\mathbb{C}^3$ with non simply-connected Milnor fibers.

If $F \in \mathbb{C}[x, y, z]$ is a homogeneous polynomial of degree $d$, an important topological invariant is its Milnor fiber. The Milnor fiber of a homogeneous singularity is a fiber of $F : \mathbb{C}^3 \setminus F^{-1}(0) \to \mathbb{C}^*$, say $F^{-1}(1)$. The restriction to $F^{-1}(1)$ of $F$ of the natural map $\mathbb{C}^3 \setminus \{0\} \to \mathbb{P}^2$ is a $d$-cyclic cover onto the complement of the tangent cone $\mathcal{C}_d$ in $\mathbb{P}^2$, defined by an epimorphism $\pi_1(\mathbb{P}^2 \setminus \mathcal{C}_d) \to \mathbb{Z}/d$.

If $\omega$ is a weight and $F \in \mathbb{C}[x, y, z]$ is an $\omega$-quasi-homogeneous polynomial of $\omega$-degree $d$, its Milnor fiber $F = 1$ can also be recovered as a $d$-cyclic orbifold cover of $\mathbb{P}^2_\omega \setminus \mathcal{C}_d$ (the complement of the *tangent* $\omega$-*quasi-cone*) defined by an epimorphism $\pi_1^{\mathrm{orb}}(\mathbb{P}^2_\omega \setminus \mathcal{C}_d) \to \mathbb{Z}/d$. If the elements of $\omega$ are pairwise coprime and the vertices are in $\mathcal{C}_d$, then the notions of $\pi_1$ and $\pi_1^{\mathrm{orb}}$ coincide. If it is not the case, the notion of orbifold fundamental groups apply.

The curves obtained via the Cremona transformation provide homogeneous singularities whose topology is not complicated and such that the Milnor fiber has non-trivial fundamental group. The following result is a direct consequence of Proposition 4.2.

**Proposition 4.6** *Let* $\omega = (\alpha_1, \alpha_2, \alpha_3)$ *be pairwise coprime and let* $\beta_1, \beta_2$ *be such that* $\alpha_1\alpha_2 + \alpha_3 = \alpha_1\beta_1 + \alpha_2\beta_2$. *Let* $S := \{F_{\omega,\beta_1,\beta_2}(x, y, z) = 0\}$, *where*

$$F_{\omega,\beta_1,\beta_2}(x, y, z) = y^{2\alpha_1}z^2 + x^{2\alpha_2}z^2 + x^{2\beta_1}y^{2\beta_2} - 2z(x^{\alpha_2}y^{\alpha_1}z + x^{\alpha_2+\beta_1}y^{\alpha_1} + x^{\alpha_2}y^{\alpha_1+\beta_2})$$

*defines a* $\omega$*-homogeneous singularity. Then, the fundamental group of the Milnor fiber of* $S$ *is cyclic of order* $\gcd(\alpha_1 + 2\beta_2, \alpha_2 + 2\beta_1)$.

More complicated fundamental groups can be obtained by choosing the orbifold variant. Let $(d_1, d_2, d_3)$ be a triple of pairwise coprime integers, $d_i > 1$, let $\omega = (d_2d_3, d_1d_3, d_1d_2)$ be a weight, and let

$$F_{d_1,d_2,d_3}(x, y, z) = x^{2d_1} + y^{2d_2} + z^{2d_3} - 2(x^{d_1}y^{d_2} + x^{d_1}z^{d_3} + y^{d_2}z^{d_3}).$$

As a direct consequence of Proposition 4.4 we obtain the following result.

**Proposition 4.7** *The fundamental group of the Milnor fiber of* $\{F_{d_1,d_2,d_3} = 0\}$ *is infinite and non-abelian.*

# 5 Weighted Lê–Yomdin Surface Singularities

In this section we study the relationship between (weighted) projective plane curves and normal surface singularities whose link is a rational (or integral) homology sphere.

## 5.1 *The Determinant of a Normal Surface Singularity*

Let $(S, 0)$ be a germ of normal surface singularity and let $K$ be its link. It is well known that $K$ is a graph manifold whose plumbing decorated graph is the dual graph $\Gamma$ of a simple normal crossing resolution. Each vertex $v$ of $\Gamma$ is decorated with two numbers $(g_v, e_v)$, where $g_v$ is the genus of the corresponding irreducible component $E_v$ and $e_v$ is its self-intersection. Let $A$ be the intersection matrix of the graph; recall that $A$ is negative definite. In a natural way, $A$ is also the presentation matrix of an abelian group yielding the following classical result.

**Proposition 5.1** *The free part of* $H_1(K; \mathbb{Z})$ *has rank* $2\sum_v g_v + \operatorname{Rank} H_1(\Gamma; \mathbb{Z})$. *The torsion part is isomorphic to* $\operatorname{coker} A$ *and, in particular, its cardinality is* $\det(-A)$.

As a direct consequence of this, the determinant $\det(-A)$ does not depend on the resolution. This justifies the definition of the determinant of a normal surface singularity.

**Definition 5.2** The *determinant* $\det S$ of a normal surface singularity $S$ is defined as $\det(-A)$, where $A$ is the intersection matrix of any resolution of $S$.

As a consequence, one has the following combinatorial criteria to detect rational (resp. integral) homology sphere singularities, that is, surface singularities whose link is a rational (resp. integral) homology sphere.

**Corollary 5.3** *The surface singularity $S$ is a rational (resp. integral) homology sphere if and only if all $g_v$'s vanish and $\Gamma$ is a tree (resp. and* $\det S = 1$*).*

## 5.2 Superisolated and Lê–Yomdin Singularities

In [15], the authors relate hypersurface singularities whose link is a rational homology sphere with rational cuspidal curves using superisolated singularities. In our search for more examples of surface singularities whose link is a rational (or integral) homology sphere, a generalization of this method will be discussed here. For the sake of completeness we present a classical result.

**Definition 5.4** Let $(S, 0) \subset (\mathbb{C}^3, 0)$ be the germ of a hypersurface singularity with equation $F = f_d + f_{d+k} + \dots$, where the previous decomposition is the decomposition in homogeneous parts. Assume $f_d \neq 0$, $k > 0$. Let $\mathcal{C}_m := V_{\mathbb{P}}(f_m)$ denote the projective zero locus in $\mathbb{P}^2$ of the homogeneous polynomial $f_m$. We say that $S$ is a *Lê–Yomdin* singularity if $\operatorname{Sing}(\mathcal{C}_d) \cap \mathcal{C}_{d+k} = \emptyset$. If $k = 1$, $S$ is called a *superisolated* singularity.

Superisolated singularities were introduced by Luengo [24]: they can be resolved by one blow-up. In [25], the authors show that the link of a superisolated singularity is a rational homology sphere if and only if all the irreducible components of $\mathcal{C}_d$ are cuspidal rational and if the curve is reducible they only intersect at one point. Besides the smooth case, no other one provides an integral homology sphere as can be deduced from the following result in [24]. We reproduce the proof since it will be generalized for other classes of singularities.

**Proposition 5.5 ([24])** *Let $S$ be a superisolated singularity with tangent cone $\mathcal{C}_d$ of degree $d$. Let $\Pi : \widehat{\mathbb{C}}^3 \to \mathbb{C}^3$ be the blow-up of $0 \in S \subset \mathbb{C}^3$ and $\pi : \hat{S} \to S$ the restriction of $\Pi$ to the strict transform of $S$. If $E \cong \mathbb{P}^2$ is the exceptional divisor of $\Pi$, then the exceptional divisor of $\pi$ is $\mathcal{C}_d = E \cap \hat{S}$.*

*Moreover, if $\mathcal{C}_{d,1}, \dots, \mathcal{C}_{d,s}$ denote the irreducible components of $\mathcal{C}_d$ and $\delta_i := \deg \mathcal{C}_{d,i}$, then*

$$(\mathcal{C}_{d,i} \cdot \mathcal{C}_{d,i})_{\hat{S}} = -\delta_i(d - \delta_i + 1), \quad (\mathcal{C}_{d,i} \cdot \mathcal{C}_{d,j})_{\hat{S},P} = (\mathcal{C}_{d,i} \cdot \mathcal{C}_{d,j})_{\mathbb{P}^2,P}, i \neq j,\ P \in \operatorname{Sing} \mathcal{C}_d.$$

***Proof*** Let us assume that $[0 : 0 : 1] \in \operatorname{Sing} \mathcal{C}_d$. We can fix the usual chart of the blowing-up. Assume that $S = \{F = 0\}$, where $F = f_d + f_{d+1} + \dots$; in the chart $(x, y, z) \mapsto (xz, yz, z)$ and $E = \{z = 0\}$, $\hat{S} = \{f_d(x, y, 1) + z(f_{d+1}(x, y, 1) + \dots) = 0\}$, i.e., $\mathcal{C}_d = \{z = f_d(x, y, 1) = 0\}$. In the neighborhood of $P$, $(E, \mathcal{C}_d)$ and $(\hat{S}, \mathcal{C}_d)$ are isomorphic. We deduce that for $i \neq j$, $(\mathcal{C}_{d,i} \cdot \mathcal{C}_{d,j})_{\hat{S},P} = (\mathcal{C}_{d,i} \cdot \mathcal{C}_{d,j})_{\mathbb{P}^2,P}$.

The surfaces $E$ and $\hat{S}$ are *generically* transversal, namely outside $\operatorname{Sing} \mathcal{C}_d$. The Euler class $e(E) = -L$, where $L$ is a line in $E$. Then,

$$(\mathcal{C}_d \cdot \mathcal{C}_{d,i})_{\hat{S}} = (e(E) \cdot \mathcal{C}_{d,i})_{\mathbb{P}^2} = -\delta_i.$$

Also

$$(\mathcal{C}_d \cdot \mathcal{C}_{d,i})_{\hat{S}} = (\mathcal{C}_{d,i} \cdot \mathcal{C}_{d,i})_{\hat{S}} + \sum_{j \neq i} (\mathcal{C}_{d,j} \cdot \mathcal{C}_{d,i})_{\hat{S}} = (\mathcal{C}_{d,i} \cdot \mathcal{C}_{d,i})_{\hat{S}} + \sum_{j \neq i} (\mathcal{C}_{d,j} \cdot \mathcal{C}_{d,i})_{\mathbb{P}^2} =$$
$$(\mathcal{C}_{d,i} \cdot \mathcal{C}_{d,i})_{\hat{S}} + \delta_i (d - \delta_i),$$

and the result follows.

Although $\pi$ is not necessarily a resolution with normal crossings, $\det S$ can be recovered from it using its intersection matrix; it is a classical result which will follow from a later proposition.

**Corollary 5.6** *If $S$ is as above, then*

$$\det S = (d + 1)^{s-1} \cdot \delta_1 \cdot \dots \cdot \delta_s.$$

*In particular, if $\mathcal{C}_d$ is irreducible, then* $\det S = d$.

***Proof*** By Proposition 5.5, the diagonal terms of the intersection matrix for $\pi$ equal $-\delta_i(d - \delta_i + 1)$ and the non-diagonal terms are $\delta_i \cdot \delta_j$. Replacing the first row by the sum of all rows, one obtains $-(\delta_1, \dots, \delta_s)$. If we add the new first row multiplied by $\delta_i$ times the $i$th-row $(i > 1)$, all the non-diagonal terms vanish and the diagonal term becomes $-\delta_i(d + 1)$.

For Lê–Yomdin singularities we follow the same strategy. If $S = \{F = 0\}$ with $F = f_d + f_{d+k} + \dots$, and we keep the notation above, the main difference is that $\hat{S}$ is no longer smooth, in general. If $P \in \operatorname{Sing} \mathcal{C}_d$, then the local equation of $\hat{S}$ at $P$ is $z^k - f(x, y) = 0$ where $f(x, y) = 0$ is the local equation of $\mathcal{C}_d$ at $P$. Intersection theory can be used also in normal surfaces, see [18, 28] for definitions and [8] for useful tips. As the following result shows the intersection form of a partial resolution is also useful.

**Lemma 5.7** *Let $(S, 0)$ be a normal surface singularity and let $\pi : (X, D) \to (S, 0)$ be a proper birational morphism which is an isomorphism outside $D = \pi^{-1}(0)$ on* *the normal surface X. Let A be the intersection matrix for D. Then,*

$$\det S = \det(-A) \prod_{P \in D} \det(X, P).$$

***Proof*** Note first that the product in the formula is finite since only a finite number of singular points may arise. Let $\sigma : (Y, E) \to (X, D)$ be a resolution of the singularities of $X$. Let $B$ the intersection matrix of $E$. Instead of expressing this matrix in terms of the irreducible components of $E$, we replace the strict transforms of the components of $D$ by their total transforms.

Then, $B$ is replaced by a matrix $\tilde{B}$, with the same determinant, which is a diagonal sum of $A$ and the intersection matrices of the singular points. Then,

$$\det S = \det(-B) = \det(-\tilde{B}) = \det(-A) \prod_{P \in \operatorname{Sing} X} \det(X, P).$$

**Proposition 5.8** *Let $S$ be a $k$-Lê–Yomdin singularity with tangent cone $\mathcal{C}_d$ of degree $d$. With the notation of Proposition 5.5,*

$$(\mathcal{C}_{d,i} \cdot \mathcal{C}_{d,i})_{\hat{S}} = -\frac{\delta_i(d - \delta_i + k)}{k}, \; (\mathcal{C}_{d,i} \cdot \mathcal{C}_{d,j})_{\hat{S},P} = \frac{(\mathcal{C}_{d,i} \cdot \mathcal{C}_{d,j})_{\mathbb{P}^2,P}}{k}, i \neq j, \; P \in \operatorname{Sing} \mathcal{C}_d.$$

***Proof*** We follow the guidelines of the proof of Proposition 5.5. Note that it is not true any more that in the neighborhood of $P \in \operatorname{Sing} \mathcal{C}_d$ the germs $(E, \mathcal{C}_d)$ and $(\hat{S}, \mathcal{C}_d)$ are isomorphic. However, the projection $\rho(x, y, z) := (x, y)$ restricts to a $k : 1$ proper map $(\hat{S}, \mathcal{C}_d) \to (E, \mathcal{C}_d)$. Since $\pi^*(\pi_* \mathcal{C}_{d,i}) = k\mathcal{C}_{d,i}$ we have that for $i \neq j$

$$(\mathcal{C}_{d,i} \cdot \mathcal{C}_{d,j})_{\hat{S},P} = \frac{1}{k^2}(\pi^*\pi_*\mathcal{C}_{d,i} \cdot \pi^*\pi_*\mathcal{C}_{d,j})_{\hat{S},P} = \frac{1}{k}(\mathcal{C}_{d,i} \cdot \mathcal{C}_{d,j})_{\mathbb{P}^2,P}.$$

For the self-intersections we apply the same ideas:

$$(\mathcal{C}_d \cdot \mathcal{C}_{d,i})_{\hat{S}} = (e(E) \cdot \mathcal{C}_{d,i})_{\mathbb{P}^2} = -\delta_i$$

$$(\mathcal{C}_d \cdot \mathcal{C}_{d,i})_{\hat{S}} = (\mathcal{C}_{d,i} \cdot \mathcal{C}_{d,i})_{\hat{S}} + \sum_{j \neq i} (\mathcal{C}_{d,j} \cdot \mathcal{C}_{d,i})_{\hat{S}} = (\mathcal{C}_{d,i} \cdot \mathcal{C}_{d,i})_{\hat{S}} + \frac{\delta_i(d - \delta_i)}{k},$$

and the result follows.

A similar proof to the one of Corollary 5.6 provides the following result.

**Corollary 5.9** *If $S$ is a $k$-Lê–Yomdin as above, then*

$$\det S = \delta_1 \cdot \ldots \cdot \delta_s \cdot \left(\frac{d + k}{k}\right)^{s-1} \prod_{P \in \operatorname{Sing} \mathcal{C}_d} \det S_{P,k},$$

*where*

$$S_{P,k} = \{z^k = f_P(x, y) \mid P \in \operatorname{Sing} \mathcal{C}_d\},$$

*and* $f_P(x, y) = 0$ *is a local equation of* $\mathcal{C}_d$ *at* $P$.

*In particular, if* $\mathcal{C}_d$ *is smooth, then* $\det S = d$.

*Example 5.10* Let $S_k$ be the singularity $\{z^k = x^2 + y^2\}$, then $\det S_k = k$. Denote by $T_k$ the singularity $\{z^k = x^2 + y^3\}$, then we have

$$\det T_k = \begin{cases} 1 & \text{if } \gcd(k, 6) = 1, 6 \\ 3 & \text{if } \gcd(k, 6) = 2 \\ 4 & \text{if } \gcd(k, 6) = 3. \end{cases}$$

Note that $T_k$ admits a $\mathbb{Q}$-resolution with only one exceptional divisor. This divisor has positive genus (equal to one) if and only if $\gcd(k, 6) = 6$.

We did not find in the literature a general formula for this determinant. >From the above computations and the periodicity properties of the Alexander invariants, the following statement may be true.

**Conjecture 1** *Let* $C : f(x, y) = 0$ *be a germ of a reduced plane curve singularity, and let* $S_k : z^k = f(x, y)$ *be a cyclic germ of surface. Let* $N$ *be the order of the semisimple factor of the monodromy of* $C$. *Then* $\det S_k$ *is a quasi-polynomial in* $k$ *of period* $N$.

**Proposition 5.11** *A k-Lê–Yomdin singularity with tangent cone* $\mathcal{C}_d$ *has as link a rational homology sphere if and only if* $\mathcal{C}_d$ *is a union of rational cuspidal curves with only one intersection point and the links of the k-cyclic singularities associated with the singular points of* $\mathcal{C}_d$ *have also a rational homology sphere as a link.*

The proof of this proposition is a direct consequence of the previous result. We have not proven that Lê–Yomdin singularities do not provide integral homology sphere links, mainly since we do not have a closed formula for the determinant of a cyclic singularity. Our experimentation leads to this conjecture.

**Conjecture 2** *No k-Lê–Yomdin singularity* $k > 1$ *has an integral homology sphere link.*

## 5.3 Weighted Lê–Yomdin Singularities

We are going to generalize these families of singularities using weighted homogeneous curves. We use the notation $\omega$, $\eta$, etc. introduced in Sect. 2.2. The following notion of weighted Lê–Yomdin singularity was introduced in [7].

**Definition 5.12** A hypersurface $(S, 0) := \{F = 0\}$ is an *$(\omega, k)$-weighted Lê–Yomdin singularity* if the following holds. Let $F := f_d + f_{d+k} + \dots$ be the decomposition in $\omega$-weighted homogeneous forms, then $\operatorname{Jac}(f_d) \cap V(f_{d+k}) = \emptyset$.

In order to relate geometrically this definition with the definition of superisolated and Lê–Yomdin singularities, let us consider the weighted blow-up $\Pi_\omega : \widehat{\mathbb{C}}^3_\omega \to \mathbb{C}^3$. In Sect. 2.3 we have described a stratification of the exceptional divisor $E_\omega \cong \mathbb{P}^2_\omega \cong \mathbb{P}^2_\eta$ according to the singularities of $\widehat{\mathbb{C}}^3_\omega$, see Proposition 2.3.

One needs to study the two curves $\mathcal{C}_d, \mathcal{C}_{d+k} \subset E_\omega$. In general, note that $f_d(x, y, z) = x^{\varepsilon_x} y^{\varepsilon_y} z^{\varepsilon_z} g(x^{d_1}, y^{d_2}, z^{d_3})$, where $\varepsilon_x, \varepsilon_y, \varepsilon_z \in \{0, 1\}$ and $g$ is $\eta$-weighted homogeneous of degree $\frac{d - e_1\varepsilon_x - e_2\varepsilon_y - e_3\varepsilon_z}{d_1 d_2 d_3}$. If we see this curve in $\mathbb{P}^2_\eta$ its equation is $x^{\varepsilon_x} y^{\varepsilon_y} z^{\varepsilon_z} g(x, y, z) = 0$.

**Proposition 5.13** *Let $S = \{F = 0\}$ be an $(\omega, k)$-weighted Lê–Yomdin singularity with $\omega$-quasi-tangent cone $\mathcal{C}_d = \{f_d = 0\}$. Let $\Pi_\omega$ be the $\omega$-blow-up, $E_\omega \cong \mathbb{P}^2_\omega \cong \mathbb{P}^2_\eta$ is the exceptional divisor and $\hat{S}$ is the strict transform (and a partial resolution) of $S$. Recall the stratification of $E_\omega = \mathcal{P} \cup \mathcal{L} \cup \mathcal{T}$ as given in Notation 2.6. The structure of $\hat{S}$ along $P \in \mathcal{C}_d = E_\omega \cap \hat{S}$ is as follows:*

(1) $P \in \mathcal{T}$.

- (a) *If $P \notin \operatorname{Sing} \mathcal{C}_d$ then $\hat{S}$ is smooth at $P$ and $E_\omega \pitchfork_P \hat{S}$.*
- (b) *If $P \in \operatorname{Sing} \mathcal{C}_d$ then $P \notin \mathcal{C}_{d+k}$. There are local coordinates $U, V, W$ centered at $P$ such that $E_\omega = \{W = 0\}$, $\mathcal{C}_d = \{W = g(U, V) = 0\}$ and $\hat{S} = \{W^k = g(U, V)\}$; in particular $\hat{S}$ is smooth at $P$ if and only if $k = 1$ (but it is not transversal to $E_\omega$).*

(2) $P \in \mathcal{L}_y$ *(a similar statement holds for $\mathcal{L}_x, \mathcal{L}_z$).*

- (a) *If $\mathcal{C}_d$ is transversal to $Y$ at $P$ then $(\hat{S}, P) \cong \frac{1}{d_2}(e_2, -1)$. In the quotient ambient space $(\widehat{\mathbb{C}}^3_\omega, P)$ the situation is similar to (1a).*
- (b) *If $(\mathcal{C}_d, P) = (Y, P)$ then $(\hat{S}, P)$ is smooth. In the quotient ambient space $(\widehat{\mathbb{C}}^3_\omega, P)$ the situation is similar to (1a).*
- (c) *If $\mathcal{C}_d \not\pitchfork_P Y$, i.e. the order of $f_d(x + t, y, 1)$ is $> 1$ ($P = [t : 0 : 1]$), then $P \notin \mathcal{C}_{d+k}$. The germ $(\hat{S}, P)$ is isomorphic to $z^k = f_d(x + t, y, 1)$ in the threefold quotient singularity $\frac{1}{d_2}(0, e_2, -1)$, where $z = 0$ is the equation of $E_\omega$.*

(3) $P = P_z$ *(a similar statement holds for $P_x, P_y$).*

- (a) *If $\mathcal{C}_d$ is extremely quasi-smooth at $P$ (i.e. the order of $f_d(x, y, 1)$ is 1) the situation is as in (1a) replacing the ambient smooth space by the threefold quotient singularity $\frac{1}{e_3}(e_1, e_2, -1)$. Let $h_1(x, y)$ be the linear part of $f_d(x, y, 1)$.*
  - (i) *If $h_1(x, y)$ is proportional to $x$, then $(\hat{S}, P) \cong \frac{1}{e_3}(e_1, -1)$.*
  - (ii) *If $h_1(x, y)$ is proportional to $y$, then $(\hat{S}, P) \cong \frac{1}{e_3}(e_2, -1)$.*
  - (iii) *Otherwise, $e_1 \equiv e_2 \bmod e_3$ and the above cases coincide.*

(b) *If $\mathcal{C}_d$ is not extremely quasi-smooth at $P$ (i.e., the order of $f_d(x, y, 1)$ is $>1$), then $P \notin \mathcal{C}_{d+k}$ and $d+k \equiv 0 \bmod e_3$. The germ $(\hat{S}, P)$ is isomorphic to $z^k = f_d(x, y, 1)$ in the threefold quotient singularity $\frac{1}{e_3}(e_1, e_2, -1)$, where $z = 0$ is the equation of $E_\omega$.*

***Proof*** The different parts of the statement will be particular cases of the following general situation. Assume $P = [0 : 0 : 1] \in \mathcal{C}_d = E_\omega \cap \hat{S}$ is a point of the strict transform $\hat{S}$ of $S$ on the exceptional divisor $E_\omega$. The total transform of $S$ is equal to $\hat{S} + dE_\omega$ and hence, its equation in the chart $\Psi_{\omega,3}$ is:

$$z^d(f_d(x, y, 1) + z^k \underbrace{(f_{d+k}(x, y, 1) + \ldots)}_{q(x,y)}) = 0.$$

By hypothesis $f_d(0, 0, 1) = 0$, let us denote by $\ell(x, y)$ the linear part of $f_d(x, y, 1)$. The following conditions are immediate

$$\begin{cases} P \notin \mathcal{C}_{d+k} \Longleftrightarrow q(x, y) \text{ is a unit,} \\ P \in \mathcal{C}_{d+k} \Longrightarrow \ell(x, y) \neq 0. \end{cases}$$

We will consider $(x_1, y_1, z_1)$ a change of coordinates where

$$(x_1, y_1, z_1) = \begin{cases} (x, y, zq(x, y)^{\frac{1}{k}}) & \text{if } \ell(x, y) = 0, \\ (\frac{1}{a}(f_d(x, y, 1) + z^k q(x, y)), y, z) & \text{if } \ell(x, y) = ax, \\ (x, \frac{1}{b}(f_d(x, y, 1) + z^k q(x, y) - ax), z) & \text{if } \ell(x, y) = ax + by, b \neq 0. \end{cases}$$

Note that the action of $\mu_{e_3}$ on $(x_1, y_1, z_1)$ reads as in $(x, y, z)$. If $P \notin \mathcal{C}_{d+k}$ then $d + k \equiv 0 \bmod e_3$.

In these coordinates $E_\omega : z_1 = 0$ and $\mathcal{C}_d : W = g(x_1, y_1) = 0$, where

$$g(x_1, y_1) = \begin{cases} f_d(x_1, y_1, 1) & \text{if } \ell(x, y) = 0, \\ \ell(x_1, y_1) & \text{otherwise.} \end{cases}$$

The local equations for $dE_\omega + \hat{S}$ are $z_1^d(z_1^k + g) = 0$. If $\ell(x, y) \neq 0$, then both look like two surfaces in a quotient ambient space whose preimages in $\mathbb{C}^3$ are smooth and transversal.

The case $P \in \mathcal{T}$ locally corresponds to $\omega = (1, 1, 1)$. Note that (1a) implies $\ell(x_1, y_1) \neq 0$ whereas (1b) implies $\ell(x_1, y_1) = 0$. The case $P \in \mathcal{L}_y$ corresponds with the choice $\omega = (d_2, e_2, d_2)$ where (2a) and (2b) refers to $\ell \neq 0$ and (2c) refers to $\ell = 0$. Finally, $P = P_z$, corresponds to the choice $\omega = (e_1, e_2, e_3)$. In this case (3a) refers to the different cases of $\ell \neq 0$ and (3b) refers to $\ell = 0$.

The divisor $\mathcal{C}_d$ has an irreducible decomposition in $s+\varepsilon_x+\varepsilon_y+\varepsilon_z$ components $\varepsilon_x X+\varepsilon_y Y+\varepsilon_z Z+\tilde{\mathcal{C}}_d$, where $\tilde{\mathcal{C}}_d=\sum_{i=1}^s \mathcal{C}_{d,i}$ and $\varepsilon_x, \varepsilon_y, \varepsilon_z \in \{0,1\}$. Recall that $d_1 d_2 d_3$ divides $\deg \mathcal{C}_{d,i}$ and we can write $\delta_i = d_1 d_2 d_3 \hat{\delta}_i$.

Consider the stratification of a weighted projective plane as above. We call a curve in a weighted projective plane *stratified smooth* if it is smooth, intersects the axes transversally and does not contain the vertices.

**Proposition 5.14** *Let $S$ be an $(\omega,k)$-weighted Lê–Yomdin singularity with* quasi-tangent cone $\mathcal{C}_d \subset \mathbb{P}^2_\omega$ *of degree $d$. With the notation of Proposition* 5.8:

(1) $(\mathcal{C}_{d,i}\cdot\mathcal{C}_{d,i})_{\hat{S}} = -\frac{\delta_i(d-\delta_i+k)}{k e_1 e_2 e_3} = -\frac{\hat{\delta}_i(d-\delta_i+k)}{k d_1 d_2 d_3 \alpha_1\alpha_2\alpha_3}$.
(2) *If* $\varepsilon_x = 1$, *then* $(X\cdot X)_{\hat{S}} = -\frac{d_1^2 e_1(d-e_1+k)}{k e_1 e_2 e_3} = -\frac{d_1^2(d-e_1+k)}{k e_2 e_3} = -\frac{d-e_1+k}{k d_2 d_3 \alpha_2\alpha_3}$. *Similar formulas hold for $Y$ and $Z$.*
(3) *If* $i\neq j$, *then* $(\mathcal{C}_{d,i}\cdot\mathcal{C}_{d,j})_{\hat{S}} = \frac{\delta_i\delta_j}{k e_1 e_2 e_3} = \frac{\hat{\delta}_i\hat{\delta}_j}{k\alpha_1\alpha_2\alpha_3}$.
(4) *If* $\varepsilon_x = 1$ *then* $(\mathcal{C}_{d,i}\cdot X)_{\hat{S}} = \frac{d_1\delta_i}{k e_2 e_3} = \frac{\hat{\delta}_i}{k\alpha_2\alpha_3}$. *Similar formulas hold for $Y$ and $Z$.*
(5) *If* $\varepsilon_x\varepsilon_y = 1$ $(X\cdot Y)_{\hat{S},P_z} = \frac{d_1 d_2}{k e_3} = \frac{1}{k\alpha_3}$. *Similar formulas hold for the other pairs involving $X$, $Y$, and $Z$.*

***Proof*** We follow the ideas in the proofs of Propositions 5.5 and 5.8 with some modifications. If $\omega\neq\eta$, the map $\pi_{\eta,\omega}^{-1}:\mathbb{P}^2_\eta\to\mathbb{P}^2_\omega$ can be considered as the *identity* where $\mathbb{P}^2_\omega$ is seen as $(\mathbb{P}^2_\eta)^{\text{orb}}$, where $(\pi_{\eta,\omega}^{-1})^*(X)=\frac{1}{d_1}X$, $(\pi_{\eta,\omega}^{-1})^*(Y)=\frac{1}{d_2}Y$, and $(\pi_{\eta,\omega}^{-1})^*(Z)=\frac{1}{d_3}Z$. Analogously, the abstract strict transform $\hat{S}$ has a natural orbifold embedded structure $\hat{S}^{\text{orb}}\subset\widehat{\mathbb{C}}^3_\omega$ where the embedding $\pi:\hat{S}\to\hat{S}^{\text{orb}}$ has the same properties for $X, Y, Z$ as $\pi_{\eta,\omega}^{-1}$ whenever $X, Y, Z$ are contained in $\mathcal{C}_d$.

The divisor $e(E_\omega)$ in $E_\omega\equiv\mathbb{P}^2_\omega$ has degree 1 and Bézout's Theorem for the $\omega$-projective plane states that the sum of the intersection numbers of two divisors equals the product of the degrees divided by $e_1e_2e_3$. Hence, we obtain the same formulas as in Proposition 5.8 with two differences: $e_1e_2e_3$ appears in the denominator and all the intersection numbers are considered in $\hat{S}^{\text{orb}}$.

When we consider the intersection numbers in $\hat{S}$, when $X$ appears ($\varepsilon_x = 1$), the formulas must be multiplied by $d_1$. A similar argument holds for $Y, Z$.

**Corollary 5.15** *If $S$ is an $(\omega,k)$-weighted Lê–Yomdin as above and $A$ is the intersection matrix of the blowing-up, then*

$$\det S = d_1^{2\varepsilon_x}\cdot d_2^{2\varepsilon_y}\cdot d_3^{2\varepsilon_z}\cdot\delta_1\cdot\ldots\cdot\delta_s\cdot\left(\frac{d+k}{k e_1 e_2 e_3}\right)^{s-1}\prod_{P\in\hat{S}}\det(\hat{S}_{k,P}),$$

*where $\hat{S}_{k,P}$ is the surface singularity at $P$ as described in Proposition* 5.13.
*In particular, if $\mathcal{C}_d$ is stratified smooth, then* $\det S = \frac{d}{e_1e_2e_3}$.

# 6 Normal Surface Singularities with Rational Homology Sphere Links

In this section we will use the results and strategies presented in Sect. 5 in order to exhibit examples of weighted Lê–Yomdin singularities whose links are rational homology spheres, generalizing the strategy in [15]. We will be using Proposition 5.11 in the context of weighted Lê–Yomdin singularities.

## *6.1 Brieskorn–Pham Singularities*

We will interpret these singularities as Lê–Yomdin singularities and study their $\mathbb{Q}$-resolution graph. Consider $\omega_0 = (n_1, n_2, n_3)$ and the Brieskorn–Pham singularity $S = \{F_{\omega_0} = x^{n_1} + y^{n_2} + z^{n_3} = 0\} \subset (\mathbb{C}^3, 0)$, where $n_1, n_2, n_3$ are not assumed to be coprime.

Denote by $e := \gcd \omega_0$, and $\alpha_k := \frac{1}{e}\gcd(n_i, n_j)$, where $\{i, j, k\} = \{1, 2, 3\}$. Note that $d_i := \frac{n_i}{e\alpha_j\alpha_k} \in \mathbb{Z}_{>0}$ are pairwise coprime. If

$$\omega = (e_1, e_2, e_3) := \frac{1}{e^2\alpha_1\alpha_2\alpha_3}(n_2n_3, n_1n_3, n_1n_2) = (\alpha_1d_2d_3, \alpha_2d_1d_3, \alpha_3d_1d_2),$$

then $F_{\omega_0}(x, y, z)$ is an $\omega$-weighted homogeneous polynomial of degree $d := \frac{n_1n_2n_3}{e^2\alpha_1\alpha_2\alpha_3}$ and hence $S$ can be viewed as an $(\omega, k)$-weighted Lê–Yomdin singularity for any $k \geq 1$. Following the general construction, $f_d = F_{\omega_0}(x, y, z) = g(x^{d_1}, y^{d_2}, z^{d_3}) = 0$ can be considered a curve in $\mathbb{P}^2_\eta \cong \mathbb{P}^2_\omega$ for $\eta = (\alpha_1, \alpha_2, \alpha_3)$ of $\eta$-degree $d_\eta = e\alpha_1\alpha_2\alpha_3$ given by the equation $g(x, y, z) = x^{e\alpha_2\alpha_3} + x^{e\alpha_1\alpha_3} + z^{e\alpha_1\alpha_2} = 0$. Its genus is

$$\frac{d_\eta(d_\eta - |\eta|)}{2\alpha_1\alpha_2\alpha_3} + 1 = \frac{e^2\alpha_1\alpha_2\alpha_3 - e(\alpha_1 + \alpha_2 + \alpha_3) + 2}{2}.$$

Since the curve $\mathcal{C}_d$ is transversal to the axes we obtain that the exceptional locus of $\hat{S}$ has (in the intersection with the axes) $e\alpha_i$ cyclic points of order $d_i$. The determinant of the singularity is

$$\frac{d}{(d_1d_2d_3)^2(\alpha_1\alpha_2\alpha_3)}\left(d_1^{\alpha_1}d_2^{\alpha_2}d_3^{\alpha_3}\right)^e = ed_1^{e\alpha_1-1}d_2^{e\alpha_2-1}d_3^{e\alpha_3-1}.$$

As a consequence of this discussion one obtains the following.

**Proposition 6.1** *The Brieskorn–Pham singularity $S = \{F_{\omega_0} = x^{n_1} + y^{n_2} + z^{n_3} = 0\} \subset (\mathbb{C}^3, 0)$ is a rational homology sphere singularity if and only if either $\alpha_1 = \alpha_2 = \alpha_3 = 1, e = 2$ or $\alpha_i = \alpha_j = e = 1$ for some $i \neq j$.*

*Moreover, it is an integral homology sphere if and only if the exponents are pairwise coprime.*

## 6.2 Examples Coming from Cremona Transformations and Kummer Covers

The purpose of this section is to provide more candidates to surface singularities with rational homology sphere links by applying the techniques used in Sect. 4. In particular, we will start with the strict transforms of the conic by the Cremona transformations.

In order to do so one needs $\omega := (\alpha_1, \alpha_2, \alpha_3)$ pairwise coprime, and $\beta_1, \beta_2 \in \mathbb{Z}_{>0}$ such that $\alpha_1\alpha_2 + \alpha_3 = \alpha_1\beta_1 + \alpha_2\beta_2$. The weighted homogeneous polynomial

$$\begin{aligned} f_\omega(x, y, z) &= f(y^{\alpha_1} z, x^{\alpha_2} z, x^{\beta_1} y^{\beta_2}) \\ &= y^{2\alpha_1} z^2 + x^{2\alpha_2} z^2 + x^{2\beta_1} y^{2\beta_2} - 2z \left( x^{\alpha_2} y^{\alpha_1} z + x^{\beta_1} y^{\alpha_1+\beta_2} + x^{\alpha_2+\beta_1} y^{\beta_2} \right) \end{aligned}$$

has $\omega$-degree $2(\alpha_1\alpha_2 + \alpha_3)$ and defines a rational curve in $\mathbb{P}^2_\omega$ which is smooth outside the vertices. Assume for simplicity that $\alpha_1\alpha_2 + \alpha_3 < \alpha_1\alpha_2\alpha_3$. Hence for any generic quasi-homogeneous polynomial $g_\omega(x, y, z)$ of degree $2\alpha_1\alpha_2\alpha_3$, $F := f_\omega + g_\omega$ defines an $(\omega, k)$-weighted Lê–Yomdin singularity, for $k = 2(\alpha_1\alpha_2\alpha_3 - \alpha_1\alpha_2 - \alpha_3)$. A partial resolution of this singularity has an exceptional locus which is a rational curve with three singular points. In most cases the link of this singularity is a rational homology sphere. For simplicity, we will prove it in a special case.

**Proposition 6.2** *With the previous notation, take $\alpha_1 = 1$, $\beta_1 = \alpha_3$, and $\beta_2 = 1$. Then, for any $\alpha_2, \alpha_3 > 1$ satisfying $\gcd(3, k) = \gcd(3, \alpha_2\alpha_3 - \alpha_2 - \alpha_3) = 1$ and a generic $g_\omega$, the equation $\{F = f_\omega + g_\omega = 0\} \subset \mathbb{C}^3$ defines a surface singularity with a rational homology sphere link.*

***Proof*** We study the strict transform of this singularity at $P_x, P_y, P_z$ after an $\omega$-weighted blow-up. At $P_x$, the ambient space is smooth and the strict transform has equation

$$0 = x^k + y^2 z^2 + z^2 + y^2 - 2yz\,(z + y + 1) = x^k + (y_1 + z)^2 z^2 + y_1^2 - 2(y_1 + z)z\,(2z + y_1)$$

if $y_1 = y - z$. This is topologically equivalent to $0 = x^k + y^2 + z^3$. Since $\gcd(3, \alpha_2\alpha_3 - \alpha_2 - \alpha_3) = 1$, by Proposition 6.1 the link of this singularity is a rational homology sphere.

At $P_y$, the ambient space is $\frac{1}{\alpha_2}(1, -1, \alpha_3)$ and the strict transform has equation

$$\begin{aligned} f_\omega(x, y, z) &= y^k + z^2 + x^{2\alpha_2} z^2 + x^{2\alpha_3} - 2z \left( x^{\alpha_2} z + x^{\alpha_3} + x^{\alpha_2+\alpha_3} \right) \\ &= y^k + z_1^2 - 2x^{\alpha_2+2\alpha_3} + \ldots \end{aligned}$$

if $z_1 = z - x^{\alpha_3}$. This change of variable is compatible with the action; this equation defines a singularity in $\mathbb{C}^3$ whose link is a rational homology sphere, and so is the case in the quotient manifold.

By symmetry arguments, the same happens for $P_z$. Hence, $F$ defines a singularity whose link is a rational homology sphere.

Let us use the orbifold approach. Given $(d_1, d_2, d_3)$ pairwise coprime consider $\omega := (d_2d_3, d_1d_3, d_1d_2)$; the normalized $\eta$ is $(1, 1, 1)$ and the isomorphism $\mathbb{P}^2_\omega \to \mathbb{P}^2$ is given by $[x : y : z]_\omega \mapsto [x^{d_1} : y^{d_2} : z^{d_3}]$, see (2.1). This isomorphism can be seen as a *weighted Kummer cover* and the homogeneous polynomial

$$f_\omega(x, y, z) = f(x^{d_1}, y^{d_2}, z^{d_3}) = x^{2d_1} + y^{2d_2} + z^{2d_3} - 2\left(y^{d_2}z^{d_3} + x^{d_1}z^{d_3} + x^{d_1}y^{d_2}\right)$$

of $\omega$-degree $2d_1d_2d_3$, which defines a rational curve in $\mathbb{P}^2_\omega \cong \mathbb{P}^2$ and it is tangent to the axes. In most cases the link of this singularity is a rational homology sphere. Let us study a special case.

**Proposition 6.3** *For any generic quasi-homogeneous polynomial $g_\omega(x, y, z)$ of degree $3d_1d_2d_3$, and $d_1, d_2, d_3$ odd numbers $\{F := f_\omega + g_\omega = 0\} \subset \mathbb{C}^3$ defines a surface singularity with a rational homology sphere link.*

***Proof*** Note that $\{F = 0\}$ defines an $(\omega, k)$-weighted Lê–Yomdin singularity, for $k = d_1d_2d_3$. A partial resolution of this singularity has an exceptional locus which is a rational curve with three singular points (corresponding to the tangencies). In most cases the link of this singularity is a rational homology sphere.

By symmetry reasons we study only the strict transform of this singularity at the tangency point with $Y$ after an $\omega$-weighted blow-up. After a change of coordinates the local equation of $F$ is

$$\begin{aligned} 0 &= z^{d_1d_2d_3} + (x+1)^{2d_1} + y^{2d_2} + 1 - 2\left(y^{d_2} + (x+1)^{d_1} + (x+1)^{d_1}y^{d_2}\right) \\ &= z^{d_1d_2d_3} + {d_1}^2x^2 - 2y^{d_2} + \dots \end{aligned}$$

and the ambient space is $\frac{1}{d_2}(0, d_1d_3, -1)$. Since $d_1, d_2, d_3$ are odd numbers, by Proposition 6.1 this equation defines a singularity in $\mathbb{C}^3$ whose link is a rational homology sphere, and so is the case in the quotient manifold.

## 6.3 Integral Homology Sphere Surface Singularities

Following ideas of the third named author, Veys, and Vos, we present an infinite family of normal surface singularities which are complete intersection in $\mathbb{C}^4$ and whose links are integral homology spheres. The splice diagram of this family is precisely the one given in [29, p. 765] and the corresponding semigroup conditions are satisfied. The examples given here can be generalized to any dimension.

Let $n_0, n_1, n_2, n_3 \in \mathbb{Z}_{>0}$ and $b_{20}, b_{21}, b_{30}, b_{31}, b_{32} \in \mathbb{Z}_{\geq 0}$. Consider $S$ the surface singularity in $(\mathbb{C}^4, 0)$ defined by

$$S = \{f_1 + f_2 = f_2 + f_3 = 0\} \subset (\mathbb{C}^4, 0), \text{ where } \begin{cases} f_1 = x_1^{n_1} - x_0^{n_0}, \\ f_2 = x_2^{n_2} - x_0^{b_{20}} x_1^{b_{21}}, \\ f_3 = x_3^{n_3} - x_0^{b_{30}} x_1^{b_{31}} x_2^{b_{32}}. \end{cases} \tag{6.1}$$

The purpose of this section is to show when the link of $S$ is a rational homology sphere as well as to characterize when it is integral. The idea is to resolve $S$ with $\mathbb{Q}$-normal crossings and apply Lemma 5.7 to compute $\det S$. In order to do so we consider the Cartier divisors of $S$ defined by $Y = \{f_1 = 0\}$ and $H_i = \{x_i = 0\}$, $i = 0, 1, 2$. This family was recently studied in Vos' PhD thesis in a more general context and we just briefly discuss here the construction of the partial resolution obtained in [27, section 5].

**Theorem 6.4** *Let $S \subset (\mathbb{C}^4, 0)$ be the surface singularity defined above. Assume $n_0, n_1, n_2, n_3 \in \mathbb{Z}_{>0}$ are pairwise coprime, then $S$ is a rational homology sphere. Moreover, in that case $S$ is an integral homology sphere singularity if and only if $m := \gcd(n_3, b_{20}n_1 + b_{21}n_0) = 1$.*

***Proof*** Let $\pi_1 : \widehat{\mathbb{C}}^4 \to \mathbb{C}^4$ be the weighted blow-up at the origin of $\mathbb{C}^4$ with weights $w_1 = (\frac{n}{n_0}, \frac{n}{n_1}, \frac{n}{n_2}, \frac{n}{n_3})$ where $n = n_0 n_1 n_2 n_3$. The exceptional divisor of $\pi_1$ is the weighted projective variety $E_1 = \mathbb{P}^3_{w_1}$. The assumption on the integers $n_i$, $i = 0, .., 3$ being pairwise coprime implies that the exceptional divisor $\mathcal{E}_1$ of the restriction $\varphi_1 = \pi_1|_{\hat{S}} : \hat{S} \to S$ is a rational irreducible curve which contains three singular points of $\hat{S}$, namely $Q_0 = \hat{H}_0 \cap \mathcal{E}_1$, $Q_1 = \hat{H}_1 \cap \mathcal{E}_1$, and $P_1 = \hat{H}_2 \cap \mathcal{E}_1 = \hat{Y} \cap \mathcal{E}_1$, see Fig. 5. The local type of the singularities at $Q_0$ and $Q_1$ are given by

$$Q_0 : \begin{cases} \hat{S} = \dfrac{1}{n_0}(n_1 n_2 n_3, -1) \\ \mathcal{E}_1 : x_1 = 0, \quad \hat{H}_0^{\text{red}} : x_0 = 0, \end{cases} \qquad Q_1 : \begin{cases} \hat{S} = \dfrac{1}{n_1}(-1, n_0 n_2 n_3) \\ \mathcal{E}_1 : x_0 = 0, \quad \hat{H}_1^{\text{red}} : x_1 = 0. \end{cases}$$

Around $P_1$ the surface $\hat{S}$ can be described inside $\frac{1}{n_2 n_3}(-1, n_0 n_1 n_3, n_0 n_1 n_2)$ as the set of zeros of $x_2^{n_2} - x_0^{b'_2} + x_3^{n_3} - x_0^{b'_3} x_2^{b_{32}} + (x_2^{n_2} - x_0^{b'_2}) R'_2(x_0, x_2)$ where $b'_i = b_{i,0}\frac{n}{n_0} + \cdots + b_{i,i-1}\frac{n}{n_{i-1}} - n$, $i = 2, 3$, and $R'_i(0, x_2) = 0$. Since the monomial with higher order will not play any role in the resolution of $S$, roughly speaking the

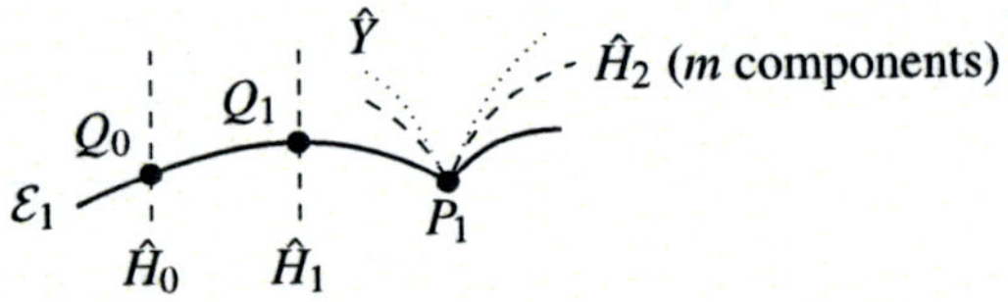


**Fig. 5** First step of the $\mathbb{Q}$-resolution of $S$

situation at $P_1$ with variables $[(x_0, x_2, x_3)]$ can be thought of as

$$P_1 : \begin{cases} \hat{S} = \{x_0^{b_2'} + x_2^{n_2} + x_3^{n_3} = 0\} \subset \frac{1}{n_2 n_3}(-1, n_0 n_1 n_3, n_0 n_1 n_2) \\ \mathcal{E}_1 : x_0 = 0, \quad \hat{H}_2^{\text{red}} : x_2 = 0, \quad \hat{Y} : x_0^{b_2'} + x_2^{n_2} = 0. \end{cases} \tag{6.2}$$

The points $Q_0$ and $Q_1$ already have $\mathbb{Q}$-normal crossings, so one does not need to blow them up anymore. Consider the previous coordinates around $P_1$ and let $\pi_2$ be the blow-up at $P_1$ with weights $w_2 = (1, \frac{b_2'}{n_2}, \frac{b_2'}{n_3})$. The exceptional divisor of $\pi_2$ is $E_2 = \mathbb{P}^2_{w_2}/G$ where $G$ is a cyclic group of order $n_2 n_3$ acting diagonally as in (6.2). The exceptional divisor $\mathcal{E}_2$ of the restriction $\varphi_2|_{\hat{S}} : \hat{S} \to \hat{S}$ is again a rational irreducible curve containing $2 + m$ cyclic quotient singular points of $\hat{S}$, namely $Q_{12} = \mathcal{E}_1 \cap \mathcal{E}_2$, $P_2 = \hat{Y} \cap \mathcal{E}_2$, and $m$ points $Q_{2j} \in \hat{H}_2 \cap \mathcal{E}_2$ (Fig. 6).

The composition $\varphi = \varphi_1 \circ \varphi_2 : \hat{S} \to S$ is a $\mathbb{Q}$-resolution of $S$ and the order of the groups at $Q_{12}$, $Q_{2j}$, and $P_2$ are $d$, $n_2$, and $\frac{n_3}{m}$, respectively. Since the $\mathbb{Q}$-resolution graph is a tree and the exceptional divisors are isomorphic to $\mathbb{P}^1$ the link of $S$ is a rational homology sphere. In order to compute $\det S$ one needs to calculate the self-intersection numbers $\mathcal{E}_i^2 = -a_i$, $i = 1, 2$, which can be done by exploiting our information on the curve $Y$ in the partial resolution of $S$. First, note that the intersection of $\mathcal{E}_2$ with $\hat{Y}$ at $P_2$ is $m$. Second,

$$\varphi^* Y = \hat{Y} + N_1 \mathcal{E}_1 + N_2 \mathcal{E}_2$$

where $N_1 = n_0 n_1 n_2 n_3 = n$ and $N_2 = \frac{b_2' + n}{m}$. Since $\mathcal{E}_i \cdot \varphi^* Y = 0$, $i = 1, 2$, one obtains that $a_1 = \frac{N_2}{N_1 d}$ and $a_2 = \frac{m + \frac{N_1}{d}}{N_2}$. Therefore the determinant of the intersection matrix is given by

$$\det(A) = \det \begin{pmatrix} -a_1 & \frac{1}{d} \\ \frac{1}{d} & -a_2 \end{pmatrix} = \frac{m}{N_1 d}.$$

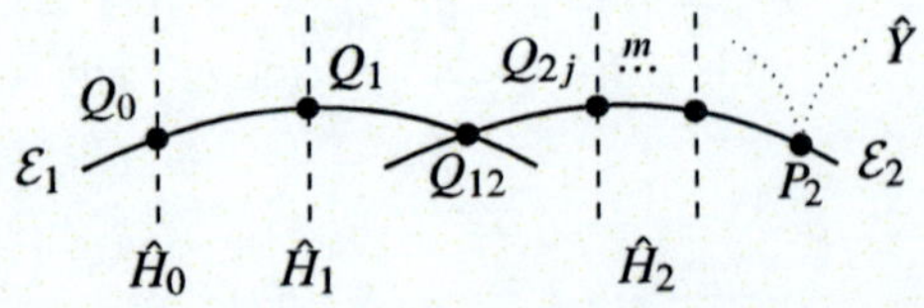


**Fig. 6** $\mathbb{Q}$-resolution of $S$

By Lemma 5.7 one has

$$\det S = \det(-A) n_0 n_1 d n_2^m \frac{n_3}{m} = n_2^{m-1}.$$

Therefore by Corollary 5.3 the link of $S$ is a integral homology sphere if and only if $\det S = 1$, or equivalently, $m = 1$ as claimed.

*Remark 6.5* If the exponents $n_i$'s are not pairwise coprime, then $\mathcal{E}_1 = \bigsqcup_j \mathcal{E}_{1j}$ has $n_{23} = \gcd(n_2, n_3)$ irreducible components and $\mathcal{E}_2$ is irreducible. They have genus

$$g(\mathcal{E}_{1j}) = \frac{1}{2}\left(\frac{n_{123}}{n_{23}} - 1\right)\left(\frac{n_{023}}{n_{23}} - 1\right) \quad \text{and} \quad g(\mathcal{E}_2) = \frac{1}{2}(n_{23} - 1)(m - 1),$$

where $m = \gcd(n_3, b)$ with $b = b_{20} n_1 + b_{21} n_0$. The determinant of $S$ can be rewritten as

$$\det S = \left(\frac{b}{m}\right)^{n_{23}-1} \left(\frac{N_1}{\alpha}\right)^{n_{123}-n_{23}} \left(\frac{N_1}{\beta}\right)^{n_{023}-n_{23}} \left(\frac{n_2}{n_{23}}\right)^{m-1}$$

where $N_1 = \operatorname{lcm}(n_0, n_1, n_2, n_3)$, $\alpha = \operatorname{lcm}(n_1, n_2, n_3)$, $\beta = \operatorname{lcm}(n_0, n_2, n_3)$, and $n_{ijk} = \frac{n_i n_j n_k}{\operatorname{lcm}(n_i, n_j, n_k)}$. From here it can easily be shown that the link of $S$ is an integral homology sphere if and only if $\gcd(n_i, n_j) = 1$, $i \neq j$, and $m = 1$. The details are left to the reader.

**Acknowledgments** We thank the anonymous referee for their valuable suggestions which have helped to improve the manuscript.

Partially supported by MTM2016-76868-C2-2-P and Gobierno de Aragón (Grupo de referencia "Álgebra y Geometría"), E22_17R and E22_20R, cofunded by Feder 2014–2020 "Construyendo Europa desde Aragón". The third author is also partially supported by FQM-333 "Junta de Andalucía".

## References

1. Artal, E.: Forme de Jordan de la monodromie des singularités superisolées de surfaces. Mem. Am. Math. Soc. **109**(525), x+84 (1994)
2. Artal, E.: Sur les couples de Zariski. J. Algebraic Geom. **3**(2), 223–247 (1994)
3. Artal, E., Carmona, J.: Zariski pairs, fundamental groups and Alexander polynomials. J. Math. Soc. Jpn **50**(3), 521–543 (1998)
4. Artal, E., Cogolludo-Agustín, J.I., Martín-Morales, J.: Cyclic branched coverings of surfaces with abelian quotient singularities. Indiana U. Math. J. (2019). arXiv:1912.08670 [math.AG]
5. Artal, E., Cogolludo-Agustín, J.I., Matei, D.: Orbifold groups, quasi-projectivity and covers. J. Singul. **5**, 33–47 (2012)
6. Artal, E., Cogolludo-Agustín, J.I., Ortigas-Galindo, J.: Kummer covers and braid monodromy. J. Inst. Math. Jussieu **13**(3), 633–670 (2014)
7. Artal, E., Fernández de Bobadilla, J., Luengo, I., Melle, A.: Milnor number of weighted-Lê-Yomdin singularities. Int. Math. Res. Not. IMRN **2010**(22), 4301–4318 (2010)

8. Artal, E., Martín-Morales, J., Ortigas-Galindo, J.: Intersection theory on abelian-quotient $V$-surfaces and $\mathbf{Q}$-resolutions. J. Singul. **8**, 11–30 (2014)
9. Blasco-García, R., Cogolludo-Agustín, J.I.: Quasi-projectivity of even Artin groups. Geom. Topol. **22**(7), 3979–4011 (2018)
10. Borodzik, M., Livingston, C.: Heegaard Floer homology and rational cuspidal curves. Forum Math. Sigma **2**, e28, 23 (2014)
11. Cogolludo-Agustín, J.I.: Fundamental group for some cuspidal curves. Bull. Lond. Math. Soc. **31**(2), 136–142 (1999)
12. Dolgachev, I.: Weighted projective varieties. In: Group actions and vector fields (Vancouver, B.C., 1981). Lecture Notes in Mathematics, vol. 956, pp. 34–71. Springer, Berlin (1982)
13. Esnault, H.: Fibre de Milnor d'un cône sur une courbe plane singulière. Invent. Math. **68**(3), 477–496 (1982)
14. Esnault, H., Viehweg, E.: Revêtements cycliques. In: Algebraic threefolds (Varenna, 1981). Lecture Notes in Mathematics, vol. 947, pp. 241–250. Springer, Berlin-New York (1982)
15. Fernández de Bobadilla, J., Luengo, I., Melle, A., Némethi, A.: On rational cuspidal projective plane curves. Proc. Lond. Math. Soc. (3) **92**(1), 99–138 (2006)
16. Flenner, H., Zaĭdenberg, M.G.: On a class of rational cuspidal plane curves. Manuscripta Math. **89**(4), 439–459 (1996)
17. Fujita, T.: On the topology of noncomplete algebraic surfaces. J. Fac. Sci. Univ. Tokyo Sect. IA Math. **29**(3), 503–566 (1982)
18. Fulton, W.: Intersection theory, 2nd ed.. Ergebnisse der Mathematik und ihrer Grenzgebiete. 3. Folge. A Series of Modern Surveys in Mathematics, vol. 2, Springer, Berlin (1998)
19. Koras, M., Palka, K.: The Coolidge-Nagata conjecture. Duke Math. J. **166**(16), 3085–3145 (2017)
20. Koras, M., Palka, K.: Complex planar curves homeomorphic to a line have at most four singular points (2019). Preprint arXiv:1905.11376 [math.AG]
21. Libgober, A.: Alexander polynomial of plane algebraic curves and cyclic multiple planes. Duke Math. J. **49**(4), 833–851 (1982)
22. Libgober, A.: Characteristic varieties of algebraic curves. In: Applications of Algebraic Geometry to Coding Theory, Physics and Computation (Eilat, 2001), pp. 215–254. Kluwer, Dordrecht (2001)
23. Loeser, F., Vaquié, M.: Le polynôme d'Alexander d'une courbe plane projective. Topology **29**(2), 163–173 (1990)
24. Luengo, I.: The $\mu$-constant stratum is not smooth. Invent. Math. **90**(1), 139–152 (1987)
25. Luengo, I., Melle, A., Némethi, A.: Links and analytic invariants of superisolated singularities. J. Algebraic Geom. **14**(3), 543–565 (2005)
26. Martín-Morales, J.: Semistable reduction of a normal crossing $\mathbb{Q}$-divisor. Ann. Mat. Pura Appl. (4) **195**(5), 1749–1769 (2016)
27. Martín-Morales, J., Veys, W., Vos, L.: The monodromy conjecture for a space monomial curve with a plane semigroup. Publ. Math. (accepted)
28. Mumford, D.: The topology of normal singularities of an algebraic surface and a criterion for simplicity. Inst. Hautes Études Sci. Publ. Math. **9**, 5–22 (1961)
29. Neumann, W.D., Wahl, J.: Complex surface singularities with integral homology sphere links. Geom. Topol. **9**, 757–811 (2005)
30. Project Jupyter, et al.: Binder 2.0—Reproducible, interactive, sharable environments for science at scale. In: Akici, F., Lippa, D., Niederhut, D., Pacer, M (eds.) Proceedings of the 17th Python in Science Conference, pp. 113–120 (2018)
31. Sabbah, C.: Modules d'Alexander et $\mathcal{D}$-modules. Duke Math. J. **60**(3), 729–814 (1990)
32. Steenbrink, J.H.M.: Mixed Hodge structure on the vanishing cohomology. Real and complex singularities. In: Proceedings of the Ninth Nordic Summer School/NAVF Sympos. Math., Oslo, 1976, pp. 525–563 (1977)
33. Stein et al., W.A.: Sage Mathematics Software (Version 8.9). The Sage Development Team (2019). http://www.sagemath.org

34. The GAP Group, GAP—Groups, Algorithms, and Programming, Version 4.10.2 (2019)
35. Vaquié, M.: Irrégularité des revêtements cycliques des surfaces projectives non singulières. Am. J. Math. **114**(6), 1187–1199 (1992)
36. Zariski, O.: On the irregularity of cyclic multiple planes. Ann. of Math. (2) **32**(3), 485–511 (1931)

# Normal Reduction Numbers of Normal Surface Singularities

**Tomohiro Okuma**

*Dedicated to Professor András Némethi on the occasion of his sixtieth birthday*

**Abstract** This article consists of two parts. The first part is a survey on the normal reduction numbers of normal surface singularities. It includes results on elliptic singularities, cone-like singularities and homogeneous hypersurface singularities. In the second part, we prove a new results on the normal reduction numbers and related invariants of Brieskorn complete intersections.

**Keywords** Normal reduction number · Normal surface singularity · Geometric genus · Elliptic singularity · Brieskorn complete intersection · Homogeneous hypersurface singularity

**Subject Classifications** Primary 14J17; Secondary 14B05, 32S25, 13B22

## 1 Introduction

In this paper, we survey results on the normal reduction numbers of normal complex surface singularities and some related topics [24, 26, 28, 29], and prove new results on the normal reduction numbers of Brieskorn complete intersections. The normal reduction number has appeared in the study of normal Hilbert polynomials from a ring-theoretic point of view (cf. [6, 14]). We study the normal reduction numbers of the local ring of normal surface singularities using resolution of singularities, and we wish to know what kind of geometric property of singularities relates to the normal reduction numbers.

---

T. Okuma (✉)
Department of Mathematical Sciences, Yamagata University, Yamagata, Japan
e-mail: okuma@sci.kj.yamagata-u.ac.jp



J. Fernández de Bobadilla et al. (eds.), *Singularities and Their Interaction with Geometry and Low Dimensional Topology*, Trends in Mathematics,
https://doi.org/10.1007/978-3-030-61958-9_8

Let us briefly recall some basic facts about integral closure and reduction of ideals in a local ring. Let $(A, \mathfrak{m})$ be a Noetherian local ring and $I$ an $\mathfrak{m}$-primary ideal (namely, $\sqrt{I} = \mathfrak{m}$). Let $\overline{I}$ denote the integral closure of $I$, that is, $\overline{I}$ is an ideal of $A$ consists of all elements $z \in A$ such that $z^n + c_1 z^{n-1} + \cdots + c_n = 0$ for some $n \geq 1$ and $c_i \in I^i$ $(i = 1, \ldots, n)$. The ideal $I$ is said to be integrally closed if $I = \overline{I}$. An ideal $Q \subset I$ is called a *reduction* of $I$ if $I^{n+1} = QI^n$ for some $n \geq 0$. It is known that an ideal $Q$ is a reduction of $I$ if and only if $I \subset \overline{Q}$ (cf. [5, 1.2.5]). For a reduction $Q$ of $I$, $\mathrm{r}_Q(I) := \min\left\{n \mid I^{n+1} = QI^n\right\}$ is called the reduction number of $I$ with respect to $Q$.

Let $(V, p)$ be a normal complex surface singularity[1] and $\mathcal{O}_{V,p}$ the local ring of the singularity with maximal ideal $\mathfrak{m}$. Let $I \subset \mathcal{O}_{V,p}$ be an $\mathfrak{m}$-primary integrally closed ideal. It is known that any minimal reduction of $I$ is generated by two elements and that two general elements of $I$ generate a minimal reduction of $I$ (see [5, 8.3.7, 8.6.6]). Suppose that $Q$ is a minimal reduction of $I$. We define two normal reduction numbers, which are analogues of the reduction number $r_Q(I)$, as follows:

$$\mathrm{nr}(I) = \min\{n \in \mathbb{Z}_{\geq 0} \mid \overline{I^{n+1}} = Q\overline{I^n}\},$$

$$\bar{\mathrm{r}}(I) = \min\{n \in \mathbb{Z}_{\geq 0} \mid \overline{I^{N+1}} = Q\overline{I^N} \text{ for every } N \geq n\}.$$

We note that $\mathrm{nr}(I)$ and $\bar{\mathrm{r}}(I)$ are independent of the choice of $Q$ (see e.g. [4, Theorem 4.5], Proposition 3.2), though $r_Q(I)$ is not independent of the choice of a minimal reduction $Q$ in general. It is obvious by the definition that $\mathrm{nr}(I) \leq \bar{\mathrm{r}}(I)$. We will show that $\bar{\mathrm{r}}(I) \leq p_g(V, p) + 1$ in general (see Proposition 3.2). We can also show that for any integer $g \geq 2$ there exists a singularity $(V, p)$ with $\mathrm{nr}(I) = 1$ and $\bar{\mathrm{r}}(I) = p_g(V, p) + 1 = g + 1$ (Example 4.5). We define

$$\mathrm{nr}(V, p) = \max\{\mathrm{nr}(J) \mid J \text{ is an } \mathfrak{m}\text{-primary integrally closed ideal of } \mathcal{O}_{V,p}\},$$

$$\bar{\mathrm{r}}(V, p) = \max\{\bar{\mathrm{r}}(J) \mid J \text{ is an } \mathfrak{m}\text{-primary integrally closed ideal of } \mathcal{O}_{V,p}\}.$$

The invariant $\bar{\mathrm{r}}(V, p)$ naturally appears in several situation as follows. For any $\mathfrak{m}$-primary integrally closed ideal $I \subset \mathcal{O}_{V,p}$, there exist a resolution $\pi : X \to V$ and a divisor $Z$ on $X$ such that $\mathcal{O}_X(-Z)$ is $\pi$-generated and $I = \pi_*\mathcal{O}_X(-Z)_p$ (see Sect. 2). Let $r := \bar{\mathrm{r}}(I)$. By the definition of $\bar{\mathrm{r}}$ and Proposition 3.2, we have the following:

(1) Briançon-Skoda type inclusion (cf. [3, 13]): $\overline{I^{r+k}} \subset Q^k$ for $k \geq 1$.
(2) The natural homomorphism $\pi_*\mathcal{O}_X(-nZ) \otimes \pi_*\mathcal{O}_X(-Z) \to \pi_*\mathcal{O}_X(-(n+1)Z)$ is surjective for $n \geq r$.

[1] In our papers [24, 26, 28, 29], we treat a singularity $(\mathrm{Spec}\, A, \mathfrak{m})$, where $(A, \mathfrak{m})$ is an excellent normal two-dimensional local ring such that the residue field $k$ is algebraically closed and $k \subset A$.

(3) The function $\phi(n) := \dim_{\mathbb{C}} H^0(\mathcal{O}_X)/H^0(\mathcal{O}_X(-nZ))$ is a polynomial function of $n$ for $n \geq r$; note that $\phi(n) = \chi(\mathcal{O}_{nZ}) + h^1(\mathcal{O}_X) - h^1(\mathcal{O}_X(-nZ))$ by Kato's Riemann-Roch Theorem ([8]).

So we expect that the normal reduction numbers can characterize good singularities. For example, we see that $(V, p)$ is a rational singularity if and only if $\bar{\mathrm{r}}(V, p) = 1$ (see Proposition 3.6). However, we can only show that $\bar{\mathrm{r}}(V, p) = 2$ if $(V, p)$ is an elliptic singularity (see Theorem 3.9, Proposition 5.13). At present, we have computed the normal reduction numbers only for some special cases, and we do not know whether those invariants are topological or not.

This paper is organized as follows. Sections 2–4 are devoted to a survey of fundamental results on the normal reduction numbers and some related topics. We refer the reader to [20] and [32] for basic facts about normal surface singularities. In Sect. 2, we set up notation and briefly recall the basic results on the cohomology groups of ideal sheaves of cycles on a resolution space. Then we mention a question about the range of the dimension of those cohomology groups. In Sect. 3, we give a relation between the normal reduction numbers and the dimension of the cohomology groups associated with an $\mathfrak{m}$-primary integrally closed ideal in $\mathcal{O}_{V,p}$ and review fundamental results on the normal reduction numbers. Then we review the results on elliptic singularities. In Sect. 4, we consider the cone-like singularities, namely, those homeomorphic to the cone over a nonsingular curve. We give an upper bound of $\bar{\mathrm{r}}$ using the genus and gonality of the curve and the self-intersection number of the fundamental cycles. Then we show a formula for the normal reduction numbers of homogeneous hypersurface singularities. In Sect. 5, we prove an explicit formula for $\bar{\mathrm{r}}$ of the maximal ideal of a Brieskorn complete intersection and apply the formula to classify elliptic singularities, which are natural generalization of the results about Brieskorn hypersurface singularities in [28].

## 2 Cycles and Cohomology

Let $(V, p)$ be a normal complex surface singularity, namely, the germ of a normal complex surface $V$ at $p \in V$. We always assume that $V$ is Stein and suitably small. Let $\pi : X \to V$ denote a resolution of the singularity $(V, p)$ with exceptional set $E = \pi^{-1}(p)$ and let $\{E_i\}_{i \in \mathcal{I}}$ denote the set of irreducible components of $E$. We call a divisor on $X$ supported in $E$ a *cycle* and denote by $\sum \mathbb{Z}E_i$ the group of cycles.

For a function $h \in H^0(\mathcal{O}_X(-E))$, we denote by $(h)_E \in \sum \mathbb{Z}E_i$ the exceptional part of the divisor $\mathrm{div}_X(h)$; so, $\mathrm{div}_X(h) - (h)_E$ is an effective divisor containing no components of $E$. We simply write $(h)_E$ instead of $(h \circ \pi)_E$ for $h \in \mathfrak{m}$.

An element of $\sum \mathbb{Q}E_i := (\sum \mathbb{Z}E_i) \otimes \mathbb{Q}$ is called a $\mathbb{Q}$-*cycle*. A $\mathbb{Q}$-cycle $D$ is said to be *nef* (resp. *anti-nef*) if $DE_i \geq 0$ (resp. $DE_i \leq 0$) for all $i \in \mathcal{I}$. Note that if $D \neq 0$ is anti-nef, then $D \geq E$.

**Definition 2.1** The *maximal ideal cycle* on $X$ is the minimum of $\{(h)_E \mid h \in \mathfrak{m}\}$ and denoted by $M_X$. There exists a $\mathbb{Q}$-cycle $Z_{K_X}$ such that $(K_X + Z_{K_X})E_i = 0$ for every $i \in \mathcal{I}$, where $K_X$ is a canonical divisor on $X$. We call $Z_{K_X}$ the *canonical cycle* on $X$.

In the following, we assume that $Z > 0$ is a cycle such that $\mathcal{O}_X(-Z)$ *has no fixed component*, namely, there exists a function $h \in H^0(\mathcal{O}_X(-Z))$ such that $(h)_E = Z$. We say that $\mathcal{O}_X(-Z)$ is *generated* if it is $\pi$-generated (i,e., $\pi^*\pi_*\mathcal{O}_X(-Z) \to \mathcal{O}_X(-Z)$ is surjective). For any coherent sheaf $\mathcal{F}$ on $X$, we write $H^i(\mathcal{F}) = H^i(X, \mathcal{F})$ and $h^i(\mathcal{F}) = \dim_{\mathbb{C}}(H^i(\mathcal{F}))$.

**Definition 2.2** The *geometric genus* of the singularity $(V, p)$ is defined by $p_g(V, p) = h^1(\mathcal{O}_X)$.

**Definition 2.3** Let $A \geq 0$ be an effective cycle on $X$ and let

$$h(A) = \max\left\{h^1(\mathcal{O}_B) \;\middle|\; B \in \sum \mathbb{Z}E_i,\ B \geq 0,\ \mathrm{Supp}(B) \subset \mathrm{Supp}(A)\right\}.$$

We put $h^1(\mathcal{O}_B) = 0$ if $B = 0$. There exists a unique minimal cycle $C$ such that $h^1(\mathcal{O}_C) = h(A)$ (cf. [32, 4.8]). We call $C$ the *cohomological cycle* of $A$. Note that $p_g(V, p) = h(E)$ and that if $(V, p)$ is Gorenstein and $\pi$ is the minimal resolution, then $Z_{K_X}$ is the cohomological cycle of $E$ ([32, 4.20]).

We define a reduced cycle $A^\perp$ to be the sum of the components $E_i \subset E$ such that $AE_i = 0$.

*Remark 2.4* Let $F_1, \ldots, F_k$ be the connected component of $Z^\perp$ and let $(V_i, p_i)$ be the normal surface singularity obtained by contracting $F_i$. If $C$ is the cohomological cycle of $Z^\perp$, we have

$$h^1(\mathcal{O}_C) = \sum_{i=1}^{k} p_g(V_i, p_i).$$

**Definition 2.5** Let $q(Z) = h^1(\mathcal{O}_X(-Z))$ and $q_Z(n) = h^1(\mathcal{O}_X(-nZ))$ for $n \geq 0$. Let $s(Z) = \min\left\{n \in \mathbb{Z}_{\geq 0} \mid q_Z(n) = q_Z(n+1)\right\}$.

**Proposition 2.6 (See [26, §3], [24, 3.6])** *We have the following.*

(1) $q_Z(n) \geq q_Z(n+1)$ *for every integer* $n \geq 0$.
(2) *If* $q_Z(1) = p_g(V, p)$, *namely,* $s(Z) = 0$, *then* $q(n) = p_g(V, p)$ *for* $n \geq 0$.
(3) *If* $\mathcal{O}_X(-Z)$ *is generated, then* $q_Z(n) = q_Z(s(Z)) = h^1(\mathcal{O}_C)$ *for* $n \geq s(Z)$, *where* $C$ *is the cohomological cycle of* $Z^\perp$.
(4) $\mathcal{O}_X(-nZ)$ *is generated for* $n > s(Z)$.

We are interested in the range of the function $q$. Let $\mathcal{A}$ (resp. $\mathcal{A}'$) denotes the set of the pairs $(Y, W)$ such that $W > 0$ is a cycle on a resolution $Y \to V$ such that

$\mathcal{O}_Y(-W)$ is generated (resp. has no fixed components). Clearly, $\mathcal{A} \subset \mathcal{A}'$. Let

$$q(\mathcal{A}) = \left\{ h^1(\mathcal{O}_Y(-W)) \mid (Y, W) \in \mathcal{A} \right\}, \; q(\mathcal{A}') = \left\{ h^1(\mathcal{O}_Y(-W)) \mid (Y, W) \in \mathcal{A}' \right\}.$$

By Proposition 2.6, we have

$$q(\mathcal{A}) \subset q(\mathcal{A}') \subset \{0, 1, \dots, p_g(V, p)\}.$$

The proof of the following theorem is included in the proof of [24, 3.12].

**Proposition 2.7** *We have the equality*

$$q(\mathcal{A}') = \{0, 1, \dots, p_g(V, p)\}.$$

**Conjecture 2.8** *For every normal complex surface singularity, the equality $q(\mathcal{A}) = q(\mathcal{A}')$ holds.*

At present, we have the equality $q(\mathcal{A}) = q(\mathcal{A}')$ only for a few cases (cf. Proposition 3.11, Example 4.5). Some results related to Conjecture 2.8 are obtained in [16].

The next lemma is used in Sect. 5. For a $\mathbb{Q}$-cycle $D$, let $\mathcal{O}_X(D) = \mathcal{O}_X(\lfloor D \rfloor)$, where $\lfloor D \rfloor$ denotes the integral part of $D$.

**Lemma 2.9** *Let $C < E$ be a reduced cycle and $\{I_n\}_{n \in \mathbb{Z}_{\geq 0}}$ a filtration of $\mathcal{O}_{V,p}$ such that $(h)_E \geq nC$ for all $n \in \mathbb{Z}_{\geq 0}$ and all $h \in I_n \setminus \{0\}$ and that $\bigoplus_{n \geq 0} I_n/I_{n+1}$ is reduced. Assume that there exists an anti-nef $\mathbb{Q}$-cycle $\widetilde{C} = \sum a_i E_i$ such that $a_i = 1$ for $E_i \leq C$ and $\widetilde{C} E_i = 0$ for every $E_i \not\leq C$. Moreover assume that there exists an integer $d > 0$ such that $d\widetilde{C} \in \sum \mathbb{Z} E_i$ and $(h)_E = d\widetilde{C}$ for some $h \in I_d$. Then $I_n = \widetilde{I}_n := \pi_* \mathcal{O}_X(-n\widetilde{C})_p$.*

***Proof*** First we show that $I_n \subset \widetilde{I}_n$ for every $n \geq 0$. Let $h \in I_n$ and $\Delta = (h)_E - n\widetilde{C}$. We write $\Delta = \Delta_1 - \Delta_2$, where $\Delta_1$ and $\Delta_2$ are effective and have no common components. Since $(h)_E \geq nC$, by the assumption on $\widetilde{C}$, we have $\mathrm{Supp}(\Delta_2) \subset \mathrm{Supp}(\widetilde{C} - C) = \mathrm{Supp}(E - C)$, and hence $\widetilde{C}\Delta_2 = 0$. If $\Delta_2 \neq 0$, then $0 < -\Delta_2^2 \leq \Delta\Delta_2 = (h)_E \Delta_2$; it contradicts that $(h)_E$ is anti-nef. Hence $\Delta = \Delta_1 \geq 0$, namely, $h \in \widetilde{I}_n$.

From the arguments in §2.2–2.4 of [36], since $\bigoplus_{n \geq 0} I_n/I_{n+1}$ is reduced, we have a $\mathbb{Q}$-cycle $D > 0$ such that $I_n = \pi_* \mathcal{O}_X(-nD)_p$ for all $n \in \mathbb{Z}_{\geq 0}$, and we may assume that $dD \in \sum \mathbb{Z} E_i$ and $\mathcal{O}_X(-dD)$ is generated. The inclusion $I_d \subset \widetilde{I}_d$ implies that $dD \geq d\widetilde{C}$. Since there exists $h \in I_d$ such that $d\widetilde{C} = (h)_E \geq dD$, we obtain $\widetilde{C} = D$.

## 3 Cohomology and Normal Reduction Numbers

Let $\mathfrak{m} \subset \mathcal{O}_{V,p}$ denote the maximal ideal. In the following, we always assume that $I \subset \mathcal{O}_{V,p}$ is an $\mathfrak{m}$-primary integrally closed ideal, namely, $I$ satisfies that $\sqrt{I} = \mathfrak{m}$ and $\bar{I} = I$. Let $Q$ be a minimal reduction of $I$. Then there exist a resolution $\pi : X \to V$ and a cycle $Z > 0$ such that

$$I = I_Z := \pi_* \mathcal{O}_X(-Z)_p$$

and $I\mathcal{O}_X = \mathcal{O}_X(-Z)$ (cf. [12, §6]). In this case, we say that $I$ is *represented* by a cycle $Z$ on $X$. We use the symbol "$I_Z$" only when $\mathcal{O}_X(-Z)$ is generated. Conversely, such an ideal $I_Z$ is $\mathfrak{m}$-primary and integrally closed. Note that $\overline{I_Z I_{Z'}} = I_{Z+Z'}$. Thus we can write

$$\mathrm{nr}(I_Z) = \min \left\{ n \in \mathbb{Z}_{>0} \mid I_{(n+1)Z} = QI_{nZ} \right\},$$
$$\bar{\mathrm{r}}(I_Z) = \min \left\{ n \in \mathbb{Z}_{>0} \mid I_{(m+1)Z} = QI_{mZ},\ m \geq n \right\}.$$

In the rest of this section, we always assume that $I$ is represented by a cycle $Z$ on $X$, namely, $I = I_Z$.

**Definition 3.1** We put $q(I) = q(Z) = h^1(\mathcal{O}_X(-Z))$; this is independent of the representation of $I$ (cf. [25, Lemma 3.4]).

**Proposition 3.2 (Cf. [26, §2])** *Let $q_I(n) := q(\overline{I^n}) = q_Z(n)$ for $n \geq 0$. We have the following.*

(1) *For any integer $n \geq 1$, we have*

$$2q_I(n) + \dim_{\mathbb{C}}(\overline{I^{n+1}}/Q\overline{I^n}) = q_I(n+1) + q_I(n-1).$$

*In particular,*

$$\mathrm{nr}(I) = \min \left\{ n \in \mathbb{Z}_{\geq 0} \mid q_I(n-1) - q_I(n) = q_I(n) - q_I(n+1) \right\}.$$

(2) *We have*

$$\bar{\mathrm{r}}(I) = \min \left\{ n \in \mathbb{Z}_{\geq 0} \mid q_I(n-1) = q_I(n) \right\}.$$

*In particular, $\bar{\mathrm{r}}(I) = s(Z) + 1 \leq p_g(V, p) + 1$ and $q_I(n) = q_I(s(Z))$ for every $n \geq s(Z)$.*

***Proof*** We write $H^i(Z) := H^i(\mathcal{O}_X(-Z))$. Let $h_1, h_2 \in H^0(Z)$ and $Q := (h_1, h_2) \subset \mathcal{O}_{V,p}$. Suppose that $h_1, h_2$ are sufficiently general so that $Q$ is a minimal reduction of $I = I_Z$ and that the following sequence is exact:

$$0 \to \mathcal{O}_X(-(n-1)Z) \xrightarrow{(h_1\ h_2)} \mathcal{O}_X(-nZ)^{\oplus 2} \xrightarrow{\binom{-h_2}{h_1}} \mathcal{O}_X(-(n+1)Z) \to 0.$$

Taking cohomology, we obtain the long exact sequence:

$$0 \to \overline{I^n}Q \to \overline{I^{n+1}} \to H^1((n-1)Z) \to H^1(nZ)^{\oplus 2} \to H^1((n+1)Z) \to 0.$$

This yields (1). We write

$$\dim_{\mathbb{C}}(\overline{I^{n+1}}/Q\overline{I^n}) = \Delta_I(n-1) - \Delta_I(n) \geq 0,$$

where $\Delta_I(n) = q_I(n) - q_I(n+1)$. By Proposition 2.6 (1), $\Delta_I(n) \geq 0$. Therefore, if $\Delta_I(n-1) = 0$, then $\Delta_I(n+k) = 0$ for $k \geq 0$. Hence we have (2).

By the argument similar to the proof of Proposition 3.2, we have

**Proposition 3.3 ([28, 2.9])** *Let* $r = \mathrm{nr}(I)$. *Then*

$$r(r-1)/2 + q(r) \leq p_g(V, p).$$

In [28, 3.13], the hypersurface $V = \{x^a + y^b + z^c = 0\} \subset \mathbb{C}^3$ with $p_g(V, o) = r(r-1)/2$ are classified.

*Remark 3.4* Let $X \to Y$ be the contraction of $Z^{\perp}$ (cf. Remark 2.4). Then we obtain that $\bar{\mathrm{r}}(I) - 1 = \min\{n \in \mathbb{Z}_{\geq 0} \mid H^1(I^n\mathcal{O}_Y) = 0\}$ (cf. [24, 3.8]).

*Remark 3.5* The ideal $I$ is called the $p_g$*-ideal* if $q(I) = p_g(V, p)$. It immediately follows from Proposition 2.6 that $\bar{\mathrm{r}}(I) = 1$ if and only if $I$ is a $p_g$-ideal. Moreover, the following are equivalent (see [25, 3.10], [26, 4.1]):

- $I$ is a $p_g$-ideal.
- $\mathcal{O}_C(-Z) \cong \mathcal{O}_C$, where $C$ is the cohomological cycle of $E$.
- The Rees algebra $\bigoplus_{n\geq 0} I^n$ is a Cohen-Macaulay normal domain.

The $p_g$-ideals have nice properties and studied in [25–27]. For example, if $I$ is a $p_g$-ideal and $J$ an $\mathfrak{m}$-primary integrally closed ideal of $\mathcal{O}_{V,p}$, then $IJ = \overline{IJ}$ and $q(IJ) = q(J)$; in particular, $p_g$-ideals form a semigroup with respect to the product (cf. [25, 2.6, 3.5]).

The singularity $(V, p)$ is said to be *rational* if $p_g(V, p) = 0$. Rational surface singularities can be characterized in many ways [1, 10, 12, 22, 27]. We have also a characterization in terms of the normal reduction numbers as follows.

**Proposition 3.6 ([29, 1.1])** *The following are equivalent:*

(1) *$(V, p)$ is a rational singularity.*
(2) *Every $\mathfrak{m}$-primary integrally closed ideal in $\mathcal{O}_{V,p}$ is a $p_g$-ideal.*

(3) $\bar{r}(V, p) = 1$.
(4) $\mathrm{nr}(V, p) = 1$.

*Remark 3.7* The singularities with $\bar{r}(\mathfrak{m}) = 1$ ($\mathfrak{m}$ is a $p_g$-ideal in this case) have been characterized in [31, 5.2]. In case $(V, p)$ is Gorenstein and $p_g(V, p) > 0$, the condition $\bar{r}(\mathfrak{m}) = 1$ implies that $(V, p)$ is an elliptic double point (see [26, 4.3], [24, 4.10]).

The elliptic singularities were introduced by P. Wagreich, and the theory of those singularities were developed by Wagreich [37], H. Laufer [11], M. Reid [32, §4], S.S.-T. Yau [38–41], M. Tomari [34, 35], and A. Némethi [21], Nagy–Némethi [17, 18].

Let $Z_f$ denote the *fundamental cycle* on $X$, namely, the minimal non-zero anti-nef cycle. The *fundamental genus* $p_f(V, p)$ is defined by $p_f(V, p) = p_a(Z_f) = 1 - \chi(\mathcal{O}_{Z_f})$. By the Riemann-Roch formula, $p_f(V, p) = Z_f(Z_f + K_X)/2 + 1$. This is independent of the choice of a resolution, and hence a topological invariant of the singularity $(V, p)$.

**Definition 3.8** The singularity $(V, p)$ is said to be *elliptic* if $p_f(V, p) = 1$.

The following are well-known:

(1) For any positive integer $m$, there exists an elliptic singularity $(V, p)$ with $p_g(V, p) = m$ (Yau [41, §2]).
(2) For any elliptic surface singularity $(V', p')$, there exists an elliptic singularity $(V, p)$ with $p_g(V, p) = 1$ such that $(V', p')$ and $(V, p)$ have the same topological type (Laufer [11, Theorem 4.1]).

**Theorem 3.9 (See [24, §3])** *If $(V, p)$ is elliptic, then $\mathrm{nr}(V, p) = \bar{r}(V, p) = 2$. In fact, $s(W) = 1$ for any $(Y, W) \in \mathcal{A}'$.*

The point of the proof of Theorem 3.9 is as follows. Using Yau's elliptic sequences and Röhr's vanishing theorem [33], we have

**Proposition 3.10 (Cf. [24, 3.11])** *If $(V, p)$ is elliptic and $W > 0$ is a cycle on $X$ such that $\mathcal{O}_X(-W)$ has no fixed component, then $h^1(\mathcal{O}_X(-W)) = h^1(\mathcal{O}_{C_W})$, where $C_W$ is the cohomological cycle of $W^{\perp}$.*

This proposition implies that $h^1(\mathcal{O}_{C_Z}) = q_Z(n)$ for $n \geq 1$ (take $W = nZ$). If $I$ is not a $p_g$-ideal, then $s(Z) = 1$, and $\bar{r}(I) = 2$ by Proposition 3.2 (2).

**Proposition 3.11 (cf. [24, 3.12])** *If $(V, p)$ is elliptic, then $q(\mathcal{A}) = q(\mathcal{A}')$.*

***Proof*** By Proposition 2.7, there exist a resolution $Y$ and cycles $W_0, \ldots, W_{p_g(V,p)}$ on $Y$ such that $q(W_i) = i$. Since $s(W_i) = 1$, Propositions 2.6 and 3.10 imply that $\mathcal{O}_Y(-2W_i)$ is generated and $q(W_i) = q(2W_i)$.

**Problem 3.1** Characterize the singularities $(V, p)$ with $\bar{r}(V, p) = 2$. Is the converse of Theorem 3.9 true?

We define a topological invariant min-$p_g(V, p)$ to be the minimum of the geometric genus $p_g$ of normal complex surface singularities homeomorphic to $(V, p)$. For example, if $(V, p)$ is elliptic, then $\bar{r}(V, p) - 1 = 1 = \text{min-}p_g(V, p)$ by Theorem 3.9 and Laufer's result mentioned above. Let us recall that $\bar{r}(V, p) \leq p_g(V, p) + 1$ (Proposition 3.2).

**Problem 3.2** For a normal complex surface singularity $(V, p)$, does the inequality $\bar{r}(V, p) \leq \text{min-}p_g(V, p) + 1$ hold? Characterize singularities which satisfy $\bar{r}(V, p) = \text{min-}p_g(V, p) + 1$.

# 4 Cone-Like Singularities

If $C$ is a nonsingular projective curve over $\mathbb{C}$ and $D$ an ample divisor on $C$, then $V(C, D) := \operatorname{Spec} \bigoplus_{n \geq 0} H^0(\mathcal{O}_C(nD))$ is a normal surface with at most an isolated singularity at the "vertex"(cf. [30]). Such a singularity is called a *cone singularity*. The exceptional set of the minimal resolution of $V(C, D)$ is isomorphic to $C$ with self-intersection number $-\deg D$. For example, if $R = \oplus_{n \geq 0} R_n$ is a two-dimensional normal graded ring generated by $R_1$ over $R_0 = \mathbb{C}$, then $\operatorname{Spec} R$ has a cone singularity.

**Definition 4.1** Let $\pi_0 \colon X_0 \to V$ be the minimal resolution of the singularity $(V, p)$ and $F$ the exceptional set of $\pi_0$. We call $(V, p)$ a *cone-like singularity* if $F$ consists of a unique smooth curve. Note that in this case $(V, p)$ is homeomorphic to the cone singularity $(V(F, -F|_F), \text{vertex})$.

In the rest of this section, we always assume that $(V, p)$ is a cone-like singularity. Let $g$ denote the genus of the exceptional curve $F$ of the minimal resolution $\pi_0 \colon X_0 \to V$ and let $d = -F^2$. Assume that $g \geq 1$. Let $\pi \colon X \to V$ be any resolution with exceptional set $E$ as in the preceding section. Then we have a natural morphism $X \to X_0$. We denote by $E_0 \subset X$ the proper transform of $F$; this is the unique irreducible exceptional curve on $X$ with positive genus. Note that $d = -Z_f^2$ because $F$ is the fundamental cycle on $X_0$; the number $d$ is sometimes called the degree of $(V, p)$.

**Definition 4.2** Let $C$ be a nonsingular projective curve. The *gonality* of the curve $C$ is the minimum of the degree of surjective morphisms from $C$ to $\mathbb{P}^1$, and denoted by $\operatorname{gon}(C)$. It is known that $\operatorname{gon}(F) \leq \lfloor (g + 3)/2 \rfloor$.

**Definition 4.3** For any $\alpha \in \mathbb{R}$, let $[[\alpha]] = \min \{m \in \mathbb{Z} \mid m > \alpha\}$. For example, $[[2]] = [[5/2]] = 3$.

We give an upper bound for $\bar{r}(V, p)$ using the invariants $g, d, \operatorname{gon}(E_0)$. Note that $g$ and $d$ are topological invariant of $(V, p)$, but $\operatorname{gon}(E_0)$ is not.

**Theorem 4.4** **([29, 3.9])** *Let $(V, p)$ be a cone-like singularity and let $I = I_Z$ be an $\mathfrak{m}$-primary integrally closed ideal represented by a cycle $Z$ on the resolution $X$. Then we have the following.*

(1) *If $ZE_0 = 0$, then $\bar{\mathrm{r}}(I) \le [[(2g-2)/d]] + 1$.*
(2) *If $ZE_0 < 0$, then $\bar{\mathrm{r}}(I) \le [[(2g-2)/\operatorname{gon}(E_0)]] + 1$.*

*In particular, $\bar{\mathrm{r}}(V, p) \le [[(2g-2)/\min\{d, \operatorname{gon}(E_0)\}]] + 1$.*

For the proof we apply Röhr's vanishing theorem (see [29, §3] for the details). The following example is a special case of [29, 3.10] (take $b = g$).

*Example 4.5* Let $C$ be a hyperelliptic curve with genus $g \ge 2$ and $D_0$ a divisor on $C$ which is the pull-back of a point via the double cover $C \to \mathbb{P}^1$. Let $D = gD_0$ and $V = \operatorname{Spec} \bigoplus_{n \ge 0} H^0(X, \mathcal{O}_C(nD))$. Then $C \cong F \subset X_0$. We have $p_g(V, p) = g$ by Pinkham [30, Theorem 5.7].

If we take a general element $h \in H^0(\mathcal{O}_{X_0}(-F))$, then $\operatorname{div}_{X_0}(h) = F + H$, where $H$ is the non-exceptional part and $F \cap H$ consists of distinct $2g$ points $P_1, \dots, P_{2g}$. We may assume that $P_1 + P_2 \sim D_0$. Let $\phi\ X \to X_0$ be the blowing-up with center $\{P_3, \dots, P_{2g}\}$ and let $Z = (h)_E$, the exceptional part of $\operatorname{div}_X(h)$. If we put $E_i = \phi^{-1}(P_i)$ for $3 \le i \le 2g$, then $Z = E_0 + 2(E_3 + \cdots + E_{2g})$. We can see that $\mathcal{O}_X(-Z)$ is generated since a general element of $H^0(\mathcal{O}_X(-2F))$ has no zero on $H$.

Then we have $h^1(\mathcal{O}_X(-(g-1)Z)) \ge h^1(\mathcal{O}_{E_0}(-(g-1)Z)) = h^1(K_C) = 1$ and $H^1(\mathcal{O}_X(-gZ)) = 0$. It follows from Proposition 2.6 (1) and Proposition 3.2 (2) that $q_Z(n) = g - n$ for $0 \le n \le g$. Hence we have $\bar{\mathrm{r}}(I_Z) = p_g(V, p) + 1 = [[(2g-2)/\operatorname{gon}(E_0)]] + 1$, $\operatorname{nr}(I_Z) = 1$, $q(\mathcal{A}) = q(\mathcal{A}')$.

## 4.1 *Homogeneous Hypersurface Singularities*

Assume that $V \subset \mathbb{C}^3$ is a hypersurface defined by a homogeneous polynomial $f \in \mathbb{C}[x, y, z]$ with degree $d \ge 3$ ($\deg x = \deg y = \deg z = 1$) having an isolated singularity at the origin $p \in \mathbb{C}^3$. Then $F \cong \{f = 0\} \subset \mathbb{P}^2$, $g = (d-1)(d-2)/2$. Let $D = -F|_F$. Then $V = \operatorname{Spec} \bigoplus_{n \ge 0} H^0(\mathcal{O}_C(nD))$. Since $\mathfrak{m} = I_F$, we have

$$q_F(n) = h^1(\mathcal{O}_Y(-nF)) = \sum_{m \ge n} h^1(\mathcal{O}_F(mD)) = \sum_{m=n}^{d-3} \binom{d-1-m}{2} = \binom{d-n}{3}.$$

Hence we have $\operatorname{nr}(\mathfrak{m}) = \bar{\mathrm{r}}(\mathfrak{m}) = d - 1$ by Proposition 3.2. By the definition, $\bar{\mathrm{r}}(V, p) \ge d - 1$. On the other hand, by Namba's theorem (Max Noether's theorem) [19, Theorem 2.3.1], we have $\operatorname{gon}(F) = d - 1$. By Theorem 4.4, we have

$$\bar{\mathrm{r}}(V, p) \le [[(2g-2)/(d-1)]] + 1 = [[d - 2 - 2/(d-1)]] + 1 = d - 1.$$

Hence we obtain

**Theorem 4.6 ([29, 4.1])** $\mathrm{nr}(\mathfrak{m}) = \bar{\mathrm{r}}(\mathfrak{m}) = \mathrm{nr}(V, p) = \bar{\mathrm{r}}(V, p) = d - 1$.

*Remark 4.7 (See [29, §4])* Suppose that $R = \oplus_{n \geq 0} R_n$ is a normal graded ring generated by $R_1$ over $R_0 = \mathbb{C}$ and $V = \operatorname{Spec} R$. Then $\mathfrak{m}^n = \overline{\mathfrak{m}^n}$. Let $a(R)$ denote the $a$-invariant of $R$ (see [2]). If $Q$ is a minimal reduction of $\mathfrak{m}$ generated by elements of $R_1$, we can see

$$\mathfrak{m}^{a(R)+2} \neq Q\mathfrak{m}^{a(R)+1} \quad \text{and} \quad \mathrm{nr}(\mathfrak{m}) = a(R) + 2 = \bar{\mathrm{r}}(\mathfrak{m}).$$

If $R = \mathbb{C}[x, y, z]/(f)$ as above, then $a(R) = d - 3$ (cf. [2, (3.1.6)]).

# 5 Brieskorn Complete Intersections

In [28], we obtained an explicit expression of $\bar{\mathrm{r}}(\mathfrak{m})$ for Brieskorn hypersurfaces using ring-theoretic arguments and gave a classification of Brieskorn hypersurfaces having elliptic singularities. In this section, we extend these results to the case of Brieskorn complete intersections, using resolution of singularities.

In the following, we assume that $V \subset \mathbb{C}^m$ is a Brieskorn complete intersection define by the following $m - 2$ polynomials:

$$q_{i1}x_1^{a_1} + \cdots + q_{im}x_m^{a_m} \quad (q_{ij} \in \mathbb{C}, \quad i = 3, \ldots, m),$$

where $a_i$ are integers such that $2 \leq a_1 \leq \cdots \leq a_m$. We also assume that $V$ has an isolated singularity at the origin $p \in \mathbb{C}^m$. Then, since every maximal minor of the matrix $(q_{ij})$ does not vanish (see [7, §7]), we may assume that

$$(q_{ij}) = \begin{pmatrix} 1 & 0 & \cdots & 0 & p_1 & q_1 \\ 0 & 1 & \cdots & 0 & p_2 & q_2 \\ \vdots & \vdots & \ddots & \vdots & \vdots & \vdots \\ 0 & 0 & \cdots & 1 & p_{m-2} & q_{m-2} \end{pmatrix}, \qquad (5.1)$$

where $p_i, q_i \neq 0$ and $p_i q_j \neq p_j q_i$ for $i \neq j$.

## 5.1 *The Maximal Ideal Cycle, the Fundamental Cycle, and the Canonical Cycle*

We summarize the results in [15] which will be used in this section; those are a natural extension of the hypersurface case obtained by Konno and Nagashima [9].

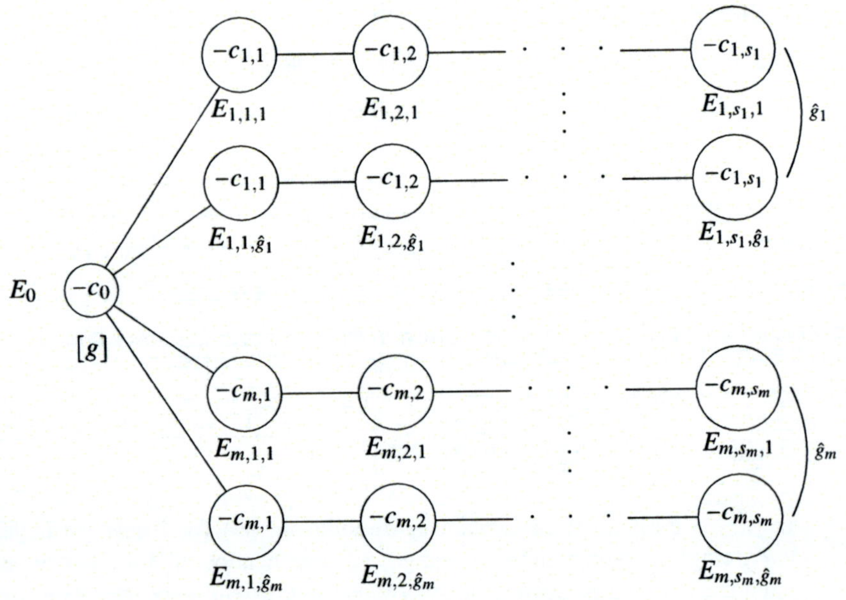


**Fig. 1** The weighted dual graph of a Brieskorn complete intersection

In the following, we assume that $\pi : X \to V$ is the minimal good resolution. Since $(V, p)$ is Gorenstein, the canonical cycle $Z_{K_X}$ is an effective cycle.

We define positive integers $\ell$, $\ell_i$, $\alpha$, $\alpha_i$, $\hat{g}$, $\hat{g}_i$, and $\lambda_i$ as follows:[2]

$$\ell := \operatorname{lcm}(a_1, \ldots, a_m), \ \ \ell_i := \operatorname{lcm}(a_1, \ldots, \hat{a_i}, \ldots, a_m), \text{ where } \hat{a_i} \text{ is omitted,}$$

$$\alpha_i := \ell/\ell_i, \ \ \alpha := \alpha_1 \cdots \alpha_m, \ \ \hat{g} := a_1 \cdots a_m/\ell, \ \ \hat{g}_i := \hat{g}\alpha_i/a_i, \ \ \lambda_i := \ell/a_i.$$

We easily see that the polynomials $x_i^{a_i} + p_i x_{m-1}^{a_{m-1}} + q_i x_m^{a_m}$ are weighted homogeneous polynomials of degree $\ell$ with respect to the weights $(\lambda_1, \ldots, \lambda_m)$. Then the weighted dual graph of the exceptional set $E$ is as in Fig. 1, where

$$E = E_0 + \sum_{w=1}^{m} \sum_{\nu=1}^{s_w} \sum_{\xi=1}^{\hat{g}_w} E_{w,\nu,\xi},$$

$g$ denotes the genus of the central curve $E_0$, $c_0 = -E_0^2$, and $c_{w,v} = -E_{w,\nu,\xi}^2$ (see [15, 4.4]).

For any $\mathbb{Q}$-cycle $B$ on $X$ and any irreducible component $F \subset E$, let $\operatorname{cff}_F(B)$ denote the coefficient of $F$ in $B$. Let $Z^{(i)} = (x_i)_E$.

[2] Using the notation of [15, §3], we have $l = d_m$, $\ell_i = d_{im}$, $\alpha_i = n_{im}$, $\lambda_i = e_{im}$, $\lambda_m = e_{mm} = e_m$.

**Theorem 5.1 ([15, 4.4])** *We have the following:*

$$Z^{(i)} = \lambda_0^{(i)} E_0 + \sum_{w=1}^{m} \sum_{\nu=1}^{s_w} \sum_{\xi=1}^{\hat{g}_w} \lambda_{w,\nu,\xi}^{(i)} E_{w,\nu,\xi} \quad (1 \le i \le m),$$

*where* $\lambda_0^{(i)}$ *and the sequence* $\{\lambda_{w,\nu,\xi}^{(i)}\}$ *are determined as follows:*

$$\lambda_0^{(i)} := \lambda_{w,0,\xi}^{(i)} := \lambda_i,$$

$$\lambda_{w,s_w+1,\xi}^{(i)} := \begin{cases} 1 & \text{if } w = i \\ 0 & \text{if } w \neq i, \end{cases}$$

$$\lambda_{w,\nu-1,\xi}^{(i)} = \lambda_{w,\nu,\xi}^{(i)} c_{w,\nu} - \lambda_{w,\nu+1,\xi}^{(i)}.$$

*The cycle* $Z^{(i)}$ *is the smallest one among the cycles* $Z > 0$ *such that* $Z$ *is anti-nef and* $\mathrm{cff}_{E_0}(Z) = \lambda_i$ *(cf. [15, 2.1]). In particular, we have* $M_X = Z^{(m)}$*, since* $\lambda_1 \ge \cdots \ge \lambda_m$.

**Theorem 5.2 ([15, 5.3])** *We have*

$$Z_{K_X} = E + \frac{(m-2)l}{\alpha} Z_0 - \sum_{w=1}^{m} Z^{(w)},$$

*where* $Z_0$ *is the anti-nef cycle such that* $\mathrm{cff}_{E_0}(Z_0) = \alpha$ *and* $Z_0(E - E_0) = 0$.

**Theorem 5.3 ([15, 5.1, 5.2, 5.4])** *If* $\lambda_m \ge \alpha$*, then* $Z_f = Z_0$ *and*

$$p_f(V, p) = \frac{1}{2}\alpha \left\{ (m-2)\hat{g} - \frac{(\alpha - 1)\hat{g}}{l} - \sum_{w=1}^{m} \frac{\hat{g}_w}{\alpha_w} \right\} + 1.$$

*If* $\lambda_m \le \alpha$*, then* $Z_f = M_X$ *and*

$$p_f(V, p) = \frac{1}{2}\lambda_m \left\{ (m-2)\hat{g} - \frac{(2\lceil \lambda_m/\alpha_m \rceil - 1)\hat{g}_m}{\lambda_m} - \sum_{w=1}^{m-1} \frac{\hat{g}_w}{\alpha_w} \right\} + 1.$$

## 5.2 The Normal Reduction Numbers

Since $M_X = (x_m)_E$ by Theorem 5.1, $\mathcal{O}_X(-M_X)$ has no fixed components; however, it is not generated in general.

Let $H = \mathrm{div}_X(x_m) - M_X$. Then $E + H$ is simple normal crossing and the set of the base points of the linear system $|\mathcal{O}_X(-M)|$ is an empty set or $\{t_1, \dots, t_{g_m}\}$, where $\{t_\xi\} = E_{m,s_m,\xi} \cap H$ (see Theorem 5.1). Let us look in detail at a point. Let $x, y$ be the local coordinates at $t_\xi \in X$ such that $E = \{x = 0\}$ and $H = \{y = 0\}$. We write $\eta_i = \lambda^{(i)}_{m,s_m,\xi}$ and $\delta = \eta_{m-1} - \eta_m$. Then $\delta \geq 0$ and $\mathfrak{m}\mathcal{O}_{X,t_\xi} = (x_{m-1}, x_m) = (x^{\eta_m} y, x^{\eta_{m-1}}) = x^{\eta_m}(y, x^\delta)$.

**Proposition 5.4 ([15, 6.4])** *The following conditions are equivalent:*

(1) $\delta = 0$
(2) *The base points of the linear system $|\mathcal{O}_X(-M)|$ on E is empty.*

*If $\delta > 0$, each base point can be resolved by a succession of $\delta$ blowing-ups at the intersection of the exceptional set and the proper transform of $H$.*

Let $\phi \colon Y \to X$ be the minimal morphism such that $\mathfrak{m}\mathcal{O}_Y$ is invertible and let $F = \phi^{-1}(E)$. Let $W_i = (x_i)_F$ $(i = 1, \dots, m)$, and let $M_Y$ denote the maximal ideal cycle on $Y$ and $H_Y$ the proper transform of $H$ on $Y$. Then

$$W_i = \phi^* Z^{(i)} \quad \text{for } i \neq m, \quad W_m = M_Y = \phi^* Z^{(m)} + K_{Y/X}, \tag{5.2}$$

where $K_{Y/X} = K_Y - \phi^* K_X$. Now, $\mathfrak{m}$ is represented by $M_Y$ and $\overline{\mathfrak{m}^n} = I_{nM_Y}$. Fix an irreducible component $F_\xi \subset F$ intersecting $H_Y$. For any cycle $W$ on $Y$, we write $\gamma(W) = \mathrm{cff}_{F_\xi}(W)$. Note that $\gamma(M_Y)$ is independent of the choice of a component intersecting $H_Y$ (see Theorem 5.1) and

$$\gamma(W_i) = \eta_i \quad \text{for } i \neq m, \quad \gamma(W_m) = \gamma(M_Y) = \eta_m + \delta = \eta_{m-1}. \tag{5.3}$$

**Lemma 5.5** *Let $(u_1, \dots, u_m) \in (\mathbb{Z}_{\geq 0})^m$. For any positive integer n,*

$$\prod_{i=1}^m x_i^{u_i} \in \overline{\mathfrak{m}^n} \quad \text{if and only if} \quad \sum_{i=1}^{m-2} \frac{u_i}{a_i} \geq \frac{n - (u_{m-1} + u_m)}{a_{m-1}}.$$

*Proof* We have $\left(\prod_{i=1}^m x_i^{u_i}\right)_F = W := \sum_{i=1}^m u_i W_i$. First we show that $\prod_{i=1}^m x_i^{u_i} \in \overline{\mathfrak{m}^n}$ if and only if $\gamma(W) \geq \gamma(nM_Y)$. Clearly, if $W \geq nM_Y$, then $\gamma(W) \geq \gamma(nM_Y)$. So we show the converse. Let $W - nM_Y = D_1 - D_2$, where $D_1$ and $D_2$ are effective cycles without common components. By the assumption, $D_2$ has no components of $F$ intersecting $H_Y$. Thus $M_Y D_2 = 0$. Then $0 \leq D_1 D_2 - D_2^2 = W D_2 \leq 0$. Hence $D_2 = 0$. We have proved the claim.

We have the following (see [9, Lemma 1.2 (4)] for the first equality):

$$\eta_i = \lambda_i / \alpha_m = \ell / a_i \alpha_m \quad (1 \leq i \leq m-1). \tag{5.4}$$

Then we have

$$\gamma(W) - \gamma(nM_Y) = \sum_{i=1}^{m-1} u_i\eta_i + u_m\eta_{m-1} - n\eta_{m-1}$$
$$= \frac{l}{\alpha_m}\left(\sum_{i=1}^{m-2}\frac{u_i}{a_i} + \frac{u_{m-1}+u_m-n}{a_{m-1}}\right).$$

This implies the assertion.

Let $P \subset A := \mathbb{C}[x_1, \dots, x_m]$ denote the ideal generated by the polynomials $\{x_i^{a_i} + p_i x_{m-1}^{a_{m-1}} + q_i x_m^{a_m} \mid i = 1, \dots, m-2\}$ defining $V \subset \mathbb{C}^m$. For simplicity, let $P$ also denote the ideal in $\mathbb{C}\{x_1, \dots, x_m\}$ generated by these polynomials; so $\mathcal{O}_{V,p} = \mathbb{C}\{x_1, \dots, x_m\}/P$. We easily see the following (cf. [23, Theorem 3.1]).

**Lemma 5.6** *For any $1 \le i \le m$, the quotient ring $A/(P + (x_i))$ is reduced.*

**Proposition 5.7** *For $n \in \mathbb{Z}_{\ge 0}$, let $I_n \subset \mathcal{O}_{V,p}$ be an ideal generated by monomials $\prod_{i=1}^m x_i^{u_i}$ such that*

$$\sum_{i=1}^{m-2}\frac{u_i}{a_i} \ge \frac{(n/\eta_{m-1}) - (u_{m-1}+u_m)}{a_{m-1}}.$$

*Then $I_{n\eta_{m-1}} = \overline{\mathfrak{m}^n}$ for $n \in \mathbb{Z}_{\ge 0}$. In particular, $\overline{\mathfrak{m}^n}$ is generated by monomials.*

***Proof*** First we show that $G := \bigoplus_{n\ge 0} I_n/I_{n+1}$ is reduced. It follows from (5.3) and (5.4) that the inequality is equivalent to the following (cf. the proof of Lemma 5.5):

$$\gamma\left(\left(\prod_{i=1}^m x_i^{u_i}\right)_F\right) = \sum_{i=1}^{m-1} u_i\eta_i + u_m\eta_{m-1} \ge n. \tag{5.5}$$

Therefore the filtration $\{I_n\}_{n\in\mathbb{Z}_{\ge 0}}$ is induced from the weight filtration of the power series ring $\mathbb{C}\{x_1, \dots, x_m\}$ with weight vector $(\eta_1, \dots, \eta_{m-1}, \eta_{m-1}) \in \mathbb{Z}^m$. Let $I \subset A = \mathbb{C}[x_1, \dots, x_m]$ denote the ideal generated by the leading form, with respect to these weights, of the polynomials $\{x_i^{a_i} + p_i x_{m-1}^{a_{m-1}} + q_i x_m^{a_m} \mid i = 1, \dots, m-2\}$. Then $A/I$ is complete intersection and isomorphic to $G$ (cf. the proof of [23, Theorem 2.6]). If $a_{m-1} = a_m$, then $G = A/P$. If $a_{m-1} < a_m$, then $G \cong (A/P + (x_m))[x_m]$, and thus $G$ is reduced by Lemma 5.6.

Let $C = \sum_{\xi=1}^{\hat{g}_m} F_\xi$, the sum of the irreducible components of $F$ intersecting $H_Y$. From (5.5), every $h \in I_n$ satisfies $(h)_F \ge nC$. Now we can apply Lemma 2.9. Since $\eta_{m-1}\widetilde{C} = M_Y$, we obtain that $I_{n\eta_{m-1}} = \overline{\mathfrak{m}^n}$.

Let $Q = (x_{m-1}, x_m) \subset \mathcal{O}_{V,p}$. Then $x_i^{a_i} \in Q$ for every $i$, and thus $Q$ is a minimal reduction of $\mathfrak{m}$ (cf. [5, 8.3.6]).

**Theorem 5.8** *We have the following.*

(1)

$$\mathrm{nr}(\mathfrak{m}) = \bar{\mathrm{r}}(\mathfrak{m}) = \left\lfloor a_{m-1} \sum_{i=1}^{m-2} \frac{a_i - 1}{a_i} \right\rfloor.$$

(2) *The image of the monomials* $\prod_{i=1}^{m-2} x_i^{u_i}$ *such that*

$$\sum_{i=1}^{m-2} \frac{u_i}{a_i} \geq \frac{n+1}{a_{m-1}} \quad \text{and} \quad 0 \leq u_i \leq a_i - 1 \quad (i = 1, \dots, m-2)$$

*in the vector space* $\overline{\mathfrak{m}^{n+1}}/Q\overline{\mathfrak{m}^n}$ *form a basis. In particular,* $\dim_{\mathbb{C}}(\overline{\mathfrak{m}^{n+1}}/Q\overline{\mathfrak{m}^n})$ *is a non-increasing function of* $n$.

***Proof*** Note that $Q\overline{\mathfrak{m}^n}$ and $\overline{\mathfrak{m}^{n+1}}$ are generated by monomials for every $n \geq 0$ by Proposition 5.7. Let $N = \left\lfloor a_{m-1} \sum_{i=1}^{m-2} (a_i - 1)/a_i \right\rfloor$. First we prove that $Q\overline{\mathfrak{m}^n} = \overline{\mathfrak{m}^{n+1}}$ for $n \geq N$. Let $v = \prod_{i=1}^{m} x_i^{u_i} \in \overline{\mathfrak{m}^{n+1}}$. By Lemma 5.5, we have

$$\sum_{i=1}^{m-2} \frac{u_i}{a_i} \geq \frac{n+1-(u_{m-1}+u_m)}{a_{m-1}} = \frac{n-(u_{m-1}+u_m-1)}{a_{m-1}}.$$

Therefore, if $u_{m-1} \geq 1$ or $u_m \geq 1$, we have $v/x_{m-1} \in \overline{\mathfrak{m}^n}$ or $v/x_m \in \overline{\mathfrak{m}^n}$, and hence $v \in Q\overline{\mathfrak{m}^n}$. We consider the case that $u_{m-1} = u_m = 0$ and $u_i \geq a_i$ for some $1 \leq i \leq m-2$; we may assume that $i = 1$. Then it follows that $x_1^{u_1} \in x_1^{u_1-a_1}(x_{m-1}^{a_{m-1}}, x_m^{a_m})$, since $x_1^{a_1} + p_1 x_{m-1}^{a_{m-1}} + q_1 x_m^{a_m} = 0$. We show that $w_1 := (x_1^{u_1-a_1} x_{m-1}^{a_{m-1}}) \prod_{i=2}^{m-2} x_i^{u_i} \in Q\overline{\mathfrak{m}^n}$. Let $w' = w_1/x_{m-1} = (x_1^{u_1-a_1} x_{m-1}^{a_{m-1}-1}) \prod_{i=2}^{m-2} x_i^{u_i}$. Since

$$\frac{u_1 - a_1}{a_1} + \sum_{i=2}^{m-2} \frac{u_i}{a_i} \geq \frac{n+1}{a_{m-1}} - 1 = \frac{n-(a_{m-1}-1)}{a_{m-1}},$$

we have $w' \in \overline{\mathfrak{m}^n}$ by Lemma 5.5. Thus $w_1 = x_{m-1} w' \in Q\overline{\mathfrak{m}^n}$. In a similar way, we also have that $w_2 := (x_1^{u_1-a_1} x_m^{a_m}) \prod_{i=2}^{m-2} x_i^{u_i} \in Q\overline{\mathfrak{m}^n}$, since $\frac{n-(a_{m-1}-1)}{a_{m-1}} \geq \frac{n-(a_m-1)}{a_{m-1}}$. Hence we obtain that $v \in (w_1, w_2) \subset Q\overline{\mathfrak{m}^n}$. Next assume that $u_{m-1} = u_m = 0$ and $u_i < a_i$ for $1 \leq i \leq m-2$. Then we have

$$\sum_{i=1}^{m-2} \frac{a_i - 1}{a_i} \geq \sum_{i=1}^{m-2} \frac{u_i}{a_i} \geq \frac{n+1}{a_{m-1}}.$$

However this implies that $n \leq N - 1$. Hence we obtain that $Q\overline{\mathfrak{m}^n} = \overline{\mathfrak{m}^{n+1}}$ for $n \geq N$.

Next we prove that $Q\overline{\mathfrak{m}^{N-1}} \neq \overline{\mathfrak{m}^N}$. Let $v := \prod_{i=1}^{m-2} x_i^{a_i-1}$. Then $v \notin Q$, because $\mathcal{O}_{V,p}/Q = \mathbb{C}\{x_1, \dots, x_m\}/(x_1^{a_1}, \dots, x_{m-2}^{a_{m-2}}, x_{m-1}, x_m)$. However, since

$$\sum_{i=1}^{m-2} \frac{a_i - 1}{a_i} \geq \frac{N}{a_{m-1}},$$

we have $v \in \overline{\mathfrak{m}^N}$ by Lemma 5.5. Hence we obtain that $\bar{r}(\mathfrak{m}) = N$.

From the arguments above, we see that (2) holds, because any non-trivial linear combinations of those monomials is not in the ideal $P + (x_{m-1}, x_m) = (x_1^{a_1}, \dots, x_{m-2}^{a_{m-2}}, x_{m-1}, x_m)$. Since $\dim_{\mathbb{C}}(\overline{\mathfrak{m}^{n+1}}/Q\overline{\mathfrak{m}^n})$ is a non-increasing function of $n$, we have $\mathrm{nr}(\mathfrak{m}) = \bar{r}(\mathfrak{m})$ (cf. Proposition 3.2).

*Example 5.9* If $m = 3$, we have

$$\mathrm{nr}(\mathfrak{m}) = \left\lfloor \frac{a_2(a_1 - 1)}{a_1} \right\rfloor,$$

$$\begin{aligned} \dim_{\mathbb{C}}(\overline{\mathfrak{m}^{n+1}}/Q\overline{\mathfrak{m}^n}) &= \sharp\left\{ u \in \mathbb{Z} \;\middle|\; \frac{a_1(n+1)}{a_2} \leq u \leq a_1 - 1 \right\} \\ &= \max\left( a_1 - \left\lceil \frac{a_1(n+1)}{a_2} \right\rceil, 0 \right). \end{aligned}$$

The formula for $q(\mathfrak{m})$ in [28, 3.8] is generalized as follows.

**Proposition 5.10** *Let* $p(n+1) = \dim_{\mathbb{C}}(\overline{\mathfrak{m}^{n+1}}/Q\overline{\mathfrak{m}^n})$ *and* $q(n) = h^1(\mathcal{O}_Y(-nM_Y))$ *for* $n \geq 0$. *Then we have the following:*

$$q(n) = p_g(V, p) + \frac{n}{2}(M_Y^2 - M_Y K_Y) + \sum_{i=1}^{n} (n + 1 - i)p(i).$$

*(Note that the same formula holds for any normal surface singularity.)*

***Proof*** It is well-known that the multiplicity of $\mathcal{O}_{V,p}$ coincides with $\dim_{\mathbb{C}} \mathcal{O}_{V,p}/Q$ (e.g., [5, 11.2.2]). Thus we have $p(1) = \dim_{\mathbb{C}}(\mathfrak{m}/Q) = -M_Y^2 - 1$. From the exact sequence

$$0 \to \mathcal{O}_Y(-M_Y) \to \mathcal{O}_Y \to \mathcal{O}_{M_Y} \to 0,$$

we have

$$q(1) - q(0) = \chi(\mathcal{O}_{M_Y}) - 1 = \chi(\mathcal{O}_{M_Y}) + M_Y^2 + p(1) = \frac{1}{2}(M_Y^2 - M_Y K_Y) + p(1).$$

For $n \geq 1$, it follows from Proposition 3.2 (1) that

$$q(n) - q(n-1) = q(1) - q(0) + \sum_{i=2}^{n} p(i) = \frac{1}{2}(M_Y^2 - M_Y K_Y) + \sum_{i=1}^{n} p(i).$$

Hence we obtain

$$q(n) - q(0) = \frac{n}{2}(M_Y^2 - M_Y K_Y) + \sum_{i=1}^{n}(n+1-i)p(i). \qquad \square$$

*Remark 5.11* The invariant $M_Y^2 - M_Y K_Y$ can be computed from $a_1, \ldots, a_m$ as follows. First we have $M_Y^2 = -\operatorname{mult}(V, p) = -\prod_{i=1}^{m-2} a_i$ (see [15, 6.3]). On the other hand, from (5.2), we have

$$M_Y^2 + M_Y K_Y = M_X^2 + M_X K_X + 2(K_{Y/X})^2 = 2p_a(M_X) - 2 - 2\delta \hat{g}_m.$$

We have seen a formula for $p_a(M_X)$ in Theorem 5.3.

## 5.3 Elliptic Singularities of Brieskorn Type

We classify the exponents $(a_1, \ldots, a_m)$ such that $(V, p)$ is elliptic, applying the formula for $\bar{r}(\mathfrak{m})$.

**Theorem 5.12** *$(V, p)$ is elliptic if and only if $(a_1, \ldots, a_m)$ is one of the following.*

(1) $(a_1, a_2, a_3) = (2, 3, a)$, $a \geq 6$.
(2) $(a_1, a_2, a_3) = (2, 4, a)$, $a \geq 4$.
(3) $(a_1, a_2, a_3) = (2, 5, a)$, $5 \leq a \leq 9$.
(4) $(a_1, a_2, a_3) = (3, 3, a)$, $a \geq 3$.
(5) $(a_1, a_2, a_3) = (3, 4, a)$, $4 \leq a \leq 5$.
(6) $(a_1, a_2, a_3, a_4) = (2, 2, 2, a)$, $a \geq 2$.

***Proof*** For the case (1)–(6) in the theorem, we can check that $\alpha \geq \lambda_m$ and obtain $p_f(V, p) = 1$ using Theorem 5.3.

Assume that $(V, p)$ is elliptic. By Theorems 3.9 and 5.8, we have

$$3 > a_{m-1} \sum_{i=1}^{m-2} \frac{a_i - 1}{a_i} \geq a_{m-1}(m-2)/2 \geq m - 2.$$

Hence $m \le 4$. We first consider the case $m = 4$. We have $a_3 < 3$, and thus $a_1 = a_2 = a_3 = 2$. Then $\alpha/\lambda_4 = a_4/2 \ge 1$ and $p_f(V, p) = 1$ by Theorem 5.3,

Next assume that $m = 3$. Then we have $\bar{r}(\mathfrak{m}) = \left\lfloor \dfrac{(a_1 - 1)a_2}{a_1} \right\rfloor \le 2$, and thus $(a_1 - 1)(a_2 - 3) \le 2$. If $a_2 = 2$, then $a_1 = 2$ and $(V, p)$ is a rational. Hence $a_2 \ge 3$ and the list of $(a_1, a_2)$ is as follows:

$$(2, 3), (2, 4), (2, 5), (3, 3), (3, 4).$$

We can see that $\alpha \ge \lambda_3$ for those cases. So it follows from Theorem 5.3 that

$$p_f(V, p) = \frac{1}{2}\{a_1a_2 - a_1 - a_2 - (2\lceil \mathrm{lcm}(a_1, a_2)/a_3\rceil - 1)\gcd(a_1, a_2)\} + 1.$$

Let us look at each case.

(1) The case where $(a_1, a_2) = (2, 3)$. We know that $(V, p)$ is rational if $a_3 \le 5$. Hence $a_3 \ge 6$. We have $\alpha/\lambda_3 = a_3/\gcd(6, a_3)$ and $p_f(V, p) = 1$.
(2) The case where $(a_1, a_2) = (2, 4)$, $a_3 \ge 4$. We have $\alpha/\lambda_3 = a_3/4$ if $4 \mid a_3$, $\alpha/\lambda_3 = a_3/2$ otherwise, and $p_f(V, p) = 1$.
(3) The case where $(a_1, a_2) = (2, 5)$, $a_3 \ge 5$. We have $\alpha/\lambda_3 = a_3/\gcd(10, a_3)$ and $p_f(V, p) = 3 - \lceil 10/a_3\rceil$. Sine $p_f(V, p) = 1$, we have $a_3 \le 9$.
(4) The case where $(a_1, a_2) = (3, 3)$, $a_3 \ge 3$. We have $\alpha/\lambda_3 = a_3/3$ and $p_f(V, p) = 1$ for all $a_3 \ge 3$.
(5) The case where $(a_1, a_2) = (3, 4)$, $a_3 \ge 4$. We have $\alpha/\lambda_3 = a_3/\gcd(12, a_3)$ and $p_f(V, p) = 4 - \lceil 12/a_3\rceil$. Sine $p_f(V, p) = 1$, we have $a_3 \le 5$.

Hence we have proved the theorem.

From the proof of Theorem 5.12, we obtain that $\bar{r}(\mathfrak{m}) = 2$ and $p_f(V, p) \ge 2$ if $(a_1, a_2, a_3) = (2, 5, a)$ with $a \ge 10$ or $(3, 4, a)$ with $a \ge 6$. For the cases $(2, 5, a)$ with $a \ge 10$ and $(3, 4, a)$ with $a \ge 8$, letting $Q = (y, z^2)$ and $I = \overline{Q}$, we have $\bar{r}(I) \ge 3$. Hence we obtain the following.

**Proposition 5.13 ([28, 4.5])** $\bar{r}(V, p) = 2$ *if and only if* $p_f(V, p) = 1$, *except for the cases* $(a_1, \ldots, a_m) = (3, 4, 6), (3, 4, 7)$.

For the reader's convenience, we put some information about the two exceptional cases above. Both singularities have $p_g = 3$ and $p_f = 2$. The weighted dual graph $\Gamma_1$ (resp. $\Gamma_2$) of $\{x^3 + y^4 + z^6 = 0\}$ (resp. $\{x^3 + y^4 + z^7 = 0\}$) is as in Fig. 2.

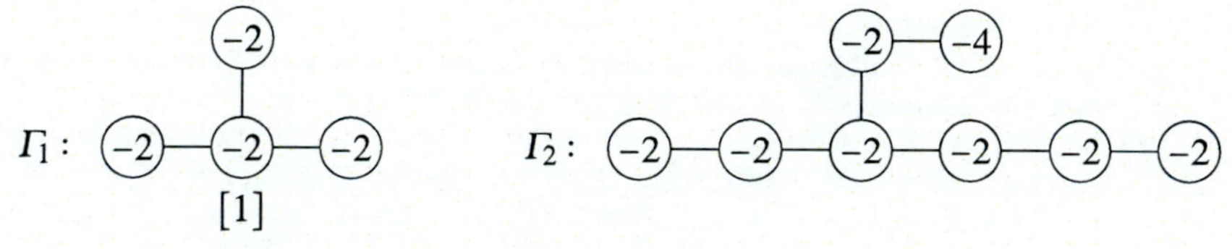


**Fig. 2** The weighted dual graphs $\Gamma_1$ and $\Gamma_2$

As we have seen above, the equality $\bar{\mathfrak{r}}(V, p) = \bar{\mathfrak{r}}(\mathfrak{m})$ does not hold in general (see also Remark 3.7).

**Problem 5.1** For a given normal surface singularity $(V, p)$, characterize $\mathfrak{m}$-primary integrally closed ideals $I \subset \mathcal{O}_{V,p}$ (or, cycles which represent $I$) such that $\bar{\mathfrak{r}}(V, p) = \bar{\mathfrak{r}}(I)$. Characterize normal surface singularities $(V, p)$ such that $\bar{\mathfrak{r}}(V, p) = \bar{\mathfrak{r}}(\mathfrak{m})$.

**Acknowledgments** The author is grateful to the referee for the careful reading of the manuscript and helpful comments.

This work was partially supported by JSPS Grant-in-Aid for Scientific Research (C) Grant Number 17K05216.

## References

1. Artin, M.: On isolated rational singularities of surfaces. Amer. J. Math. **88**, 129–136 (1966)
2. Goto, S., Watanabe, K.-i.: On graded rings. I. J. Math. Soc. Jpn. **30**(2), 179–213 (1978)
3. Hochster, M., Huneke, C.: Tight closure, invariant theory, and the Briançon-Skoda theorem. J. Amer. Math. Soc. **3**(1), 31–116 (1990)
4. Huneke, C.: Hilbert functions and symbolic powers. Michigan Math. J. **34**(2), 293–318 (1987)
5. Huneke, C., Swanson, I.: Integral Closure of Ideals, Rings, and Modules. London Mathematical Society Lecture Note Series, vol. 336, Cambridge University Press, Cambridge (2006)
6. Itoh, S.:Integral closures of ideals generated by regular sequences. J. Algebra **117**(2), 390–401 (1988)
7. Jankins, M., Neumann, W.D.: Lectures on Seifert Manifolds. Brandeis Lecture Notes, vol. 2, Brandeis University, Waltham (1983)
8. Kato, M.: Riemann-Roch theorem for strongly pseudoconvex manifolds of dimension 2. Math. Ann. **222**(3), 243–250 (1976)
9. Konno, K., Nagashima, D.: Maximal ideal cycles over normal surface singularities of Brieskorn type. Osaka J. Math. **49**(1), 225–245 (2012)
10. Laufer, H.B.: On rational singularities. Amer. J. Math. **94**, 597–608 (1972)
11. Laufer, H.B.: On minimally elliptic singularities. Amer. J. Math. **99**(6), 1257–1295 (1977)
12. Lipman, J.: Rational singularities, with applications to algebraic surfaces and unique factorization. Inst. Hautes Études Sci. Publ. Math. **36**, 195–279 (1969)
13. Lipman, J., Teissier, B.: Pseudorational local rings and a theorem of Briançon-Skoda about integral closures of ideals. Michigan Math. J. **28**(1), 97–116 (1981)
14. Mandal, M., Masuti, S., Verma, J.K.: Normal Hilbert polynomials: A survey. In: Commutative Algebra and Algebraic Geometry (CAAG-2010). Lecture Notes Series, vol. 17, pp. 139–166. Ramanujan Mathematical Society, Mysore (2013)
15. Meng, F.-N., Okuma, T.: The maximal ideal cycles over complete intersection surface singularities of Brieskorn type. Kyushu J. Math. **68**(1), 121–137 (2014)
16. Nagy, J.: Holes in possible values of $h^1$ and geometric genus (2019). arXiv:1911.07300
17. Nagy, J., Némethi, A.: On the topology of elliptic singularities (2019). arXiv:1901.06224
18. Nagy, J., Némethi, A.: The Abel map for surface singularities III. Elliptic germs (2019). arXiv:1902.07493
19. Namba, M.: Families of Meromorphic Functions on Compact Riemann Surfacesl. Lecture Notes in Mathematics, vol. 767. Springer, Berlin (1979)
20. Némethi, A.: Five lectures on normal surface singularities. In: Low Dimensional Topology (Eger, 1996/Budapest, 1998), Bolyai Society Mathematical Studies, vol. 8, pp. 269–351. János Bolyai Mathematical Society, Budapest (1999); With the assistance of Ágnes Szilárd and Sándor Kovács

21. Némethi, A.: "Weakly" elliptic Gorenstein singularities of surfaces. Invent. Math. **137**(1), 145–167 (1999)
22. Némethi, A.: Links of rational singularities, L-spaces and LO fundamental groups. Invent. Math. **210**(1), 69–83 (2017)
23. Neumann, W.D., Wahl, J.: Complete intersection singularities of splice type as universal abelian covers. Geom. Topol. **9**, 699–755 (2005) (electronic)
24. Okuma, T.: Cohomology of ideals in elliptic surface singularities. Illinois J. Math. **61**(3–4), 259–273 (2017)
25. Okuma, T., Watanabe, K.-i., Yoshida, K.-i.: Good ideals and $p_g$-ideals in two-dimensional normal singularities. Manuscripta Math. **150**(3–4), 499–520 (2016)
26. Okuma, T., Watanabe, K.-i., Yoshida, K.-i.: Rees algebras and $p_g$-ideals in a two-dimensional normal local domain. Proc. Amer. Math. Soc. **145**(1), 39–47 (2017)
27. Okuma, T., Watanabe, K.-i., Yoshida, K.-i.: A characterization of two-dimensional rational singularities via core of ideals. J. Algebra **499**, 450–468 (2018)
28. Okuma, T., Watanabe, K.-i., Yoshida, K.-i.: Normal reduction numbers for normal surface singularities with application to elliptic singularities of Brieskorn type. Acta Math. Vietnam. **44**(1), 87–100 (2019)
29. Okuma, T., Watanabe, K.-i., Yoshida, K.-i.: The normal reduction number of two-dimensional cone-like singularities (2019). arXiv:1909.13190
30. Pinkham, H.: Normal surface singularities with $C^*$ action. Math. Ann. **227**(2), 183–193 (1977)
31. Puthenpurakal, T., Watanabe, K., Yoshida, K.-i.: The strong Rees property of powers of the maximal ideal and Takahashi-Dao's question. J. Algebra, In Press. arXiv:1708.06090
32. Reid, M.: Chapters on Algebraic Surfaces, Complex Algebraic Geometry. IAS/Park City Mathematics Series, vol. 3, pp. 3–159. American Mathematical Society, Providence (1997)
33. Röhr, A.:A vanishing theorem for line bundles on resolutions of surface singularities. Abh. Math. Sem. Univ. Hamburg **65**, 215–223 (1995)
34. Tomari, M.: A geometric characterization of normal two-dimensional singularities of multiplicity two with $p_a \leq 1$. Publ. Res. Inst. Math. Sci. **20**(1), 1–20 (1984)
35. Tomari, M.: A $p_g$-formula and elliptic singularities. Publ. Res. Inst. Math. Sci. **21**(2), 297–354 (1985)
36. Tomari, M., Watanabe, K.-i.: Filtered rings, filtered blowing-ups and normal two-dimensional singularities with "star-shaped" resolution. Publ. Res. Inst. Math. Sci. **25**(5), 681–740 (1989)
37. Wagreich, P.: Elliptic singularities of surfaces. Amer. J. Math. **92**, 419–454 (1970)
38. Yau, S.S.T.: Gorenstein singularities with geometric genus equal to two. Amer. J. Math. **101**(4), 813–854 (1979)
39. Yau, S.S.T.: Hypersurface weighted dual graphs of normal singularities of surfaces. Amer. J. Math. **101**(4), 761–812 (1979)
40. Yau, S.S.T.: Normal two-dimensional elliptic singularities. Trans. Amer. Math. Soc. **254**, 117–134 (1979)
41. Yau, S.S.T.: On maximally elliptic singularities. Trans. Amer. Math. Soc. **257**(2), 269–329 (1980)

# Motivic Chern Classes of Cones

László M. Fehér

**Abstract** We study motivic Chern classes of cones. First we show examples of projective cones of smooth curves such that their various $K$-classes (sheaf theoretic, push-forward and motivic) are all different. Then we show connections between the torus equivariant motivic Chern class of a projective variety and of its affine cone, generalizing results on projective Thom polynomials.

**Keywords** Motivic Chern classes of varieties · K-theory · K-theory class of a subvariety · Projective Thom polynomial

**Subject Classifications** 19E15

## 1 Introduction

We study two different topics in this paper. The common technical issue is to understand the motivic properties of cones. Equivariant motivic classes of cones were studied in [17] previously. Our results are related, but the philosophy is somewhat different. We try to stay in $K$-theory without using the transition to cohomology using the Chern character. We hope to convince the reader that some of the arguments are more transparent in $K$-theory.

In the first part we introduce three different notions of the $K$-class of a projective subvariety, and show by examples that they are different. We explain their connection with classical algebraic geometric invariants as the Hilbert function and polynomial, and the arithmetic genus. We discuss the equivariant version, too, study the transversality properties and how these properties connected to $K$-theoretic Thom polynomials.

L. M. Fehér (✉)
Institute of Mathematics, Eötvös University, Budapest, Hungary

 
J. Fernández de Bobadilla et al. (eds.), *Singularities and Their Interaction with Geometry and Low Dimensional Topology*, Trends in Mathematics,
https://doi.org/10.1007/978-3-030-61958-9_9

The second part is an attempt to introduce the motivic version of the projective Thom polynomial. The cohomological projective Thom polynomial was introduced by András Némethi, Richárd Rimányi and the author in [7] and used later in other projects. I hope that this motivic version will be just as useful in applications.

## 2 What Is the $K$-Class of a Subvariety?

There are several candidates for the $K$-class of a subvariety of an ambient smooth variety $M$. We show that they are different and have different functorial properties.

### 2.1 *Algebraic K-Theory and the Sheaf K-Class*

First we recall the basic constructions in algebraic $K$-theory following [11, §15.1]:

For any scheme $X$, $K^0X$ denotes the Grothendieck group of vector bundles (locally free sheaves) on $X$. Each vector bundle $E$ determines an element, denoted by $[E]$, in $K^0X$. $K^0X$ is the free abelian group on the set of isomorphism classes of vector bundles, modulo the relations

$$[E]=[E']+[E''],$$

whenever $E'$ is a subbundle of a vector bundle $E$, with quotient bundle $E''=E/E'$. The tensor product makes $K^0X$ a ring. For any morphism $f:Y\to X$ there is an induced pull-back homomorphism

$$f^*:K^0X\to K^0Y,$$

taking $[E]$ to $[f^*E]$, where $f^*E$ is the pull-back bundle; this makes $K^0$ a contravariant functor from schemes to commutative rings.

The Grothendieck group of coherent sheaves on $X$, denoted by $K_0X$, is defined to be the free abelian group on the isomorphism classes of coherent sheaves on X, modulo the relations

$$[\mathcal{F}]=[\mathcal{F}']+[\mathcal{F}''],$$

for each exact sequence

$$0\to\mathcal{F}'\to\mathcal{F}\to\mathcal{F}''\to 0$$

of coherent sheaves.

For any proper morphism $f:X\to Y$, there is a push-forward homomorphism

$$f_*:K_0X\to K_0Y,$$

which takes $[\mathcal{F}]$ to $\sum_{i\geq 0}(-1)^i[R^i f_*\mathcal{F}]$, where $R^i f_*\mathcal{F}$ is Grothendieck's higher direct image sheaf.

On any X there is a canonical "duality" homomorphism:

$$K^0X \to K_0X$$

which takes a vector bundle to its sheaf of sections. When $X$ is non-singular, this duality map is an isomorphism. The reason for this is that a coherent sheaf $\mathcal{F}$ on a non-singular X has a finite resolution by locally free sheaves, i.e., there is an exact sequence

$$0 \to E_n \to E_{n-1} \to \cdots \to E_1 \to E_0 \to \mathcal{F} \to 0$$

with $E_0, \ldots, E_n$ locally free. The inverse homomorphism from $K_0$ to $K^0$ takes $[\mathcal{F}]$ to $\sum_{i=0}^n(-1)^i[E_i]$, for such a resolution.

In this paper we only study the case when $X$ is non-singular, so we identify $K_0$ with $K^0$ and denote the pushforward by $f_!$. We can define the *sheaf* $K$*-class* of a subvariety $Y$ by $[\mathcal{O}_Y] \in K_0(X)$.

## 2.2 Topological K-Theory

Topological $K$-theory is a complex oriented cohomology theory, which has several consequences. Any complex vector bundle $E \to X$ has an Euler class $e(E) \in K_{\text{top}}(X)$. (In this notation we incorporated the fact that $K_{\text{top}}$ is 2-periodic.) The Euler class of a line bundle $L$ is given by $e(L) = 1 - [L^*]$. Similarly to ordinary cohomology a complex submanifold $Y$ of the complex manifold $M$ represents a class $[Y \subset M] \in K_{\text{top}}(M)$. Given a complex vector bundle $E$ with a section $\sigma : M \to E$ transversal to the zero section we have $e(E) = [\sigma^{-1}(0) \subset M]$. We have an obvious map from algebraic $K$-theory to topological $K$-theory, which is an isomorphism for $\mathbb{P}^n$, so in the remaining of the paper we identify these rings and also drop the upper and lower 0 indices.

**Theorem 2.1** *The forgetful map $K \to K_{top}$ respects pushforward,*

by Atiyah and Hirzebruch [1], in particular for a complex submanifold $Y$ of the complex manifold $M$ we have

$$[Y \subset M] = [\mathcal{O}_Y].$$

The main goal of this section is to explore how to define this class for non smooth subvarieties.

## 2.3 The K-Theory of $\mathbb{P}^n$

$K(\mathbb{P}^n) = \mathbb{Z}[t]/((1-t)^{n+1})$, where $t = [\gamma]$, the class of the tautological line bundle. The corresponding sheaf is $\mathcal{O}(-1)$. The dual bundle is $L = \mathcal{O}(1)$. $L$ has sections transversal to the zero section so we get that

$$e(L) = 1 - t = [\mathbb{P}^{n-1} \subset \mathbb{P}^n].$$

We will also denote this class by $H$, the class of the hyperplane. Therefore we also have the description $K(\mathbb{P}^n) = \mathbb{Z}[H]/(H^{n+1})$.

## 2.4 Hilbert Polynomial

We show now that for the subvariety $X \subset \mathbb{P}^n$ the class $[\mathcal{O}_X]$ contains the same information as the Hilbert polynomial of $X$.

For $X \subset \mathbb{P}^n$ let $S = \mathbb{C}[x_0, \dots, x_n]$ denote the ring of polynomials, $I_X \lhd S$ the ideal of $X$ and $S(X) := S/I_X$ the homogeneous coordinate ring of $X$. The coordinate ring is a graded ring $S(X) = \bigoplus S^j(X)$ and we would like to encode the dimensions $h_j(X) := \dim S^j(X)$ (i.e. $j \mapsto h_j(X)$ is the Hilbert function of $X$). Notice that the embedding of $X$ is encoded in the grading of $S(X)$. It is well-known that there is a unique polynomial $p_X(x)$—the Hilbert polynomial of $X$—such that $h_j(X) = p_X(j)$ for $j \gg 0$. For us it will be more convenient to use the Hilbert series

$$HS(X) := \sum_{j=0}^{\infty} h_j(X) t^j.$$

For example for $X = \mathbb{P}^n$ we have $h_j(X) = \binom{n+j}{n}$ which is clearly a polynomial of degree $n$ in $j$. The coefficients are certain Stirling numbers. On the other hand the Hilbert series has a particularly simple form:

$$HS(\mathbb{P}^n) = \sum \binom{n+j}{n} t^j = \frac{1}{(1-t)^{n+1}}. \tag{1}$$

The key property of the Hilbert series is (see e.g. in [14])

**Theorem 2.2** *There is a unique polynomial (the $\mathcal{K}$-polynomial) $\mathcal{K}_X(t)$, such that*

$$HS(X) = \frac{\mathcal{K}_X(t)}{(1-t)^{n+1}}.$$

Now we can state the proposed connection of the sheaf $K$-class with the Hilbert polynomial (see e.g. [4, §21]):

**Proposition 2.1** $[\mathcal{O}_X] = \mathcal{K}_X(t)$ *in* $K(\mathbb{P}^n) = \mathbb{Z}[t]/((1-t)^{n+1})$.

Notice that adding a multiple of $(1-t)^{n+1}$ to $\mathcal{K}_X(t)$ changes only finitely many $h_j$'s, so the Hilbert polynomial doesn't change. $\mathcal{K}_X(t)$ encodes the Hilbert function and $[\mathcal{O}_X]$ encodes the Hilbert polynomial. In this sense the natural generalization of the Hilbert polynomial for $X \subset M$ is the $K$-class $[\mathcal{O}_X] \in K(M)$.

*Example 2.1* It is not difficult to calculate (see [12]), that for three generic points $X \subset \mathbb{P}^2$ we have $h_X(j) = 3$ for all $j > 0$ (notice that $h_X(0) = 1$ for all nonempty $X$!), and for three collinear points $Y \subset \mathbb{P}^2$ we have $h_Y(1) = 2$ and $h_Y(j) = 3$ for all $j > 1$. Therefore

$$HS(X) = \left(3\sum t^j\right) - 2 = \frac{3}{1-t} - 2 = \frac{3(1-t)^2 - 2(1-t)^3}{(1-t)^3}$$

and

$$HS(Y) = HS(X) - t = \frac{3(1-t)^2 - (t+2)(1-t)^3}{(1-t)^3},$$

so their Hilbert polynomial i.e. their sheaf $K$-class agrees, but their $\mathcal{K}$-polynomial is different. We will see later that the $\mathcal{K}$-polynomial can be interpreted as the GL(1)-equivariant sheaf $K$-class of the cone.

Using standard resolution techniques (Koszul complex) one can show that

**Corollary 2.1** *Let* $X = (f_1, \dots, f_k) \subset \mathbb{P}^n$ *be a complete intersection. Then the sheaf-theoretic* $K$*-class*

$$[\mathcal{O}_X] = \prod_{i=1}^{k}(1 - t^{d_i}),$$

*where* $d_i$ *is the degree of the generator* $f_i$.

*Remark 2.1* Theorem 2.2 and Proposition 2.1 implies that the assignment $X \mapsto p_X(m)$ induces a map of $K(\mathbb{P}^n)$ to $\mathbb{Q}(m)/(m^{n+1})$, the truncated ring of the possible Hilbert polynomials, but this is only an additive homomorphism. According to (1) the base change is given by

$$H^n \mapsto 1,\ H^{n-1} \mapsto m+1,\ H^{n-2} \mapsto \frac{1}{2}m^2+\frac{3}{2}m+1,\ H^{n-3} \mapsto \frac{1}{6}m^3+m^2+\frac{11}{6}m+1,$$

etc.

The dimension and the degree of $X$ can be read off from the Hilbert polynomial. In fact some people use this fact to define the dimension and the degree. Translating this connection to the $H$-variable we get

**Corollary 2.2** *The sheaf $K$-class of $X \subset \mathbb{P}^n$ has the form $[\mathcal{O}_X] = \sum_{i=d}^{n} q_i H^i$, where $d$ is the codimension of $X$ and $q_d = \deg(X)$.*

*Remark* 2.2 If the ideal of $X \subset \mathbb{P}^n$ is known, then the 'hilbert_numerator' command of Sage (Singular) calculates $[\mathcal{O}_X]$. Also Maple has the 'HilbertSeries' which has to be multiplied by the denominator $(1-t)^{n+1}$. These calculations are feasible only for small examples.

## 2.5 The Pushforward K-Class

This is also only defined for closed subvarieties: Take a resolution $\varphi : \tilde{X} \to X \subset Y$, where $Y$ is smooth.

$$[X] := \varphi_*[\mathcal{O}_{\tilde{X}}].$$

It is a non-trivial fact that this class is independent of the resolution. By Theorem 2.1 we have

**Proposition 2.2**

$$[X] = \varphi_! 1,$$

*for the $K$-theory pushforward.*

For singular $X$ this is not easy to calculate. Even if we know a resolution, the $K$-group of $\tilde{X}$ can be complicated. However for $X = \bigcup_{i=1}^{k} X_i$ where $X_i$ are the irreducible components of $X$, we clearly have

$$[X] = \sum_{i=1}^{k} [X_i], \tag{2}$$

which helps if the components are smooth. Almost by definition we have

**Theorem 2.3** *If $X$ is irreducible with only rational singularities, then*

$$[\mathcal{O}_X] = [X].$$

Indeed, in this case $R^i f_* \mathcal{O}_{\tilde{X}} = 0$ for $i > 0$, and $R^0 f_* \mathcal{O}_{\tilde{X}} = \mathcal{O}_X$ always holds for $X$ normal.

## 2.6 Motivic Invariants and the Motivic K-Class

The motivic $K$-class is defined as the constant term of the motivic Chern class:

**Definition 2.1** ([2]) Suppose that $X \subset Y$, with $Y$ smooth, then the motivic $K$-class of $X$ is

$$\mathrm{mC}_0(X) := \mathrm{mC}(X)_{y=0},$$

where the ambient manifold $Y$ is not denoted if it is clear from the context.

A simple consequence of the definition is that $\mathrm{mC}_0(X) = [X]$ for smooth $X$.

Let us recall what motivic invariants are. The surprising (and not so easy to prove) fact (Fulton) is that $\chi(W) = \chi(W \setminus U) + \chi(U)$ for any $U \subset W$ for (quasiprojective) varieties over $\mathbb{C}$. Over the reals this is not true: e.g. $W = \mathbb{R}$, $U = \{0\}$. We will call these type of invariants *motivic*.

Main examples are the Chern–Schwartz–MacPherson (CSM) and the motivic Chern class. There are several variations, but I prefer now the following setup: Let $h^*$ be a complex oriented cohomology theory (ordinary cohomology for CSM and $K$-theory for motivic Chern class). Then a motivic class for $h^*$ is a functor

$$m(U \subset M) \in h^*(M)$$

(or in $h^*(M)[y]$ for the motivic Chern class) for pairs of varieties (note that $U$ in not necessarily closed in $M$, it is a constructible subset) if it has the *motivic property*: $m(W) = m(W \setminus U) + m(U)$ for any $U \subset W$, and a property I will call *homology property*: Suppose that $f : M \to N$ is proper and $f$ is an isomorphism restricted to $U \subset M$, then

$$f_! m(U \subset M) = m(f(U) \subset N). \tag{!}$$

This property means that $m(U \subset M)$ essentially depends only on $U$ and the dependence on the embedding is very simple. For example the fundamental cohomology class of $U$ has this property, and the reason for that is that there is a fundamental homology class as well. We restrict ourselves to $M$ smooth though it is not necessary. This property is also called covariant functoriality.

We claim that any motivic class in the above sense is determined by its value on closed submanifolds. Indeed, for any pair $U \subset M$ with $U, M$ smooth but $U$ not necessarily closed in $M$ there is a proper map $f : \tilde{M} \to M$ such that

1. $f|_{f^{-1}(U)} : f^{-1}(U) \to U$ is an isomorphism,
2. The „divisor" $D := \tilde{M} \setminus f^{-1}(U)$ is the union of closed submanifolds $D_i$, $i = 1, \ldots, s$ such that for all $I \subset \underline{s} = \{1, 2, \ldots, s\}$ the intersection $D_I := \bigcap_{i \in I} D_i$ is a submanifold.

By the weak factorization theorem such a map (called proper normal crossing extension) always exists. Then, by the motivic and the homology property we have:

$$m(U \subset M) = \sum_{I \subset \underline{s}} (-1)^{|I|} f_!^I m(D_I), \tag{3}$$

where $f^I = f|_{D_I}$, and we use the convention that $m(M) := m(M \subset M)$.

For a smooth variety $M$ we define $\mathrm{mC}(M) := \lambda_y(T^*M)$, where for any complex vector bundle $\lambda_y(E) = \sum_{i=0}^{\mathrm{rank}\, E} [\Lambda^i E] y^i$. Notice that $\lambda_y(E)$ is a natural K-theory analogue of the total Chern class of cohomology theory. The existence of the motivic Chern class is proved in [2]. The definition of the equivariant version is again straightforward, see in [10].

The motivic property of mC is inherited by $\mathrm{mC}_0$.

*Example 2.2* Let $X$ be a projective cubic plane curve. Then its sheaf-class is $1 - t^3 = 3H - 3H^2$. For the smooth cubics all three classes agree. For the others see Table 1. The calculations are straightforward for the reducible ones. For the nodal and cuspical ones we use the fact that they have a rational resolution $\varphi : \mathbb{P}^1 \to \mathbb{P}^2$ of degree 3. First we need to calculate $\varphi_! 1$. Since $\varphi_! 1$ depends only on the degree, we have

$$\varphi_! 1 = i_! f_! 1,$$

where $i : \mathbb{P}^1 \to \mathbb{P}^2$ is the linear inclusion and $f : \mathbb{P}^1 \to \mathbb{P}^1$ has degree 3. Since $i$ is injective, it is easy to see that $i_! 1 = H$ and $i_! H = H^2$. Less obvious is to calculate $f_! 1$:

**Lemma 2.1** *Let $f_d : \mathbb{P}^1 \to \mathbb{P}^1$ has degree $d$. Then*

$$f_{d!} 1 = d - (d-1)H.$$

**Table 1** $K$-classes of plane cubics

| Description of $X$ | Shape | Representant | $[\mathcal{O}_X]$ | $[X]$ | $\mathrm{mC}_0(X)$ |
|---|---|---|---|---|---|
| Nodal cubic |  | $x^3 + y^3 + xyz$ | $3H - 3H^2$ | $3H - 2H^2$ | $3H - 3H^2$ |
| Cuspidal cubic |  | $x^3 + y^2 z$ | $3H - 3H^2$ | $3H - 2H^2$ | $3H - 2H^2$ |
| Conic and intersecting line |  | $x^3 + xyz$ | $3H - 3H^2$ | $3H - H^2$ | $3H - 3H^2$ |
| Conic and tangent line |  | $x^2 y + y^2 z$ | $3H - 3H^2$ | $3H - H^2$ | $3H - 2H^2$ |
| Three nonconcurrent lines | 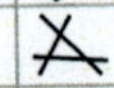 | $xyz$ | $3H - 3H^2$ | $3H$ | $3H - 3H^2$ |
| Three concurrent lines |  | $x^2 y + xy^2$ | $3H - 3H^2$ | $3H$ | $3H - 2H^2$ |

***Proof*** Let $v_d : \mathbb{P}^1 \to \mathbb{P}^d$ be the degree $d$ Veronese embedding. The image $C_d$ is smooth, so $v_{d!}1 = [C_d \subset \mathbb{P}^d]$. Using the Hilbert polynomial or Corollary 2.5 we can see, that $[C_d \subset \mathbb{P}^d] = H^{d-1}\big(d-(d-1)H\big)$. Let $i_d : \mathbb{P}^1 \to \mathbb{P}^d$ be the linear embedding. Then $i_d f_d$ is homotopic to $v_d$, so $v_{d!}1 = i_{d!}f_{d!}1$. On the other hand $i_{d!}H^j = H^{j+d-1}$, implying our result. □

Therefore $\varphi_!1 = i_!f_!1 = i_!(3-2H) = 3H-2H^2 =$ [] = []. Using the definition (3) of the motivic class (notice that these resolutions are evidently normal crossing) we get that mC$_0$() $= \varphi_!1$ since $\varphi$ injective in this case, and mC$_0$(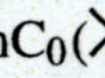) $= \varphi_!1 - H^2$ since $\varphi$ has a double point in this case.

**Theorem 2.4 ([2])** *If $X$ has only Du Bois singularities then* $[\mathcal{O}_X] = \mathrm{mC}_0(X)$.

Also the definition of Du Bois singularity can be found in [2]. Additional information can be found in [17]. Important cases of Du Bois singularities are rational singularities, transversal union of smooth varieties (these are not rational) and cone hypersurfaces in $\mathbb{C}^n$ of degree $d \le n$. We can see from our calculations that ,  and  are not Du Bois.

*Example 2.3* Let $X = E^{n-k} \cup E^{n-l} \subset \mathbb{P}^n$, where $E^j$ is a $j$-dimensional projective subspace. We assume that $E^{n-k}$ and $E^{n-l}$ are in general position and $k, l > 0$. Then $X$ has only Du Bois singularities, therefore $[\mathcal{O}_X] = \mathrm{mC}_0(X)$. The latter can be easily calculated by the motivic property:

$$\mathrm{mC}_0(X) = (1-t)^k + (1-t)^l - (1-t)^{k+l},$$

where the last term is 0 in the $K$-group if $k + l > n$, i.e. $E^{n-k} \cap E^{n-l} = \emptyset$. On the other hand

$$[E^{n-k} \cup E^{n-l}] = [E^{n-k}] + [E^{n-l}] = (1-t)^k + (1-t)^l.$$

The calculation of $[\mathcal{O}_X]$ is usually done using the Hilbert syzygy theorem by calculating a resolution. For non obvious examples this is difficult, even with computers. Let us demonstrate this on $X = E^1 \cup E^1 \subset \mathbb{P}^3$. Then $I_X = (x, y) \cdot (w, z) = (xw, xz, yw, yz)$. The four relations are

$$\begin{array}{llllll} g_1 : & z & -w & 0 & 0 \\ g_2 : & 0 & 0 & z & -w \\ g_3 : & -y & 0 & x & 0 \\ g_4 : & 0 & -y & 0 & x \end{array}$$

Finally we have a relation among the relations: $yg_1 - xg_2 + zg_3 - wg_4$. This gives us

$$[\mathcal{O}_X] = 1 - 4t^2 + 4t^3 - t^4,$$

which is indeed congruent to $2(1-t)^2$ modulo $(1-t)^4$.

## 2.7 The Todd Genus

We have already seen that understanding pushforward is essential in our calculations. Compared to cohomology, $K$-theory has a new feature. Let $\mathrm{co}_X : X \to *$ denote the collapse map of $X$ for $X$ projective and smooth. Then the *Todd genus*

$$\mathrm{Td}(X) := \mathrm{co}_{X!}1 = \int_X 1 = \chi(X, \mathcal{O}_X) \in \mathbb{Z}$$

is a non-trivial invariant. (In this paper the integral sign will always denote the $K$-theory pushforward to the point.) This is a genus in the sense that it defines a ring homomorphism from the complex cobordism ring to the ring of integers.

As being a genus suggests it is enough to calculate $\mathrm{Td}(\mathbb{P}^n)$ to be able to calculate more involved examples. It is a key result in $K$-theory, in particular the proof of Theorem 2.1 uses it. It is usually proved using the topological Grothendieck–Riemann–Roch theorem.

**Theorem 2.5 ([1])** *For any $n \in \mathbb{N}$ the Todd genus of the projective space $\mathbb{P}^n$ is 1:*

$$\mathrm{Td}(\mathbb{P}^n) = 1.$$

Then by basic properties of pushforward immediately yields:

**Corollary 2.3** *Let $X \subset \mathbb{P}^n$ be smooth with $K$-theory fundamental class*

$$[X] = \sum_{i=0}^{n} q_i H^i = q(H).$$

*Then* $\mathrm{Td}(X) = \sum_{i=0}^{n} q_i = q(1)$.

***Proof*** Notice first that $\mathrm{Td}(X) = \int_{\mathbb{P}^n} [X]$, then we can apply the integral formula

$$\int_{\mathbb{P}^n} \sum_{i=0}^{n} q_i H^i = q(1). \tag{4}$$

□

Recalling the connection of $[X] = [\mathcal{O}_X]$ with the Hilbert polynomial $p_X$ we have that $q(1) = p_X(0)$. Recall that the *arithmetic genus* is defined as

$$p_a(X) := (-1)^{\dim X}(p_X(0) - 1),$$

so we see that it is essentially the same as the Todd genus for smooth $X$. In other words the Todd genus of $X$ is equal to its *holomorphic Euler characteristics* $\chi(X, \mathcal{O}_X)$.

## 2.8 Genus of Smooth Hypersurfaces

For a smooth degree $d$ hypersurface $X_d \subset \mathbb{P}^n$ we have $[X] \equiv 1 - t^d$, so a simple binomial identity implies that

**Corollary 2.4** *The arithmetic genus of the smooth degree d hypersurface* $X_d \subset \mathbb{P}^n$ *is* $\binom{d-1}{n}$.

Notice that by definition $\binom{d-1}{n} = 0$ for $d \le n$. This is the first sign that hypersurfaces of degree higher that $n$ behaves very differently then the low degree ones.

**Corollary 2.5** *For the degree d rational normal curve* $X_d \subset \mathbb{P}^d$ *we have* $[X_d \subset \mathbb{P}^d] = dH^{d-1} - (d-1)H^d$.

Indeed, by Corollary 2.2. we have $[X_d \subset \mathbb{P}^d] = dH^{d-1} + q_d H^d$. But $X_d \cong \mathbb{P}^1$ so $1 = \mathrm{Td}(X_d) = d + q_d$. □

## 2.9 The $\chi_y$ Genus

A straightforward extension of the Todd genus is the $\chi_y$-genus of Hirzebruch:

$$\chi_y(X) := \int_X \lambda_y(T^*X),$$

for $X$ projective and smooth. In general if $X \subset M$ for $M$ projective and smooth we can define

$$\chi_y(X) := \int_M \mathrm{mC}(X \subset M), \tag{5}$$

which is independent of the embedding of $X$ (the reader is encouraged to check this), providing a motivic extension of the $\chi_y$ genus. Clearly, substituting $y = 0$ into the $\chi_y$ genus we obtain the Todd genus if $X$ is smooth, and the holomorphic Euler characteristics is general.

*Example 2.4* It is instructive to calculate $\chi_y(\mathbb{P}^n)$. Since we have the short exact sequence of vector bundles

$$0 \to \mathbb{C} \to L^{n+1} \to T\mathbb{P}^n \to 0,$$

and $\lambda_y$ is multiplicative, we have

$$\mathrm{mC}(\mathbb{P}^n) = \frac{(1+yt)^{n+1}}{1+y} \equiv \frac{(1+yt)^{n+1} - (-y)^{n+1}(1-t)^{n+1}}{1+y}$$
$$= \sum_{i=0}^{n} (1+yt)^i (y(t-1))^{n-i}.$$

Applying the integral formula (4) by substituting $t = 0$ we arrive at

$$\chi_y(\mathbb{P}^n) = 1 - y + y^2 - \cdots \pm y^n,$$

which was already calculated by Hirzebruch.

*Remark 2.3* It is interesting to write $\mathrm{mC}(\mathbb{P}^n)$ in the form $\sum_{i=0}^n q_i H^i$ for $q_i \in \mathbb{Z}[y]$ (See e.g. in [4, §22]):

$$\mathrm{mC}(\mathbb{P}^n) = \sum_{i=0}^{n} \binom{n+1}{i} (-y)^i (1+y)^{n-i} H^i.$$

For the next example we recall the "divisor trick", the multiplicativity of $\lambda_y$ implies the following:

**Corollary 2.6** *Suppose that $Y$ is the zero locus of a section of a vector bundle $E \to M$, which is transversal to the zero-section. Then the motivic Chern class is*

$$\mathrm{mC}(Y \subset M) = e(E)\lambda_y(-E)\,\mathrm{mC}(M).$$

*Example 2.5* A smooth degree $d$ hypersurface $Z_d \subset \mathbb{P}^n$ is the zero locus of a section of the line bundle $(\gamma^*)^d$. Using the divisor trick we get

$$\mathrm{mC}(Z_d) = \frac{(1+yt)^{n+1}}{1+y} \cdot \frac{1-t^d}{1+yt^d}.$$

A closed formula for the $\chi_y$ genus gets complicated, so we give the answer for small $n$ only:

$$n = 2: \quad \chi_y(Z_d) = \left(\binom{d-1}{2} - 1\right)(y-1),$$

$$n = 3: \quad \chi_y(Z_d) = \left(\binom{d-1}{3} + 1\right)(y-1)^2 + \left(2\binom{d-1}{3} - 4\binom{d}{3} - d + 2\right)y.$$

Substituting $y = 0$ we can see that it is consistent with 2.4.

## 3 Cones

The simplest singularities in some sense are the conical ones. We study two closely related cases.

### 3.1 The Projective Case

**Proposition 3.1** *Let $X \subset \mathbb{P}^n$ be smooth and denote its cone in $\mathbb{P}^{n+1}$ by $\hat{X}$. Then for the unique polynomials $q_i \in \mathbb{Z}[y]$ and $\hat{q}_i \in \mathbb{Z}[y]$ such that* $\mathrm{mC}(X \subset \mathbb{P}^n) = \sum_{i=0}^{n} q_i H^i$ *and* $\mathrm{mC}(\hat{X} \subset \mathbb{P}^{n+1}) = \sum_{i=0}^{n+1} \hat{q}_i H^i$ *we have*

$$\hat{q}_i = (1+y)q_i - yq_{i-1} \ \textit{for } i = 0, \dots, n \quad \textit{and} \ \hat{q}_{n+1} = 1 - yq_n - (1+y)\chi_y(X),$$

*where $q_{-1} = 0$ and as we mentioned before $\chi_y(X) = \sum_{i=0}^{n} q_i$.*

***Proof*** First notice that $j : \mathbb{P}^n \to \mathbb{P}^{n+1}$ is transversal to $\hat{X}$ and intersect in $X$. We assumed that $X$ is smooth, so we have the pullback formula:

$$\frac{\mathrm{mC}(X)}{\mathrm{mC}(\mathbb{P}^n)} = j^* \left( \frac{\mathrm{mC}(\hat{X})}{\mathrm{mC}(\mathbb{P}^{n+1})} \right). \tag{6}$$

**Definition 3.1** The motivic Chern class has its Segre version just as the Chern–Schwartz–MacPherson class:

$$\mathrm{mS}(X \subset M) := \frac{\mathrm{mC}(X \subset M)}{\mathrm{mC}(M)}.$$

Multiplicativity of $\lambda_y$ implies that the motivic Segre class behaves nicely with respect to transversal maps:

**Proposition 3.2** *Let $f : A \to M$ be a proper (in the topological sense) map of smooth varieties such that $f$ is transversal to $X \subset M$ in the sense that it is transversal to the smooth part of $X$ and does not intersect the singular part of $X$. Then*

$$\mathrm{mS}(f^{-1}(X) \subset A) = f^* \mathrm{mS}(X \subset M).$$

Using that $j^*(H) = H$ and $\frac{j^* \mathrm{mC}(\mathbb{P}^{n+1})}{\mathrm{mC}(\mathbb{P}^n)} = 1 + yt$ we get that

$$\hat{q}_i = (1+y)q_i - yq_{i-1} \ \text{ for } i = 0, \dots, n. \tag{7}$$

To calculate $\hat{q}_{n+1}$ we consider the blowup at the vertex 0 of the cone. Restricting to the preimage of the cone we get a normal crossing resolution $\varphi : Y \to \hat{X} \subset \mathbb{P}^{n+1}$, where $Y$ is a fiber bundle over $X$ with fiber $\mathbb{P}^1$. From the definition (3) we have

$$\mathrm{mC}(\hat{X} \setminus 0) = \varphi_! \lambda_y(Y) - \chi_y(X) H^{n+1}. \tag{8}$$

The first term is difficult to calculate directly so we push forward (8) to a point:

$$\chi_y(\hat{X} \setminus 0) = \chi_y(Y) - \chi_y(X). \tag{9}$$

Now we use that $Y \to X$ is the projective bundle of a vector bundle, so

$$\chi_y(Y) = \chi_y(\mathbb{P}^1)\chi_y(X) = (1-y)\chi_y(X),$$

implying

$$\chi_y(\hat{X}) = 1 - y\chi_y(X).$$

(This product property of $\chi_y$ was already known by Hirzebruch, see e. g. [13].) On the other hand $\chi_y(\hat{X}) = \sum_{i=0}^{n+1} \hat{q}_i$ and $\chi_y(X) = \sum_{i=0}^{n} q_i$, so we can express $\hat{q}_{n+1}$ using (7). □

*Remark 3.1* Proposition 3.1. immediately generalises to $X$ being a constructible set, if we use the motivic extension of the $\chi_y$ genus, i.e. we take (5) as the definition of the $\chi_y$ genus for any constructible set.

Substituting $y = 0$ we get the following:

**Corollary 3.1** *Express* $\mathrm{mC}_0(\hat{X} \subset \mathbb{P}^{n+1})$ *and* $\mathrm{mC}_0(X \subset \mathbb{P}^n)$ *in their reduced form, i.e as polynomials of degree at most $n+1$ and $n$, respectively, in the variable $H$. Then*

$$\mathrm{mC}_0(\hat{X} \subset \mathbb{P}^{n+1}) = \mathrm{mC}_0(X \subset \mathbb{P}^n) + (1 - \mathrm{Td}(X))H^{n+1}.$$

*Remark 3.2* The other two $K$-classes of a cone can also be calculated. For the pushforward $K$-class we can use the same resolution as above:

$$[\hat{X}] = \varphi_! 1,$$

so Proposition 3.1 implies that

$$[\hat{X}] = \mathrm{mC}_0(\hat{X}) + (\mathrm{Td}(X) - 1)H^{n+1} = [X].$$

So the three $K$-classes of the cone differs only in the top coefficient, which is 0 for the pushforward class and $(1 - \mathrm{Td}(X))$ for the motivic $K$-class.

For the sheaf theoretic $K$-class of $\hat{X}$ we need more than the corresponding one for $X$. It follows from the definition, that $\mathcal{K}_{\hat{X}}(t) = \mathcal{K}_X(t)$. Write $\mathcal{K}_X(t) = \sum p_i H^i$ as a polynomial (of degree possibly much higher than $n+1$) of $H = 1 - t$. Then the reduced form of the sheaf theoretic $K$-class of $X$ and $\hat{X}$ are $\sum_{i=0}^{n} p_i H^i$, and $\sum_{i=0}^{n+1} p_i H^i$, respectively.

*Remark 3.3* For CSM classes the calculation is simpler: for $c^{sm}(X) := \sum_{i=0}^{n} q_i H^i \in H^*(\mathbb{P}^n)$ and $c^{sm}(\hat{X}) := \sum_{i=0}^{n+1} \hat{q}_i H^i \in H^*(\mathbb{P}^{n+1})$ we get $\hat{q}_i = q_i + q_{i-1}$ for $i \leq n$ and $\hat{q}_{n+1} = \chi(\hat{X}) = q_n + 1 = \chi(X) + 1$.

*Example 3.1* Corollary 3.1 allows us to find irreducible examples of varieties for which all 3 $K$-classes are different. Let $X = Z_d \subset \mathbb{P}^2$ a smooth curve of degree $d$. Then $\mathrm{mC}_0(Z_d) \equiv 1 - t^d \equiv dH - \binom{d}{2} H^2$. Then

$$\mathrm{mC}_0(\hat{Z}_d) = dH - \binom{d}{2} H^2 + \binom{d-1}{2} H^3.$$

On the other hand

$$[\mathcal{O}_{\hat{Z}_d}] \equiv 1 - t^d \equiv dH - \binom{d}{2} H^2 + \binom{d}{3} H^3,$$

and

$$[\hat{Z}_d] = \mathrm{mC}_0(Z_d \subset \mathbb{P}^n) + H^{n+1} = dH - \binom{d}{2} H^2$$

by the previous remark.

Therefore all these 3 classes of $\hat{Z}_d$ are different if $d \geq 4$. It implies that these hypersurfaces are not Du Bois. In fact it is known that $\hat{Z}_d \subset \mathbb{P}^{n+1}$ is Du Bois if and only if $d \leq n + 1$, so this calculation detects all the non Du Bois cases among the $\hat{Z}_d$'s.

*Remark 3.4* Pushing forward the three $K$-classes to the point we get 3 different extensions of the Todd genus to singular varieties. $\int_{\mathbb{P}^n} [\mathcal{O}_X]$ is the holomorphic Euler characteristics. $\int_{\mathbb{P}^n} [X] = \mathrm{Td}(\tilde{X})$ is the Todd genus of the resolution (note that this is independent of the resolution!). We can call $\chi_{y=0} := \int_{\mathbb{P}^n} \mathrm{mC}_0(X)$ the *motivic Todd genus* of $X$. Our calculations show that for $X = \hat{Z}$, the projective cone of the smooth variety $Z$, $\int_{\mathbb{P}^n} [X] = \mathrm{Td}(\tilde{X}) = \mathrm{Td}(Z)$ and $\chi_{y=0}(X) = 1$, so even these three Todd genus extensions are different for the projective cone of a smooth curve of degree $d$ if $d \geq 4$. A similar example was discovered in [17, ex 14.1].

## 4 Equivariant Classes

If an algebraic linear group $G$ acts on $M$, then we can define the Grothendieck group $K_0^G(M)$ of coherent $G$-sheaves. Also we can define the Grothendieck group $K_G^0(M)$ of $G$-vector bundles. For smooth $M$ they are isomorphic. For $M = \mathbb{C}^n$ and

$\mathbb{P}^n$ the forgetful map to $K^G_{top}(M)$—the Grothendieck group of topological $G$-vector bundles—is also isomorphism. The ring $K(B_G M)$ is much bigger, we will not use it. The definition of the equivariant motivic Chern classes and the various equivariant $K$-classes are straightforward, one can repeat the same definitions equivariantly. For the existence of the equivariant motivic Chern class and more details about equivariant $K$-theory see e.g. [10]. Notice that compared to cohomology the transition to the equivariant theory is much smoother, we do not need classifying spaces and approximation of the classifying spaces by algebraic varieties.

There are several reasons to introduce equivariant theory in this context. One is that implicitly we are already using scalar equivariant objects: vector bundles admit a canonical scalar action therefore they admit equivariant Euler class, which in cohomology can be identified with the total Chern class and in $K$-theory with $\lambda_y$ of the vector bundle, which is the starting point of building the motivic Chern class.

The second reason is that the definition of the Hilbert function is based on a scalar action: the homogeneous coordinate ring of $X \subset \mathbb{P}^n$ is graded according to the natural scalar action on it. This implies the following reformulation:

Consider first the scalar $\Gamma := \mathrm{GL}(1)$-action on $\mathbb{C}^{n+1}$. For any $X \subset \mathbb{P}^n$ the cone $CX \subset \mathbb{C}^{n+1}$ is $\Gamma$-invariant, so its structure sheaf $\mathcal{O}_{CX}$ is a coherent $\Gamma$-sheaf.

**Proposition 4.1** *The Grothendieck group $K_0^\Gamma(\mathbb{C}^{n+1})$ is isomorphic to $\mathbb{Z}[t, t^{-1}]$ via the restriction map to the origin, and*

$$[\mathcal{O}_{CX}]_\Gamma = \mathcal{K}_X(t).$$

The first statement is proved in [3, 5.4.17]. The statement on the $\mathcal{K}$-polynomial is folklore, see [14, p. 172]. It is essentially equivalent to [3, 6.6.8]. Consequently the Kirwan-type homomorphism $K_0^\Gamma(\mathbb{C}^{n+1}) \to K_0(\mathbb{P}^n)$ with $t \mapsto t$ maps $\mathcal{K}_X(t) = [\mathcal{O}_{CX}]_\Gamma$ to $[\mathcal{O}_X]$, i.e. assigns the Hilbert polynomial to the Hilbert function.

*Remark 4.1* Notice, that $t$ as an element in the representation ring of $\Gamma := \mathrm{GL}(1)$ is the inverse of the standard representation. It looks awkward first but this is the choice which reflects that the hyperplane is the zero locus of a section of the *dual* of the tautological bundle. A more conceptual explanation will be given in 4.3.1.

The third reason to introduce the equivariant theory might be the most important: Equivariant motivic, $K$, etc classes on $G$-invariant subvarieties are *universal classes for degeneracy loci*. Let us explain this statement in more details. An introduction into the cohomological theory can be found e.g. in [5, §2.] and for the CSM case in [6] and [15], so we concentrate on the $K$-theory cases here.

## 4.1 Universal Classes in K-Theory

Let $G$ be a connected linear algebraic group and suppose that $\pi_P : P \to M$ is a principal $G$-bundle over the smooth $M$ and $A$ is a smooth $G$-variety. Then we can define a map

$$a : K_G(A) \to K(P \times_G A)$$

by association: For any $G$-vector bundle $E$ over $A$ the associated bundle $P \times_G E$ is a vector bundle over $P \times_G A$.

**Proposition 4.2** *Let $Y \subset A$ be $G$-invariant. Then*

$$\mathrm{mS}(P \times_G Y \subset P \times_G A) = a\big(\mathrm{mS}_G(Y \subset A)\big).$$

The proof can be found in [9, Pr 8.7]. Substituting $y = 0$ we get the corresponding statement for the motivic $K$-class:

**Proposition 4.3** *Let $Y \subset A$ be $G$-invariant. Then*

$$\mathrm{mC}_0(P \times_G Y \subset P \times_G A) = a\big(\mathrm{mC}_0^G(Y \subset A)\big).$$

Similarly one can show the analogous statement for the push forward $K$-class:

**Proposition 4.4** *Let $Y \subset A$ be a $G$-invariant subvariety. Then*

$$[P \times_G Y \subset P \times_G A] = a[Y \subset A]_G.$$

For the proof you need to check that for a $G$-equivariant resolution $\varphi : \tilde{Y} \to A$ of $Y$ the induced resolution $\hat{\varphi} : P \times_G \tilde{Y} \to P \times_G A$ of $P \times_G Y$ has the property $\hat{\varphi}_! 1 = a\varphi_! 1$.

For the sheaf $K$-class we have

**Proposition 4.5** *Let $Y \subset A$ be a $G$-invariant subvariety. Then*

$$[\mathcal{O}_{P \times_G Y}] = a[\mathcal{O}_Y]_G.$$

In most applications $A$ is a vector space and a section $\sigma : M \to P \times_G A$ sufficiently transversal to $P \times_G Y$ is given. For example

**Corollary 4.1** *Suppose that $\sigma : M \to P \times_G A$ is a section motivically transversal to $P \times_G Y$. Then*

$$\mathrm{mS}(Y(\sigma) \subset M) = a\big(\mathrm{mS}_G(Y \subset A)\big),$$

*where $Y(\sigma) = \sigma^{-1}(P \times_G Y)$ is the $Y$-locus of the section $\sigma$.*

*If $A$ is a vector space then we identify the $K$-theory of $M$ with the $K$-theory of $P \times_G A$ via $\sigma^*$.*

For the proof and the definition of motivically transversal see [9, §8]. The corollary implies the analogous statement for $\mathrm{mC}_0$.

For the push forward $K$-class a weaker transversality condition is sufficient: we only need that the pullback of the resolution $\varphi : \tilde{Y} \to A$ by $\sigma$ is a resolution of $\sigma^{-1}(Y)$.

Recently Rimanyi and Szenes studied the $K$-theoretical Thom polynomial of the singularity $A_2$ in [16]. They choose the push forward $K$-class which means that for a reasonably wide class of maps their $K$-theoretical Thom polynomial calculates the push forward $K$-class of the $A_2$-locus. It would be interesting to study the motivic version of their $K$-theoretical Thom polynomial.

For the sheaf $K$-class the conditions are more complicated. $Y$ has to be Cohen-Macaulay of pure dimension (in many applications like [16] this is not satisfied). If $\sigma^{-1}(Y)$ is also of pure dimension and its codimension agrees with the codimension of $Y$ then $[\mathcal{O}_{Y(\sigma)}] = a[\mathcal{O}_Y]_G$, where $Y(\sigma)$ is the pull back scheme. To ensure that $Y(\sigma)$ is reduced we need further transversality conditions.

## 4.2 Equivariant Classes of Cones in Cohomology: The Projective Thom Polynomial

Earlier we explained the connection between the motivic Chern class of $X \subset \mathbb{P}^n$ and of its projective cone. Just as interesting is the case of the affine cone, however we are forced to use equivariant setting, otherwise there is not enough information in the class of the affine cone.

Suppose that a complex torus $\mathbb{T}$ of rank $k$ acts on $\mathbb{C}^{n+1}$ linearly, i.e. a homomorphism $\rho : \mathbb{T} \to \mathrm{GL}(n+1)$ is given. We assume that the action *contains the scalars*: there is a non zero integer $q$ and a homomorphism $\varphi : \mathrm{GL}(1) \to \mathbb{T}$ such that $\rho\varphi(z) = z^q I$ for all $z \in \mathrm{GL}(1)$. Suppose that $X \subset \mathbb{P}^n$ is $\mathbb{T}$-invariant. Then $CX \subset \mathbb{C}^{n+1}$ is also $\mathbb{T}$-invariant and we can compare their various classes. The first such connection was found about the equivariant cohomology class in [7], what we recall now.

After reparamerization of $\mathbb{T}$ we can assume that $\varphi(z) = \mathrm{diag}(z^{w_1}, \dots, z^{w_k})$, where the integers $w_1, \dots, w_k$ are the weights of $\varphi$. Then we have the following:

**Proposition 4.6** *The* $\mathbb{T}$*-equivariant cohomology class*

$$[X \subset \mathbb{P}^n] \in H^*_{\mathbb{T}}(\mathbb{P}^n) = \mathbb{Z}[a_1, \dots, a_k][x]/(\prod_{i=1}^{n+1}(b_i - x)),$$

*where* $x = c_1^{\mathbb{T}}(\gamma)$ *is the equivariant first Chern class of the tautological bundle with the induced* $\mathbb{T}$*-action and* $b_i$ *are the weights of the* $\mathbb{T}$*-action on* $\mathbb{C}^{n+1}$*, can be expressed from the* $\mathbb{T}$*-equivariant cohomology class*

$$[CX \subset \mathbb{C}^{n+1}] \in H^*_{\mathbb{T}}(\mathbb{C}^{n+1}) = \mathbb{Z}[a_1, \dots, a_k]$$

*by the substitution*

$$[X] = \mathrm{sub}([CX], a_i \mapsto a_i - \frac{w_i}{q}x).$$

This formula has several useful applications, in particular it helps to calculate the degree of certain subvarieties (see e.g. [8]). There is a counterpart which is quite obvious in cohomology:

**Proposition 4.7** *The* $\mathbb{T}$*-equivariant cohomology class*

$$[CX \subset \mathbb{C}^{n+1}]_{\mathbb{T}} \in H^*_{\mathbb{T}}(\mathbb{C}^{n+1}) = \mathbb{Z}[a_1, \dots, a_k]$$

*can be expressed from the* $\mathbb{T}$*-equivariant cohomology class*

$$[X \subset \mathbb{P}^n]_{\mathbb{T}} \in H^*_{\mathbb{T}}(\mathbb{P}^n) = \mathbb{Z}[a_1, \dots, a_k][x]/(\prod_{i=1}^{n+1}(b_i - x))$$

*by the substitution*

$$[CX \subset \mathbb{C}^{n+1}]_{\mathbb{T}} = \mathrm{sub}([X]_{\mathbb{T}}, x \mapsto 0),$$

*where the substitution is done into the reduced form of* $[X]_{\mathbb{T}}$*, the unique polynomial of* $x$ *degree at most* $n$ *representing* $[X]_{\mathbb{T}}$ *in* $\mathbb{Z}[a_1, \dots, a_k][x]$.

## 4.3 Projective Thom Polynomial for the Motivic Chern Class

There is an analogous result for the motivic Chern class. First we need to understand $K_{\mathbb{T}}(\mathbb{P}^n)$.

### 4.3.1 The Kirwan Map in $K$-Theory

We have a Kirwan-type surjective map $\kappa : K_{\Gamma\times\mathbb{T}}(\mathbb{C}^{n+1}) \to K_{\mathbb{T}}(\mathbb{P}^n)$. More generally let $V$ be a $\Gamma \times \mathbb{T}$-vector space for a connected algebraic group $\Gamma$ and assume that $P \subset V$ is an open $\Gamma \times \mathbb{T}$-invariant subset such that $\pi : P \to P/\Gamma$ is a principal $\Gamma$-bundle over the smooth $M := P/\Gamma$.

Strictly speaking $\Gamma$ acts on $V$ from the left, and on $P$ on the right, so we need to define

$$pg := g^{-1}p \tag{10}$$

for all $g \in \Gamma$ and $p \in P \subset V$.

Notice first that we have a restriction map $r : K_{\Gamma\times\mathbb{T}}(V) \to K_{\Gamma\times\mathbb{T}}(0)$. Then for a $\Gamma \times \mathbb{T}$-representation $W$ we can apply the association map $W \to P \times_{\Gamma} W$ to induce a map $a : K_{\Gamma\times\mathbb{T}}(0) \to K_{\mathbb{T}}(M)$, and we can define $\kappa := ar$.

Specializing to $V = \mathbb{C}^{n+1}$, $\Gamma = \mathrm{GL}(1)$ acting as scalar multiplication, $P = \mathbb{C}^{n+1} \setminus 0$ we obtain $\kappa : K_{\Gamma\times\mathbb{T}}(\mathbb{C}^{n+1}) \to K_{\mathbb{T}}(\mathbb{P}^n)$. Notice that the switch between left and right action explained above implies that $\kappa(t) = [\gamma]_{\mathbb{T}}$ for $t$ denoting the inverse of the standard representation of $\Gamma = \mathrm{GL}(1)$, explaining the convention in Remark 4.1.

The $\mathbb{T}$-bundle $\mathrm{Hom}(\gamma, \mathbb{C}^{n+1})$ over $\mathbb{P}^n$ has a nowhere zero $\mathbb{T}$-equivariant section (the inclusion of $\gamma$ into the trivial bundle $\mathbb{P}^n \times \mathbb{C}^{n+1}$), therefore

$$e_{\mathbb{T}}\big(\mathrm{Hom}(\gamma, \mathbb{C}^{n+1})\big) = \prod_{i=1}^{n+1}(1 - t/\beta_i) = 0.$$

It can be checked that this is the only relation, therefore

$$K_{\mathbb{T}}(\mathbb{P}^n) \cong \mathbb{Z}[\alpha_1, \alpha_1^{-1}, \ldots, \alpha_k, \alpha_k^{-1}][t, t^{-1}]/(\prod_{i=1}^{n+1}(1 - t/\beta_i)).$$

The relation can be rewritten as $\prod_{i=1}^{n+1}(t-\beta_i) = 0$ which implies that any element $\omega \in K_{\mathbb{T}}(\mathbb{P}^n)$ can be written uniquely as a polynomial of degree at most $n$ in $t$ with coefficients in $K_{\mathbb{T}}$ (i.e. $\mathbb{P}^n$ is equivariantly formal in $K$-theory for linear $\mathbb{T}$-actions). We call this polynomial the *reduced form* of $\omega$.

### 4.3.2 The Affine to Projective Formula

The analogue of Proposition 4.6 for motivic classes is similar, and the proof is essentially the same:

**Theorem 4.1** *The $\mathbb{T}$-equivariant motivic Segre class*

$$\mathrm{mS}_{\mathbb{T}}(X \subset \mathbb{P}^n) \in K_{\mathbb{T}}(\mathbb{P}^n)[y] = \mathbb{Z}[\alpha_1, \alpha_1^{-1}, \ldots, \alpha_k, \alpha_k^{-1}][t, t^{-1}]/(\prod_{i=1}^{n+1}(1 - t/\beta_i))[y],$$

*where $t = [\gamma]_{\mathbb{T}}$ is the class of the tautological bundle with the induced $\mathbb{T}$-action and $\beta_i$ are the characters of the $\mathbb{T}$-action on $\mathbb{C}^{n+1}$, can be expressed from the $\mathbb{T}$-equivariant motivic Segre class*

$$\mathrm{mS}_{\mathbb{T}}(C_0X \subset \mathbb{C}^{n+1}) \in K_{\mathbb{T}}(\mathbb{C}^{n+1})[y] = \mathbb{Z}[\alpha_1, \alpha_1^{-1}, \ldots, \alpha_k, \alpha_k^{-1}][y]$$

*by the substitution*

$$\mathrm{mS}_{\mathbb{T}}(X) = \mathrm{sub}(\mathrm{mS}_{\mathbb{T}}(C_0X), \alpha_i \mapsto \alpha_i \cdot t^{-\frac{w_i}{q}}),$$

*where $C_0X = CX \setminus 0$.*

It is natural to use the motivic Segre class because it has the transversal pull back property. We need to use $C_0X$ instead of $CX$, because $C_0X$ is the preimage of $X$ under the quotient map $\mathbb{C}^{n+1} \setminus 0 \to \mathbb{P}^n$.

Strictly speaking the motivic Segre class lives in the completion of $K_{\mathbb{T}}(\mathbb{P}^n)[y]$ because of the division with $\mathrm{mC}(\mathbb{P}^n)$, but we will not denote this completion. At the end we are mainly interested in the motivic Chern class where the completion is not needed.

***Proof*** First notice that by a simple change of variables we have

**Proposition 4.8** *If the $\mathbb{T}$-action contains the scalars as above then*

$$\mathrm{mS}_{\Gamma\times\mathbb{T}}(Z) = \mathrm{sub}(\mathrm{mS}_{\mathbb{T}}(Z), \alpha_i \mapsto \alpha_i \cdot t^{-\frac{w_i}{q}})$$

*for any $\mathbb{T}$-invariant (therefore $\Gamma \times \mathbb{T}$-invariant) constructible subset $Z \subset \mathbb{C}^{n+1}$.*

On the other hand we have

**Proposition 4.9**

$$\mathrm{mS}_{\mathbb{T}}(X) = \kappa\, \mathrm{mS}_{\Gamma\times\mathbb{T}}(C_0X),$$

as a special case of [9, Thm 8.12]. □

We can translate the result to motivic Chern class easily:

**Theorem 4.2** *The $\mathbb{T}$-equivariant motivic Chern class of $X \subset \mathbb{P}^n$ can be calculated via the substitution*

$$\mathrm{mC}_{\mathbb{T}}(X) = \frac{1}{1+y}\, \mathrm{sub}(\mathrm{mC}_{\mathbb{T}}(C_0X), \alpha_i \mapsto \alpha_i \cdot t^{-\frac{w_i}{q}}).$$

This formula probably can be used to calculate the Hilbert polynomial of certain subvarieties of the projective space.

**Corollary 4.2** *Applying Proposition 4.9 to the trivial torus we get*

$$\mathrm{mS}(X) = \kappa\, \mathrm{mS}_{\Gamma}(C_0X),$$

*implying that*

$$\mathrm{mC}(X) = \frac{1}{1+y}\kappa\, \mathrm{mC}_{\Gamma}(C_0X).$$

## 4.4 The Projective to Affine Formula

Interestingly, the calculation of $\mathrm{mC}_{\mathbb{T}}(C_0X)$ from $\mathrm{mC}_{\mathbb{T}}(X)$ is more involved than the obvious formula of Proposition 4.7 for the corresponding cohomology classes. The projective cone formula 3.1 already indicates the subtleties ahead of us. Finding such formula is important since some motivic Chern class calculations are simpler in the projective case than in the affine case, as unpublished works of B. Kőműves show. In this section we give an "inverse" to Theorem 4.2:

**Theorem 4.3** *Suppose that the torus $\mathbb{T}$ acts linearly on $\mathbb{C}^{n+1}$ and $X \subset \mathbb{P}^n$. Then*

$$\mathrm{mC}_{\Gamma\times\mathbb{T}}(C_0X) = (1+y)\big(\mathrm{mC}_{\mathbb{T}}(X) - \chi_y(X)[0]_{\Gamma\times\mathbb{T}}\big),$$

*where $[0]_{\Gamma\times\mathbb{T}}$ is the $\Gamma\times\mathbb{T}$-equivariant $K$-class of the origin, and $\mathrm{mC}_{\mathbb{T}}(X)$ is written in the reduced form in the variable $t$, the $\mathbb{T}$-equivariant class of the tautological bundle $\gamma$.*

Notice that the $\chi_y$ genus has no $\mathbb{T}$-equivariant version. This is called the *rigidity* of the $\chi_y$ genus, see e.g. [17, pr. 7.2].

Before proving the theorem let us have a look at the case when $\mathbb{T}$ is the trivial torus.

**Corollary 4.3** *Suppose that $X \subset \mathbb{P}^n$ is a constructible subset. Then the affine cone minus the origin $C_0X \subset \mathbb{C}^{n+1}$ is invariant for the scalar action of $\Gamma = \mathrm{GL}(1)$, and*

$$\mathrm{mC}_{\Gamma}(C_0X) = (1+y)\big(\mathrm{mC}(X) - \chi_y(X)[0]\big),$$

*where $[0] = (1-t)^{n+1}$ is the $\Gamma$-equivariant $K$-class of the origin, and $\mathrm{mC}(X)$ is written in the reduced form in the variable $t$, the $K$-theory class of the tautological bundle $\gamma$.*

An other way to express Corollaries 4.9 and 4.3 together is that written in the variable $H = 1-t$ the coefficients of $(1+y)\,\mathrm{mC}(X)$ (in the reduced form) and $\mathrm{mC}_{\Gamma}(C_0X)$ are the same, except $\mathrm{mC}_{\Gamma}(C_0X)$ has also an $n+1$'st coefficient to assure that the sum of the coefficients is zero.

Comparing with Proposition 4.1 we can see an important difference between the sheaf $K$-class and the motivic $K$-class: The scalar-equivariant motivic $K$-class of the cone of $X$ contains no additional information than the motivic $K$-class of $X$, on the other hand the scalar-equivariant sheaf $K$-class of the cone of $X$ determines the Hilbert function not just the Hilbert polynomial of $X$.

*Example 4.1* Let us study the case $X = \mathbb{P}^k \subset \mathbb{P}^n$. Then by Remark 2.3 and the simple fact that $i_!(H^j) = H^{j+n-k}$ for the inclusion $i : X \to \mathbb{P}^n$ we have

$$\mathrm{mC}(\mathbb{P}^k \subset \mathbb{P}^n) = \sum_{i=0}^{k}\binom{k+1}{i}(-y)^i(1+y)^{k-i}H^{n-k+i}.$$

For the cone we have the product formula (see [9, §2.7])

$$\mathrm{mC}_{\Gamma}(\mathbb{C}^{k+1} \subset \mathbb{C}^{n+1}) = (1-t)^{n-k}(1+yt)^{k+1},$$

so removing the origin we get

$$\mathrm{mC}_{\Gamma}(C_0X \subset \mathbb{C}^{n+1}) = (1-t)^{n-k}(1+yt)^{k+1} - (1-t)^{n+1}.$$

We know from Example 2.4 that $\chi_y(X) = 1 - y + y^2 - \cdots \pm y^k = \frac{1-(-y)^{k+1}}{1+y}$ so Corollary 4.3 gives the identity

$$\left((1+y)\sum_{i=0}^{k}\binom{k+1}{i}(-y)^i(1+y)^{k-i}H^{n-k+i}\right) - (1-(-y)^{k+1})H^{n+1} =$$
$$= (1-t)^{n-k}(1+yt)^{k+1} - (1-t)^{n+1},$$

where $H = 1-t$, which of course can be checked directly.

The idea of the proof of Theorem 4.3 is that we first prove it for the special case of $X$ being a projective space, and show how this result implies the result for general $X$.

*Example 4.2* Suppose that the torus $\mathbb{T}$ acts on $\mathbb{C}^{n+1}$ with characters $\beta_1, \ldots, \beta_{n+1}$. Let $X = \mathbb{P}^k \subset \mathbb{P}^n$ be invariant for the induced $\mathbb{T}$-action on $\mathbb{P}^n$. Without loss of generality we can assume that $CX$ is spanned by the first $k+1$ eigenvectors. Then

$$\mathrm{mC}_{\Gamma\times\mathbb{T}}(\mathbb{C}^{k+1} \setminus 0 \subset \mathbb{C}^{n+1}) = M - R,$$

where

$$M := \prod_{i=1}^{k+1}\left(1+\frac{yt}{\beta_i}\right)\prod_{i=k+2}^{n+1}\left(1-\frac{t}{\beta_i}\right),$$

and

$$R := [0]_{\Gamma\times\mathbb{T}} = \prod_{i=1}^{n+1}\left(1-\frac{t}{\beta_i}\right),$$

where $R$ is also the relation in $K_{\mathbb{T}}(\mathbb{P}^n)$ after identifying $t$ with the $\mathbb{T}$-equivariant class of $\gamma$. On the other hand we have

$$(1+y)\,\mathrm{mC}_{\Gamma\times\mathbb{T}}(\mathbb{P}^k \subset \mathbb{P}^n) \equiv M,$$

but this is not the reduced form yet, the coefficient of $t^{n+1}$ is not zero. Comparing $M$ and $R$ we can see that the reduced form is

$$(1+y)\,\mathrm{mC}_{\mathbb{T}}(\mathbb{P}^k \subset \mathbb{P}^n) = M - (-y)^{k+1}R,$$

so the right hand side of Theorem 4.3 becomes

$$(1+y)(\mathrm{mC}_{\mathbb{T}}(\mathbb{P}^k \subset \mathbb{P}^n) - \chi_y(\mathbb{P}^k)R) = M - (-y)^{k+1}R - (1-(-y)^{k+1})R,$$

since $(1+y)\chi_y(\mathbb{P}^k) = 1-(-y)^{k+1}$. Consequently we see that Theorem 4.3 holds for these examples.

The next step is to prove Theorem 4.3 for $X$ being a $\mathbb{T}$-invariant smooth subvariety of $\mathbb{P}^n$. In this case the blowup of $\mathbb{C}^{n+1}$ at the origin provides a $\varphi : Y \to \mathbb{C}^{n+1}$ proper normal crossing extension for $C_0X$, where $Y$ is the total space of the restriction of the tautological bundle $\gamma \to \mathbb{P}^n$ to $X$. The resolution factors as

$$Y \xrightarrow{\ j\ } \mathbb{P}^n \times \mathbb{C}^{n+1} \xrightarrow{\ \pi\ } \mathbb{C}^{n+1} \ ,$$

which implies that

$$\mathrm{mC}_{\Gamma\times\mathbb{T}}(C_0X) = \int_{\mathbb{P}^n} \mathrm{mC}_{\mathbb{T}}(X)\lambda_y(\gamma^*)e(\mathbb{C}^{n+1}/\gamma) \ - \ e(\mathbb{C}^{n+1})\int_{\mathbb{P}^n} \mathrm{mC}_{\mathbb{T}}(X), \tag{11}$$

where the $\lambda_y$ class and the Euler classes are $\Gamma \times \mathbb{T}$-equivariant.

It is quite difficult to use (11) for calculations. Luckily we do not need it. We only need to notice that (11) implies that the left hand side can be calculated from the reduced form of $\mathrm{mC}_{\mathbb{T}}(X)$ providing a $K_{\mathbb{T}}[y]$-module homomorphism. This implies that it is enough to check 4.3 for a basis of the space of polynomials of degree at most $n$ in the variable $t$ and coefficients in $K_{\mathbb{T}}[y]$. We claim that the cases of Example 4.2 will give such a basis. Indeed, substituting $\beta_i = 1$ and $y = 0$ we get $\mathrm{mC}_0(\mathbb{P}^k \subset \mathbb{P}^n) = (1-t)^{n-k}$.

The last step is to extend the result to all $\mathbb{T}$-invariant constructible subsets of $\mathbb{P}^n$. For that we just have to notice that all 3 components of the formula are motivic and we finished the proof of Theorem 4.3. □

Forgetting the scalar action we still get a nontrivial statement:

**Theorem 4.4** *Suppose that the torus $\mathbb{T}$ acts linearly on $\mathbb{C}^{n+1}$. Then*

$$\mathrm{mC}_{\mathbb{T}}(C_0X) = (1+y)\big(\mathrm{mC}_{\mathbb{T}}(X)|_{t=1} - \chi_y(X)[0]_{\mathbb{T}}\big),$$

*where $[0]_{\mathbb{T}}$ is the $\mathbb{T}$-equivariant $K$-class of the origin, and $\mathrm{mC}_{\mathbb{T}}(X)$ is written in the reduced form in the variable $t$.*

**Acknowledgments** I am grateful for Andrzej Weber for patiently explaining me the intricacies of the motivic Chern class. I had inspiring conversations on the topic with Richárd Rimányi, András Némethi, Ákos Matszangosz and Balázs Kőműves. A special case of Theorem 4.4 was first proved by the latter. I thank Anders Buch for explaining the role of the Cohen-Macaulay condition in pulling back the sheaf theory $K$-class.

The author László M. Fehér was partially supported by NKFI 112703 and 112735 and NKFIH KKP 126683 as well as ERC Advanced Grant LTDBud and enjoyed the hospitality of the Rènyi Institute.

## References

1. Atiyah, M.F., Hirzebruch, F.: The Riemann-Roch theorem for analytic embeddings. Topology **1**, 151–166 (1962)
2. Brasselet, J.-P., Schürmann, J., Yokura, S.: Hirzebruch classes and motivic chern classes for singular spaces. J. Topol. Anal. **2**(1), 1–55 (2010)
3. Chriss, N., Ginzburg, V.: Representation Theory and Complex Geometry. Birkhäuser, Boston (1997)
4. Dugger, D.: A geometric introduction to K-theory (2014). https://pages.uoregon.edu/ddugger/kgeom.pdf
5. Fehér, L.M., Rimányi, R.: Thom series of contact singularities. Ann. Math. **176**(3), 1381–1426 (2012)
6. Fehér, L.M., Rimányi, R.: Chern-Schwartz-MacPherson classes of degeneracy loci. Geom. Topol. **22**(6), 3575–3622 (2018)
7. Fehér, L.M., Némethi, A., Rimányi, R.: Degeneracy of 2-forms and 3-forms. Can. Math. Bull. **48**(4), 547–560 (2005)
8. Fehér, L.M., Némethi, A., Rimányi, R.: The degree of the discriminant of irreducible representations. J. Algebraic Geom. **17**(4), 751–780 (2008)
9. Fehér, L.M., Rimányi, R., Weber, A.: Characteristic classes of orbit stratifications, the axiomatic approach. In: Schubert Calculus and its Applications in Combinatorics and Representation Theory. Springer, Berlin (2018)
10. Fehér, L.M., Rimányi, R., Weber, A.: Motivic Chern classes and $K$-theoretic stable envelopes. In: Proceedings of London Math. Soc. (2020). https://doi.org/10.1112/plms.12374
11. Fulton, W.: Intersection theory. In: Ergebnisse der Mathematik und ihrer Grenzgebiete. 3. Folge. A Series of Modern Surveys in Mathematics [Results in Mathematics and Related Areas. 3rd Series. A Series of Modern Surveys in Mathematics], vol. 2, 2nd edn. Springer, Berlin (1998)
12. Harris, J.: Algebraic Geometry: A First Course, vol. 133. Springer, Berlin (2013)
13. Hirzebruch, F.: Topological methods in algebraic geometry. In: Die Grundlehren der Mathematischen Wissenschaften, Band, vol. 131, Third enlarged edition. Springer, New York (1966)
14. Miller, E., Sturmfels, B.: Combinatorial Commutative Algebra, vol. 227. Springer, Berlin (2004)
15. Ohmoto, T.: Singularities of maps and characteristic classes. In: School on Real and Complex Singularities in São Carlos, 2012. Advanced Studies in Pure Mathematics, vol. 68, pp. 191–265. Mathematical Society Japan, Tokyo (2016)
16. Rimányi, R., Szenes, A.: Residues, Grothendieck polynomials and $K$-theoretic Thom polynomials (2018). arXiv:1811.02055
17. Weber, A.: Equivariant Hirzebruch class for singular varieties. Sel. Math. **22**(3), 1413–1454 (2016)

# Semicontinuity of Singularity Invariants in Families of Formal Power Series

**Gert-Martin Greuel and Gerhard Pfister**

*Dedicated to András Némethi on the occasion of his sixtieth birthday*

**Abstract** The problem we are considering came up in connection with the classification of singularities in positive characteristic. Then it is important that certain invariants like the determinacy can be bounded simultaneously in families of formal power series parametrized by some algebraic variety. In contrast to the case of analytic or algebraic families, where such a bound is well known, the problem is rather subtle, since the modules defining the invariants are quasi-finite but not finite over the base space. In fact, in general the fibre dimension is not semicontinuous and the quasi-finite locus is not open. However, if we pass to the completed fibres in a family of rings or modules we can prove that their fibre dimension is semicontinuous under some mild conditions. We prove this in a rather general framework by introducing and using the completed and the Henselian tensor product, the proof being more involved than one might think. Finally we apply this to the Milnor number and the Tjurina number in families of hypersurfaces and complete intersections and to the determinacy in a family of ideals.

**Keywords** Formal power series · Completed tensor product · Henselian tensor product · Semicontinuity · Milnor number · Tjurina number · Determinacy

**Subject Classifications** 13B35, 13B40, 14A15, 14B05, 14B07

---

G.-M. Greuel (✉) · G. Pfister
Department of Mathematics, University of Kaiserslautern, Kaiserslautern, Germany
e-mail: greuel@mathematik.uni-kl.de; pfister@mathematik.uni-kl.de



J. Fernández de Bobadilla et al. (eds.), *Singularities and Their Interaction with Geometry and Low Dimensional Topology*, Trends in Mathematics,
https://doi.org/10.1007/978-3-030-61958-9_10

# 1 Introduction

In connection with the classification of singularities defined by formal power series over a field a fundamental invariant is the modality of the singularity (with respect to some equivalence relation like right or contact equivalence). To determine the modality one has to investigate adjacent singularities that appear in nearby fibres. This cannot be done by considering families over complete local rings but one has to consider families of power series parametrized by some algebraic variety in the neighbourhood of a given point. To determine potential adjacencies, an important tool is the semicontinuity of certain singularity invariants like, for example, the Milnor or the Tjurina number. Another basic question is if the determinacy of an ideal can be bounded by a semicontinuous invariant. In the complex analytic situation the answer to these questions is well known and positive, but for formal power series the problem is much more subtle than one might think at the first glance. This is mainly due to the fact that ideals or modules that define the invariants are quasi-finite but not finite over the base space.

The modality example shows that the questions treated in this paper are rather natural and appear in important applications. Moreover, the semicontinuity in general is a very basic property with numerous applications in many other contexts. Therefore we decided to choose a rather general framework with families of modules presented by matrices of power series and parametrized by the spectrum of some Noetherian ring. It is not difficult to see that the fibre dimension is in general not semicontinuous and that the quasi-finite locus is in general not open (in contrast the case of ring maps of finite type, where the quasi-finite locus is open by Zariski's Main Theorem, cf. Proposition 2.7), see Examples 2.3 and 2.4. It turns out that the situation is much more satisfactory if we consider not the fibres but the completed fibres and we prove the desired semicontinuity for the completed fibre dimension under some conditions on the family. To guarantee that the completed fibre families behave well under base change we introduce the notion of a (partial) completed tensor product and study its properties in Sects. 2.1 and 2.2.

Unfortunately, we cannot prove the semicontinuity of the completed fibre dimension in full generality. We prove it if either the base ring has dimension one (in Sect. 2.3), or if the base ring is complete local containing a field, or if the presentation matrix has polynomials or algebraic power series as entries (in Sect. 2.5). Together, these cases cover most applications (see Corollary 2.7 for a summary). To treat the latter case, we use Henselian rings and the Henselian tensor product, for which we give a short account in Sect. 2.4. It would be interesting to know, if the result holds for presentation matrices with arbitrary power series as entries or if there are counterexamples. In Sect. 2.6 we consider also the case of families of finite type over the base ring and prove a version of Zariski's main theorem for modules. Moreover, we compare the completed fibre with the usual fibre.

In Sect. 3 we apply our results to singularity invariants. We discuss and compare first the notions of regularity and smoothness (over a field) and show that both notions coincide for the completed fibres (Lemma 3.3). Under the restrictions mentioned above, we prove the semicontinuity of the Milnor number and Tjurina number for hypersurfaces (Sect. 3.2) and the Tjurina number for complete intersections (Sect. 3.4) as well as an upper bound for the determinacy of an ideal (Sect. 3.3). Since the base ring may be the integers, our results are of some interest for computational purposes. For example, if a power series has integer coefficients then the Milnor number over the rationals is bounded by the Milnor number modulo just one (possibly unlucky) prime number if this is finite (see Corollary 3.1 and, more generally, Corollary 2.4 and Remark 2.5).

We assume all rings to be associative, commutative and with unit. Throughout the paper $\Bbbk$ denotes an arbitrary field, $A$ a ring, $R = A[[x]]$, $x = (x_1, \cdots, x_n)$, the formal power series ring over $A$ and $M$ an $R$-module. For our main results we will assume that $A$ is Noetherian and that $M$ is finitely generated as $R$-module.

## 2 Quasi-Finite Modules and Semicontinuity

### 2.1 The Completed Tensor Product

Let $A$ be a ring, $R = A[[x]]$ and $M$ an $R$-module. For any prime ideal $\mathfrak{p}$ of $A$ let $k(\mathfrak{p}) = A_\mathfrak{p}/\mathfrak{p}A_\mathfrak{p}$ be the residue field of $\mathfrak{p}$. $k(\mathfrak{p}) = \mathrm{Quot}(A/\mathfrak{p})$ is the quotient field of $A/\mathfrak{p}$ and hence $k(\mathfrak{p}) = A/\mathfrak{p}$ if $\mathfrak{p}$ is a maximal ideal. We consider $M$ via the canonical map $A \hookrightarrow R$ as an $A$-module and set

$$M(\mathfrak{p}) := M_\mathfrak{p} \otimes_{A_\mathfrak{p}} k(\mathfrak{p}) = (M \otimes_A A_\mathfrak{p}) \otimes_{A_\mathfrak{p}} k(\mathfrak{p}) = M \otimes_A k(\mathfrak{p}),$$

which is called the *fibre* of $M$ over $\mathfrak{p}$. $M(\mathfrak{p})$ is a vector space over $k(\mathfrak{p})$ and its dimension is denoted by

$$d_\mathfrak{p}(M) := \dim_{k(\mathfrak{p})} M(\mathfrak{p}).$$

$M$ is called *quasi-finite*[1] over $\mathfrak{p}$ if $d_\mathfrak{p}(M) < \infty$. We are interested in the behavior of $d_\mathfrak{p}(M)$ as $\mathfrak{p}$ varies in Spec $A$, in particular in finding conditions under which $d_\mathfrak{p}(M)$ is semicontinuos on Spec $A$.

We say that a function $d : \mathrm{Spec}\, A \to \mathbb{R}$, $\mathfrak{p} \mapsto d_\mathfrak{p}$, is *(upper) semicontinuous* at $\mathfrak{p}$ if $\mathfrak{p}$ has an open neighbourhood $U \subset \mathrm{Spec}\, A$ such that $d_\mathfrak{q} \le d_\mathfrak{p}$ for all $\mathfrak{q} \in U$. $d$ is semicontinuous on Spec $A$ if it is semicontinuous at every $\mathfrak{p} \in \mathrm{Spec}\, A$.

For finitely presented $A$-modules $M$ the semicontinuity of $\mathfrak{p} \mapsto d_\mathfrak{p}(M)$ is true and well known (cf. Lemma 2.1). However, in many applications $M$ is not

[1] For $M = R/I$, $I$ an ideal, this is the original definition of Grothendieck. Nowadays most authors (e.g. [15]) require in addition that $R$ is of finite type over $A$.

finitely generated over $A$ but finite over some $A$-algebra $R$. Such a situation appears naturally in algebraic geometry, when one considers families of schemes or of coherent sheaves over Spec $A$. But then it is usually assumed that the ring $R$ is either of (essentially) finite type over $A$ (in algebraic geometry) or an analytic $A$-algebra (in complex analytic geometry). When we study families of singularities defined by formal power series (cf. Sect. 3), we have to consider $R = A[[x]]$, which is not of finite type over $A$. As far as we know, this situation has not been systematically studied and it leads to some perhaps unexpected results. For example, $d_{\mathfrak{p}}(M)$ is in general not semicontinuous on Spec $A$ (cf. Examples 2.3 and 2.4).

It turns out that the situation is much more satisfactory if we pass from the usual fibres to the completed fibres, that is, we consider the *completed fibre dimension*

$$\hat{d}_{\mathfrak{p}}(M) := \dim_{k(\mathfrak{p})} M(\mathfrak{p})^{\wedge},$$

where $M(\mathfrak{p})^{\wedge}$ is the $\langle x\rangle$-adic completion of the $R(\mathfrak{p})$-module $M(\mathfrak{p})$. To guarantee that the completed fibres behave well when $\mathfrak{p}$ varies in Spec $A$, we introduce the notion of a completed tensor product below.

For a finitely presented $A$-module $M$ the semicontinuity of $\mathfrak{p} \to d_{\mathfrak{p}}(M)$ is well known:

**Lemma 2.1** *If $M$ is a finitely presented $A$-module then $d_{\mathfrak{p}}(M)$ is semicontinuous on* Spec $A$. *Moreover, if $M$ is $A$-flat, then $d_{\mathfrak{p}}(M)$ is locally constant on* Spec $A$.

***Proof*** Fix $\mathfrak{p} \in$ Spec $A$ and consider a presentation of $M$,

$$A^p \xrightarrow{P} A^q \to M \to 0,$$

with matrix $P = (p_{ij}),\ p_{ij} \in A$. Applying $\otimes_A k(\mathfrak{p})$ to this sequence we get the exact sequence of vector spaces

$$k(\mathfrak{p})^p \xrightarrow{\overline{P}_{\mathfrak{p}}} k(\mathfrak{p})^q \to M(\mathfrak{p}) \to 0,$$

with entries of $\overline{P}_{\mathfrak{p}}$ being the images of $p_{ij}$ in $k(\mathfrak{p})$. Then $d_{\mathfrak{p}}(M)$ is finite and $d_{\mathfrak{p}}(M) = q - rank(\overline{P}_{\mathfrak{p}})$. Since $rank(\overline{P}_{\mathfrak{p}}) \leq rank(\overline{P}_{\mathfrak{q}})$ for all $\mathfrak{q}$ in some neighbourhood $U$ of $\mathfrak{p}$, the claim follows.

If $M$ is flat, then $M_{\mathfrak{p}}$ is free over the local ring $A_{\mathfrak{p}}$ for a given $\mathfrak{p} \in$ Spec $A$. By [12], Theorem 4.10 (ii) (and its proof) there exists an $f \notin \mathfrak{p}$ such that $M_f$ is a free $A_f$-module of some rank $r$ and hence $d_{\mathfrak{q}}(M) = r$ for $\mathfrak{q}$ in the open neighbourhood $D(f)$ of $\mathfrak{p}$.

We introduce now the completed tensor product. Let us denote by

$$\langle x\rangle := \langle x_1, \ldots, x_n\rangle_R$$

the ideal in $R$ generated by $x_1, \ldots, x_n$. More generally, if $S$ is an $R$-algebra, then $\langle x\rangle_S$ denotes the ideal in $S$ generated by (the images of) $x_1, \ldots, x_n$.

For an $R$-module $N$ denote by

$$N^\wedge := \varprojlim N/\langle x\rangle^m N$$

the $\langle x\rangle$-adic completion of $N$. If $N$ is also an $S$-module for some $R$-algebra $S$, then $\langle x\rangle^m N = (\langle x\rangle_S)^m N$, and hence the $\langle x\rangle$-adic completion and the $\langle x\rangle_S$-adic completion of $N$ coincide.

**Definition 2.1** Let $A$ be a ring, $R = A[[x]]$, $B$ an $A$-algebra and $M$ an $R$-module. We define the *completed tensor product* of $R$ and $B$ over $A$ as the ring

$$R\hat{\otimes}_A B := \varprojlim \big((R/\langle x\rangle^m) \otimes_A B\big)$$

and the *completed tensor product* of $M$ and $B$ over $A$ as the module

$$M\hat{\otimes}_A B := \varprojlim \big((M/\langle x\rangle^m M) \otimes_A B\big).$$

If $N$ is an $A$-module, we define the $R$-module

$$M\hat{\otimes}_A N := \varprojlim \big((M/\langle x\rangle^m M) \otimes_A N\big)$$

and call it the *completed tensor product* of $M$ and $N$ over $A$.

One reason why we consider the completed tensor product is that it provides the right base change property in the category of rings of the form $A[[x]]$ by the following Proposition 2.1.1.

**Proposition 2.1** *The completed tensor product has the following properties (assumptions as in Definiton 2.1).*

1. $A[[x]]\hat{\otimes}_A B = (R \otimes_A B)^\wedge = B[[x]]$.
2. $M\hat{\otimes}_A N = (M \otimes_A N)^\wedge$.
3. *If $M$ is finitely presented over $R$ and $N$ is a finitely presented $B$-module, then*

$$M\hat{\otimes}_A N \cong (M \otimes_A N) \otimes_{R\otimes_A B} \mathrm{fl}(R\hat{\otimes}_A B).$$

4. *The canonical map $M \otimes_A N \to M\hat{\otimes}_A N$ is injective if $A$ is Noetherian, $M$ finite over $R$ and $N$ finite over $A$.*
5. *If $\langle x\rangle^m \subset \mathrm{Ann}_R(M)$ for some $m$ then $M\hat{\otimes}_A N = M \otimes_A N$ for every $A$-module $N$.*

***Proof***

1. We have $\varprojlim \big(A[[x]]/\langle x\rangle^m \otimes_A B\big) = \varprojlim \big(A[x]/\langle x\rangle^m \otimes_A B\big) = \varprojlim B[x]/\langle x\rangle^m = B[[x]]$, showing the second equality. The first follows since $(R/\langle x\rangle^m) \otimes_A B = (R \otimes_A B)/\langle x\rangle^m(R \otimes_A B)$.

2. Since $(M/\langle x\rangle^m M)\otimes_A N = (M\otimes_A N)/\langle x\rangle^m(M\otimes_A N)$ the equality follows and that $(M\otimes_A N)^\wedge$ is the $\langle x\rangle_R$ as well as the $\langle x\rangle_{R\otimes_A B}$-adic completion of $M\otimes_A N$.
3. If $M$ resp. $N$ are finitely presented over $R$ resp. $B$, then $M\otimes_A N$ is finitely presented over $R\otimes_A B$. Hence we can apply (the proof of) [1, Proposition 10.13] and use 1. to show the isomorphism.
4. If $A$ is Noetherian then $R$ is Noetherian. If $M$ is finitely generated over $R$ and $N$ finitely generated over $A$ then $M\otimes_A N$ is finitely generated over $R$. The injectivity follows from 2. and [1, Theorem 10.17 and Corollary 10.19], since $\langle x\rangle$ is contained in the Jacobson radical of $R$ by Lemma 2.4.
5. If $\langle x\rangle^m M = 0$ for some $m$, then $M\hat{\otimes}_A N = M\otimes_A N$ by definition of the completed tensor product.

**Corollary 2.1** *The completed tensor product is right-exact on the category of finitely presented modules. That is, let*

$$M' \to M \to M'' \to 0, \textit{ resp.}$$
$$N' \to N \to N'' \to 0$$

*be exact sequences of finitely presented $R$-modules resp. $B$-modules. Then the sequences of finitely presented $R\hat{\otimes}_A B$-modules*

$$M'\hat{\otimes}_A N \to M\hat{\otimes}_A N \to M''\hat{\otimes}_A N \to 0, \textit{ resp.}$$
$$M\hat{\otimes}_A N' \to M\hat{\otimes}_A N \to M\hat{\otimes}_A N'' \to 0$$

*are exact.*

***Proof*** The sequences $M'\otimes_A N \to M\otimes_A N \to M''\otimes_A N \to 0$ and $M\otimes_A N' \to M\otimes_A N \to M\otimes_A N'' \to 0$ are exact. Now tensor these sequences with $R\hat{\otimes}_A B$ over $R\otimes_A B$ and apply Proposition 2.1.3.

**Corollary 2.2**

*(i)* $R\hat{\otimes}_A A = R$.
*(ii)* *If $S$ is multiplicatively closed in $A$ then* $A[[x]]\hat{\otimes}_A(S^{-1}A) = (S^{-1}A)[[x]]$.
*(iii)* *For any $R$-module $M$ we have* $M\hat{\otimes}_A A = M^\wedge$.
*(iv)* *If $M$ is finitely presented over $R$ then $M = M^\wedge$. If moreover $N$ is finitely presented over $A$, then* $M\hat{\otimes}_A N = M\otimes_A N$.

***Proof*** (i) and (ii) follow immediately from Proposition 2.1.1, (iii) is a special case of Proposition 2.1.2 and (iv) follows from (i) and Proposition 2.1.3 with $B = A$.

Applying Corollary 2.1 and Proposition 2.1.1 we get

**Corollary 2.3** *If* $A[[x]]^p \xrightarrow{T} A[[x]]^q \to M \to 0$ *is an $A[[x]]$-presentation of $M$ and $B$ an $A$-algebra, then*

$$M\hat{\otimes}_A B = \operatorname{coker}\big(B[[x]]^p \xrightarrow{T} B[[x]]^q\big).$$

*Remark 2.1* Let $A$ be Noetherian and $B$ an $A$-algebra. If $\langle x\rangle$ is contained in the Jacobson radical of $R \otimes_A B$, then $R\hat{\otimes}_A B$ is faithfully flat over $R \otimes_A B$, by Proposition 2.1.1 and [12, Theorem 8.14]. Note however that although $\langle x\rangle$ is contained in the Jacobson radical of $R$ by Lemma 2.4 below, it need not be in the Jacobson radical of $R \otimes_A B$ (cf. Example 2.2).

*Example 2.1* Let $\langle f_1, \ldots, f_k\rangle \subset R = A[[x]]$ be an ideal and $M = A[[x]]/\langle f_1, \ldots, f_k\rangle$. If $\mathfrak{p}$ is a prime ideal in $A$ then $R\hat{\otimes}_A A_\mathfrak{p} = A_\mathfrak{p}[[x]]$ and form Corollary 2.3 we get $M\hat{\otimes}_A A_\mathfrak{p} = A_\mathfrak{p}[[x]]/\langle f_1, \ldots, f_k\rangle$. If $k(\mathfrak{p})$ is the residue field of $A$ at $\mathfrak{p}$ then $M\hat{\otimes}_A k(\mathfrak{p}) = k(\mathfrak{p})[[x]]/\langle f_1, \ldots, f_k\rangle$, something what one expects as fibre of $M$ over $\mathfrak{p}$. While $A_\mathfrak{p}[[x]]$ and $k(\mathfrak{p})[[x]]$ are nice local rings, the subrings $R \otimes_A A_\mathfrak{p} \subset A_\mathfrak{p}[[x]]$ and $R \otimes_A k(\mathfrak{p}) \subset k(\mathfrak{p})[[x]]$ are in general not local if $\mathfrak{p}$ is not a maximal ideal (see Example 2.2).

*Remark 2.2* Proposition 2.1.1 with $B = A[[y]]$, $y = (y_1, \ldots, y_m)$, implies

$$A[[x]]\hat{\otimes}_A A[[y]] = A[[x, y]].$$

Now let $A$ be Noetherian. If $I$ resp. $J$ are ideals in $A[[x]]$ resp. $A[[y]]$, we get from Corollary 2.1

$$A[[x]]/I\hat{\otimes}_A A[[y]]/J = A[[x, y]]/\langle I, J\rangle A[[x, y]].$$

We call an $A$-algebra a *formal $A$-algebra* if it is isomorphic to an $A$-algebra $A[[x]]/I$. For two formal $A$-algebras $B = A[[x]]/I$ and $C = A[[y]]/J$ the completed tensor product can be defined as $B\hat{\otimes}_A C = A[[x, y]]/\langle I, J\rangle A[[x, y]]$. It has the usual universal property of the tensor product in the category of formal $A$-algebras, analogous to the analytic tensor product for analytic algebras (cf. [5, Chapter III.5]). Thus, Definition 2.1 generalizes the completed tensor product of formal $A$-algebras.

## 2.2 *Fibre and Completed Fibre*

Let again $A$ be a ring and $M$ an $R = A[[x]]$-module. We introduce the completed fibre $\hat{M}(\mathfrak{p})$ and the completed fibre dimension $\hat{d}_\mathfrak{p} M$ of $M$ for $\mathfrak{p} \in \operatorname{Spec} A$ and compare it with the usual fibre $M(\mathfrak{p})$ and the usual fibre dimension $d_\mathfrak{p} M$.

At the end of this section we give examples, showing that semicontinuity of $d_\mathfrak{p}(M)$ does not hold in general on $\operatorname{Spec} A$, even if $A = \mathbb{C}[t]$ or $A = \mathbb{Z}$ (Examples 2.3 and 2.4). However, we show in the next Sects. 2.3 and 2.5 that, under some conditions, semicontinuity holds for the completed fibre dimension $\hat{d}_\mathfrak{p}(M)$.

**Notation 2.1** We have canonical maps

$$A \overset{j}{\hookrightarrow} R \overset{\pi}{\twoheadrightarrow} R/\langle x\rangle \underset{\cong}{\overset{i}{\to}} A,$$

with $i \circ \pi \circ j = id$ and for an ideal $I \subset R$ we set $\overline{I} := \pi(I)$. On the level of schemes we have the maps $\mathrm{Spec}\, A \underset{\cong}{\overset{i^*}{\to}} V(\langle x\rangle) \overset{\pi^*}{\hookrightarrow} \mathrm{Spec}\, R \overset{j^*}{\twoheadrightarrow} \mathrm{Spec}\, A$, with $i^*(\mathfrak{p}) = \langle \mathfrak{p}, x\rangle$, $j^*(\langle \mathfrak{p}, x\rangle) = \langle \mathfrak{p}, x\rangle \cap A = \mathfrak{p}$ for $\mathfrak{p} \in \mathrm{Spec}\, A$. We denote by

$$\mathfrak{n}_\mathfrak{p} := \langle \mathfrak{p}, x\rangle = \langle \mathfrak{p}, x_1, \ldots, x_n\rangle_R$$

the ideal in $R$ generated by $\mathfrak{p} \in \mathrm{Spec}\, A$ and $x_1, \ldots, x_n$. The family

$$f := j^* : \mathrm{Spec}\, R \to \mathrm{Spec}\, A$$

has the trivial section $\sigma = (i \circ \pi)^* : \mathrm{Spec}\, A \to \mathrm{Spec}\, R$, $\mathfrak{p} \mapsto \mathfrak{n}_\mathfrak{p}$, and the composition $h := \pi \circ j : A \cong R/\langle x\rangle$ induces an isomorphism

$$h^* : V(\langle x\rangle) \overset{\cong}{\to} \mathrm{Spec}\, A,$$

the restriction of $f$ to $V(\langle x\rangle)$. We call $R_\mathfrak{p} := R \otimes_A A_\mathfrak{p}$ the *stalk* of $R$ over $\mathfrak{p}$. $R_\mathfrak{p}$ is not a local ring, its local ring at $\mathfrak{n}_\mathfrak{p}$ is $(R_\mathfrak{p})_{\mathfrak{n}_\mathfrak{p}} = R_{\mathfrak{n}_\mathfrak{p}}$ with residue field $k(\mathfrak{n}_\mathfrak{p}) = k(\mathfrak{p})$ (by Lemma 2.2 below).

If $M$ is an $R$-module, we call $M_\mathfrak{p} = M \otimes_A A_\mathfrak{p}$ the *stalk* of $M$ over $\mathfrak{p}$ and we are interested in the behavior of $M$ along the section $\sigma$. However, we are not interested in the $R(\mathfrak{p})$-modules $M(\mathfrak{p})$ since $R(\mathfrak{p})$ is not a power series ring (and does not behave nicely). We are interested in the completed stalk $\hat{M}_\mathfrak{p}$ and in the completed fibres $\hat{M}(\mathfrak{p})$, which we introduce now.

**Definition 2.2** Let $A$ be a ring, $R = A[[x]]$, $M$ an $R$-module and $\mathfrak{p} \in \mathrm{Spec}\, A$.

1. We set $\hat{R}_\mathfrak{p} := R\hat{\otimes}_A A_\mathfrak{p}$, a local ring isomorphic to $A_\mathfrak{p}[[x]]$ (Proposition 2.1.1), and call the $\hat{R}_\mathfrak{p}$-module

$$\hat{M}_\mathfrak{p} := M\hat{\otimes}_A A_\mathfrak{p}$$

the *completed stalk* of $M$ over $\mathfrak{p}$.

2. The ring $\hat{R}(\mathfrak{p}) := R\hat{\otimes}_A k(\mathfrak{p})$ is called the *completed fibre of $R$ over* $\mathfrak{p}$. It is a local ring isomorphic to $k(\mathfrak{p})[[x]]$ (Proposition 2.1.1). The $\hat{R}(\mathfrak{p})$-module

$$\hat{M}(\mathfrak{p}) := M\hat{\otimes}_A k(\mathfrak{p}) = \hat{M}_\mathfrak{p} \otimes_{A_\mathfrak{p}} k(\mathfrak{p})$$

is called the *completed fibre of $M$ over* $\mathfrak{p}$.

3. $\hat{M}(\mathfrak{p})$ is a $k(\mathfrak{p})$-vector space and we call its dimension

$$\hat{d}_\mathfrak{p}(M) := \dim_{k(\mathfrak{p})} \hat{M}(\mathfrak{p})$$

the *completed fibre dimension of $M$ over* $\mathfrak{p}$.

4. $M$ is called *quasi-completed-finite over* $\mathfrak{p}$ if $\hat{d}_\mathfrak{p}(M) < \infty$.

The map $A \to R$ induces a map of local rings $A_\mathfrak{p} \to R_{\mathfrak{n}_\mathfrak{p}}$ and for an $R$-module $M$ we have the fibre $M(\mathfrak{p}) = M \otimes_A k(\mathfrak{p})$ of $M$ w.r.t. $A \to R$ and the fibre

$$M_{\mathfrak{n}_\mathfrak{p}}(\mathfrak{p}) = M_{\mathfrak{n}_\mathfrak{p}} \otimes_{A_\mathfrak{p}} k(\mathfrak{p}) = M_{\mathfrak{n}_\mathfrak{p}}/\mathfrak{p}M_{\mathfrak{n}_\mathfrak{p}}$$

of $M_{\mathfrak{n}_\mathfrak{p}}$ w.r.t. $A_\mathfrak{p} \to R_{\mathfrak{n}_\mathfrak{p}}$. The fibres are in general different but the completed fibres coincide by Lemma 2.3 if $M$ is finitely $R$-presented.

Let us first compare the fibre $M(\mathfrak{p})$ with its completed fibre $\hat{M}(\mathfrak{p})$.

**Lemma 2.2** *For any R-module M the following holds.*

*(i)* $\hat{M}_\mathfrak{p} = (M_\mathfrak{p})^\wedge$ *and* $\hat{M}(\mathfrak{p}) = M(\mathfrak{p})^\wedge$.
*(ii)* $\mathfrak{n}_\mathfrak{p}$ *is a prime ideal in R with* $\mathfrak{n}_\mathfrak{p} \cap A = \mathfrak{p}$ *and the residue field of* $\mathfrak{n}_\mathfrak{p}$ *in R satisfies* $k(\mathfrak{n}_\mathfrak{p}) = k(\mathfrak{p})$.
*(iii) If* $\mathfrak{n}$ *is any prime ideal in R containing* $\langle x\rangle$, *then* $\mathfrak{n} = \mathfrak{n}_\mathfrak{p}$ *with* $\mathfrak{p} = \mathfrak{n} \cap A \in \operatorname{Spec} A$.

***Proof*** Statement (i) follows from Proposition 2.1.1. The first statement of (ii) follows since $R/\mathfrak{n}_\mathfrak{p} = A/\mathfrak{p}$ is an integral domain. Since $R/\mathfrak{n}_\mathfrak{p} = A/\mathfrak{p}$ we have $k(\mathfrak{n}_\mathfrak{p}) = \operatorname{Quot}(R/\mathfrak{n}_\mathfrak{p}) = \operatorname{Quot}(A/\mathfrak{p}) = k(\mathfrak{p})$. (iii) is obvious.

*Remark 2.3* We have strict flat inclusions $A_\mathfrak{p} \subsetneqq R_\mathfrak{p} \subsetneqq R_{\mathfrak{n}_\mathfrak{p}} \subsetneqq A_\mathfrak{p}[[x]]$ of rings that are Noetherian if $A$ is Noetherian.

The strictness is easy to see. E.g. $g_0 + \sum_{|\alpha|\geq 1}(g_\alpha/h_\alpha)x^\alpha$, $g_0 \notin \mathfrak{p}$, with arbitrary $h_\alpha \in R \smallsetminus \mathfrak{n}_\mathfrak{p}$, is a unit in $A_\mathfrak{p}[[x]]$ but it is not contained in $R_{\mathfrak{n}_\mathfrak{p}}$, where only finitely many different denominators are allowed. We have $R_\mathfrak{p} = S^{-1}R$, with $S$ the multiplictive set $A \smallsetminus \mathfrak{p}$ and $R_{\mathfrak{n}_\mathfrak{p}} = (R_\mathfrak{p})_{\mathfrak{n}_\mathfrak{p}}$. Since localization preserves flatness ([1, Corollary 3.6]) and the Noether property ([1, Proposition 7.3]), the inclusions $A_\mathfrak{p} \subset R_\mathfrak{p} \subset R_{\mathfrak{n}_\mathfrak{p}}$ are flat and the rings are Noetherain if $A$ is Noetherain. The flatness of $A_\mathfrak{p}[[x]]$ over $R_{\mathfrak{n}_\mathfrak{p}}$ follows, since the first is the $\langle x\rangle$-adic completion of the second by Lemma 2.2 (i). Since both rings are local, $R_{\mathfrak{n}_\mathfrak{p}} \subset A_\mathfrak{p}[[x]]$ is faithfully flat.

The rings $R_\mathfrak{p}$ and $R_{\mathfrak{n}_\mathfrak{p}}$ are "strange" subrings of $A_\mathfrak{p}[[x]]$. The ring $A_\mathfrak{p}[[x]]$ is of interest in applications (cf. Sect. 3), while the rings $R_\mathfrak{p}$ and $R_{\mathfrak{n}_\mathfrak{p}}$ are of minor interest. By the following Lemma 2.3 we have $(R_\mathfrak{p})^\wedge = (R_{\mathfrak{n}_\mathfrak{p}})^\wedge = A_\mathfrak{p}[[x]]$.

*Example 2.2* As an example let $A = \mathbb{k}[t]$ and $R = A[[x]]$ with $t$ and $x$ one variable, $\mathfrak{p} = \langle 0\rangle \in \operatorname{Spec} A$. We have $A_\mathfrak{p} = k(\mathfrak{p}) = \mathbb{k}(t)$ and

$$R_\mathfrak{p} = \mathbb{k}[t][[x]] \otimes_{\mathbb{k}[t]} \mathbb{k}(t) = \{g/h \mid g \in \mathbb{k}[t][[x]], h \in \mathbb{k}[t] \smallsetminus 0\},$$

$g = g_0 + \sum_{i\geq 1} g_i x^i$, $g_i \in \mathbb{k}[t]$, a subring strictly contained in $R\hat{\otimes}_A A_\mathfrak{p} = \mathbb{k}(t)[[x]]$.

- $\langle x\rangle$ is contained in the Jacobson radical of $R\hat{\otimes}_A A_\mathfrak{p}$ by Lemma 2.4.
- $\langle x\rangle$ is not contained in the Jacobson radical of $R_\mathfrak{p} = R \otimes A_\mathfrak{p}$.

To see this, note that the element $t-x$ is a unit in $R\hat{\otimes}_A A_\mathfrak{p}$, since $1/(t-x) = 1/t\sum_{i\geq 0}(x/t)^i$, but $1/(t-x)$ is not an element in $R_\mathfrak{p}$. The ideal $\langle t-x\rangle$ is a maximal ideal in $R_\mathfrak{p}$, since $R_\mathfrak{p}/\langle t-x\rangle \cong \Bbbk((t))$ (see Example 2.3.2), but $x \notin \mathfrak{m}$ since otherwise $t \in \mathfrak{m}$, contradicting the fact that $t$ is a unit $R_\mathfrak{p}$.

- The rings $R_\mathfrak{p}$ and $R(\mathfrak{p})$ are in general not local.
  Since $R_\mathfrak{p}/\langle x\rangle = \Bbbk(t)$, the ideals $\langle x\rangle$ and $\langle t-x\rangle$ are two different maximal ideals and $R_\mathfrak{p} = R(\mathfrak{p})$ ($\mathfrak{p} = \langle 0\rangle$) is not local.

**Lemma 2.3** *Let $M$ be a finitely presented $R$-module and* $\mathfrak{p} \in \operatorname{Spec} A$.

1. *We have isomorphisms*

$$\hat{M}_\mathfrak{p} \cong M_{\mathfrak{n}_\mathfrak{p}}\hat{\otimes}_{A_\mathfrak{p}} A_\mathfrak{p} = M_{\mathfrak{n}_\mathfrak{p}}\hat{\otimes}_A A = (M_{\mathfrak{n}_\mathfrak{p}})^\wedge.$$

2. $\hat{M}(\mathfrak{p}) \cong (M_{\mathfrak{n}_\mathfrak{p}}/\mathfrak{p}M_{\mathfrak{n}_\mathfrak{p}})^\wedge$.
3. *If* $M = \operatorname{coker}\left(A[[x]]^p \xrightarrow{T} A[[x]]^q\right)$ *then*

$$\begin{aligned}\hat{M}_\mathfrak{p} = M\hat{\otimes}_A A_\mathfrak{p} &= \operatorname{coker}\left(A_\mathfrak{p}[[x]]^p \xrightarrow{T} A_\mathfrak{p}[[x]]^q\right),\\ \hat{M}(\mathfrak{p}) &= \operatorname{coker}\left(k(\mathfrak{p})[[x]]^p \xrightarrow{T} k(\mathfrak{p})[[x]]^q\right).\end{aligned}$$

*Note that* $\hat{R}_\mathfrak{p} = A_\mathfrak{p}[[x]] = (R_\mathfrak{p})^\wedge \cong (R_{\mathfrak{n}_\mathfrak{p}})^\wedge$ *and* $\hat{R}(\mathfrak{p}) = k(\mathfrak{p})[[x]] = R(\mathfrak{p})^\wedge \cong (R_{\mathfrak{n}_\mathfrak{p}}/\mathfrak{p}R_{\mathfrak{n}_\mathfrak{p}})^\wedge$ *are local rings but* $R_\mathfrak{p} \not\cong R_{\mathfrak{n}_\mathfrak{p}}$ *and* $R(\mathfrak{p}) \not\cong R_{\mathfrak{n}_\mathfrak{p}}/\mathfrak{p}R_{\mathfrak{n}_\mathfrak{p}}$, *since* $R_\mathfrak{p}$ *and* $R(\mathfrak{p})$ *are in general not local.*

***Proof***

1. The natural inclusion $R_{\mathfrak{n}_\mathfrak{p}} = A[[x]]_{\mathfrak{n}_\mathfrak{p}} \hookrightarrow A_\mathfrak{p}[[x]]$ is given as follows. Let $h/g \in R_{\mathfrak{n}_\mathfrak{p}}$ with $h, g \in R$, $g \notin \mathfrak{n}_\mathfrak{p}$ and write $g = g_0 - g_1$ with $g_0 \in A$ and $g_1 \in \langle x\rangle R$. Then $g \notin \mathfrak{n}_\mathfrak{p} = \langle \mathfrak{p}, x\rangle$ iff $g_0 \notin \mathfrak{p}$ and $g$ is a unit in $R_{\mathfrak{n}_\mathfrak{p}}$ iff its image in $A_\mathfrak{p}[[x]]$ is a unit. We get

$$h/g = \frac{g_0^{-1}h}{(1-g_1/g_0)} = g_0^{-1}h\sum_{i\geq 0}(g_1/g_0)^i \in A_\mathfrak{p}[[x]].$$

Now it is not difficult to see that the induced map $A[[x]]_{\mathfrak{n}_\mathfrak{p}}/\langle x\rangle^m \to A_\mathfrak{p}[[x]]/\langle x\rangle^m$ is bijective (a finite sum $\sum_{|\alpha|=0}^{m-1}(a_\alpha/b_\alpha)x^\alpha$, $a_\alpha, b_\alpha \in A$, $b_\alpha \notin \mathfrak{p}$ in $A_\mathfrak{p}[[x]]$ can be written as $1/b\sum_{|\alpha|=0}^{m-1}(a_\alpha b'_\alpha)x^\alpha$ with $b = \prod b_\alpha \notin \mathfrak{n}_\mathfrak{p}$, $b'_\alpha = b/b_\alpha \in A$, and hence is in $A[[x]]_{\mathfrak{n}_\mathfrak{p}}$). We get

$$R_{\mathfrak{n}_\mathfrak{p}}\hat{\otimes}_A A = \varprojlim A[[x]]_{\mathfrak{n}_\mathfrak{p}}/\langle x\rangle^m \otimes_A A = \varprojlim A_\mathfrak{p}[[x]]/\langle x\rangle^m = A_\mathfrak{p}[[x]]$$

and also $R_{\mathfrak{n}_\mathfrak{p}}\hat{\otimes}_{A_\mathfrak{p}} A_\mathfrak{p} = A_\mathfrak{p}[[x]] = R^\wedge_{\mathfrak{n}_\mathfrak{p}}$. Now apply Corollary 2.1 to the presentation of $M$ and deduce the claim for $M\hat{\otimes}_A A_\mathfrak{p}$.

2. $\hat{M}(\mathfrak{p}) = M\hat{\otimes}_A(A_\mathfrak{p}/\mathfrak{p}A_\mathfrak{p}) = (M\hat{\otimes}_A A_\mathfrak{p})/\mathfrak{p}(M\hat{\otimes}_A A_\mathfrak{p}) = M^\wedge_{\mathfrak{n}_\mathfrak{p}}/\mathfrak{p}M^\wedge_{\mathfrak{n}_\mathfrak{p}}$ by Corollary 2.2 (iv) and the first statement of this lemma. Since $M_{\mathfrak{n}_\mathfrak{p}}$ is finitely presented over $R_{\mathfrak{n}_\mathfrak{p}}$ we have $M^\wedge_{\mathfrak{n}_\mathfrak{p}} = M_{\mathfrak{n}_\mathfrak{p}} \otimes_{R_{\mathfrak{n}_\mathfrak{p}}} R^\wedge_{\mathfrak{n}_\mathfrak{p}}$, which implies the result.
3. This follows from Corollary 2.3.

Over maximal ideals the fibre and the completed fibre coincide:

**Lemma 2.4** *Let $A$ be Noetherian and $M$ a finitely generated $R$-module. For $\mathfrak{a} \subset A$ a* maximal *ideal the following holds.*

*(i)*

$$\hat{M}(\mathfrak{a}) = M(\mathfrak{a}),\ \ \hat{d}_\mathfrak{a}(M) = d_\mathfrak{a}(M).$$

*(ii)* $R/\mathfrak{a}R = R(\mathfrak{a}) = \hat{R}(\mathfrak{a}) = k(\mathfrak{a})[[x]]$ *and* $\mathfrak{a}R$ *is a prime ideal in* $R$.
*(iii)* $\mathfrak{n}_\mathfrak{a}$ *is a maximal ideal of $R$ and any maximal ideal of $R$ is of the form $\mathfrak{n}_\mathfrak{a}$ for some $\mathfrak{a} \in \operatorname{Max} A$. Hence $\langle x \rangle$ is contained in the Jacobson radical of $R$.*
*(iv)* $M(\mathfrak{a}) = M/\mathfrak{a}M \cong M_{\mathfrak{n}_\mathfrak{a}}/\mathfrak{a}M_{\mathfrak{n}_\mathfrak{a}}$.

***Proof***

(i) Since $\mathfrak{a}$ is maximal, $k(\mathfrak{a}) = A/\mathfrak{a}$ is a finite $A$-module. Corollary 2.2 (iv) implies $\hat{M}(\mathfrak{a}) = M\hat{\otimes}_A A/\mathfrak{a} = M \otimes_A A/\mathfrak{a} = M(\mathfrak{a})$.
(ii) This follows from (i) and the fact that $R/\mathfrak{a}R = k(\mathfrak{a})[[x]]$ is integral.
(iii) Cf. [12, §1, Example 1] and [1, Chapter 1, Exercise 5]).
(iv) $M/\mathfrak{a}M = M \otimes_A A/\mathfrak{a} = M(\mathfrak{a}) = \hat{M}(\mathfrak{a}) = M\hat{\otimes}_A A/\mathfrak{a} = \operatorname{coker}(R^p_{\mathfrak{n}_\mathfrak{a}}\hat{\otimes}_A A/\mathfrak{a} \to R^q_{\mathfrak{n}_\mathfrak{a}}\hat{\otimes}_A A/\mathfrak{a}) = M_{\mathfrak{n}_\mathfrak{a}}/\mathfrak{a}M_{\mathfrak{n}_\mathfrak{a}}$, as in the proof of Lemma 2.3.

As a first step to semicontinuity we show below (Lemma 2.5) that the vanishing locus of $\hat{d}_\mathfrak{p}(M)$ is open. For an arbitrary $A$-module $M$

$$\operatorname{Supp}_A(M) := \{\mathfrak{p} \in \operatorname{Spec} A \mid M_\mathfrak{p} \neq 0\}$$

denotes the support of $M$ and $\operatorname{Ann}_A(M) = \{f \in A \mid fM = 0\}$ the annihilator ideal of $M$.

In general $\operatorname{Supp}_A(M)$ is not closed in $\operatorname{Spec} A$, but if $M$ a finitely generated $A$-module, then it is well known that $\operatorname{Supp}_A(M) = V(\operatorname{Ann}_A(M))$, which is closed in $\operatorname{Spec} A$. More generally we have:

*Remark 2.4* For any $A$-module $M$ we have

$$\operatorname{Supp}_A(M) \subset V(\operatorname{Ann}_A(M)).$$

If $R$ is an $A$-algebra and $M$ an $R$-module, then $\operatorname{Ann}_A(M) = \operatorname{Ann}_R(M) \cap A$. If $M$ is a finite $R$-module then

$$\operatorname{Supp}_A(M) = V(\operatorname{Ann}_A(M))$$

and hence $\mathrm{Supp}_A(M)$ is closed in Spec $A$.

To see the first claim, let $\mathfrak{p} \in \mathrm{Spec}\, A$. Note that $\mathfrak{p} \notin \mathrm{Supp}_A M \Leftrightarrow M_\mathfrak{p} = 0 \Leftrightarrow \forall m \in M\ \exists f \in A,\ f \notin \mathfrak{p},\ fm = 0$ and that $\mathfrak{p} \notin V(\mathrm{Ann}_A(M)) \Leftrightarrow \mathrm{Ann}_A(M) \not\subset \mathfrak{p} \Leftrightarrow \exists f \notin \mathfrak{p}, fM = 0$. Hence $\mathfrak{p} \notin V(\mathrm{Ann}_A(M))$ implies $\mathfrak{p} \notin \mathrm{Supp}_A(M)$, i.e. $\mathrm{Supp}_A(M) \subset V(\mathrm{Ann}_A(M))$.

Now let $M$ be generated over $R$ by $m_1, \ldots, m_k \in M$. If $M_\mathfrak{p} = 0$, choose $f_i \in A$, $f_i \notin \mathfrak{p}$, $f_i m_i = 0$. Then $f = f_1 \cdots f_k \notin \mathfrak{p}$ satisfies $fM = 0$ and hence $f \in \mathrm{Ann}_A(M)$. We get $\mathfrak{p} \notin V(\mathrm{Ann}_A(M))$ and hence the other inclusion $\mathrm{Supp}_A(M) \supset V(\mathrm{Ann}_A(M))$.

In our situation for $R = A[[x]]$ and $M$ finitely $R$-presented we have (possibly strict) inclusions (cf. Lemma 2.2)

$$\{\mathfrak{p} \in \mathrm{Spec}\, A \mid \hat{d}_\mathfrak{p}(M) \neq 0\} \subset \{\mathfrak{p} \mid d_\mathfrak{p}(M) \neq 0\} \subset \mathrm{Supp}_A(M),$$

where the first (Lemma 2.5) and the last (Remark 2.4) sets are closed in Spec $A$, while the middle set may not be closed (Example 2.3.4).

**Lemma 2.5** *Let $M$ be a finitely presented $R = A[[x]]$-module.*

1. *We have (possibly strict) inclusions*

$$\{\mathfrak{p} \in \mathrm{Spec}\, A \mid \hat{d}_\mathfrak{p}(M) \neq 0\} \subset \{\mathfrak{p} \mid d_\mathfrak{p}(M) \neq 0\} \subset \mathrm{Supp}_A(M),$$

*where the first and the last sets are closed in* Spec $A$, *while the middle set may not be closed.*

2. *The map* $\mathrm{Supp}_R(M) \to \mathrm{Supp}_A(M), \mathfrak{n} \mapsto \mathfrak{n} \cap A$, *is dominant and induces a homeomorphism*

$$V(\langle x \rangle) \cap \mathrm{Supp}_R(M) \xrightarrow{\approx} \{\mathfrak{p} \in \mathrm{Spec}\, A \mid \hat{d}_\mathfrak{p}(M) \neq 0\}.$$

*Hence* $\{\mathfrak{p} \in \mathrm{Spec}\, A \mid \hat{d}_\mathfrak{p}(M) = 0\}$, *the* vanishing locus *of* $\hat{d}_\mathfrak{p}(M)$, *is open in* Spec $A$.

3. *Let* $A' = A/\mathrm{Ann}_A(M)$, $R' = A'[[x]]$ *and denote by* $M'$ *the module* $M$ *considered as* $A'$*-module. Then* $M'$ *is a finitely presented* $R'$*-module and for* $\mathfrak{p} \in \mathrm{Spec}(A') \subset \mathrm{Spec}(A)$ *we have* $M_\mathfrak{p} = M'_\mathfrak{p}$, $M(\mathfrak{p}) = M'(\mathfrak{p})$, $\hat{M}_\mathfrak{p} = \hat{M}'_\mathfrak{p}$, *and* $\hat{M}(\mathfrak{p}) = \hat{M}'(\mathfrak{p})$. *For* $\mathfrak{p} \in \mathrm{Spec}(A) \setminus \mathrm{Spec}(A')$ *the modules* $M_\mathfrak{p}, M(\mathfrak{p}), \hat{M}_\mathfrak{p}$, *and* $\hat{M}(\mathfrak{p})$ *vanish.*

   In particular, we may consider $M$ as an $A'$-module, whenever we study the fibres or the completed fibres of $M$.

***Proof***

1. The first inclusion follows from Lemma 2.2(i), the second from the definition. For an example where these inclusions are strict, see Example 2.3.2, 3 and 2.3.4.

The first set is closed by item 2. and the third by Remark 2.4. In Example 2.3.4 the middle set is not closed.

2. Since $\mathrm{Ann}_A(M) = \mathrm{Ann}_R(M) \cap A$, the map $A/\mathrm{Ann}_A(M) \to R/\mathrm{Ann}_R(M)$ is injective and induces therefore a dominant morphism $\mathrm{Spec}(R/\mathrm{Ann}_R(M)) \to \mathrm{Spec}(A/\mathrm{Ann}_A(M))$. The first claim follows hence from Remark 2.4.

   For the second claim consider $A[[x]]^p \xrightarrow{T} A[[x]]^q \to M \to 0$, a presentation of $M$ with $T = (t_{ij})$, $t_{ij} \in A[[x]]$, and let $\mathfrak{p} \in \mathrm{Spec}\, A$. Then $k(\mathfrak{p})[[x]]^p \xrightarrow{T'} k(\mathfrak{p})[[x]]^q \to \hat{M}(\mathfrak{p}) \to 0$ is a presentation of $\hat{M}(\mathfrak{p})$ with $T' = (t'_{ij})$, $t'_{ij} \in k(\mathfrak{p})[[x]]$, the induced map (Corollary 2.3).

   Now $\hat{M}(\mathfrak{p}) = 0$ iff $T'$ is surjective, i.e., iff the 0-th Fitting ideal (the ideal of $q$-minors) of $T'$ contains a unit $u' \in k(\mathfrak{p})[[x]]$. Write $u'$ as $u' = u'_0 + u'_1$ with $u'_0 \in k(\mathfrak{p}) \setminus \{0\}$, $u'_1 \in \langle x\rangle k(\mathfrak{p})[[x]]$. Since Fitting ideals are compatible with base change, the 0-th Fitting ideal $F_0 \subset A[[x]]$ of $M$ contains an element $u = u_0 + u_1 \in A[[x]]$ with $u_0 \in A$, $u_1 \in \langle x\rangle A[[x]]$, that maps to $u'$ under $A[[x]] \to A_{\mathfrak{p}}[[x]] \to A_{\mathfrak{p}}/\mathfrak{p}A_{\mathfrak{p}}[[x]]$. Hence $u'$ is a unit iff $u_0 \notin \mathfrak{p}$, i.e., iff $\langle x, \mathfrak{p}\rangle \notin V(F_0 + \langle x\rangle)$. The result follows since $\mathrm{Supp}_R(M) = V(F_0)$.

3. Any $R$-presentation of $M$ induces obviously an $R'$-presentation of $M'$. The equalities $M_{\mathfrak{p}} = M'_{\mathfrak{p}}$ and $M(\mathfrak{p}) = M'(\mathfrak{p})$ for $\mathfrak{p} \in \mathrm{Spec}(A')$ are clear, the equalities $\hat{M}_{\mathfrak{p}} = \hat{M}'_{\mathfrak{p}}$ and $\hat{M}(\mathfrak{p}) = \hat{M}'(\mathfrak{p})$ follow from this and from Lemma 2.2(i). Since $\mathrm{Supp}_A(M) = \mathrm{Spec}(A')$ by Remark 2.4, $M_{\mathfrak{p}} = M(\mathfrak{p}) = 0$ for $\mathfrak{p} \notin \mathrm{Spec}(A')$ and Lemma 2.2(i) implies then $\hat{M}_{\mathfrak{p}} = \hat{M}(\mathfrak{p}) = 0$.

At the end of this section we give two examples where $d_{\mathfrak{p}}(M)$ is not semicontinuous on Spec $A$ while $\hat{d}_{\mathfrak{p}}(M)$ is. The examples show also that $\hat{M}(\mathfrak{p}) = 0$ may happen for $M(\mathfrak{p}) \neq 0$.

*Example 2.3* Let $A = K[t]$, $K$ an algebraically closed field, $R = A[[x]]$, and $M = R/\langle t - x\rangle \cong K[[t]]$ as $A$-module via $f(x,t) \mapsto f(t,t)$, with $t$ and $x$ one variable each. The following properties illustrate the difference between the fibres and the completed fibres. Let $\mathfrak{a} = \langle t - c\rangle$, $c \in K$, denote the maximal ideals in $A$. By Lemma 2.4 $\hat{M}(\mathfrak{a}) = M(\mathfrak{a}) = M/\mathfrak{a}M$ which is isomorphic to $K[[t]]/\langle t - c\rangle$. Hence $M(\langle t\rangle) = K$ and $M(\langle t - c\rangle) = 0$ for $c \neq 0$.

1. $M$ is not finitely generated over $A$, $d_{\mathfrak{a}}(M) = \hat{d}_{\mathfrak{a}}(M) < \infty$ for $\mathfrak{a} \in \mathrm{Max}\, A$ and $d_{\mathfrak{a}}(M)$ is semicontinuous on Max $A$:
   $K[[t]]$ is not finite over $K[t]$ and $d_{\mathfrak{a}}(M) = 1$ if $c = 0$ and $0$ if $c \neq 0$.
2. $d_{\mathfrak{p}}(M)$ is not semicontinuous on Spec $A$:
   The prime ideal $\langle 0\rangle$ is contained in every neighbourhood of $\mathfrak{a} = \langle t\rangle$ in Spec $A$. It satisfies $k(\langle 0\rangle) = K(t)$ and we get $M(\langle 0\rangle) \cong K[[t]] \otimes_A K(t) = K((t))$, the field of formal Laurent series. Since $\dim_{K(t)} K((t)) = \infty$, $d_{\langle 0\rangle}(M) = \infty$, while $d_{\mathfrak{a}}(M) \leq 1$ for $\mathfrak{a} \in \mathrm{Spec}\, A \setminus \langle 0\rangle$.
3. $\hat{d}_{\mathfrak{p}}(M)$ is semicontinuous on Spec $A$:
   We have $\hat{M}(\langle 0\rangle) = K(t)[[x]]/\langle t - x\rangle$ by Corollary 2.3. Since $t$ is a unit in $K(t)$, $\hat{d}_{\langle 0\rangle}(M) = 0$.

4. $M(\mathfrak{a}) = M_\mathfrak{a}/\mathfrak{a}M_\mathfrak{a} = 0$ does not imply $M_\mathfrak{a} = 0$:
In fact, we have $M_{\langle t-c\rangle} \cong K[[t]]_{\langle t-c\rangle}$ as a $K[t]$-module. For $c \neq 0$ we get $K[[t]]_{\langle t-c\rangle} = K((t))$ (since $t \notin \langle t-c\rangle$) while $K[[t]]_{\langle t-c\rangle}/\langle t-c\rangle K[[t]]_{\langle t-c\rangle} = K((t))/\langle t-c\rangle = 0$. We have $\{\mathfrak{p} \mid \hat{d}_\mathfrak{p}(M) \neq 0\} = \{\langle t\rangle\} \subsetneqq \{\mathfrak{p} \mid d_\mathfrak{p}(M) \neq 0\} = \{\langle t\rangle, \langle 0\rangle\} \subsetneqq \operatorname{Supp}_A(M) = \operatorname{Spec} A$.
5. $M$ is flat over $A$. By 1. and 3. we cannot expect any continuity of $d_\mathfrak{p}(M)$ or $\hat{d}_\mathfrak{p}(M)$ on Max $A$ or on Spec $A$ for flat $A$-modules.
6. The quasi-finite locus of $A \to A[[x]]/\langle t-x\rangle$ is not open in Spec $A$:
The quasi-finite locus $\{\mathfrak{p} \in \operatorname{Spec} A \mid d_\mathfrak{p}(M) < \infty\}$ is $\operatorname{Spec} A \setminus \langle 0\rangle$ by 1. and 2. Recall that if $B$ is a ring of finite type over $A$, then the quasi-finite locus of $A \to B$ is open by Zariski's main Theorem (cf. [15, 10.122]).
7. The quasi-completed-finite locus of $A \to A[[x]]/\langle t-x\rangle$ is open in Spec $A$:
Let us call $\{\mathfrak{p} \in \operatorname{Spec} A \mid \hat{d}_\mathfrak{p}(M) < \infty\}$ the *quasi-completed-finite locus*. It is Spec $A$ in our example.
In general, if semicontinuity of $\hat{d}_\mathfrak{p}(M)$ holds (Corollary 2.7), then the quasi-completed-finite locus is open.

*Example 2.4* The following example may be of interest for arithmetic and computational purposes. It goes along similar lines as Example 2.3.

Let $A = \mathbb{Z}$, $R = \mathbb{Z}[[x]]$, and $M = R/\langle x-p\rangle$, $p \in \mathbb{Z}$ a prime number. Since $R = \varprojlim \mathbb{Z}[x]/\langle x\rangle^n$ we obtain $M = \varprojlim \mathbb{Z}/p^n = \hat{\mathbb{Z}}_{\langle p\rangle}$, the ring of $p$-adic integers.

Now let $\langle q\rangle \in \operatorname{Max} \mathbb{Z}$.
If $q \neq p$ then $q$ is a unit in $\mathbb{Z}_{\langle p\rangle}$ hence in $\hat{\mathbb{Z}}_{\langle p\rangle}$ and $M \otimes_\mathbb{Z} \mathbb{Z}/q = M/\langle q\rangle M = \hat{\mathbb{Z}}_{\langle p\rangle}/q\hat{\mathbb{Z}}_{\langle p\rangle} = 0$.
If $q = p$ then $M \otimes_\mathbb{Z} \mathbb{Z}/p = \hat{\mathbb{Z}}_{\langle p\rangle}/p\hat{\mathbb{Z}}_{\langle p\rangle} = \mathbb{Z}/p$.
Hence $d_{\langle q\rangle}(M) = \dim_{\mathbb{Z}/q} M \otimes_\mathbb{Z} \mathbb{Z}/q$ is 0 if $q \neq p$ and 1 if $q = p$.

On the other hand, looking at the prime ideal $\langle 0\rangle$ we get

$$\hat{M}(\langle 0\rangle) = M\hat{\otimes}_\mathbb{Z}\mathbb{Q} = \mathbb{Q}[[x]]/\langle x-p\rangle = 0,$$

while

$$M(\langle 0\rangle) = M \otimes_\mathbb{Z} \mathbb{Q} = \hat{\mathbb{Z}}_{\langle p\rangle} \otimes_\mathbb{Z} \mathbb{Q} = Quot(\hat{\mathbb{Z}}_{\langle p\rangle})$$

has dimension $d_{\langle 0\rangle}(M) = \dim_\mathbb{Q} Quot(\hat{\mathbb{Z}}_{\langle p\rangle}) = \infty$.

To see the last equality in the formula for $M(\langle 0\rangle)$ one checks that the following diagram has the universal property of the tensor product:

$$\begin{array}{ccc} \hat{\mathbb{Z}}_{\langle p\rangle} & \xrightarrow{i_1} & Quot(\hat{\mathbb{Z}}_{\langle p\rangle}) = (\hat{\mathbb{Z}}_{\langle p\rangle})_p \\ \uparrow{\scriptstyle i_2} & & \uparrow{\scriptstyle j_1} \\ \mathbb{Z} & \xrightarrow[j_2]{} & \mathbb{Q} \end{array}.$$

Here $i_1$, $i_2$ and $j_2$ are the canonical inclusions and $j_1$ is given as follows: if $\alpha, \beta \in \mathbb{Z}$, $p \nmid \beta$, then $j_1(\frac{\alpha}{p^m \beta}) = \frac{1}{p^m}\frac{\alpha}{\beta}$, $\frac{\alpha}{\beta} \in \hat{\mathbb{Z}}_{\langle p \rangle}$ since $p \nmid \beta$. The universality of the diagram is easily seen. If $T$ is a $\mathbb{Z} - algebra$ and $\phi : \hat{\mathbb{Z}}_{\langle p \rangle} \to T$ and $\psi : \mathbb{Q} \to \mathbb{Z}$ are $\mathbb{Z}$-algebra homomorphisms then the morphism $\sigma : (\hat{\mathbb{Z}}_{\langle p \rangle})_p \to T$, given as $\sigma(\alpha/p^m) = \phi(\alpha)\psi(1/p^m)$, $p \nmid \alpha$ is the unique one, making the obvious diagram commutative.

## 2.3 Semicontinuity Over a 1-Dimensional Ring

In this section $A$ will be Noetherian and $M$ a finitely generated $R$-module (unless we say otherwise). Then $R = A[[x]]$ is Noetherian and $M$ is finitely presented as $R$-module. At the moment we can prove the semicontinuity of $d_{\mathfrak{q}}(M)$ on Spec $A$ in full generality only under certain assumptions on the irreducible components of $\mathrm{Supp}_R(M)$. This includes the case $\dim A = 1$ where $\dim A$ denotes the Krull dimension of $A$. The case of arbitrary Noetherain $A$ is treated in the next section under the assumption that the presentation matrix of $M$ is algebraic.

In an important special case semicontinuity holds for arbitrary $A$:

**Proposition 2.2** *Let $A$ be Noetherian and $M$ a finitely generated $R$-module.*

1. *If $\mathrm{Supp}_R(M) \subset V(\langle x \rangle)$ then $M$ is finitely generated over $A$ and $\hat{M}(\mathfrak{q}) = M(\mathfrak{q})$ for all $\mathfrak{q} \in \mathrm{Spec}\, A$. In particular semicontinuity of $\hat{d}_{\mathfrak{p}}(M) = d_{\mathfrak{p}}(M)$ holds at any $\mathfrak{p} \in \mathrm{Spec}\, A$.*
2. *If $M$ is finitely generated as an $A$-module (in particular $d_{\mathfrak{p}}(M) < \infty$ for $\mathfrak{p} \in \mathrm{Spec}\, A$), then $\hat{d}_{\mathfrak{p}}(M) \le d_{\mathfrak{p}}(M)$ and $\hat{d}_{\mathfrak{p}}(M)$ (as well as $d_{\mathfrak{p}}(M)$) is semicontinuous at any $\mathfrak{p} \in \mathrm{Spec}\, A$.*

*Proof*

1. Since $V(\mathrm{Ann}_R(M)) = \mathrm{Supp}_R(M) \subset V(\langle x \rangle)$, we have $\langle x \rangle \subset \sqrt{\mathrm{Ann}_R(M)}$ and there exists an $m$ such that $\langle x \rangle^m \subset \mathrm{Ann}_R(M)$. We get a surjection

$$A[[x]]/\langle x \rangle^m \to R/\mathrm{Ann}_R(M)$$

and since $A[[x]]/\langle x \rangle^m$ is finitely generated over $A$ this holds for $R/\mathrm{Ann}_R(M)$ too. Since $M$ is finitely generated over $R/\mathrm{Ann}_R(M)$ it is finitely generated over $A$ and hence finitely presented. By Lemma 2.1 there is an open neighborhood $U$ of $\mathfrak{p}$ in Spec $A$ such that $d_{\mathfrak{q}}(M) \le d_{\mathfrak{p}}(M)$, $\mathfrak{q} \in U$. By Proposition 2.1.5, $\hat{M}(\mathfrak{q}) = M(\mathfrak{q})$ for all $\mathfrak{q} \in \mathrm{Spec}\, A$, showing the claim.
2. Let $m < m'$ be two strictly positive integers and consider the natural surjective maps

$$R \longrightarrow R/\langle x \rangle^{m'} \longrightarrow R/\langle x \rangle^m.$$

By the right exactness of the tensor product, they induce surjective maps

$$M \longrightarrow M/\langle x\rangle^{m'}M \longrightarrow M/\langle x\rangle^{m}M.$$

and

$$M(\mathfrak{p}) = M \otimes_A k(\mathfrak{p}) \longrightarrow M/\langle x\rangle^{m'}M \otimes_A k(\mathfrak{p}) \xrightarrow{\pi_{m,m'}(\mathfrak{p})} M/\langle x\rangle^{m}M \otimes_A k(\mathfrak{p}).$$

Since $M$ is finitely generated over $A$, all the modules appearing in the last sequence are finite-dimensional $k(\mathfrak{p})$-vector spaces, with $\dim_{k(\mathfrak{p})}(M(\mathfrak{p})) = d_{\mathfrak{p}}(M)$ by definition. Since $\pi_{m,m'}(\mathfrak{p})$ is surjective for all $m < m'$, the elements of the inverse system $\{M/\langle x\rangle^m M \otimes_A k(\mathfrak{p})\}_m$ of finite-dimensional $k(\mathfrak{p})$-vector spaces have dimensions increasing with $m$ and bounded above by $d_{\mathfrak{p}}(M)$. Thus $\dim_{k(\mathfrak{p})}\big(M/\langle x\rangle^m M \otimes_A k(\mathfrak{p})\big)$ stabilizes for large $m$. Hence, for $m$ large

$$M/\langle x\rangle^m M \otimes_A k(\mathfrak{p}) = \varprojlim_{m'}(M/\langle x\rangle^{m'}M \otimes_A k(\mathfrak{p})) = M\hat{\otimes}_A k(\mathfrak{p}) = \hat{M}(\mathfrak{p}).$$

This implies $\hat{d}_{\mathfrak{p}}(M) = \dim_{k(\mathfrak{p})}\hat{M}(\mathfrak{p}) \le d_{\mathfrak{p}}(M)$.

To see the semicontinuity of $\hat{d}_{\mathfrak{p}}(M)$, we use the semicontinuity of $d_{\mathfrak{p}}(M)$ (by Lemma 2.1). It follows that there exists an open neighbourhood $U$ of $\mathfrak{p}$ such that the sequence $\{\dim_{k(\mathfrak{q})}\big(M/\langle x\rangle^m M \otimes_A k(\mathfrak{q})\big)\}_m$ is bounded by $d_{\mathfrak{p}}(M)$ simultaneously for all $\mathfrak{q} \in U$. Hence,

$$d_{\mathfrak{q}}(M/\langle x\rangle^m M) = \dim_{k(\mathfrak{q})}(M/\langle x\rangle^m M \otimes_A k(\mathfrak{q})) = \dim_{k(\mathfrak{q})}(\hat{M}(\mathfrak{q})) = \hat{d}_{\mathfrak{q}}(M)$$

for a fixed large $m$ and $\mathfrak{q} \in U$. Since $M/\langle x\rangle^m M$ is finitely $A$-generated, Lemma 2.1 implies that $d_{\mathfrak{p}}(M/\langle x\rangle^m M)$ is semicontinuous, and so is $\hat{d}_{\mathfrak{p}}(M)$.

The inequality $\hat{d}_{\mathfrak{p}}(M) \le d_{\mathfrak{p}}(M)$ of item 2. and its proof, as well as Example 2.6, were suggested to us by the anonymous referee. Note that $\hat{d}_{\mathfrak{p}}(M) = d_{\mathfrak{p}}(M)$ for $\mathfrak{p}$ a maximal ideal (by Lemma 2.4), but that $\hat{d}_{\mathfrak{p}}(M) < d_{\mathfrak{p}}(M)$ may happen by Example 2.7 for $\mathfrak{p}$ not maximal.

Before we formulate the next result, we introduce some notations to be used throughout this section. Consider a minimal primary decomposition of $\mathrm{Ann}_R(M)$,

$$\mathrm{Ann}_R(M) = \bigcap_{i=1}^{r} Q_i \subset R.$$

Since $M$ is finitely generated over $R$, $\mathrm{Supp}_R(M) = V(\mathrm{Ann}_R(M)) = \bigcup_{i=1}^{r} V(Q_i)$ and $\dim M = \dim \mathrm{Supp}_R(M)$.

Let $P_1, \ldots P_s \subset R$ be the minimal associated primes of $\langle x \rangle$. Since they correspond via $h : A \cong R/\langle x \rangle$ to the minimal associated primes $\bar{P}_1, \ldots \bar{P}_s$ of $A$, we have $\dim V(P_j) \leq \dim A$.

**Lemma 2.6** *For $A$ Noetherian, $M$ finitely generated over $R$ and $\mathfrak{p} \in \operatorname{Spec} A$ the following holds:*

1. *Let $A'$ be the reduction of $A$, $R' = A'[[x]]$ and $M'$ the $R'$-module $M \otimes_R R'$. Then $\hat{M}'(\mathfrak{p}) \cong \hat{M}(\mathfrak{p})$ and hence $\hat{d}_{\mathfrak{p}}(M') = \hat{d}_{\mathfrak{p}}(M)$.*
2. *Let $Q \subset R$ be an ideal. Then $\hat{d}_{\mathfrak{p}}(M/QM) \leq \hat{d}_{\mathfrak{p}}(M)$.*
3. *If $Q_i \not\subset \mathfrak{n}_{\mathfrak{p}}$ for some $1 \leq i \leq r$, then $\hat{d}_{\mathfrak{q}}(M) = \hat{d}_{\mathfrak{q}}(M/QM)$, with $Q = \cap_{j \neq i} Q_j$, for $\mathfrak{q}$ in some neighbourhood of $\mathfrak{p}$ in $\operatorname{Spec} A$.*
4. *If $Q_i \subset \mathfrak{n}_{\mathfrak{p}}$ and $\dim V(Q_i) > \dim(A/Q_i \cap A)_{\mathfrak{p}}$ for some $1 \leq i \leq r$, then $\hat{d}_{\mathfrak{p}}(M) = \infty$.*
5. *Let $U = \operatorname{Spec} B \subset \operatorname{Spec} A$ be an affine open neighbourhood of $\mathfrak{p}$ and $M_B = M \otimes_A B$ the restriction of $M$ to $U$. Then $\hat{M}_B(\mathfrak{q}) = \hat{M}(\mathfrak{q})$ for all $\mathfrak{q} \in U$.*

***Proof***

1. Since $\mathfrak{p} \in \operatorname{Spec} A$ contains the nilpotent elements, $A'/\mathfrak{p}' = A/\mathfrak{p}$, where $\mathfrak{p}'$ is the image of $\mathfrak{p}$ in $A'$, and hence the residue field does not change if we pass from $A$ to $A'$. By Proposition 2.1.1 we have $\hat{R}'(\mathfrak{p}') = R'\hat{\otimes}_{A'}k(\mathfrak{p}) = k(\mathfrak{p})[[x]] = \hat{R}(\mathfrak{p})$. Consider a presentation $R^p \xrightarrow{T} R^q \to M \to 0$ of $M$. Applying $\otimes_R R'$, we get a presentation of $M'$, $R'^p \xrightarrow{T} R'^q \to M' \to 0$. Apply $\hat{\otimes}_A k(\mathfrak{p})$ to the first resp. $\hat{\otimes}_{A'} k(\mathfrak{p})$ to the second exact sequence above. The sequences stay exact by Corollary 2.1. Since $(\hat{R}(\mathfrak{p}))^k = (\hat{R}'(\mathfrak{p}'))^k$ it follows that the canonical morphism $M \to M'$ induces an isomorphism $\hat{M}(\mathfrak{p}) \cong \hat{M}'(\mathfrak{p})$.
2. Since $(M/QM)\hat{\otimes}_A k(\mathfrak{p}) = M\hat{\otimes}_A k(\mathfrak{p})/Q(M\hat{\otimes}_A k(\mathfrak{p}))$ by Corollary 2.1, the result follows.
3. $Q_i \not\subset \mathfrak{n}_{\mathfrak{p}}$ means that $\mathfrak{n}_{\mathfrak{p}}$ is not a point of $V(Q_i)$. Hence $\mathfrak{n}_{\mathfrak{q}} \notin V(Q_i)$ and $M_{\mathfrak{n}_{\mathfrak{q}}} = (M/Q)_{\mathfrak{n}_{\mathfrak{q}}}$ for $\mathfrak{n}_{\mathfrak{q}}$ in some neighbourhood of $\mathfrak{n}_{\mathfrak{p}}$ in $V(\langle x \rangle)$. The result follows from Lemma 2.3.
4. Set $\bar{R} := R/Q_i$, $\bar{A} := A/Q_i \cap A$ and $\bar{M} := M/Q_i M$. Then $Q_i \subset \operatorname{Ann}_R(\bar{M}) \subset \sqrt{(Q_i)}$ and $\dim \bar{R}_{\mathfrak{n}_{\mathfrak{p}}} = \dim \bar{M}_{\mathfrak{n}_{\mathfrak{p}}} = \dim V(Q_i) > \dim \bar{A}_{\mathfrak{p}}$ by assumption. Considering $\bar{M}$ as $R$-module, we have $\bar{M}\hat{\otimes}_A k(\mathfrak{p}) = \bar{M}(\mathfrak{p})^\wedge = (\bar{M}_{\mathfrak{n}_{\mathfrak{p}}}/\mathfrak{p}\bar{M}_{\mathfrak{n}_{\mathfrak{p}}})^\wedge$ by Lemma 2.3. Since the $\bar{R}_{\mathfrak{n}_{\mathfrak{p}}}$-modules $\bar{M}(\mathfrak{p})$ and its $\langle x \rangle$-adic completion $\bar{M}(\mathfrak{p})^\wedge$ have the same Hilbert-Samuel function w.r.t. $\mathfrak{n}_{\mathfrak{p}}$, their dimension coincides (cf. [7, Corollary 5.6.6]). Moreover, $\mathfrak{p}\bar{R}_{\mathfrak{n}_{\mathfrak{p}}}$ is the annihilator of $\bar{M}_{\mathfrak{n}_{\mathfrak{p}}}/\mathfrak{p}\bar{M}_{\mathfrak{n}_{\mathfrak{p}}}$ and therefore $\dim \bar{M}(\mathfrak{p})^\wedge = \dim \bar{R}_{\mathfrak{n}_{\mathfrak{p}}}/\mathfrak{p}\bar{R}_{\mathfrak{n}_{\mathfrak{p}}}$.

   We apply now [12, Theorem 15.1] to the map of local rings $\bar{A}_{\mathfrak{p}} \to \bar{R}_{\mathfrak{n}_{\mathfrak{p}}}$ and get that $\dim \bar{R}_{\mathfrak{n}_{\mathfrak{p}}}/\mathfrak{p}\bar{R}_{\mathfrak{n}_{\mathfrak{p}}} \geq \dim \bar{R}_{\mathfrak{n}_{\mathfrak{p}}} - \dim \bar{A}_{\mathfrak{p}} > 0$ and hence $\dim_{k(\mathfrak{p})} \bar{M}(\mathfrak{p})^\wedge = \infty$. Then $\hat{d}_{\mathfrak{p}}(M) = \dim_{k(\mathfrak{p})} M(\mathfrak{p})^\wedge = \infty$ by 2. of this lemma.

5. We may assume that $B = A_f$ for some $f \notin \mathfrak{p}$. Since $A_\mathfrak{q} = (A_f)_\mathfrak{q}$ for $\mathfrak{q} \in U = D(f)$, we have $k(\mathfrak{q}) = A_f \otimes_A k(\mathfrak{q})$. Now Proposition 2.1.1 implies $\hat{M}_{A_f}(\mathfrak{q}) = (M \otimes_A A_f)\hat{\otimes}_A k(\mathfrak{q}) = (M \otimes_A A_f \otimes_A k(\mathfrak{q}))^\wedge = M(\mathfrak{q})^\wedge = \hat{M}(\mathfrak{q})$.

**Proposition 2.3** *Let $A$ be Noetherian, $M$ a finitely generated $R$-module and fix $\mathfrak{p} \in \operatorname{Spec} A$. Let $Q_1, \ldots, Q_r$ be the primary components of $\operatorname{Ann}_R(M)$, which we renumerate such that*

*I. $V(Q_i) \subset V(\langle x\rangle)$ for $1 \le i \le k$,*
*II. $V(Q_i) \not\subset V(\langle x\rangle)$ for $k+1 \le i \le r$,*

*and set $Q_I := \bigcap_{i=1}^{k} Q_i$ and $Q_{II} := \bigcap_{i=k+1}^{r} Q_i$. Assume that either (a) $V(Q_{II}) = \emptyset$ (i.e., $k = r$), or (b) $\dim V(Q_{II}) > \dim V(Q_{II} \cap A)$ or (c) $\mathfrak{n}_\mathfrak{p}$ is an isolated point of $V(\langle x\rangle) \cap V(Q_{II})$.*

*Then there is an open neighbourhood $U$ of $\mathfrak{p}$ in $\operatorname{Spec} A$ such that $\hat{d}_\mathfrak{q}(M) \le \hat{d}_\mathfrak{p}(M)$ for all prime ideals $\mathfrak{q} \in U$.*

***Proof*** We set $M_I := M/Q_I M$ and $M_{II} := M/Q_{II}M$. Then $\operatorname{Ann}_R(M_I) = Q_I$ and $\operatorname{Ann}_R(M_{II}) = Q_{II}$. By Lemma 2.6.3 we may assume that $Q_i \subset \mathfrak{n}_\mathfrak{p}$ for all $1 \le i \le r$.

We have $\operatorname{Supp}_R(M_I) = V(Q_I) \subset V(\langle x\rangle)$. By Proposition 2.2 there is an open neighborhood $U_1$ of $\mathfrak{p}$ in $\operatorname{Spec} A$ such that

$$\hat{d}_\mathfrak{q}(M_I) \le \hat{d}_\mathfrak{p}(M_I), \quad \mathfrak{q} \in U_1. \tag{2.1}$$

(a) If $V(Q_{II}) = \emptyset$, then $M = M_I$ and the claim follows from (2.1).
(b) If $\dim V(Q_{II}) > \dim V(Q_{II} \cap A)$ then $\dim V(Q_i) > \dim(A/Q_i \cap A)_\mathfrak{p}$ for some $i$ and hence $\hat{d}_\mathfrak{p}(M) = \infty$ by Lemma 2.6.4, implying the claim.
(c) Now let $\mathfrak{n}_\mathfrak{p}$ be an isolated point of $V(\langle x\rangle) \cap V(Q_{II})$. Then there exists an open neighbourhood $U_2 \subset \operatorname{Spec} A$ of $\mathfrak{p}$ such that $M_{I,\mathfrak{n}_\mathfrak{q}} = M_{\mathfrak{n}_\mathfrak{q}}$ if $\mathfrak{q} \in U_2 \setminus \{\mathfrak{p}\}$. Since $\hat{d}_\mathfrak{q}(M) = \dim_\Bbbk (M_{\mathfrak{n}_\mathfrak{q}}/\mathfrak{q}M_{\mathfrak{n}_\mathfrak{q}})^\wedge$ we get

$$\hat{d}_\mathfrak{q}(M) = \hat{d}_\mathfrak{q}(M_I), \quad \mathfrak{q} \in U_2 \setminus \{\mathfrak{p}\}. \tag{2.2}$$

Using (2.1) and (2.2), we have $\hat{d}_\mathfrak{q}(M) \le \hat{d}_\mathfrak{p}(M_I)$ for $\mathfrak{q} \in U_1 \cap U_2 \setminus \{\mathfrak{p}\}$ and by Lemma 2.6.2

$$\hat{d}_\mathfrak{p}(M_I) \le \hat{d}_\mathfrak{p}(M). \tag{2.3}$$

Hence $\hat{d}_\mathfrak{q}(M) \le \hat{d}_\mathfrak{p}(M)$ for $\mathfrak{q} \in U_1 \cap U_2$.

As a corollary we get the following theorem, which was already proved for maximal ideals in $A = \Bbbk[t]$ in [6].

**Theorem 2.1** *Let $A$ be Noetherian, $M$ a finitely generated $R$-module and $\mathfrak{p} \in \operatorname{Spec} A$. If $\dim_{\mathfrak{p}}(\operatorname{Supp}_A(M)) \leq 1$ then there is an open neighbourhood $U$ of $\mathfrak{p}$ in $\operatorname{Spec} A$ such that*

$$\hat{d}_{\mathfrak{q}}(M) \leq \hat{d}_{\mathfrak{p}}(M) \text{ for all } \mathfrak{q} \in U.$$

***Proof*** By Lemma 2.5.3 we may assume that $\operatorname{Ann}_A(M) = 0$, such that $\operatorname{Supp}_A(M) = A$. We may further assume that $\dim A = \dim V(\langle x \rangle) = 1$ and $\hat{d}_{\mathfrak{p}}(M) < \infty$. Using the notations from Proposition 2.3, we have $\dim V(Q_{II} \cap A) \leq 1$ and by the proof of Proposition 2.3(b) that $\dim V(Q_{II}) \leq 1$. Hence, either $V(Q_{II}) = \emptyset$, or $\mathfrak{n}_{\mathfrak{p}}$ is an isolated point of $V(\langle x \rangle) \cap V(Q_{II})$. The result follows from Proposition 2.3.

**Corollary 2.4** *Let $A = \mathbb{Z}$ and let $M$ be a finitely generated $\mathbb{Z}[[x]]$-module, $x = (x_1 \cdots x_n)$, given by a presentation*

$$\mathbb{Z}[[x]]^r \to \mathbb{Z}[[x]]^s \to M \to 0.$$

*Denote by*

$$M_p := \hat{M}(\langle p \rangle) = \operatorname{coker}\left(\mathbb{F}_p[[x]]^r \xrightarrow{\bar{T}} \mathbb{F}_p[[x]]^s\right)$$

*if $p \in \mathbb{Z}$ is a prime number and by*

$$M_0 := \hat{M}(\langle 0 \rangle) = \operatorname{coker}\left(\mathbb{Q}[[x]]^r \xrightarrow{\bar{T}} \mathbb{Q}[[x]]^s\right)$$

*the induced modules.*

1. *Fix a prime number $p$. If $\dim_{\mathbb{F}_p} M_p < \infty$ then $\dim_{\mathbb{F}_p} M_p \geq \dim_{\mathbb{Q}} M_0$. Moreover, for all except finitely many prime numbers $q \in \mathbb{Z}$, $\dim_{\mathbb{F}_p} M_p \geq \dim_{\mathbb{F}_q} M_q$.*
2. *If $\dim_{\mathbb{Q}} M_0 < \infty$ then $\dim_{\mathbb{Q}} M_0 \geq \dim_{\mathbb{F}_q} M_q$ for all except finitely many prime numbers $q \in \mathbb{Z}$, and hence "$=$" for all except finitely many prime numbers $q \in \mathbb{Z}$.*

The first part of statement 1. follows, since $\langle 0 \rangle$ is in every neighbourhood of $p$. In particular $\dim_{\mathbb{Q}} M_0$ is finite if $\dim_{\mathbb{F}_p} M_p$ is finite for some prime number $p$.

*Remark 2.5* The corollary is important for practical computations in computer algebra systems. For simplicity let $I$ be an ideal in $\mathbb{Z}[[x]]$ generated by polynomials, $M = \mathbb{Z}[[x]]/I$, and $I_p$ the image of $I$ in $\mathbb{F}_p[[x]]$. The dimension of $\mathbb{Q}[[x]]/I$ resp. of $\mathbb{F}_p[[x]]/I_p$, if finite, is equal to the dimension of $\mathbb{Q}[x]_{\langle x \rangle}/I$ resp. of $\mathbb{F}_p[x]_{\langle x \rangle}/I_p$. These dimensions can be computed in the localizations $\mathbb{Q}[x]_{\langle x \rangle}$ resp. $\mathbb{F}_p[x]_{\langle x \rangle}$ by computing a Gröbner or standard basis of $I$ resp. of $I_p$ w.r.t. a local monomial ordering (cf. [7]). Such algorithms are implemented e.g. in SINGULAR [4]. Usually the computations over $\mathbb{Q}$ are very time consuming or do not finish, due to extreme coefficient growths, and therefore often modular methods are used. The above corollary says that for all except finally many prime numbers $p$ we have equality $\dim_{\mathbb{Q}} \mathbb{Q}[x]_{\langle x \rangle}/I = \dim_{\mathbb{F}_p} \mathbb{F}_p[x]_{\langle x \rangle}/I_p$, and if this holds $p$ is sometimes called a "lucky" prime. This fact can also be proved by Gröbner basis methods. More interesting is however that $\dim_{\mathbb{Q}} \mathbb{Q}[x]_{\langle x\rangle}/I < \infty$ if there exists just one $p$ (lucky or not) such that $\dim_{\mathbb{F}_p} \mathbb{F}_p[x]_{\langle x\rangle}/I_p < \infty$ and that the first dimension is bounded by the latter. This was stated in [13] without proof.

## 2.4 Henselian Rings and Henselian Tensor Product

In this section we recall some basic facts about Henselian rings and introduce similarly to the complete tensor product a Henselian tensor product. For details about Henselian rings see [15] or [9]. The Henselian tensor product is needed in Sect. 2.5 for algebraically presented modules. We start with some basic facts about étale ring maps.

**Definition 2.3**

1. A ring map $\phi : A \longrightarrow B$ is called *étale* if it is flat, unramified and of finite presentation.[2]
2. $\phi$ is called *standard étale* if $B = (A[T]/F)_G$, $F, G \in A[T]$, the univariate polynomial ring, $F$ monic and $F'$ a unit in $B$.
3. $\phi$ is called *étale at* $\mathfrak{q} \in \operatorname{Spec}(B)$ if there exist $g \in B \setminus \mathfrak{q}$ such that $A \longrightarrow B_g$ is étale.

The following proposition lists some basic properties of étale maps. The results can be found in section 10.142 of [15].

**Proposition 2.4**

1. *The map $A \longrightarrow A_f$ is étale.*
2. *A standard étale map is an étale map.*
3. *The composition of étale maps is étale.*
4. *A base change of étale maps is étale.*
5. *An étale map is open.*
6. *An étale map is quasi-finite.*
7. *Given $\phi : A \longrightarrow B$ and $g_1, \ldots, g_m \in B$ generating the unit ideal[3] such that $A \longrightarrow B_{g_i}$ is étale for all $i$ then $\phi : A \longrightarrow B$ is étale.*
8. *Let $\phi : A \longrightarrow B$ be étale. Then there exist $g_1, \ldots, g_m \in B$ generating the unit ideal such that $A \longrightarrow B_{g_i}$ is standard étale for all $i$.*
9. *Let $S \subset A$ be a multiplicatively closed subset and assume that $\phi' : S^{-1}A \longrightarrow B'$ is étale. Then there exists an étale map $\phi : A \longrightarrow B$ such that $B' = S^{-1}B$ and $\phi' = S^{-1}\phi$.*

---

[2] $\phi$ is unramified if it is of finite type and the module of Kähler differentials $\Omega_{B/A}$ vanishes. $\phi$ is of finite presentation if $B \cong A[x_1, \ldots, x_n]/\langle f_1, \ldots, f_m\rangle$ as $A$-algebras.

[3] I.e., $\operatorname{Spec}(A) = \cup D(g_i)$.

10. *Let $\phi' : A/I \longrightarrow B'$ be étale for some ideal $I \subset A$. Then there exist an étale map $\phi : A \longrightarrow B$ such that $B' = B/IB$ and the obvious diagram commutes.*

**Definition 2.4** Let $A$ be a ring and $I \subset A$ an ideal. $A$ is called *Henselian with respect to $I$* if the following holds[4] (Univariate Implicit Function Theorem):
Let $F \in A[T]$, the univariate polynomial ring, such that $F(0) \in I$ and $F'(0)$ is a unit modulo $I$. Then there exists $a \in I$ such[5] that $F(a) = 0$.

Next we associate to any pair $(A, I)$, $I \subset A$ an ideal, the Henselization $A_I^h$, i.e. the "smallest" Henselian ring with respect to $I$, such that $A_I^h \subset \hat{A}_I = \varprojlim \big(A/I^n\big)$ the $I$-adic completion.

**Definition 2.5**

1. Let $A$ be a ring and $I \subset A$ an ideal. The ring

$$A_I^h = \varinjlim \big(B \mid A \longrightarrow B \text{ an étale ring map inducing } A/I \cong B/IB\big)$$

is called the *Henselization of $A$ with respect to $I$.*

2. The Henselization of $A[x]$, $A$ any ring, $x = (x_1, \ldots, x_n)$, with respect to $I = \langle x \rangle$ is denoted by $A\langle x\rangle$. We call $A\langle x\rangle$ the *ring of albegraic power series over $A$.*

The Henselization has the following properties (cf. section 15.11 and 15.12 of [15]):

**Proposition 2.5** *Let $A$ be ring and $I \subset A$ an ideal.*

1. *$A_I^h$ is Henselian with respect to $I^h = IA_I^h$ and $A/I^m = A_I^h/(I^h)^m$ for all $m$.*
2. *$A$ is Henselian with respect to $I$ if and only if $A = A_I^h$.*
3. *If $A$ is Noetherian then the canonical map $A \longrightarrow A_I^h$ is flat.*
4. *If $A$ is Noetherian then the canonical map $A_I^h \longrightarrow \hat{A}_I$ is faithfully flat and $\hat{A}_I$ is the $I^h$-adic completion of $A_I^h$.*

*Remark 2.6* The definition of the Henselization implies that $A_I^h$ is contained in the algebraic closure of $A$ in $\hat{A}_I$. If $A$ is excellent[6] then $A_I^h$ is the algebraic closure of $A$ in $\hat{A}_I$. This is even true under milder conditions, see [9]. In this situation $C\langle x\rangle$ is called the *ring of algebraic power series* of $C[[x]]$.

Next we prove a lemma which we need later in the applications.

---

[4]Note, that (similarly to the $I$-adic completion) the condition implies that $I$ is contained in the Jacobson radical of $A$. If we start with an ideal contained in the Jacobson radical then it is enough to consider monic polynomials $F$ in the definition.

[5]Note that $a$ is uniquely determined by the condition $a \in I$, [9].

[6]For the definition of excellence see 15.51 [15]).

**Lemma 2.7** *Let $A$ be a ring and $\mathfrak{p} \in \operatorname{Spec}(A)$ a prime ideal. Let $C = A_{\mathfrak{p}}$, $I = \mathfrak{p}C$ and $f_1, \ldots, f_m \in C_I^h$. Then there exists an étale map $A \longrightarrow B$ such that*

1. $f_1, \ldots, f_m \in B$,
2. *there exists a prime ideal $\mathfrak{q} \in \operatorname{Spec}(B)$ such that $\mathfrak{q} \cap A = \mathfrak{p}$.*

***Proof*** By definition we have

$$C_I^h = \varinjlim \big(D \mid C \longrightarrow D \text{ étale inducing } C/I = D/ID\big).$$

We choose $D$ from the inductive system above such that $f_1, \ldots, f_m \in D$. Since $(C, I)$ is a local ring and $C/I = D/ID$, the ideal $ID$ is a maximal ideal in $D$ and we have $ID \cap C = I$. Using Proposition 2.4 (9) for the multiplikatively closed system $S = A \setminus \mathfrak{p}$ we find an étale map $A \longrightarrow B'$ such that $D = S^{-1}B'$. This implies that $f_1, \ldots, f_m \in B'_g$ for a suitable $g \in S$. Let $B = B'_g$ and $\mathfrak{q} = ID \cap B$ then $A \longrightarrow B$ is étale having the properties 1. and 2.

Next we define the Henselization of an $A$-module $M$ with respect to an ideal $I \subset A$ similarly to the definition of the Henselization of $A$ with respect to $I$.

**Definition 2.6** Let $A$ be a ring, $I \subset A$ an ideal and $M$ an $A$-module. The module

$$M_I^h = \varinjlim \big(M \otimes_A B \mid A \longrightarrow B \text{ étale inducing } A/I = B/IB\big)$$

is called the *Henselization of $M$ with respect to $I$*.

**Lemma 2.8** $M_I^h = M \otimes_A A_I^h$.

***Proof*** The lemma follows since the direct limit commutes with the tensor product (cf. 10.75.2 [15]).

**Definition 2.7** Let $A$ be a ring, $R = A\langle x\rangle$, $B$ an $A$-algebra and $M$ an $R$-module. We define the *henselian tensor product* of $R$ and $B$ over $A$ as the ring

$$R \otimes_A^h B := \varinjlim \big(C \mid B[x] \longrightarrow C \text{ étale inducing } B = C/\langle x\rangle C\big) = B\langle x\rangle\big).$$

$$M \otimes_A^h B := \varinjlim \big(M \otimes_A C \mid B[x] \longrightarrow C \text{ étale inducing } B = C/\langle x\rangle C\big) = M \otimes_A B\langle x\rangle\big).$$

The Henselian tensor product has similar properties as the complete tensor product. Especially we obtain the following lemma.

**Lemma 2.9** *If $A\langle x\rangle^p \xrightarrow{T} A\langle x\rangle^q \to M \to 0$ is an $A\langle x\rangle$-presentation of $M$ then*

$$M \otimes_A^h B = \operatorname{coker}\big(B\langle x\rangle^p \xrightarrow{T} B\langle x\rangle^q\big).$$

*In particular $R \otimes_A^h k(\mathfrak{p}) = k(\mathfrak{p})\langle x\rangle$ for $\mathfrak{p} \in \operatorname{Spec} A$.*

**Definition 2.8** Let $A$ be a ring, $R = A\langle x\rangle$ and $M$ an $R$-module. We define for $\mathfrak{p} \in \operatorname{Spec} A$ the $R \otimes_A^h k(\mathfrak{p}) = k(\mathfrak{p})\langle x\rangle$-module

$$M^h(\mathfrak{p}) := M \otimes_A^h k(\mathfrak{p})$$

and call it the *Henselian fibre* of $M$ over $\mathfrak{p}$. Moreover, we set

$$d_{\mathfrak{p}}^h(M) := \dim_{k(\mathfrak{p})} M^h(\mathfrak{p}).$$

## 2.5 Semicontinuity for Algebraically Presented Modules

Let $A$ be Noetherian and $M$ finitely generated as $R = A[[x]]$-module. Then $M$ is finitely $R$-presented and in this section we assume that $M$ has an algebraic presentation matrix. That is, there exists a presentaion

$$R^p \xrightarrow{T} R^q \to M \to 0$$

with $T = (t_{ij})$ a $q \times p$ matrix such that $t_{ij} \in A\langle x\rangle$, $x = (x_1, \ldots, x_n)$, the ring of algebraic power series over $A$ (cf. Definition 2.5), e.g. $t_{ij} \in A[x]$. Under this assumption we shall prove the semicontinuity of $\hat{d}_{\mathfrak{b}}(M)$ for $\mathfrak{b} \in \operatorname{Spec} A$.

We set $R_0 = A\langle x\rangle$ and $M_0 = \operatorname{coker}(R_0^p \xrightarrow{T} R_0^p)$. Then using the $\langle x\rangle$-adic completion we obtain $R_0^{\wedge} = R$ and $M_0^{\wedge} = M$.

**Lemma 2.10** *Let $B \supset A$ be an $A$-algebra, $\mathfrak{b} \in \operatorname{Spec} B$ and $\mathfrak{a} = \mathfrak{b} \cap A$. Then*

$$\hat{d}_{\mathfrak{a}}(M) < \infty \Leftrightarrow \hat{d}_{\mathfrak{b}}(M \hat{\otimes}_A B) < \infty \Leftrightarrow d_{\mathfrak{b}}^h(M_0 \otimes_A^h B) < \infty$$

*and*

$$\hat{d}_{\mathfrak{a}}(M) = \hat{d}_{\mathfrak{b}}(M \hat{\otimes}_A B) = d_{\mathfrak{b}}^h(M_0 \otimes_A^h B).$$

***Proof*** $M \hat{\otimes}_A B$ is considered as an $R \hat{\otimes}_A B = B[[x]]$-module and $M_0 \otimes_A^h B$ as $R_0 \otimes_A^h B = B\langle x\rangle$-module. Therefore we have

$$\begin{aligned}
\hat{d}_{\mathfrak{a}}(M) &= dim_{k(\mathfrak{a})}(M \hat{\otimes}_A k(\mathfrak{a}))\\
\hat{d}_{\mathfrak{b}}(M \hat{\otimes}_A B) &= dim_{k(\mathfrak{b})}(M \hat{\otimes}_A B \hat{\otimes}_B k(\mathfrak{b}))\\
d_{\mathfrak{b}}^h(M_0 \otimes_A^h B) &= dim_{k(\mathfrak{b})}(M_0 \otimes_A^h B \otimes_B^h k(\mathfrak{b}))
\end{aligned}$$

and

$$\begin{aligned}
M\hat{\otimes}_A k(\mathfrak{a}) &= \operatorname{coker}(k(\mathfrak{a})[[x]]^p \xrightarrow{\bar{T}} k(\mathfrak{a})[[x]]^q) \\
M\hat{\otimes}_A B\hat{\otimes}_B k(\mathfrak{b}) &= \operatorname{coker}(k(\mathfrak{b})[[x]]^p \xrightarrow{\bar{T}} k(\mathfrak{b})[[x]]^q) \\
M_0 \otimes_A^h B \otimes_B^h k(\mathfrak{b}) &= \operatorname{coker}(k(\mathfrak{b})\langle x\rangle^p \xrightarrow{\bar{T}} k(\mathfrak{b})\langle x\rangle^q)
\end{aligned}$$

with $\bar{T} = (\bar{t}_{ij})$ and $\bar{t}_{ij}$ the induced elements in $k(\mathfrak{a})[x]$ resp. $k(\mathfrak{b})[x]$.

If $\hat{d}_{\mathfrak{b}}(M\hat{\otimes}_A B) < \infty$ there exists an $N_0$ such that $\langle x\rangle^N M\hat{\otimes}_A B\hat{\otimes}_B k(\mathfrak{b}) = 0$ for $N \geq N_0$ and hence

$$\begin{aligned}
M\hat{\otimes}_A B\hat{\otimes}_B k(\mathfrak{b}) &= \operatorname{coker}(k(\mathfrak{b})[[x]])^p/\langle x\rangle^N \xrightarrow{\bar{T}} (k(\mathfrak{b})[[x]])^q/\langle x\rangle^N \\
&= \left(\operatorname{coker}(k(\mathfrak{a})[[x]]/\langle x\rangle^N)^p \xrightarrow{\bar{T}} (k(\mathfrak{a})[[x]]/\langle x\rangle^N)^q\right) \otimes_{k(\mathfrak{a})} k(\mathfrak{b}).
\end{aligned}$$

Since this holds for every $N \geq N_0$, we obtain $\hat{d}_{\mathfrak{a}}(M) < \infty$. Similarly we can see that $\hat{d}_{\mathfrak{a}}(M) < \infty$ implies $\hat{d}_{\mathfrak{b}}(M \otimes_A B) < \infty$ and in both cases we obtain $\hat{d}_{\mathfrak{a}}(M) = \hat{d}_{\mathfrak{b}}(M \otimes_A B)$. This gives the first equality in the Lemma. Since $B\langle x\rangle/\langle x\rangle^N = B[[x]]/\langle x\rangle^N$ we get the remaining claims.

**Lemma 2.11** *Let $(A, \mathfrak{m}, \Bbbk)$ be a local Noetherian Henselian ring and $R$ a local quasi-finite (i.e. $\dim_{\Bbbk} R/\mathfrak{m}R < \infty$) and finite type $A$-algebra in the Henselian sense.*[7] *Then $R$ is a finite $A$-algebra, i.e., finitely generated as an $A$-module.*

***Proof*** This is an immediate consequence of Proposition 1.5 of [10].

**Corollary 2.5** *Let $(A, \mathfrak{m}, \Bbbk)$ be a local Noetherian Henselian ring and $R$ a local and finite type $A$-algebra in the Henselian sense. If $M$ is a finitely generated and quasi-finite (i.e. $\dim_{\Bbbk} M/\mathfrak{m}M < \infty$) $R$-module, then $M$ is a finitely generated $A$-module.*

***Proof*** Passing from $R$ to $R/\operatorname{Ann}_R(M)$ we may assume that $\operatorname{Ann}_R(M) = 0$. In this case $\dim_{\Bbbk} M/\mathfrak{m}M < \infty$ implies $\dim_{\Bbbk} R/\mathfrak{m}R < \infty$. Lemma 2.11 implies that $R$ is a finitely generated $A$-module. Since $M$ is finitely generated over $R$ it follows that $M$ is a finitely generated $A$-module.

**Theorem 2.2** *Let $A$ be a Noetherian ring, $R = A[[x]]$, $x = (x_1, \ldots, x_n)$, and $M$ a finitely generated $R$-module admitting a presentation*

$$R^p \xrightarrow{T} R^q \to M \to 0$$

[7] $R$ is an $A$-algebra of finite type in the Henselian sense if $R = A\langle t_1, \ldots, t_s\rangle$ for suitable $t_1, \ldots, t_s \in R$.

*with algebraic presentation matrix* $T = (t_{ij})$, $t_{ij} \in A[x]$ *or, more generally,* $\in A\langle x\rangle$. *Fix* $\mathfrak{p} \in \operatorname{Spec} A$ *with* $\hat{d}_{\mathfrak{p}}(M) < \infty$. *Then there is an open neighbourhood* $U$ *of* $\mathfrak{p}$ *in* $\operatorname{Spec} A$ *such that*

$$\hat{d}_{\mathfrak{q}}(M) \leq \hat{d}_{\mathfrak{p}}(M) \text{ for all } \mathfrak{q} \in U.$$

***Proof*** Recall that $R_0 = A\langle x\rangle$ is the Henselization of $A[x]$ with respect to $\langle x\rangle$ and $M_0 = \operatorname{coker}(R_0^p \xrightarrow{T} R_0^p)$. Denote by $A^h$ the henselization of the local ring $A_{\mathfrak{p}}$ with respect to its maximal ideal. We set $R^h := A^h\langle x\rangle$ and[8] $M^h := \operatorname{coker}\big((R^h)^p \xrightarrow{T} (R^h)^q\big) = M_0 \otimes^h R^h$. Then Lemma 2.10 implies $\hat{d}_{\mathfrak{p}}(M) = d^h_{\mathfrak{p}}(M^h)$ and Corollary 2.5 that $M^h$ is a finitely generated $A^h$-module ($R^h$ is a finite type $A^h$-algebra in the Henselian sense). Lemma 2.7 implies that there is an étale neighbourhood $\pi : \operatorname{Spec} B \to \operatorname{Spec} A$ of $\mathfrak{p}$ such that $M_0 \otimes^h_A B = \operatorname{coker}\big((R_0 \otimes^h_A B)^p \xrightarrow{T} (R_0 \otimes^h_A B)^q\big)$ is a finitely generated $B$-module and $M_0 \otimes^h_A B \otimes^h_B A^h = M^h$. Choose $\mathfrak{b} \in \operatorname{Spec} B$ such that $\mathfrak{b} \cap A = \mathfrak{p}$. This is possible because of Lemma 2.7. Corollary 2.5 and Lemma 2.1 imply that there is an open neigbourhood $\tilde{U} \subset \operatorname{Spec} B$ of $\mathfrak{b}$ such that for $\mathfrak{c} \in \tilde{U}$ we have $d_{\mathfrak{c}}(M_0 \otimes^h_A B) \leq d_{\mathfrak{b}}(M_0 \otimes^h_A B)$. Since $\pi$ is étale it is open (Proposition 2.4), $U := \pi(\tilde{U})$ is an open neighbourhood of $\mathfrak{p}$ in $\operatorname{Spec} A$ and for any $\mathfrak{q} \in U \cap \operatorname{Spec} A$ there exists a $\mathfrak{c} \in \tilde{U} \cap \operatorname{Spec} B$ with $\mathfrak{c} \cap A = \mathfrak{q}$. From Lemma 2.10 we obtain $\hat{d}_{\mathfrak{q}}(M) = d_{\mathfrak{c}}(M_0 \otimes^h_A B) \leq d_{\mathfrak{b}}(M_0 \otimes^h_A B) = \hat{d}_{\mathfrak{p}}(M)$.

The important property of Henselian local rings is that quasi-finite implies finite (in the sense of Corollary 2.5). Examples of Henselian local rings are quotient rings of the algebraic power series rings $A = \Bbbk\langle y\rangle/I$ over some field $\Bbbk$, and analytic $\Bbbk$-algebras.[9]

If $A$ is a complete local ring containing a field, then any finitely generated $R$-module $M$ can be polynomially presented and semicontinuity of $\hat{d}_{\mathfrak{p}}(M)$ holds, as we show now. We start with the following proposition, based on the Weierstrass division theorem.

**Proposition 2.6** *Let* $(A, \mathfrak{m}, \Bbbk)$ *be a Noetherian complete local ring containing* $\Bbbk$, $R = A[[x]]$, $x = (x_1, \ldots, x_n)$, *and* $M$ *a finitely generated* $R$*-module such that* $\dim_{\Bbbk} M/\mathfrak{m}M < \infty$. *Let* $J = \operatorname{Ann}_R(M)$. *Then there exist* $f_1, \ldots, f_s$, $s \geq n$, *with the following properties:*

1. $f_i \in A[x]$ *for all* $i$.
2. $f_{n-i+1} \in A[x_1, \ldots, x_i]$ *is a Weierstrass polynomial with respect to* $x_i$ *for* $i = 1, \ldots, n$.
3. $J = \langle f_1, \ldots, f_s\rangle A[[x]]$.

[8] Note that $R^h$ is the Henselization of $A_{\mathfrak{p}}[x]$ with respect to the maximal ideal $\langle \mathfrak{p}, x\rangle$.

[9] An analytic $\Bbbk$-algebra is the quotient $\Bbbk\{y\}/I$, $y = (y_1, \ldots, y_s)$, of a convergent power series ring over a complete real-valued field $\Bbbk$ (cf. [8]). E.g., if $\Bbbk$ is any field with the trivial valuation, then $\Bbbk\{y\} = \Bbbk[[y]]$ is the formal power series ring; if $\Bbbk \in \{\mathbb{R}, \mathbb{C}\}$, then $\Bbbk\{y\}$ is the usual convergent power series ring.

***Proof*** To prove the statements we use induction on $n$, the number of the variables $x$. The assumption implies that $\dim_{\Bbbk}(R/J + \mathfrak{m}R) < \infty$, i.e. the ideal $J + \mathfrak{m}R$ is primary to the maximal ideal $\langle x\rangle + \mathfrak{m}R$ of $R$. This implies that $x_n^b \in J + \mathfrak{m}R$ for some $b$. Therefore there exists $g \in J$, $g = x_n^b + f$ with $f \in \mathfrak{m}R$. We know by Cohen's structure theorem that $A = \Bbbk[[y]]/I$ for suitable variables $y$ and an ideal $I \subset \Bbbk[[y]]$. We can apply in the following the Weierstrass preparation and division theorem to representatives in $\Bbbk[[y, x]]$ and then take residue classes mod $I$. Obviously $g$ is $x_n$-general. The Weierstrass preparation theorem implies $g = uh$, u a unit in $R$, and $h \in A[[x_1, \dots, x_{n-1}]][x_n]$ a Weierstrass polynomial with respect to $x_n$. To simplify the notation we assume that $g$ is already a Weierstrass polynomial with respect to $x_n$. Setting $R_0 = A[[x_1, \dots, x_{n-1}]]$, the Weierstrass division theorem (cf. [8, Theorem I.1.8]) says that for any $f$ in $R$ there exist unique $h \in R$ and $r \in R_0[x_n]$ such that $f = hg + r$, $\deg_{x_n}(r) \le b - 1$. In other words, as $R_0$ modules we have

$$R = R \cdot g \oplus R_0 \cdot x_n^{b-1} \oplus R_0 \cdot x_n^{b-2} \oplus \cdots \oplus R_0. \tag{*}$$

We may thus assume that $J = \langle g_1, \dots, g_r\rangle$ with $g_1 = g$ and $g_i \in R_0[x_n]$ with $\deg_{x_n}(g_i) \le b - 1$.

If $n = 1$ then $R_0 = A$ and the claim follows from (*). If $n \ge 2$ then $M$ is a finitely generated $R_0$-module since

- $R/\langle g\rangle$ is finite over $R_0$ and
- $g \in \mathrm{Ann}_R(M)$, i.e. $M$ is a finitely generated $R/\langle g\rangle$-module.

Now let $J_0 = \mathrm{Ann}_{R_0}(M)$. By induction hypothesis there are $f_2, \dots, f_l$, $l \ge n$, such that

1. $f_i \in A[x_1, \dots, x_{n-1}]$ for all $i$.
2. $f_{n-i+1} \in A[x_1, \dots, x_i]$ is a Weierstrass polynomial with respect to $x_i$ for $i = 1, \dots, n-1$.
3. $J_0 = \langle f_2, \dots, f_l\rangle R_0$.

Now denote by $f_1$ be the remainder of the division of $g$ successively by $f_2, \dots, f_n$ and by $f_{l+i}$ the remainder of $g_i$ by $f_2, \dots, f_n$ for $i > 1$. These are polynomials in $x_1, \dots, x_n$. Then $f_1, \dots, f_s$ satisfy the conditions 1. to 3. of the proposition.

**Corollary 2.6** *Let $(A, \mathfrak{m}, \Bbbk)$ be a Noetherian complete local ring containing $\Bbbk$, $R = A[[x]]$ and $M$ a finitely generated $R$-module such that $dim_{\Bbbk} M/\mathfrak{m}M < \infty$. Then $M$ is polynomially presented.*

***Proof*** Assume $M$ has a presentation matirx $T = (g_{ij})$, $g_{ij} \in A[[x]]$. Let $J = \mathrm{Ann}_R(M)$. The assumption implies that $\dim_{\Bbbk} R/(J + \mathfrak{m}R) < \infty$. Using Proposition 2.6 we obtain that $R/J$ is a $A$-finite and $J = \langle f_1, \dots, f_s\rangle$ with $f_{n-i+1} \in A[x_1, \dots, x_i]$ a Weierstrass polynomial with respect to $x_i$ for $i = 1, \dots, n, n \le s$. This implies that $M$ has a presentation as $R/J$-module with presentation matrix $T$ having entries in $R/J$. Now we can divide representatives in $R$ of the entries of $T$ successively by the Weierstrass polynomials $f_{n-i+1}$,

$i = 1, \dots, n$. The remainders are polynomials in $A[x]$ representing the entries of $T$, which proves the claim.

Let us collect the cases for which we proved that semicontinuity of $\hat{d}_{\mathfrak{p}}(M)$ holds.

**Corollary 2.7** *Let $A$ be Noetherian and $M$ a finitely generated $R$-module. Let $\mathfrak{p} \in \operatorname{Spec} A$ and assume that one of the following conditions is satisfied:*

1. *$M$ is finitely generated as $A$-module, e.g. $\operatorname{Supp}_R(M) \subset V(\langle x \rangle)$, or*
2. *$\dim A = 1$, or*
3. *$M$ is algebraically $R$-presented, or*
4. *$(A, \mathfrak{m}, \Bbbk)$ is a complete local ring containing a field.*[10]

*Then there is an open neighbourhood $U \subset \operatorname{Spec} A$ of $\mathfrak{p}$ such that $\hat{d}_{\mathfrak{q}}(M) \le \hat{d}_{\mathfrak{p}}(M)$ for all $\mathfrak{q} \in U$. In particular, the quasi-completed-finite locus $\{\mathfrak{p} \in \operatorname{Spec} A \mid \hat{d}_{\mathfrak{p}}(M) < \infty\}$ is open.*

***Proof*** Statement 1. follows from Proposition 2.2, statement 2. from Theorem 2.1 and 3. from Theorem 2.2. Statement 4. follows from 3. and Corollary 2.6.

We do not know if semicontinuity of $\hat{d}_{\mathfrak{p}}(M)$ holds in general for $A$ Noetherian of any dimension and $M$ finitely but not necessarily algebraically presented over $R$.

*Remark 2.7* For completeness we recall cases where semicontinuity of the usual fibre dimension $d_{\mathfrak{p}}(M)$ on $\operatorname{Spec} A$ holds if $M$ is an arbitrary finitely presented $R$-module, for different (local) rings $A$ and $R$.

- $(A, \mathfrak{m}, \Bbbk)$ local Noetherian Henselian, $R$ a finite type $A$-algebra in the Henselian sense (by Corollary 2.5 and Lemma 2.1).
- $A = \Bbbk\{y\}/I$ an analytic $\Bbbk$-algebra and $R = \Bbbk\{y, x\}/J$ with $I\Bbbk\{y, x\} \subset J$ (by Greuel et al. [8, Theorem I.1.10]).
- $A$ a Noetherian complete local ring containing a field, $R = A[[x]]$. This is a special case of the previous item. We mention it, since $R$ is of the form considered in this paper.
- In the complex analytic situation with $A = \mathbb{C}\{y\}/I$, $y = (y_1, \dots, y_s)$, and $R = \mathbb{C}\{y, x\}/J$, $I\mathbb{C}\{y, x\} \subset J$, $x = (x_1, \dots, x_n)$, we have the following stronger statement: $A \to R$ induces a morphism of complex germs $f : (X, 0) \to (Y, 0)$, $(X, 0) = V(J) \subset (\mathbb{C}^n \times \mathbb{C}^s, 0)$, $(Y, 0) = V(I) \subset (\mathbb{C}^s, 0)$ and $f$ the projection. For a sufficiently small representative $f : X \to Y$, $M$ induces a coherent $\mathcal{O}_X$-module $\mathcal{F}$ on $X$ and $d_{\mathfrak{m}}(M) < \infty$, $\mathfrak{m}$ the maximal ideal of $A = \mathcal{O}_{Y,0}$, means that the fibre dimension over $0 \in Y$ is finite, i.e. $d_0(\mathcal{F}) := \dim_{\mathbb{C}} \mathcal{F}_0/\mathfrak{m}\mathcal{F}_0 < \infty$. Then, for sufficiently small suitable $X$ and $Y$, is $f|\operatorname{Supp}\mathcal{F}$ is a finite morphism and $f_*\mathcal{F}$ is a coherent $\mathcal{O}_Y$-module (cf. [8, Theorem I.1.67]). It follows that

$$d_y(\mathcal{F}) := \dim_{\mathbb{C}} f_*\mathcal{F} \otimes_{\mathcal{O}_{Y,y}} \mathbb{C} = \sum_{z \in f^{-1}(y)} \dim_{\mathbb{C}} \mathcal{F}_z/\mathfrak{m}_y\mathcal{F}_z$$

[10]By Cohen's structure theorem this is equivalent to $A \cong \Bbbk[[y]]/I$, $\Bbbk$ a field.

is upper semicontinuous at $0 \in Y$, i.e. 0 has an open neighbourhood $U \subset Y$ such that $d_y(\mathcal{F}) \leq d_0(\mathcal{F})$ for all $y \in U$.

In the above cases $M$ is finite over $A$ if it is quasi-finite over $A$ and hence semicontinuity of $d_{\mathfrak{p}}(M)$ holds by Lemma 2.1. Example 2.3 shows that for $A$ an affine ring and $R = A[[x]]$ semicontinuity of $d_{\mathfrak{p}}(M)$ does in general not even hold for polynomially presented modules.

## 2.6 Related Results

Instead of families of power series let us now consider families of algebras of finite type, a situation which is quite common in algebraic geometry. We treat the more general case of families of modules.

Let $A$ be a ring, $R = A[x]/I$ of finite type over $A$ and $M$ a finitely presented $R$-module. $M$ is called *quasi-finite at* $\mathfrak{n} \in \operatorname{Spec} R$ over $A$ if $\dim_{k(\mathfrak{p})} M_{\mathfrak{n}}/\mathfrak{p}M_{\mathfrak{n}} < \infty$ with $\mathfrak{p} \in \operatorname{Spec} A$ lying under $\mathfrak{n}$. $M$ is called *quasi-finite over* $\mathfrak{p} \in \operatorname{Spec} A$ if it is quasi-finite at all primes $\mathfrak{n} \in \operatorname{Spec} R$ lying over $\mathfrak{p}$, and $M$ is *quasi-finite over* $A$ if it is quasi-finite at all primes $\mathfrak{n} \in \operatorname{Spec} R$. The following proposition is a generalization of results from [15], where the case of ring maps is treated.

**Proposition 2.7** *Let $A$ be a ring, $R$ an $A$-algebra of finite type over $A$, $M$ a finitely presented $R$-module and $f : \operatorname{Spec} R \to \operatorname{Spec} A$ the induced map of schemes.*

1. *The following are equivalent:*

   a. *$M$ is quasi-finite over $A$,*
   b. *$d_{\mathfrak{p}}(M) = \dim_{k(\mathfrak{p})} M(\mathfrak{p}) = \sum_{\mathfrak{n} \in f^{-1}(\mathfrak{p})} \dim_{k(\mathfrak{p})} M_{\mathfrak{n}}/\mathfrak{p}M_{\mathfrak{n}} < \infty \ \forall \mathfrak{p} \in \operatorname{Spec} A$,*
   c. *The induced map $A \to S := R/\operatorname{Ann}_R(M)$ is quasi-finite.*

2. *(Zariski's main theorem for modules). The* quasi-finite locus of $M$

$$\{\mathfrak{n} \in \operatorname{Spec} R \mid M \textit{ is quasi-finite at } \mathfrak{n}\}$$

   *is open in* $\operatorname{Spec} R$.

***Proof***

1. $(a) \Rightarrow (b)$: We have to show that the support of $M(\mathfrak{p})$ is finite. By [15, Lemma 29.19.10], if $R = A[x]/I$ is a ring of finite type and quasi-finite over $A$, the induced map $f : \operatorname{Spec} R \to \operatorname{Spec} A$ has finite fibres $R(\mathfrak{p}) = R \otimes_A k(\mathfrak{p}) = k(\mathfrak{p})[x]/I(\mathfrak{p})$. It follows that 2. holds if $M$ is a ring of finite type over $A$.

   In the general case let $I = \operatorname{Ann}_R(M)$. Then $S = R/I$ is of finite type over $A$, $M$ is finitely presented over $S$ and hence $\operatorname{Supp}(S) = \operatorname{Supp}(M)$. Moreover, let $J(\mathfrak{p})$ be the annihilator of the finitely generated $R(\mathfrak{p})$-module $M(\mathfrak{p})$ satisfying $V(J(\mathfrak{p})) = \operatorname{Supp}(M(\mathfrak{p}))$. Since $R(\mathfrak{p})$ is Noetherian and $\dim_{k(\mathfrak{p})} M_{\mathfrak{n}}(\mathfrak{p}) < \infty$ by assumption, we have $\mathfrak{n}^N M_{\mathfrak{n}}(\mathfrak{p}) = 0$ for some $N$ by Nakayama's lemma. Hence $\mathfrak{n}^N \subset J(\mathfrak{p})R_\mathfrak{n}(\mathfrak{p})$ and $\dim_{k(\mathfrak{p})} R_\mathfrak{n}\mathrm{fl}(\mathfrak{p})/J(\mathfrak{p})R_\mathfrak{n}\mathrm{fl}(\mathfrak{p}) < \infty$. In general, the annihilator is not compatible with base change, hence $I(\mathfrak{p})$ is in general different from $J(\mathfrak{p})$. But for a finitely presented module the annihilator coincides up to radical with a Fitting ideal, and Fitting ideals are compatible with base change. It follows that $\sqrt{J(\mathfrak{p})R_\mathfrak{n}(\mathfrak{p})} = \sqrt{I(\mathfrak{p})R_\mathfrak{n}(\mathfrak{p})}$ and therefore $\dim_{k(\mathfrak{p})} R_\mathfrak{n}(\mathfrak{p})/I(\mathfrak{p})R_\mathfrak{n}(\mathfrak{p}) = \dim_{k(\mathfrak{p})} S_\mathfrak{n}(\mathfrak{p}) < \infty$, which means that $A \to S$ is quasi-finite at $\mathfrak{n}$ ([15, Definition 10.121.3]). Since this holds for each $\mathfrak{n} \in \operatorname{Spec} R$, the map $A \to S$ is quasi-finite and by [15, Lemma 29.19.10] the set $\operatorname{Supp}(M(\mathfrak{p})) = \operatorname{Supp}(S(\mathfrak{p}))$ is finite.

$(b) \Rightarrow (c)$: If $d_\mathfrak{p}(M) < \infty$ for all $\mathfrak{p} \in \operatorname{Spec} A$, then $\dim_{k(\mathfrak{p})} M_\mathfrak{n}/\mathfrak{p}M_\mathfrak{n} < \infty$ for all $\mathfrak{n}$ and $\mathfrak{p}$ under $\mathfrak{n}$. Then $A \to S$ is quasi-finite by the previous step.

$(c) \Rightarrow (a)$: If $A \to S$ is quasi-finite then $\dim_{k(\mathfrak{p})} S_\mathfrak{n}(\mathfrak{p}) < \infty$ for all $\mathfrak{p}$ and $\mathfrak{n}$ over $\mathfrak{p}$. Since $M_\mathfrak{n}(\mathfrak{p})$ is finitely presented as $S_\mathfrak{n}(\mathfrak{p})$-module, $\dim_{k(\mathfrak{p})} M_\mathfrak{n}(\mathfrak{p}) < \infty$ for all $\mathfrak{p}$ and $\mathfrak{n}$ over $\mathfrak{p}$ and $M$ is quasi-finite over $A$.

2. In the first and third step of 1. we proved $\dim_{k(\mathfrak{p})} S_\mathfrak{n}(\mathfrak{p}) < \infty$ if and only if $\dim_{k(\mathfrak{p})} M_\mathfrak{n}(\mathfrak{p}) < \infty$, and hence $M$ is quasi-finite at $\mathfrak{n}$ iff $A \to S$ is quasi-finite at $\mathfrak{n}$. It follows from a version of Zarisk's main theorem as proved in [15, Lemma 10.122.13] that the set $\{\mathfrak{n} \in \operatorname{Spec} S \mid A \to S \text{ is quasi-finite at } \mathfrak{n}\}$ is open in $\operatorname{Spec} S$ and thus of the form $U \cap S$ with $U$ open in $\operatorname{Spec} R$. If $\mathfrak{n} \in V = \operatorname{Spec} R \setminus \operatorname{Spec} S$ then $M_\mathfrak{n}(\mathfrak{p}) = 0$, hence $M$ is quasi-finite at $\mathfrak{n} \in V$. Thus, the quasi-finite locus of $M$ is the open set $U \cup V$.

*Example 2.5* In the situation of Proposition 2.7, although the quasi-finite locus of $M$ is open in $\operatorname{Spec} R$, we cannot expect semicontinuity of $d_\mathfrak{p}(M)$ on $\operatorname{Spec} A$. We give an example showing that the vanishing locus of $d_\mathfrak{p}(M)$ is not open in $\operatorname{Spec} A$: Let $K$ be an algebraically closed field, $A = K[y]$, $R = A[x]$ and $M = R/\langle xy - 1\rangle$. Then $M$ is quasi-finite over $A$ but $d_\mathfrak{p}(M)$ is not semicontinuous since $d_\mathfrak{p}(M) = 0$ if $\mathfrak{p} = \langle y\rangle$ and $d_\mathfrak{p}(M) = 1$ otherwise.

By Corollary 2.7.4, semicontinuity of $\hat{d}_\mathfrak{p}(M)$ holds for $M$ a finitely generated $A[[x]]$-module if $(A, \mathfrak{m})$ is a complete Noetherian local ring containing a field under the assumption that $\operatorname{Supp}_A(M) = A$ but $\operatorname{Supp}_R(M) \not\subset V(\langle x\rangle)$ (the difficult case). The question arose whether completeness was necessary. The following example shows that this is not the case.

*Example 2.6* We give an example of a non-complete local ring $(A, \mathfrak{m})$ and a finitely presented $R = A[[x]]$-module $M$ which is also a finitely presented $A$-module with $\operatorname{Supp}_R(M)$ being not contained in $V(\langle x\rangle)$ and $\operatorname{Ann}_A(M) = 0$.
Let $\Bbbk$ be a field and $t_1, t_2$ independent variables. Let $A = \Bbbk[t_1]_{\langle t_1\rangle}[[t_2]]$ and $R = A[[x]]$ with x a single variable. The ring $A$ is local with maximal ideal $\mathfrak{m} = \langle t_1, t_2\rangle A$ and not complete. Let $M = R/\langle x - t_2\rangle$. Then $\operatorname{Ann}_R(M) = \langle x - t_2\rangle$ and $\operatorname{Supp}_R(M) = V(\langle x - t_2\rangle \not\subset V(\langle x\rangle)$ and $\operatorname{Ann}_A(M) = \langle x - t_2\rangle \cap A = (0)$. Since $M$ is polynomially presented, $\hat{d}_\mathfrak{p}(M)$ is semicontinuous.

Let $M$ be a finitely presented $R = A[[x]]$-module and also finitely presented as an $A$-module with $A$ Noetherian. In Proposition 2.2 we have shown the semicontinuity of $d_\mathfrak{p}(M)$ and $\hat{d}_\mathfrak{p}(M)$ on Spec $A$, as well as the inequality $\hat{d}_\mathfrak{p}(M) \le d_\mathfrak{p}(M)$. The following example shows that $\hat{d}_\mathfrak{p}(M) < d_\mathfrak{p}(M)$ may happen.

*Example 2.7* A modification of Example 2.3 shows that $\hat{d}_\mathfrak{p}(M) \neq d_\mathfrak{p}(M)$ may happen for $A$ a power series ring. Let $A = K[[t]]$, $K$ a field, $R = A[[x]]$, and let $M = R/\langle t - x\rangle \cong K[[t]]$. For the two prime ideals $\langle 0\rangle$ and $\langle t\rangle$ of $A$ we get:

$k(\langle 0\rangle) = K((t))$, $k(\langle t\rangle) = K$, $M(\langle 0\rangle) \cong K[[t]] \otimes_{K[[t]]} K((t)) = K((t))$, $M(\langle t\rangle) \cong K$ and $d_{\langle 0\rangle}(M) = d_{\langle t\rangle}(M) = 1$. Hence $d_\mathfrak{p}(M)$ is semicontinuous (even continuous) on Spec $A$ as predicted in Remark 2.7, third item.

$\hat{M}(\langle 0\rangle) \cong K((t))[[x]]/\langle t - x\rangle = 0$, $\hat{M}(\langle t\rangle) \cong K$ and $\hat{d}_{\langle 0\rangle}(M) = 0$, $\hat{d}_{\langle t\rangle}(M) = 1$. Hence $d_\mathfrak{p}(M)$ is semicontinuous on Spec $A$ as predicted by Corollary 2.7. Note that $M$ is finitely presented as $A$-module and we have $\hat{d}_{\langle 0\rangle}(M) < d_{\langle 0\rangle}(M)$.

## 3 Singularity Invariants

### 3.1 *Isolated Singularities and Flatness*

Recall that a local Noetherian ring $(A, \mathfrak{m})$ is said to be *regular* if $\mathfrak{m}$ can be generated by dim $A$ elements. A Noetherian ring $A$ is said to be regular if the local ring $A_\mathfrak{p}$ is regular for all $\mathfrak{p} \in$ Spec $A$. For arbitrary Noetherian rings the *regular locus* Reg $A :=$ $\{\mathfrak{p} \in \text{Spec } A \mid A_\mathfrak{p} \text{ is regular}\}$ need not be open in Spec $A$. However, Reg $A$ is open if $A$ is a complete Noetherian local ring ([12, Corollary of Theorem 30.10]) and the *non-regular locus* $\{\mathfrak{p} \in \text{Spec } A \mid A_\mathfrak{p} \text{ is not regular}\}$ is closed.

However, in our situation of families of power series, the notion of formal smoothness is more appropriate than that of regularity. Formal smoothness is a relative notion and refers to a morphism, while regularity is an absolute property of the ring. The notions are related as follows. Let $(A, \mathfrak{m})$ be a local ring containing a field $\Bbbk$. If $A$ is formally smooth over $\Bbbk$ (w.r.t. the $\mathfrak{m}$-adic topology) then $A$ is regular and the converse holds if the residue field $A/\mathfrak{m}$ is separable over $\Bbbk$ (see Remark 3.1). Hence formal smoothness of $A$ over $\Bbbk$ coincides with regularity if $\Bbbk$ is a perfect field. The notions also coincide for arbitrary $\Bbbk$ if $A$ is the quotient ring of a formal power series ring over $\Bbbk$ by an ideal (cf. Lemma 3.2).

We recall now basic facts about formal smoothness. For details and proofs see [12] and [11].

**Definition 3.1** Let $A$ be a ring, $B$ an $A$-algebra defined by $\phi : A \longrightarrow B$ and $I$ an ideal in $B$. The $A$-algebra $B$ is called *formally smooth with respect to the $I$-adic topology* (for short $B$ *is $I$-smooth over* $A$) if for any $A$-algebra $C$ and any continuous[11] $A$-algebra homomorphism $u : B \longrightarrow C/N$, $N$ an ideal in $C$ with

[11]Here we consider $B$ with the $I$-adic topology and $C/N$ with the discrete topology; $u$ is continuous if $u(I^m) = 0$ for some $m$.

$N^2 = 0$, there exist $\sigma : B \longrightarrow C$ such that $\pi\sigma = u$.

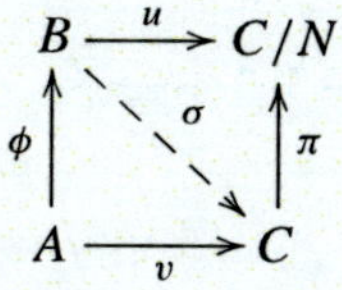


If $I = 0$ then $B$ is called a *formally smooth* $A$-algebra.

*Remark 3.1*

1. A formally smooth map of finite presentation is smooth ([15] Proposition 10.137.13).
2. $A[x]$, $x = (x_1, \ldots, x_n)$, is smooth over $A$ ([15] Lemma 10.137.4).
3. $A[[x]]$ is $\langle x \rangle$-smooth over $A$ ([12] page 215).
4. Let $(A, \mathfrak{m})$ be a local ring containing a field $\mathbb{k}$.

   a. $A$ is $\mathfrak{m}$-smooth over $\mathbb{k}$ iff $A$ is geometrically regular, i.e. $A \otimes_{\mathbb{k}} \mathbb{k}'$ is a regular ring for every finite extension field $\mathbb{k}'$ of $\mathbb{k}$ ([12, Theorem 28.7 ]).
   b. Assume that $A/\mathfrak{m}$ is separable over $\mathbb{k}$. Then $A$ is $\mathfrak{m}$-smooth over $\mathbb{k}$ iff $A$ is regular ([12] Lemma 1, page 216).

   We now generalize example 1 on page 215 of [12].

**Lemma 3.1** *Let $A$ be a ring, $B$ a $A$-algebra, $I$ an ideal in $B$ and $\widehat{B}$ the $I$-adic completion of $B$. $\phi : A \longrightarrow B$ is $I$-smooth iff $\hat{\phi} : A \longrightarrow \widehat{B}$ is $I\widehat{B}$-smooth.*

***Proof*** Assume that $B$ is $I$-smooth over $A$ and consider the following commutative diagram:

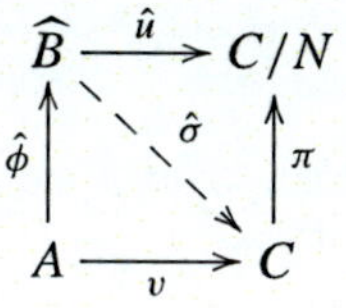


with $N^2 = 0$. We have to prove that there exists $\hat{\sigma}$ such that $\pi\hat{\sigma} = \hat{u}$. Since $\hat{u}$ is continuous there exist $m$ such that $\hat{u}(I^n\widehat{B}) = 0$. Let $i : B \longrightarrow \widehat{B}$ be the cononical map such that $\hat{\phi} = i\phi$. The $I$-smoothness of $B$ implies that there exists $\sigma : B \longrightarrow C$ such that $\sigma\pi = \hat{u}i$. $\hat{u}(I^n\widehat{B}) = 0$ implies $\sigma(I^m) \subset N$. Since $N^2 = 0$ we obtain

$\sigma(I^{2m}) = 0$. We obtain the following commutative diagram:

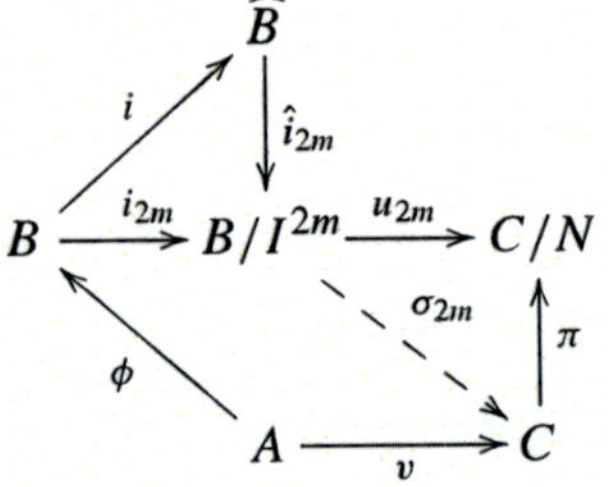


Now we define $\hat{\sigma} = \sigma_{2m}\hat{i}_{2m}$. This proves that $\hat{\phi} : A \longrightarrow \widehat{B}$ is $I\widehat{B}$-smooth.

Now assume that $\hat{\phi} : A \longrightarrow \widehat{B}$ is $I\widehat{B}$-smooth. Consider the following commutative diagram:

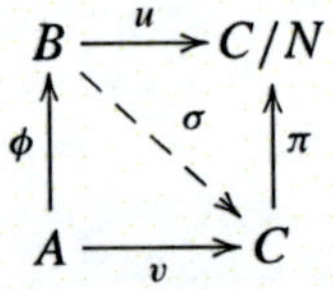


with $N^2 = 0$. We have to prove that there exists $\sigma$ such that $\pi\sigma = u$. Since $\hat{\phi} : A \longrightarrow \widehat{B}$ is $I\widehat{B}$-smooth there exists $\hat{\sigma} : \widehat{B} \longrightarrow C$ with $\pi\hat{\sigma} = \hat{u}$. Now we define $\sigma = \hat{\sigma} i$ and obtain $\pi\sigma = u$.

The following important theorem is due to Grothendieck ([12] Theorem 28.9).

**Theorem 3.1** *Let $(A, \mathfrak{m})$ be a local ring and $(B, \mathfrak{n})$ a local $A$-algebra. Let $\bar{B} = B/\mathfrak{m}B$ and $\bar{\mathfrak{n}} = \mathfrak{n}/\mathfrak{m}B$. Then $B$ is $\mathfrak{n}$-smooth over $A$ iff $\bar{B}$ is $\bar{\mathfrak{n}}$-smooth over $A/\mathfrak{m}$ and $B$ is flat over $A$.*

**Definition 3.2** Let $A$ be a ring and $B$ an $A$-algebra defined by $\phi : A \longrightarrow B$. We define the *smooth locus* of $\phi$ by

$$\mathrm{Sm}(\phi) := \{P \in \mathrm{Spec}(B) | A_{\phi^{-1}(P)} \longrightarrow B_P \text{ is } P\text{-smooth}\}.$$

and the *singular locus* of $\phi$ by

$$\mathrm{Sing}(\phi) := \mathrm{Spec}(B) \setminus \mathrm{Sm}(\phi).$$

*Remark 3.2* Let $A$ be a ring and $B$ an $A$-algebra defined by $\phi : A \longrightarrow B$. The Theorem 3.1 of Grothendieck implies that

$$\begin{aligned}\mathrm{Sm}(\phi) = \{P \in \mathrm{Spec}(B) \mid\ & A_Q \longrightarrow B_P,\ Q = \phi^{-1}(P), \text{ is flat} \\ & \text{and } B_P/QB_P \text{ is } PB_P/QB_P - \text{smooth over } k(Q)\}.\end{aligned}$$

Now let $\mathbb{k}$ be a field, $\mathbb{k}[[x]]$, $x = (x_1, \cdots, x_n)$, the formal power series ring over $\mathbb{k}$ and $I$ an ideal in $\langle x\rangle\mathbb{k}[[x]]$. If $I$ is generated by $f_1, \ldots, f_m$ we denote by $Jac(I)$ the Jacobian matrix $(\partial f_j/\partial x_i)$ and by $I_k(Jac(I))$ the ideal generated by the $k \times k$-minors of $Jac(I)$ (which is independent of the chosen generators $f_j$). The following lemma gives equivalent conditions for the maximal ideal $\langle x\rangle \in B = \mathbb{k}[[x]]/I$ to be contained in the smooth locus $\mathrm{Sm}(\phi)$ of the map $\phi : \mathbb{k} \to B$ (Remark 3.2).

**Lemma 3.2** *If* $\dim \mathbb{k}[[x]]/I = d$ *the following are equivalent.*

1. $\mathbb{k}[[x]]/I$ *is* $\langle x\rangle$*-smooth over* $\mathbb{k}$.
2. $\mathbb{k}[[x]]/I$ *is regular.*
3. $I_d(Jac(I)) = \mathbb{k}[[x]]$ (Jacobian criterion).
4. $\mathbb{k}[[x]]/I \cong \mathbb{k}[[y_1, \ldots, y_d]]$.

***Proof*** The equivalence of 1. and 2. follows from [12, Lemma 1, p. 216], the equivalence of 3. and 4. is the inverse mapping theorem for formal power series.[12]
Obviously 4. implies 2. From [12, Theorem 29.7, p. 228 in] we deduce that 2. implies 4.

*Remark 3.3* Part of the lemma can be generalized by extending the proof of Theorem 30.3 in [12] as follows:
Let $P$ be a prime ideal in $\mathbb{k}[[x]]$ containing $I = \langle f_1, \ldots, f_m\rangle$ and $\mathfrak{m}$ the maximal ideal of $A = \mathbb{k}[[x]]_P/I\mathbb{k}[[x]]_P$. Then $I_d(Jac(I)) = \mathbb{k}[[x]]_P$ implies that $A$ is $\mathfrak{m}$-smooth over $\mathbb{k}$ (or geometric regular by Remark 3.1).

We use the Jacobian criterion to define the singular locus of ideals in power series rings over a field.

**Definition 3.3**

1. If $B = \mathbb{k}[[x]]/I$ is pure $d$-dimensional (i.e. $\dim B/P = d$ for all minimal primes $P \in \operatorname{Spec} B$) we define the *singular locus of* $B$ *(or of* $I$*)* as

$$\mathrm{Sing}(B) = V(I + I_d(Jac(I))).$$

2. If $B$ is not pure dimensional we consider the minimal primes $P_1, \ldots, P_r$ of $B$. Then $B/P_i$ is pure dimensional and we define the singular locus of $B$ as

$$\mathrm{Sing}(B) = \bigcup_{i=1}^{r} \mathrm{Sing}(B/P_i) \cup \bigcup_{i\neq j} V(P_i) \cap V(P_j),$$

   which is a closed subscheme of $\operatorname{Spec} B$. The points in $\operatorname{Spec} B \setminus \mathrm{Sing}(B)$ are called *non-singular* points of $B$.
3. We say that $\mathbb{k}[[x]]/I$ (or $I$) has an *isolated singularity* (at 0) if the maximal ideal $\langle x\rangle$ is an isolated point of $\mathrm{Sing}(\mathbb{k}[[x]]/I)$ or if $\langle x\rangle$ is a non-singular point.

[12]Given $f_1, \ldots, f_n \in \mathbb{k}[[x_1, \ldots, x_n]]$ then $det(\frac{\partial f_i}{\partial x_j})$ is a unit iff $\mathbb{k}[[x_1, \ldots, x_n]] = \mathbb{k}[[f_1, \ldots, f_n]]$ ([8, Theorem I.1.18]).

*Remark 3.4* Let $i : \mathbb{k} \longrightarrow B$ be the obvious inclusion. Then $\text{Sing}(B) = \text{Sing}(i)$ whenever $\mathbb{k}$ is perfect, but not in general. A counterexample is given for instance, by letting $\text{char}(\mathbb{k}) = p > 0$, $a \in \mathbb{k} \setminus \mathbb{k}^p$ and $B = \mathbb{k}[[x_1, x_2]]/\langle x_1^p - ax_2^p \rangle$.

Note that $\text{Sing}(B)$ carries a natural scheme structure given by the Fitting ideal $I + I_d(Jac(I)) \subset \mathbb{k}[[x]]$ if $B$ is pure $d$-dimensional. In general we endow $\text{Sing}(B)$ with its reduced structure.

Now let us consider families. Let $A$ be a Noetherian ring, $F_1, \ldots, F_m \in \langle x \rangle A[[x]]$, $I \subset A[[x]]$ the ideal generated by $F_1, \ldots, F_m$ and set $B := A[[x]]/I$. We describe now the smooth locus of the map $\phi : A \to B$ along the section $\sigma : \text{Spec}\, A \to \text{Spec}\, B$, $\mathfrak{p} \mapsto \mathfrak{n}_\mathfrak{p} = \langle x, \mathfrak{p} \rangle$, of $\text{Spec}\, \phi$.

For $\mathfrak{p} \in \text{Spec}\, A$ denote by $F_i(\mathfrak{p})$ the image of $F_i$ in $k(\mathfrak{p})[[x]]$. Note that $F_1(\mathfrak{p}), \ldots, F_m(\mathfrak{p})$ generate the ideal $\hat{I}(\mathfrak{p}) \subset k(\mathfrak{p})[[x]]$, and that we have (by Lemma 2.3.3) for the *completed fibre* of $\phi$ over $\mathfrak{p}$

$$\hat{B}(\mathfrak{p}) = (B_{\mathfrak{n}_\mathfrak{p}}/\mathfrak{p}B_{\mathfrak{n}_\mathfrak{p}})^\wedge = k(\mathfrak{p})[[x]]/\hat{I}(\mathfrak{p}).$$

The maximal ideals of the local rings of the fibre $B_{\mathfrak{n}_\mathfrak{p}}/\mathfrak{p}B_{\mathfrak{n}_\mathfrak{p}}$ and the completed fibre $\hat{B}(\mathfrak{p})$ are generated by $\mathfrak{n}_\mathfrak{p}/\mathfrak{p} = \langle x \rangle$. Assume that $\phi : A \to B$ is flat. Then the theorem of Grothendieck says

$$\begin{aligned} \mathfrak{n}_\mathfrak{p} \in \text{Sm}(\phi) &\Leftrightarrow B_{\mathfrak{n}_\mathfrak{p}} \text{ is } \mathfrak{n}_\mathfrak{p}\text{-smooth over } A_\mathfrak{p} \\ &\Leftrightarrow B_{\mathfrak{n}_\mathfrak{p}}/\mathfrak{p}B_{\mathfrak{n}_\mathfrak{p}} \text{ is } \langle x \rangle\text{-smooth over } k(\mathfrak{p}). \end{aligned}$$

**Lemma 3.3** *With the above notations assume that $\phi : A \to B$ is flat. Denote by*

$$\text{Sing}_\sigma(\phi) := \{\mathfrak{n}_\mathfrak{p} \in \text{Spec}\, B \mid B_{\mathfrak{n}_\mathfrak{p}} \textit{ is not } \mathfrak{n}_\mathfrak{p}\textit{-smooth over } A_\mathfrak{p}\}$$

*the* singular locus of $\phi$ along the section $\sigma$. *Then*

$$\text{Sing}_\sigma(\phi) = \{\mathfrak{n}_\mathfrak{p} \in \text{Spec}\, B \mid \hat{B}(\mathfrak{p}) \textit{ is not regular}\}.$$

***Proof*** $B_{\mathfrak{n}_\mathfrak{p}}/\mathfrak{p}B_{\mathfrak{n}_\mathfrak{p}}$ is $\langle x \rangle$-smooth over $k(\mathfrak{p})$ iff $(B_{\mathfrak{n}_\mathfrak{p}}/\mathfrak{p}B_{\mathfrak{n}_\mathfrak{p}})^\wedge = k(\mathfrak{p})[[x]]/\hat{I}(\mathfrak{p})$ is $\langle x \rangle$-smooth over $k(\mathfrak{p})$ by Lemma 3.1. The claim follows from Lemma 3.2.

Since we assumed $B$ to be flat over $A$, we have $\dim \hat{B}(\mathfrak{p}) = \dim B_{\mathfrak{n}_\mathfrak{p}} - \dim A_\mathfrak{p}$ (by Matsumura [12, Theorem 15.1]). If $\phi$ is of pure relative dimension $d$ (i.e. $\hat{B}(\mathfrak{p})$ is pure $d$-dimensional for all $\mathfrak{p}$) then Lemma 3.2 implies

$$\text{Sing}_\sigma(\phi) = \{\mathfrak{n}_\mathfrak{p} \in \text{Spec}\, B \mid \hat{I}_d(Jac(I))(\mathfrak{p}) \text{ is a proper ideal of } k(\mathfrak{p})[[x]]\},$$

where $Jac(I)$ is the Jacobian matrix $(\partial F_j/\partial x_i)$ and $I_d(Jac(I)) \subset A[[x]]$ the ideal defined by the $d \times d$-minors.

## 3.2 Milnor Number and Tjurina Number of Hypersurface Singularities

Let $\mathbb{k}$ be a field and $f \in \mathbb{k}[[x]], x = (x_1, \cdots, x_n)$ a formal power series. The most important invaraints are the *Milnor number* $\mu(f)$ and the *Tjurina number* $\tau(f)$, defined as

$$\mu(f) = \dim_{\mathbb{k}} \mathbb{k}[[x]]/j(f),$$
$$\tau(f) = \dim_{\mathbb{k}} \mathbb{k}[[x]]/\langle f, j(f)\rangle,$$

where $j(f) = \langle \partial f/\partial x_1, \ldots, \partial f/\partial x_n\rangle$ is the *Jacobian ideal* of $f$. We say that $f$ has an *isolated critical point* (at 0) resp. an *isolated singularity* (at 0) if $\mu(f) < \infty$ resp. $\tau(f) < \infty$. Note that $\tau(f) < \infty$ iff $\mathbb{k}[[x]]/\langle f\rangle$ has an isolated singularity in the sense of Definition 3.3.

*Remark 3.5* Let char($\mathbb{k}$) $= 0$. It is proved in [3, Theorem 2] that for $f \in \langle x\rangle$, $\mu(f) < \infty \Leftrightarrow \tau(f) < \infty$ but it is easy to see that this is not true in positive characteristic. We have always $\tau(f) \leq \mu(f)$ and $\tau(f) = \mu(f) \Leftrightarrow f \in j(f)$. If $\mathbb{k} = \mathbb{C}$ and if $f \in \langle x\rangle^2$ has an isolated singularity, this is equivalent to $f$ being quasi homogeneous by a theorem of K. Saito (see [14]). His proof generalises to any algebraically closed field of characteristic zero (cf. [2, Theorem 2.1]).

We consider now families of singularities. Let $A$ be a Noetherian ring and $F \in R = A[[x]]$. Set

$$j(F) := \langle \partial F/\partial x_1, \ldots, \partial F/\partial x_n\rangle$$

and for $\mathfrak{p} \in \operatorname{Spec} A$ denote by $F(\mathfrak{p})$ the image of $F$ in $k(\mathfrak{p})[[x]]$. Then the Milnor number

$$\mu(F(\mathfrak{p})) = \dim_{k(\mathfrak{p})} k(\mathfrak{p})[[x]]/j(F(\mathfrak{p}))$$

and the Tjurina number

$$\tau(F(\mathfrak{p})) = \dim_{k(\mathfrak{p})} k(\mathfrak{p})[[x]]/\langle F(\mathfrak{p}), j(F(\mathfrak{p}))\rangle$$

are defined, and we deduce now the semicontinuity of $\mu(F(\mathfrak{p}))$ and $\tau(F(\mathfrak{p}))$.

**Proposition 3.1** *Let $A$ be Noetherian, $F \in R = A[[x]]$ and $\mathfrak{p} \in \operatorname{Spec} A$. Assume that $V(j(F)) \subset V(\langle x\rangle)$ resp. $V(\langle F, j(F)\rangle) \subset V(\langle x\rangle)$ (as sets), or $\dim A = 1$, or $F \in A[x]$, or $A$ is a complete local ring containing a field. Then $\mu(F(\mathfrak{p}))$ and $\tau(F(\mathfrak{p}))$ are semicontinuous at $\mathfrak{p} \in \operatorname{Spec} A$.*

***Proof*** Set $M = R/j(F)$ resp. $M = R/\langle F, j(F)\rangle$, then $\operatorname{Supp}_R(M) = V(j(F))$ resp. $\operatorname{Supp}_R(M) = V(\langle F, j(F)\rangle)$. Using Lemma 2.3 we get $\hat{d}_{\mathfrak{q}}(M) = \mu(F(\mathfrak{q}))$ resp. $\hat{d}_{\mathfrak{q}}(M) = \tau(F(\mathfrak{q}))$ for $\mathfrak{q} \in \operatorname{Spec} A$. The result follows from Corollary 2.7.

**Corollary 3.1** *Let $F \in \mathbb{Z}[x]$, $p \in \mathbb{Z}$ a prime number and denote by $F_p$ the image of $F$ in $\mathbb{F}_p[[x]]$ and by $F_0$ the image of $F$ in $\mathbb{Q}[[x]]$.*

*If $\mu(F_p)$ is finite, then $\mu(F_p) \geq \mu(F_0)$ and $\mu(F_p) \geq \mu(F_q)$ for all except finitely many prime numbers $q \in \mathbb{Z}$. In particular, if $\mu(F_p)$ is finite for some p then $\mu(F_0)$ is finite.*

*If $\mu(F_0)$ is finite, then $\mu(F_0) \geq \mu(F_q)$ (and hence "=") for all except finitely many prime numbers $q \in \mathbb{Z}$.*

*The same holds for the Tjurina number.*

*Example 3.1* We illustrate the corollary by a simple example. Let $F = F_0 = x^p + x^{(p+1)} + y^q$ with $p, q$ prime numbers. Then $\mu(F_0) = (p-1)(q-1)$, $\mu(F_p) = p(q-1) \geq \mu(F_0)$ while $\mu(F_q) = \infty$. Moreover, for any prime number $r \neq p, q$ we have $\mu(F_r) = \mu(F_0)$.

### 3.3 Determinacy of Ideals

Let $I$ be a proper ideal of $\Bbbk[[x]]$ and $f_1, \ldots, f_m$ a minimal set of generators of $I$. $I$ is called *contact k-determined* if for every ideal $J$ of $\Bbbk[[x]]$ that can be generated by $m$ elements $g_1, \ldots, g_m$ with $g_i - f_i \in \langle x \rangle^{k+1}$ for $i = 1, \ldots, m$, the local $\Bbbk$-algebras $\Bbbk[[x]]/I$ and $\Bbbk[[x]]/J$ are isomorphic. $I$ is called *finitely contact determined* if $I$ is contact $k$-determined for some $k$. It is easy to see (cf. [6, Proposition 4.3]) that these notions depend only on the ideal and not on the set of generators.

The ideal $I$ or the ring $\Bbbk[[x]]/I$ is called a *complete intersection* if $\dim \Bbbk[[x]]/I = n - m$ and an *isolated complete intersection singularity (ICIS)* if it has moreover an isolated singularity.

Set $f = (f_1, \ldots, f_m) \in \Bbbk[[x]]^m$ and denote by $\langle \partial f/\partial x_1, \ldots, \partial f/\partial x_n \rangle$ the submodule of $\Bbbk[[x]]^m$, generated by the m-tuples $\partial f/\partial x_i = (\partial f_1/\partial x_i, \ldots, \partial f_m/\partial x_i)$, $i = 1, \ldots, n$. We define

$$T_I := \Bbbk[[x]]^m \Big/ I\Bbbk[[x]]^m + \langle \partial f/\partial x_1, \ldots, \partial f/\partial x_n \rangle.$$

If $I$ is a complete intersection, then $\tau(I) := \dim_{\Bbbk} T_I$ is called the *Tjurina number of I*. For a complete intersection $T_I$ is concentrated on the singular locus of $\Bbbk[[x]]/I$ (Definiton 3.3) and $\tau(I)$ is finite iff $I$ has an isolated singularity. This follows from [6, Lemma 3.1], where it is shown that the ideals $I + I_{n-m}(Jac(I))$ and $\mathrm{Ann}_{\Bbbk[[x]]}(T_I)$ have the same radical.

The module $T_I$ is used in the following theorem.

**Theorem 3.2 ([6], Theorem 4.6)** *Let $I \subset \Bbbk[[x]]$ be a proper ideal and $\Bbbk$ infinite. Then the following are equivalent:*

*(i) $I$ is finitely contact determined.*
*(ii) $\dim_{\Bbbk} T_I < \infty$.*
*(iii) $R/I$ is an isolated complete intersection singularity.*

*If one of these condition is satisfied then* $I$ *is contact* $(2\dim_{\Bbbk} T_I - ord(I) + 2)$*-determined, where* $ord(I) = \max\{k \mid I \subset \langle x \rangle^k\}$*. The implications* $(iii) \Leftrightarrow (ii) \Rightarrow (i)$ *hold for any field* $\Bbbk$*, as well as* $(i) \Rightarrow (ii)$ *for hypersurfaces.*

**Proposition 3.2** *Let* $A$ *be Noetherian,* $F_1, \ldots, F_m \in \langle x \rangle A[[x]]$*. Let* $I \subset A[[x]]$ *be the ideal generated by* $F_1, \ldots, F_m$ *and* $\hat{I}(\mathfrak{p}) \subset k(\mathfrak{p})[[x]]$*,* $\mathfrak{p} \in \operatorname{Spec} A$*, the ideal generated by* $F_1(\mathfrak{p}), \ldots, F_m(\mathfrak{p}) \in k(\mathfrak{p})[[x]]$*.*

*Assume that* $V(I + I_{n-m}(Jac(I))) \subset V(\langle x \rangle)$ *(as sets), or* $\dim A = 1$*, or* $F_i \in \langle x \rangle A[x]$ *for* $i = 1, \ldots, m$*, or* $A$ *is a complete local ring containing a field.*

*Then any* $\mathfrak{p} \in \operatorname{Spec} A$ *has an open neighbourhood* $U \subset \operatorname{Spec} A$ *such that for all* $\mathfrak{q} \in U$ $\dim_{k(\mathfrak{p})} T_{\hat{I}(\mathfrak{p})} \geq \dim_{k(\mathfrak{q})} T_{\hat{I}(\mathfrak{q})}$*.*

***Proof*** By Greuel and Pham [6, Lemma 3.1] $\operatorname{Supp}(T_I) = V(I + I_{n-m}(Jac(I)))$. The claim follows from Corollary 2.7.

## *3.4 Tjurina Number of Complete Intersection Singularities*

We show first that being a regular sequence in a flat family of power series in $R = A[[x]]$ is an open property.

**Proposition 3.3** *Let* $A$ *be a Noetherian ring,* $F_i \in \langle x \rangle R$*,* $i = 1, \ldots, m$ *and* $M$ *a finitely generated* $R$*-module. For* $\mathfrak{p} \in \operatorname{Spec} A$ *we denote by* $F_i(\mathfrak{p})$ *the image of* $F_i$ *in* $\hat{R}(\mathfrak{p}) = k(\mathfrak{p})[[x]]$ *and by* $F_{i\mathfrak{n}_\mathfrak{p}}$ *the image of* $F_i$ *in* $R_{\mathfrak{n}_\mathfrak{p}}(\mathfrak{p})$ *(cf. Definition 2.2).*

(i) *If* $\mathfrak{p} \in \operatorname{Spec} A$ *then* $F_1(\mathfrak{p}), \ldots, F_m(\mathfrak{p})$ *is an* $\hat{M}(\mathfrak{p})$*-sequence iff* $F_{1\mathfrak{n}_\mathfrak{p}}, \ldots, F_{m\mathfrak{n}_\mathfrak{p}}$ *is an* $M_{\mathfrak{n}_\mathfrak{p}}(\mathfrak{p})$*-sequence.*
(ii) *Let* $F_1, \ldots, F_m$ *be an* $M$*-sequence and let* $M/\langle F_1, \ldots, F_m \rangle M$ *be* $A$*-flat. Then* $F_1(\mathfrak{p}), \ldots, F_m(\mathfrak{p})$ *is an* $\hat{M}(\mathfrak{p})$*-sequence for all* $\mathfrak{p} \in \operatorname{Spec} A$*.*
(iii) *Let* $\mathfrak{p} \in \operatorname{Spec} A$ *and* $F_1(\mathfrak{p}), \ldots, F_m(\mathfrak{p})$ *an* $\hat{M}(\mathfrak{p})$*-sequence. If* $M/\langle F_1, \ldots, F_m \rangle M$ *is flat over* $A$*, then there exists an open neighbourhood* $U$ *of* $\mathfrak{p}$ *in* $\operatorname{Spec} A$ *such that* $F_1(\mathfrak{q}), \ldots, F_m(\mathfrak{q})$ *is a* $\hat{M}(\mathfrak{q})$*-sequence for all* $\mathfrak{q}$ *in* $U$*.*

***Proof*** Set $M_0 = M$, $M_i = M/\langle F_1, \ldots, F_i \rangle M$ and consider for $i = 1, \ldots, m$ the exact sequence

$$0 \to K_{i-1} \to M_{i-1} \xrightarrow{F_i} M_{i-1} \to M_i \to 0, \qquad (*)$$

with $K_{i-1}$ the kernel of $F_i$.

(i) By Lemma 2.3.2 $\hat{R}(\mathfrak{p}) = R_{\mathfrak{n}_\mathfrak{p}}(\mathfrak{p})^\wedge$ and $\hat{M}_i(\mathfrak{p}) = M_{i,\mathfrak{n}_\mathfrak{p}}(\mathfrak{p})^\wedge$ for all $i$ and hence

$$F_i(\mathfrak{p}) = F_{i\mathfrak{n}_\mathfrak{p}}^\wedge : (M_{i-1,\mathfrak{n}_\mathfrak{p}}(\mathfrak{p}))^\wedge \to (M_{i-1,\mathfrak{n}_\mathfrak{p}}(\mathfrak{p}))^\wedge.$$

Since $M_{i,\mathfrak{n}_\mathfrak{p}}(\mathfrak{p})$ is a finite $R_{\mathfrak{n}_\mathfrak{p}}$-module we have (by Atiyah and Macdonald [1, Theorem 10.13]) $\hat{M}_i(\mathfrak{p}) = M_{i,\mathfrak{n}_\mathfrak{p}}(\mathfrak{p}) \otimes_{R_{\mathfrak{n}_\mathfrak{p}}} R^{\wedge}_{\mathfrak{n}_\mathfrak{p}}$. Moreover $R^{\wedge}_{\mathfrak{n}_\mathfrak{p}}$ is faithfully flat over the local ring $R_{\mathfrak{n}_\mathfrak{p}}$ ([12, Theorem 8.14]). Hence $F_{i\mathfrak{n}_\mathfrak{p}} : M_{i-1,\mathfrak{n}_\mathfrak{p}}(\mathfrak{p}) \to M_{i-1,\mathfrak{n}_\mathfrak{p}}(\mathfrak{p})$ is injective iff $F_i(\mathfrak{p}) : \hat{M}_{i-1}(\mathfrak{p}) \to \hat{M}_{i-1}(\mathfrak{p})$ is injective.

(ii) By assumption $K_{i-1} = 0$ for $i = 1, \ldots, m$ and $M/\langle F_1, \ldots, F_m\rangle M$ is $A$-flat. By Lemma 2.4 the Jacobson radical of $R$ contains $\langle x\rangle$ and $F_1, \ldots, F_m$ is a regular sequence contained in the Jacobson radical. Hence $M_i$ is $A$-flat and $M_{i,\mathfrak{n}_\mathfrak{p}}$ is $A_\mathfrak{p}$-flat for all $i$ (repeated application of [12, Theorem 22.2]). Tensoring $0 \to M_{i-1,\mathfrak{n}_\mathfrak{p}} \to M_{i-1,\mathfrak{n}_\mathfrak{p}} \to M_{i,\mathfrak{n}_\mathfrak{p}} \to 0$ with $\otimes_{A_\mathfrak{p}} k(\mathfrak{p})$ we get an exact sequence $0 \to M_{i-1,\mathfrak{n}_\mathfrak{p}}(\mathfrak{p}) \to M_{i-1,\mathfrak{n}_\mathfrak{p}}(\mathfrak{p})) \to M_{i,\mathfrak{n}_\mathfrak{p}}(\mathfrak{p}) \to 0$ for $i = 1, \ldots m$ by [12, Theorem 22.3]. Now apply (i).

(iii) Localizing the exact sequence (*) at $\mathfrak{n}_\mathfrak{p}$ we get an exact sequence of finite $R_{\mathfrak{n}_\mathfrak{p}}$-modules. Taking the $\langle x\rangle$-adic completion, the sequence stays exact and we see that $(K_{i-1,\mathfrak{n}_\mathfrak{p}})^\wedge = \ker\left(F^{\wedge}_{i\mathfrak{n}_\mathfrak{p}} : (M_{i-1,\mathfrak{n}_\mathfrak{p}})^\wedge \to (M_{i-1,\mathfrak{n}_\mathfrak{p}})^\wedge\right)$. By Lemma 2.3 $\hat{M}_{i-1}(\mathfrak{p}) = (M_{i-1,\mathfrak{n}_\mathfrak{p}})^\wedge \otimes_{A_\mathfrak{p}} k(\mathfrak{p})$ and $F_i(\mathfrak{p}) = F^{\wedge}_{i\mathfrak{n}_\mathfrak{p}} \otimes_{A_\mathfrak{p}} k(\mathfrak{p})$, and by assumption $F_i(\mathfrak{p})$ is injective. We apply now repeatedly [12, Theorem 22.5 ] to $A_\mathfrak{p} \to \hat{R}_{\mathfrak{n}_\mathfrak{p}} = A_\mathfrak{p}[[x]]$ and to $F^{\wedge}_{i\mathfrak{n}_\mathfrak{p}}$ to get that $(K_{i-1,\mathfrak{n}_\mathfrak{p}})^\wedge = K_{i-1,\mathfrak{n}_\mathfrak{p}} \otimes_{R_{\mathfrak{n}_\mathfrak{p}}} R^{\wedge}_{\mathfrak{n}_\mathfrak{p}} = 0$ and that $(M_{i,\mathfrak{n}_\mathfrak{p}})^\wedge = M_{i,\mathfrak{n}_\mathfrak{p}} \otimes_{R_{\mathfrak{n}_\mathfrak{p}}} R^{\wedge}_{\mathfrak{n}_\mathfrak{p}}$ is flat over $A_\mathfrak{p}$ for all $i$. Since $R^{\wedge}_{\mathfrak{n}_\mathfrak{p}}$ is faithfully flat over $R_{\mathfrak{n}_\mathfrak{p}}$ this implies $K_{i-1,\mathfrak{n}_\mathfrak{p}} = 0$ and that $M_{i,\mathfrak{n}_\mathfrak{p}}$ is flat over $A_\mathfrak{p}$.

The support of the $R$-module $K_{i-1}$ is closed and hence $(K_{i-1})^{\wedge}_{\mathfrak{n}_\mathfrak{q}} = 0$ for $\mathfrak{q}$ in an open neighbourhood $U$ of $\mathfrak{p}$ in Spec $A$. Moreover the flatness of $M/\langle F_1, \ldots, F_m\rangle M$ implies that $M^{\wedge}_{\mathfrak{n}_\mathfrak{q}}/\langle F_1, \ldots, F_m\rangle M^{\wedge}_{\mathfrak{n}_\mathfrak{q}}$ is $A_\mathfrak{q}$-flat. Applying [12, Theorem 22.5 ] now to $F^{\wedge}_{i\mathfrak{n}_\mathfrak{q}} : (M_{i-1})^{\wedge}_{\mathfrak{n}_\mathfrak{q}} \to (M_{i-1})^{\wedge}_{\mathfrak{n}_\mathfrak{q}}$ we get that $\hat{M}_{i-1}(\mathfrak{q}) \to \hat{M}_{i-1}(\mathfrak{q})$ is injective and that $F_1(\mathfrak{q}), \ldots, F_m(\mathfrak{q})$ is an $\hat{M}(\mathfrak{q})$-sequence.

**Proposition 3.4** *Let $A$ be a Noetherian ring and $I \subset \langle x\rangle A[[x]]$ an ideal generated by $F_1, \ldots, F_m$, such that $A[[x]]/IA[[x]]$ is $A$-flat. For $\mathfrak{p} \in$ Spec $A$ denote by $\hat{I}(\mathfrak{p}) \subset k(\mathfrak{p})[[x]]$ the ideal generated by $F_1(\mathfrak{p}), \ldots, F_m(\mathfrak{p})$.*

1. *If $\hat{I}(\mathfrak{p})$ is a complete intersection, then $\hat{I}(\mathfrak{q})$ is a complete intersection for $\mathfrak{q}$ in an open neighbourhood of $\mathfrak{p}$ in* Spec $A$.
2. *Assume that $\hat{I}(\mathfrak{p})$ is an ICIS and that the hypotheses of Proposition 3.2 are satisfied. Then $\hat{I}(\mathfrak{q})$ is an ICIS with $\tau(I(\mathfrak{p})) \geq \tau(I(\mathfrak{q}))$ for $\mathfrak{q}$ in an open neighbourhood of $\mathfrak{p}$ in* Spec $A$.

***Proof***

1. We may assume that $F_1(\mathfrak{p}), \ldots, F_m(\mathfrak{p})$ is a $k(\mathfrak{p})[[x]]$-sequence. By Proposition 3.3 $F_1(\mathfrak{q}), \ldots, F_m(\mathfrak{q})$ is a $k(\mathfrak{q})[[x]]$-sequence, hence $\hat{I}(\mathfrak{q})$ is a complete intersection, for $\mathfrak{q}$ in an open neighbourhood of $\mathfrak{p}$ in Spec $A$.
2. follows from Proposition 3.2 since for $\hat{I}(\mathfrak{q}) \subset k(\mathfrak{q})[[x]]$ a complete intersection $\dim_{k(\mathfrak{q})} T_{\hat{I}(\mathfrak{q})}$ is the Tjurina number of $\hat{I}(\mathfrak{q})$.

**Acknowledgments** We would like to thank the anonymous referee for their careful proofreading and in particular for providing answers to two of our original questions (cf. the comment after Proposition 2.2).

## References

1. Atiyah, M.F., Macdonald I.G.: Introduction to Commutative Algebra. Addison-Wesley, Boston (1969)
2. Boubakri, Y., Greuel, G.-M., Markwig, T.: Normal forms of hypersurface singularities in positive characteristic. Mosc. Math. J. **11**(4), 657–683 (2011)
3. Boubakri, Y., Greuel, G.-M., Markwig, T.: Invariants of hypersurface singularities in positive characteristic. Rev. Mat. Complut. **25**(1), 61–85 (2012)
4. Decker, W., Greuel, G.-M., Pfister, G., Schönemann, H.: SINGULAR 4-1-2 – A computer algebra system for polynomial computations (2019). http://www.singular.uni-kl.de
5. Grauert, H., Remmert, R.: Analytische Stellenalgebren. Grundlehren der Mathematischen Wissenschaften. Springer, Berlin (1971)
6. Greuel, G.-M., Pham, T.H.: Finite determinacy of matrices and ideals in arbitrary characteristics. J. Algebra **530**, 195–214 (2019)
7. Greuel, G.-M., Pfister, G.: A **Singular** Introduction to Commutative Algebra, 2nd edn. Springer, Berlin (2008). With contributions by O. Bachmann, C. Lossen and H. Schönemann
8. Greuel, G.-M., Lossen, C., Shustin, E.: Introduction to Singularities and Deformations. Springer, Berlin (2007)
9. Kurke, H., Pfister, G., Roczen, M.: Henselsche Ringe und Algebraische Geometrie. Mathematische Monographien. Band, vol. 11, pp. 223. VEB Deutscher Verlag der Wissenschaften, Berlin S. M 57.00 (1975)
10. Kurke, H., Pfister, G., Popescu, D., Roczen, M., Mostowski T.: Die Approximationseigenschaft Lokaler Ringe. Lecture Notes in Mathematics, vol. 634. Springer, Berlin (1978)
11. Majadas, J., Rodicio, A.G.: Smoothness, Regularity and Complete Intersection. London Mathematical Society Lecture Note Series, vol. 373. Cambridge University Press, Cambridge (2010)
12. Matsumura, H.: Commutative Ring Theory. Cambridge University Press, Cambridge (1986)
13. Pfister, G.: On modular computation of standard basis. An. Stiint. Univ. Ovidius Constanţa, Ser. Mat. **15**(1), 129–138 (2007)
14. Saito, K.: Quasihomogene isolierte Singularitäten von Hyperflächen. Invent. Math. **14**, 123–142 (1971)
15. The Stacks project authors. The Stacks project (2019). https://stacks.math.columbia.edu

# Lattices and Correction Terms

**Kyle Larson**

**Abstract** Let $L$ be a nonunimodular definite lattice, $L^*$ its dual lattice, and $\lambda$ the discriminant form on $L^*/L$. Using a theorem of Elkies we show that whether $L$ embeds in the standard definite lattice of the same rank is completely determined by a collection of lattice correction terms, one for each metabolizing subgroup of $(L^*/L, \lambda)$. As a topological application this gives a rephrasing of the obstruction for a rational homology 3-sphere to bound a rational homology 4-ball coming from Donaldson's theorem on definite intersection forms of 4-manifolds. Furthermore, from this perspective it is easy to see that if the obstruction to bounding a rational homology ball coming from Heegaard Floer correction terms vanishes, then (under some mild hypotheses) the obstruction from Donaldson's theorem vanishes too.

**Keywords** Lattice · Correction term

**Subject Classifications** Primary 57M27; Secondary 11H06

## 1 Introduction

In [5] Elkies showed that every unimodular positive definite lattice $L$ of rank $n$ contains characteristic vectors with square less than or equal to $n$, and if there are no characteristic vectors with square strictly less than $n$ then $L$ is isomorphic to the standard lattice $(\mathbb{Z}^n, I)$. One can define a *lattice correction term*

$$d_L = \min \left\{ \frac{\chi^2 - n}{4} \right\}, \tag{1.1}$$

K. Larson (✉)
Department of Mathematics, University of Georgia, Athens, GA, USA
e-mail: kyle.larson@uga.edu



J. Fernández de Bobadilla et al. (eds.), *Singularities and Their Interaction with Geometry and Low Dimensional Topology*, Trends in Mathematics,
https://doi.org/10.1007/978-3-030-61958-9_11

where the minimum is over all characteristic vectors $\chi \in L$ (see [8]). This is well-defined for all positive definite unimodular lattices, and Elkies' result translates to the statement that $d_L \leq 0$ and $d_L = 0$ if and only if $L$ is isomorphic to the standard lattice. Our first goal is to generalize this to the case where $L$ is not unimodular. In this setting one can ask whether a definite lattice *embeds* in the standard lattice of the same rank.

Recall that we have a sequence $0 \to L \to L^* \xrightarrow{\pi} L^*/L \to 0$, and $L$ is nonunimodular if and only if the discriminant group $L^*/L$ is non-trivial. There is a one-to-one correspondence between metabolizers $M < L^*/L$ and unimodular lattices $U$ with $L \subset U \subset L^*$, given by $U := \pi^{-1}(M)$ (Proposition 2.1). (Recall that a metabolizer is a subgroup $M$ with $|L^*/L| = |M|^2$ and such that the discriminant form $\lambda$ is identically zero on $M$.) For a metabolizer $M$ we denote the corresponding unimodular lattice $U(M)$. Then $U(M)$ will necessarily be positive definite of rank $n$, and hence we have a lattice correction term $d_{U(M)}$. We derive the following as a corollary of Elkies' theorem.

**Theorem 1.1** *For $L$ a positive definite lattice of rank $n$, consider the set $D := \{d_{U(M_i)}\}$ of lattice correction terms, where we range over all metabolizers $M_i < L^*/L$. Then $L$ embeds in the standard lattice of rank $n$ if and only if $D$ contains 0.*

Note that $D$ is a finite set since $L^*/L$ is a finite group, and $D$ is empty (and hence $L$ does not embed in the standard lattice) if there do not exist any metabolizers. Our main interest in this result is in application to the following question in low-dimensional topology:

*Question 1.1 ([14], Problem 4.5)* When does a rational homology 3-sphere bound a rational homology 4-ball?

We are interested in the relationship between two obstructions to a rational homology 3-sphere $Y$ smoothly bounding a rational homology 4-ball. We suppose that $Y$ bounds a smooth positive definite 4-manifold $X$, and we will make the simplifying assumption that $H_1(X) = 0$. If $Y$ bounds a smooth rational homology ball $W$ as well, we can form a smooth, closed, definite 4-manifold $Z = X \cup_Y -W$. By Donaldson's theorem [3, 4] on definite intersection forms of smooth, closed 4-manifolds, the lattice $(H_2(Z), Q_Z)$ must be isomorphic to the standard lattice. It then follows that the lattice $(H_2(X), Q_X)$ must embed in the standard lattice of the same rank. This is what we call *the obstruction to $Y$ bounding a rational homology ball coming from Donaldson's theorem.*

By Theorem 1.1, the obstruction coming from Donaldson's theorem is completely determined by a collection of lattice correction terms, one for each metabolizing subgroup of $(H_1(Y), \lambda)$ (in this context $\lambda$ is known as the linking form). Then work of Ozsváth and Szabó [20] shows that the Heegaard Floer corrections terms of $Y$ put bounds on the values of these lattice correction terms (indeed, the definition (1.1) of lattice correction terms is motivated by properties of Heegaard Floer correction terms, see Sect. 3). We use these bounds to show that the vanishing of the Heegaard Floer correction terms $d(Y, \mathfrak{t})$ on a metabolizing subgroup (in fact a slightly weaker condition) implies that $(H_2(X), Q_X)$ embeds in the standard lattice of the same rank.

**Theorem 1.2** *Suppose $Y$ is a rational homology 3-sphere such that $d(Y, \mathfrak{t}) \geq 0$ for all spin$^c$ structures $\mathfrak{t}$ with $\mathrm{PD}(c_1(\mathfrak{t}))$ belonging to some fixed metabolizer $M$ of $H_1(Y)$. If $Y$ bounds a smooth positive definite 4-manifold $X$ with $H_1(X) = 0$, then the lattice $(H_2(X), Q_X)$ must embed in the standard lattice of the same rank.*

Furthermore, if the correction terms $d(Y, \mathfrak{t})$ corresponding to a metabolizer $M$ as above are all strictly positive, then $Y$ cannot bound a positive definite 4-manifold $X$ with $H_1(X) = 0$ (Proposition 3.2). Recall that if $Y$ bounds a rational homology ball, then $d(Y, \mathfrak{t}) = 0$ for all spin$^c$ structures $\mathfrak{t}$ that extend over the rational homology ball. This can be interpreted in a convenient way if we further assume that $Y$ is a $\mathbb{Z}/2\mathbb{Z}$-homology sphere (so $|H_1(Y)|$ is odd). In particular, if such a $Y$ bounds a rational homology ball then there exists a metabolizer $M$ such that $d(Y, \mathfrak{t}) = 0$ for all spin$^c$ structures $\mathfrak{t}$ with $\mathrm{PD}(c_1(\mathfrak{t})) \in M$ (see, for example, [12]). This is what we call *the obstruction to a $\mathbb{Z}/2\mathbb{Z}$-homology sphere $Y$ bounding a rational homology ball coming from correction terms*, and hence Theorem 1.2 shows that (in this context) the obstruction to bounding a rational homology ball coming from Heegaard Floer correction terms is always at least as strong as that coming from Donaldson's theorem. Note that if $d(Y, \mathfrak{t}) = 0$ for all spin$^c$ structures $\mathfrak{t}$ with $\mathrm{PD}(c_1(\mathfrak{t})) \in M$, then $-Y$ also satisfies the conditions of Theorem 1.2 (since $d(-Y, \mathfrak{t}) = -d(Y, \mathfrak{t})$), and so by reversing orientation we get a statement that also applies to *negative* definite fillings of $Y$.

**Corollary 1.3** *If $Y$ is a $\mathbb{Z}/2\mathbb{Z}$-homology sphere on which the correction term obstruction to bounding a rational homology ball vanishes, then for any smooth (positive or negative) definite 4-manifold $X$ with $H_1(X) = 0$ and $\partial X = Y$, the lattice $(H_2(X), Q_X)$ must embed in the standard lattice of the same rank.*

When $Y$ is not a $\mathbb{Z}/2\mathbb{Z}$-homology sphere, the first Chern class mapping is no longer a bijection, and Theorem 1.2 is less useful. For example, the lens space $L(4, 1)$ (which does bound a rational homology ball) does not satisfy the hypotheses of Theorem 1.2, since $\mathrm{PD}(c_1(\mathfrak{t}))$ belongs to the unique metabolizer for *each* spin$^c$ structure on $L(4, 1)$, and there exists a spin$^c$ structure whose corresponding correction term is negative.

That there is a close relationship between Donaldson's theorem and the correction terms of Heegaard Floer homology was already established in the original paper defining correction terms [20]. Indeed, using the theorem of Elkies mentioned above, Ozsváth and Szabó gave a new proof of Donaldson's theorem using properties of the unique correction term $d(N)$ for an integral homology sphere $N$ (this mirrored another proof in Seiberg–Witten Floer theory [6]). Furthermore, they showed that the obstruction to an *integral* homology 3-sphere bounding an *integral* homology 4-ball coming from correction terms is at least as strong as that coming from Donaldson's theorem. More precisely, suppose $N$ bounds a positive definite 4-manifold $X$. If $d(N) = 0$ (which must be the case if $N$ bounds an integral homology ball), then $Q_X$ must be isomorphic to the standard form [20] (note that Corollary 1.3 is a generalization of this statement).

If we are dealing with a rational rather than an integral homology sphere, the two obstructions are slightly more complicated, as we described above. The extra complication in the case of Donaldson's theorem is because we have to consider embeddings, rather than isomorphisms, of lattices; in the case of correction terms it is because there is no longer a unique correction term, but rather a collection of correction terms corresponding to the set of spin$^c$ structures on the 3-manifold.

Nonetheless, relations between these two obstructions for rational homology spheres have appeared previously in the literature, usually in more specific contexts. In [10] Greene and Jabuka showed that in the application of these obstructions to showing that certain types of knots (e.g. alternating knots) are not slice, one can view the correction term obstruction as a second-order obstruction after the vanishing of the obstruction coming from Donaldson's theorem (see [10] Theorem 3.6 and the preceding exposition). More recently, Greene [9] showed that in certain special cases these two obstructions can be used to achieve the same purpose. For example, he showed that either obstruction is sufficient to classify which lens spaces $L(p, q)$ with odd $p$ bound rational homology balls (which had been carried out by Lisca [15] using the obstruction from Donaldson's theorem, including those with even $p$). Indeed, the proof of Theorem 1.2 is very similar to the ideas presented in [9, Proposition 2.1]. In particular, one direction of Greene's argument gives Corollary 1.3 when $H_1(Y)$ is *cyclic*. Hence the present note can be thought of as a companion to that paper, where here we take a more general and elementary perspective.

## 2 Lattices

First we develop the necessary terminology about lattices (cf. [8, Section 2]). In this paper a *lattice* $(L, Q)$ is a finite rank free abelian group $L$ together with a symmetric, bilinear form $Q\colon L \times L \to \mathbb{Q}$. We will assume that $Q$ is *nondegenerate*, i.e., for every non-zero $x \in L$ there exists some $y \in L$ such that $Q(x, y) \neq 0$. Usually the form will be understood and we will just say $L$ is a lattice. If the image of the form lies in $\mathbb{Z}$, then the lattice will be called *integral*. We will always use $L$ to denote an integral lattice. An *isomorphism* of lattices is an isomorphism of the free abelian groups that preserves the forms, and an *embedding* of lattices is a monomorphism that preserves the forms.

We say $L$ is *positive definite* if the rank of $Q$ equals its signature, and *negative definite* if the rank of $Q$ equals $-1$ times the signature. The standard positive definite lattice, or more simply, the *standard lattice* (of rank $n$), is $(\mathbb{Z}^n, I)$. This means that in a chosen basis the form is represented by the identity matrix.

The form $Q$ extends to a rational valued form on $L \otimes \mathbb{Q}$, and the *dual lattice* $L^*$ is defined as the subset $L^* = \{x \in L \otimes \mathbb{Q} \mid Q(x, y) \in \mathbb{Z}, \forall y \in L\}$. The quotient $L^*/L$ is called the *discriminant group*, and its order is the *discriminant* of

$L$, denoted $\mathrm{disc}(L)$. If $\mathrm{disc}(L) = 1$, then we say $L$ is *unimodular*. Note that we have a sequence $0 \to L \to L^* \xrightarrow{\pi} L^*/L \to 0$. We can define a symmetric, bilinear form $\lambda\colon (L^*/L) \times (L^*/L) \to \mathbb{Q}/\mathbb{Z}$, called the *discriminant form*, as follows. For any $x, y \in L^*/L$, take lifts $\bar{x}, \bar{y} \in L^*$ (so $\pi(\bar{x}) = x$ and $\pi(\bar{y}) = y$), and define $\lambda(x, y) = -Q(\bar{x}, \bar{y}) \pmod 1$. (The minus sign is to make this definition agree with a more geometric one in the case that the discriminant form arises as the linking form of a rational homology sphere. See Sect. 3 and [7] Exercises 4.5.12(c) and 5.3.13(g).) As mentioned in the introduction, a subgroup $M < L^*/L$ satisfying $\mathrm{disc}(L) = |M|^2$ and $\lambda|_{M\times M} \equiv 0$ is called a *metabolizer*. The following proposition is well-known in various forms (see [13, Lemma 2.5]), and is central to our argument.

**Proposition 2.1** *There is a one-to-one correspondence between metabolizers of $(L^*/L, \lambda)$ and unimodular integral lattices $U$ with $L \subset U \subset L^*$, given by the assignment $U(M) := \pi^{-1}(M)$, for each metabolizer $M$.*

***Proof*** The map $\pi$ induces a bijection between subgroups of $L^*$ containing $L$ and subgroups of $L^*/L$. For such a subgroup $U \subset L^*$ containing $L$, the rational valued form $Q$ on $L^*$ restricts to an integral form on $U$ if and only if $\lambda$ vanishes on $\pi(U) \times \pi(U)$. To see this, recall that for $\bar{x}, \bar{y} \in U$, $Q(\bar{x}, \bar{y}) \equiv -\lambda(\pi(\bar{x}), \pi(\bar{y})) \pmod 1$. Finally, $U$ is unimodular if and only if $U = U^*$, or equivalently, if $[U^* : U] = 1$. Now

$$\mathrm{disc}(L) = [L^* : L] = [L^* : U^*][U^* : U][U : L],$$

and since $[U : L] = [L^* : U^*]$ by Lemma 2.2 below, we have

$$\mathrm{disc}(L) = [U^* : U]([U : L])^2 = [U^* : U]|\pi(U)|^2.$$

Hence $U$ is unimodular if and only if $|\pi(U)|^2 = \mathrm{disc}(L)$.

**Lemma 2.2** *Let $L'$ be an integral lattice with $L \subset L' \subset L^*$. Then $[L' : L] = [L^* : (L')^*]$.*

***Proof*** Let $H = \pi(L')$, so $H \cong L'/L$. Furthermore let $H^\circ$ denote its *annihilator*, that is, the subgroup of $L^*/L$ consisting of all elements that pair trivially with every element of $H$ under $\lambda$. Observe that $H^\circ = \pi((L')^*)$. We claim that $(L^*/L)/H^\circ \cong H$. To see this, note that the map $\psi\colon L^*/L \to \mathrm{Hom}(L^*/L, \mathbb{Q}/\mathbb{Z})$ given by $\psi(x) = \lambda(x, \cdot)$ is an isomorphism since $\lambda$ is nondegenerate (see [22], especially Sections 1 and 7). Indeed, for each $x \in L^*/L$ there exists some $y \in L^*/L$ such that $\lambda(x, y) = 1/n \in \mathbb{Q}/\mathbb{Z}$, where $n$ is the order of $x$. We get a map $\psi'\colon L^*/L \to \mathrm{Hom}(H, \mathbb{Q}/\mathbb{Z}) \cong H$ by restricting the domain of each $\psi(x)$ to $H$. Then $\psi'$ has image isomorphic to $H$ and kernel $H^\circ$, giving $(L^*/L)/H^\circ \cong H$. It then follows that

$$[L' : L] = |H| = |(L^*/L)/H^\circ| = [L^* : (L')^*],$$

completing the proof.

We introduce some additional terminology. A *characteristic covector* $\chi \in L^*$ is an element such that $Q(\chi, y) \equiv Q(y, y) \pmod 2$ for all $y \in L$. Let $\mathrm{Char}(L)$ denote the set of characteristic covectors. If a characteristic covector $\chi$ actually lies in $L$ (as will always be the case when $L$ is unimodular), we can simply call $\chi$ a characteristic *vector*. As in the introduction, if $L$ is unimodular and positive definite, we have a well-defined *lattice correction term*

$$d_L = \min_{\chi \in \mathrm{Char}(L)} \left\{ \frac{\chi^2 - \mathrm{rk}(L)}{4} \right\}.$$

(Greene [8] contains an extended discussion of this invariant.) In this language we can state the result of Elkies as follows.

**Theorem 2.3 ([5])** *For L a unimodular positive definite lattice, $d_L \leq 0$ and $d_L = 0$ if and only if L is isomorphic to the standard lattice.*

Hence the lattice correction term completely determines when a unimodular positive definite lattice is isomorphic to the standard lattice. We can combine Theorem 2.3 and Proposition 2.1 to characterize which nonunimodular positive definite lattices embed in the standard lattice. Let $L$ be such a lattice and $M < L^*/L$ be a metabolizer. Since $L$ is positive definite, $U(M)$ is positive definite as well, since $L \subset U(M)$ and both lattices have the same rank. Hence we can define a set $D := \{d_{U(M_i)}\}$ of lattice correction terms, where we range over all metabolizers $M_i < L^*/L$. Recall that Theorem 1.1 from the introduction states that $L$ embeds in the standard lattice of the same rank if and only if $D$ contain 0. We prove this theorem now.

***Proof of Theorem 1.1*** If $D$ contains 0, then some $U(M)$ satisfies $d_{U(M)} = 0$. By Theorem 2.3, $U(M)$ is isomorphic to the standard lattice. Since $L \subset U(M)$, one direction of the proof is finished.

In the other direction, suppose $L$ embeds in the standard lattice $(\mathbb{Z}^n, I)$ of the same rank. Hence we can suppose $L \subset \mathbb{Z}^n$, and tensoring with $\mathbb{Q}$ shows that $(\mathbb{Z}^n, I) \subset L^* \subset L \otimes \mathbb{Q}$. By Proposition 2.1, $(\mathbb{Z}^n, I) = U(M)$ for some metabolizer $M$, and Theorem 2.3 implies that $d_{U(M)} = 0$. This completes the other direction of the proof.

Let $n$ denote the rank of $L$ (and $U(M)$). Recall that $d_{U(M)}$ is defined as

$$d_{U(M)} = \min_{\chi \in \mathrm{Char}(U(M))} \left\{ \frac{\chi^2 - n}{4} \right\}. \tag{2.1}$$

Since $L \subset U(M) \subset L^*$, $\mathrm{Char}(U(M)) \subset \mathrm{Char}(L)$. Indeed, $\mathrm{Char}(U(M))$ is a subset of those characteristic covectors of $L^*$ that map to elements of $M$ under the projection $\pi$. Hence from (2.1) we obtain

$$d_{U(M)} \geq \min_{\substack{\chi \in \mathrm{Char}(L) \\ \pi(\chi) \in M}} \left\{ \frac{\chi^2 - n}{4} \right\}. \tag{2.2}$$

This will be useful in the next section. Note that it is possible to show we have equality in (2.2) if $\mathrm{disc}(L)$ is odd.

## 3 Rational Homology Spheres and Correction Terms

We now turn to the topological application discussed in the introduction. Let $Y$ be a rational homology 3-sphere that bounds a smooth positive definite 4-manifold $X$ with $H_1(X) = 0$. By the long exact sequence of the pair $(X, Y)$ we get the presentation

$$0 \to H_2(X) \to H_2(X, Y) \to H_1(Y) \to 0. \tag{3.1}$$

Under suitable choices of bases, the map $H_2(X) \to H_2(X, Y)$ is given by the matrix representing the intersection form $Q_X$ (see, for example, [7, Exercise 5.3.13 (f)]). Furthermore, if we let $L$ denote the lattice $(H_2(X), Q_X)$, the dual lattice $L^*$ is identified with $(H_2(X, Y), Q_X^{-1})$, and (3.1) becomes

$$0 \to L \xrightarrow{Q_X} L^* \xrightarrow{\pi} L^*/L \to 0. \tag{3.2}$$

In this context the discriminant form is called the *linking form* $\lambda$ on $H_1(Y) \cong L^*/L$, and is defined by $\lambda(x, y) = -(Q_X)^{-1}(\pi^{-1}(x), \pi^{-1}(y)) \pmod 1$. Note that this is independent of the choice of the 4-manifold $X$.

As explained in the introduction, a consequence of Donaldson's theorem is that if $Y$ smoothly bounds a rational homology ball, then the lattice $L = (H_2(X), Q_X)$ embeds in the standard lattice of the same rank. By Theorem 1.1, this condition is completely determined by the collection of lattice correction terms $\{d_{U(M_i)}\}$, where we range over metabolizers of $(H_1(Y), \lambda)$. These lattice correction terms are in turn bounded by the Heegaard Floer correction terms of $Y$, as we now describe. Recall that in Ozsváth and Szabó's Heegaard Floer homology, correction terms are rational valued invariants of spin$^c$ rational homology spheres that are preserved under spin$^c$ rational homology cobordism. For $Y$ with spin$^c$ structure $\mathfrak{t}$, the corresponding correction term is denoted $d(Y, \mathfrak{t})$. We have the following important inequality.

**Fig. 1** A commutative diagram

$$\begin{array}{ccccc}
 & & 0 & & 0 \\
 & & \uparrow & & \uparrow \\
\mathrm{Spin}^c(Y) & \xrightarrow{c_1} & H^2(Y) & \xrightarrow{\mathrm{PD}} & H_1(Y) \\
\uparrow & & \uparrow & & \uparrow \\
\mathrm{Spin}^c(X) & \xrightarrow{c_1} & H^2(X) & \xrightarrow{\mathrm{PD}} & H_2(X,Y) \\
 & & \uparrow & & \uparrow \\
 & & H^2(X,Y) & \xrightarrow{\mathrm{PD}} & H_2(X)
\end{array}$$

**Theorem 3.1 ([20])** *Let $Y$ be a rational homology sphere that bounds a positive definite 4-manifold $X$. If $\mathfrak{s}$ is a spin$^c$ structure on $X$ with $\mathfrak{s}|_Y = \mathfrak{t}$, then*

$$\frac{1}{4}(c_1(\mathfrak{s})^2 - \mathrm{rk}(H_2(X))) \geq d(Y, \mathfrak{t}). \tag{3.3}$$

Now we relate this to lattices. The first Chern class mapping and Poincaré duality provide a bijection between spin$^c$ structures on $X$ and characteristic covectors in $H_2(X, Y)$ ([7, Proposition 2.4.16]). Under this bijection, a spin$^c$ structure $\mathfrak{s}$ on $X$ that extends a spin$^c$ structure $\mathfrak{t}$ on $Y$ corresponds to a characteristic covector $\chi$ in $H_2(X, Y) = L^*$, such that $\pi(\chi) = \mathrm{PD}(c_1(\mathfrak{t}))$. (See Fig. 1.) Then (3.3) implies that $\frac{1}{4}(\chi^2 - \mathrm{rk}(H_2(X))) \geq d(Y, \mathfrak{t})$ for each such $\chi$. (Other applications of these bounds can be found in [17] and [11].) For a metabolizer $M$, we can combine this with the inequality (2.2) and Theorem 2.3 to obtain

$$0 \geq d_{U(M)} \geq \min_{\substack{\chi \in \mathrm{Char}(L) \\ \pi(\chi) \in M}} \left\{ \frac{\chi^2 - \mathrm{rk}(H_2(X))}{4} \right\} \geq \min_{\substack{\mathfrak{t} \in \mathrm{Spin}^c(Y) \\ \mathrm{PD}(c_1(\mathfrak{t})) \in M}} \left\{ d(Y, \mathfrak{t}) \right\}. \tag{3.4}$$

Note that we would have a contradiction if there exists a metabolizer $M$ for $H_1(Y)$ with

$$\min_{\substack{\mathfrak{t} \in \mathrm{Spin}^c(Y) \\ \mathrm{PD}(c_1(\mathfrak{t})) \in M}} \left\{ d(Y, \mathfrak{t}) \right\} > 0,$$

and so such a $Y$ cannot bound a smooth positive definite 4-manifold $X$ with $H_1(X) = 0$. We record this here as a proposition. Note that this generalizes a theorem for integral homology spheres [20, Corollary 9.8], and for rational homology spheres there are similar results by Owens and Strle [18] (see Theorem 2 and Proposition 5.2).

**Proposition 3.2** *Suppose a rational homology sphere $Y$ has a metabolizer $M$ for $(H_1(Y), \lambda)$ for which $d(Y, \mathfrak{t}) > 0$ for each spin$^c$ structure $\mathfrak{t}$ with $\mathrm{PD}(c_1(\mathfrak{t})) \in M$. Then $Y$ cannot bound a smooth positive definite 4-manifold $X$ with $H_1(X) = 0$.*

We can now prove the second theorem from the introduction.

***Proof of Theorem* 1.2** Recall we are assuming that $Y$ is a rational homology 3-sphere that bounds a positive definite 4-manifold $X$ with $H_1(X) = 0$, and that there exists a metabolizer $M$ of $H_1(Y)$ such that $d(Y, \mathfrak{t}) \geq 0$ for all spin$^c$ structures $\mathfrak{t}$ with $\mathrm{PD}(c_1(\mathfrak{t})) \in M$. By Proposition 3.2, there must be at least one such $\mathfrak{t}$ such that $d(Y, \mathfrak{t}) = 0$, and hence

$$\min_{\substack{\mathfrak{t} \in \mathrm{Spin}^c(Y) \\ \mathrm{PD}(c_1(\mathfrak{t})) \in M}} \left\{ d(Y, \mathfrak{t}) \right\} = 0.$$

Then Eq. (3.4) implies that $d_{U(M)} = 0$, and by Theorem 1.1 the lattice $(H_2(X), Q_X)$ must embed in the standard lattice of the same rank.

## 4 Examples

Finally we give a couple of examples to illustrate these ideas. First we consider $(+9)$-surgery on the left-handed trefoil, $S^3_9(T_{-2,3})$. Let $Y$ denote this 3-manifold. The correction terms of surgeries on torus knots can be computed readily by combining work of [16] and [2] (see, for example, [1]). Then for the unique metabolizer of $H_1(Y)$, one can check that the corresponding correction terms are $\{2, 0, 0\}$. Since one of these is nonzero, the correction term obstruction shows that $S^3_9(T_{-2,3})$ does not bound a rational homology ball. On the other hand, Theorem 1.2 states that for every positive definite 4-manifold $X$ with $H_1(X) = 0$ and $\partial X = Y$, the lattice $(H_2(X), Q_X)$ must embed in the standard lattice of the same rank. Indeed it easy to check this condition for the two obvious positive definite 4-manifolds bounded by $Y$: the 2-handlebody given by the trace of the surgery, and the canonical definite plumbing associated to $S^3_9(T_{-2,3})$ as a Seifert fibered space. However, $Y$ also bounds a *negative* definite 4-manifold with trivial first homology (see [19]), with intersection form

$$\begin{bmatrix} -1 & 0 & 0 & 0 & 0 & 0 & 0 & 0 & 0 \\ 0 & -2 & 1 & 0 & 0 & 0 & 0 & 0 & 0 \\ 0 & 1 & -2 & 1 & 0 & 0 & 0 & 0 & 0 \\ 0 & 0 & 1 & -2 & 1 & 0 & 0 & 0 & 0 \\ 0 & 0 & 0 & 1 & -2 & 1 & 0 & 0 & 0 \\ 0 & 0 & 0 & 0 & 1 & -2 & 1 & 0 & 0 \\ 0 & 0 & 0 & 0 & 0 & 1 & -2 & 1 & 0 \\ 0 & 0 & 0 & 0 & 0 & 0 & 1 & -2 & 1 \\ 0 & 0 & 0 & 0 & 0 & 0 & 0 & 1 & -2 \end{bmatrix}.$$

Since the corresponding lattice does not embed in the standard negative definite lattice of rank 9, we see that the obstruction coming from Donaldson's theorem can also be used to show that $Y$ does not bound a rational homology ball. This suggests the following question.

*Question 4.1* Does there exist a rational homology sphere $Y$ for which the correction term obstruction does not vanish, but for any positive *or negative* definite 4-manifold bounded by $Y$, the associated lattice must embed in the standard lattice of the same rank?

Next we use Proposition 3.2 to show that the connected sum of the Poincaré homology sphere $\Sigma(2,3,5)$ (oriented as the boundary of the *negative* $E_8$ plumbing) and the Seifert fibered space $Y(1; \frac{3}{2}, \frac{21}{4}, \frac{50}{7})$ does not bound a positive definite 4-manifold with trivial first homology. We label this manifold $Z := \Sigma(2,3,5)\#Y(1; \frac{3}{2}, \frac{21}{4}, \frac{50}{7})$. Note that we chose this example because the similar obstructions of Owens and Strle mentioned above ([18] Theorem 2 and Proposition 5.2) do not apply to $Z$. The correction terms for each of these manifolds can be computed algorithmically using results of [21]. Using the fact that correction terms add over connected sums, we compute that the correction terms for $Z$ are $\{2, -\frac{4}{9}, \frac{2}{9}, 2, \frac{8}{9}, \frac{8}{9}, 2, \frac{2}{9}, -\frac{4}{9}\}$, where this set is identified with $H_1(Z) \cong \mathbb{Z}/9\mathbb{Z}$ in the obvious way. Hence the correction terms corresponding to the metabolizer are $\{2, 2, 2\}$, and so Proposition 3.2 implies that $Z$ cannot bound a positive definite 4-manifold with trivial first homology. We do not know if $Z$ bounds a negative definite 4-manifold. The same argument shows that, for any positive $n$, $\Sigma(2,3,6n-1)\#Y(1; \frac{3}{2}, \frac{21}{4}, \frac{50}{7})$ does not bound a positive definite 4-manifold with trivial first homology.

**Acknowledgments** The author is pleased to thank Marco Golla, Joshua Greene, Brendan Owens, and Sašo Strle for helpful conversations and comments. In addition, we thank Marco Golla for sharing Proposition 2.1 with the author, and a referee for helpful suggestions. This research was partially supported by NKFIH Grant K112735.

## References

1. Aceto, P., Golla, M.: Dehn surgeries and rational homology balls. Algebr. Geom. Topol. **17**(1), 487–527 (2017)
2. Borodzik, M., Livingston, C.: Heegaard Floer homology and rational cuspidal curves. Forum Math. Sigma **2**, e28, 23 (2014)
3. Donaldson, S.K.: An application of gauge theory to four-dimensional topology. J. Differ. Geom. **18**(2), 279–315 (1983)
4. Donaldson, S.K.: The orientation of Yang-Mills moduli spaces and 4-manifold topology. J. Differ. Geom. **26**(3), 397–428 (1987). http://projecteuclid.org/euclid.jdg/1214441485
5. Elkies, N.D.: A characterization of the $\mathbf{Z}^n$ lattice. Math. Res. Lett. **2**(3), 321–326 (1995)
6. Frøyshov, K.A.: The Seiberg-Witten equations and four-manifolds with boundary. Math. Res. Lett. **3**(3), 373–390 (1996)
7. Gompf, R.E., Stipsicz, A.I.: 4-manifolds and Kirby calculus. In: Graduate Studies in Mathematics, vol. 20. American Mathematical Society, Providence (1999)
8. Greene, J.E.: Lattices, graphs, and Conway mutation. Invent. Math. **192**(3), 717–750 (2013)
9. Greene, J.E.: A note on applications of the $d$-invariant and Donaldson's theorem. J. Knot Theory Ramifications **26**(2), 1740006, 8 (2017)
10. Greene, J., Jabuka, S.: The slice-ribbon conjecture for 3-stranded pretzel knots. Am. J. Math. **133**(3), 555–580 (2011)

11. Greene, J.E., Watson, L.: Turaev torsion, definite 4-manifolds, and quasi-alternating knots. Bull. Lond. Math. Soc. **45**(5), 962–972 (2013)
12. Hedden, M., Livingston, C., Ruberman, D.: Topologically slice knots with nontrivial Alexander polynomial. Adv. Math. **231**(2), 913–939 (2012)
13. Jabuka, S.: The signature of an even symmetric form with vanishing associated linking form (2012). Preprint https://arxiv.org/abs/1204.4965
14. Kirby, R.: Problems in low dimensional manifold theory. In: Algebraic and Geometric Topology (Proceedings of the Symposium on Pure Math., Stanford Univ., Stanford, Calif., 1976), Part 2, Proceedings Symposium Pure Math., XXXII, pp. 273–312. Amer. Math. Soc., Providence (1978)
15. Lisca, P.: Lens spaces, rational balls and the ribbon conjecture. Geom. Topol. **11**, 429–472 (2007)
16. Ni, Y., Wu, Z.: Cosmetic surgeries on knots in $S^3$. J. Reine Angew. Math. **706**, 1–17 (2015)
17. Owens, B., Strle, S.: Rational homology spheres and the four-ball genus of knots. Adv. Math. **200**(1), 196–216 (2006)
18. Owens, B., Strle, S.: A characterization of the $\mathbb{Z}^n \oplus \mathbb{Z}(\delta)$ lattice and definite nonunimodular intersection forms. Am. J. Math. **134**(4), 891–913 (2012)
19. Owens, B., Strle, S.: Dehn surgeries and negative-definite four-manifolds. Sel. Math. (N.S.) **18**(4), 839–854 (2012)
20. Ozsváth, P., Szabó, Z.: Absolutely graded Floer homologies and intersection forms for four-manifolds with boundary. Adv. Math. **173**(2), 179–261 (2003)
21. Ozsváth, P., Szabó, Z.: On the Floer homology of plumbed three-manifolds. Geom. Topol. **7**, 185–224 (2003)
22. Wall, C.T.C.: Quadratic forms on finite groups, and related topics. Topology **2**, 281–298 (1963)

# Complex Surface Singularities with Rational Homology Disk Smoothings

**Jonathan Wahl**

*To András Némethi on his 60th birthday*

**Abstract** A cyclic quotient singularity of type $p^2/pq-1$ $(0 < q < p, (p,q) = 1)$ has a smoothing whose Milnor fibre is a $\mathbb{Q}$HD, or rational homology disk (i.e., the Milnor number is 0). In the 1980s, we discovered additional examples of such singularities: three triply-infinite and six singly-infinite families, all weighted homogeneous. Later work of Stipsicz, Szabó, Bhupal, and the author proved that these were the only weighted homogeneous examples. In his UNC PhD thesis, our student Jacob Fowler completed the analytic classification of these singularities, and counted the number of smoothings in each case, except for types $\mathcal{W}$, $\mathcal{N}$, and $\mathcal{M}$. In this paper, we describe his results, and settle these remaining cases; there is a unique $\mathbb{Q}$HD smoothing component except in the cases of an obvious symmetry of the resolution dual graph. The method involves study of configurations of rational curves on projective rational surfaces.

**Keywords** Rational homology disk smoothings · Smoothing surface singularities

**Subject Classifications** 14J17, 32S30, 14B07

## 1 Introduction

Let $(X, 0)$ be the germ of a complex normal surface singularity. A *smoothing of* $(X, 0)$ is a morphism $f : (\mathcal{X}, 0) \to (\mathbb{C}, 0)$, where $(\mathcal{X}, 0)$ is an isolated Cohen-Macaulay singularity, equipped with an isomorphism $(f^{-1}(0), 0) \simeq (X, 0)$. The Milnor fibre $M$ of a smoothing is a general fibre $f^{-1}(\delta)$, a four-manifold whose

J. Wahl (✉)
Department of Mathematics, The University of North Carolina, Chapel Hill, NC, USA
e-mail: jmwahl@email.unc.edu



J. Fernández de Bobadilla et al. (eds.), *Singularities and Their Interaction with Geometry and Low Dimensional Topology*, Trends in Mathematics,
https://doi.org/10.1007/978-3-030-61958-9_12

boundary is the link of $(X, 0)$. The first betti number of $M$ is 0 [3]. We say $f$ is a $\mathbb{Q}$HD (or *rational homology disk*) smoothing if the second betti number of $M$ is 0 as well (the *Milnor number* $\mu = 0$). In such a case, $(X, 0)$ must be a rational singularity.

The basic examples are smoothings of the cyclic quotient singularities of type $p^2/pq-1$, where $0 < q < p$, $(p, q) = 1$ ([8, (2.7)]). For $f(x, y, z) = xz - y^p$, one has that $f : \mathbb{C}^3 \to \mathbb{C}$ is a smoothing of the $A_{p-1}$ singularity, whose Milnor fibre $M$ has Euler characteristic $p$. Now consider the cyclic subgroup $G \subset GL(3, \mathbb{C})$ generated by the diagonal matrix $[\zeta, \zeta^q, \zeta^{-1}]$, where $\zeta = e^{2\pi i/p}$. $G$ acts freely on $\mathbb{C}^3 - \{0\}$ and $f$ is $G$-invariant; so there is a map $f : \mathbb{C}^3/G \equiv \mathcal{X} \to \mathbb{C}$, a smoothing of the cyclic quotient singularity $A_{p-1}/G$, which has type $p^2/pq - 1$. The new Milnor fibre is the free quotient $M/G$, of Euler characteristic 1, hence Milnor number 0.

More examples can be produced by a similar "quotient construction"([9, 5.9.2]). For instance, let $f(x, y, z) = xy^{p+1} + yz^{q+1} + zx^{r+1}$, $N = (p + 1)(q + 1)(r + 1) + 1$, and $G \subset GL(3)$ the diagonal cyclic subgroup generated by $[\zeta, \zeta^{(q+1)(r+1)}, \zeta^{-(r+1)}]$, where $\zeta = e^{2\pi i/N}$. The resulting class of examples was later named $\mathcal{W}(p, q, r)$ in [7, (8.3)]. Another class $\mathcal{N}(p, q, r)$ was obtained by replacing $(\mathbb{C}^3, 0)$ by a hypersurface $(V, 0) \subset (\mathbb{C}^4, 0)$, $G$ by a group of automorphisms acting freely off the origin, and an appropriate $f$. Three more families are constructed in [10].

However, the major way to produce examples uses H. Pinkham's general method of "smoothing with negative weight" [6] for a weighted homogeneous singularity $(X, 0)$. Writing $X$ =Spec $A$, where $A$ is a graded ring, form the $\mathbb{C}^*$-compactification $\bar{X}$ =Proj $A[t]$ (where $t$ has weight 1). $\bar{X}$ has a smooth curve $\bar{C}$=Proj $A$ at infinity (which we assume is rational), along which are several cyclic quotient singularities. Resolving those singularities yields $\bar{X}'$ with a star-shaped collection of curves $\bar{E}$, consisting of $\bar{C}$ plus chains of rational curves. The associated graph $\Gamma$ of these curves is "dual" to the star-shaped resolution graph $\Gamma'$ of the singularity; it is non-degenerate, of signature $(1, s)$. (See e.g.[7, (8.1)] for details.) (Beware: $\Gamma'$ is itself sometimes called the "dual resolution graph.") A *smoothing of negative weight of* $X$ is a smoothing which can be extended to a smoothing of $\bar{X}'$ to which $\bar{E}$ lifts and is deformed trivially. The general fibre is a smooth projective surface $Z$, with $H^1(\mathcal{O}_Z) = 0$, containing a curve $E$ isomorphic to $\bar{E}$, which supports an ample divisor. The Milnor fibre $M$ of this smoothing may be identified with the affine variety $Z - E$ ([9, (2.2)]). Thus, $M$ is a $\mathbb{Q}$HD if and only if the curves of $E$ are rational and form a rational basis of Pic $Z$.

Conversely, Pinkham shows how to construct a smoothing of negative weight of $(X, 0)$ by starting with certain surface pairs $(Z, E)$ satisfying some cohomological vanishing. The author used this method to compile a large list of (only partially published) examples of $\Gamma$ which led to $\mathbb{Q}$HD smoothings. The paper [7] limited greatly the possible resolution graphs $\Gamma'$ of *any* singularity admitting a $\mathbb{Q}$HD smoothing, and gave names to the author's families of examples (modified slightly in [2]). This work culminated in the Bhupal-Stipsicz theorem [1], showing that the author's list of resolution dual graphs was complete for the weighted homogeneous case.

**Bhupal-Stipsicz Theorem** *The resolution graphs (or dual graphs) of weighted homogeneous surface singularities admitting a* $\mathbb{Q}$*HD smoothing are exactly those of the following types:* $p^2/(pq-1)$ *cyclic quotients;* $\mathcal{W}(p,q,r)$; $\mathcal{N}(p,q,r)$; $\mathcal{M}(p,q,r)$; $\mathcal{B}^3_2(p)$; $\mathcal{C}^3_2(p)$; $\mathcal{C}^3_3(p)$; $\mathcal{A}^4(p)$; $\mathcal{B}^4(p)$; $\mathcal{C}^4(p)$.

Resolution graphs and dual graphs for these singularities are listed at the end of this paper in Tables A.1 and A.2, from the thesis of Jacob Fowler [2]. A node (or bullet) with no decoration is always assumed to be a $-2$ curve. For the remainder of the paper, we disregard the well-understood cyclic quotients.

Previous work by H. Laufer [4] shows that the examples above with a central curve of valency 3 are taut, i.e., have a unique analytic type, necessarily weighted homogeneous; further, all deformations (in particular, smoothings) are of negative weight. In case the valency is 4 (the last 3 families), [4] implies that the only analytic invariant is the cross-ratio of the central curve, and again all deformations are of negative weight.

Fowler's Ph.D. Thesis [2] attacked the key questions remaining for these $\mathbb{Q}$HD smoothings:

- Show the cross-ratios in the three infinite families of valency 4 examples are uniquely determined, as in [10].
- Determine the number of $\mathbb{Q}$HD smoothing components in each case.
- Calculate the fundamental groups of the Milnor fibres.

We fix some language and notation. Let $(X,0)$ be a weighted homogeneous surface singularity, of resolution dual graph $\Gamma$, admitting $\mathbb{Q}$HD smoothings.

**Definition 1.1** A $\Gamma$ *surface* is a pair $(Z,E)$ consisting of a smooth rational surface and rational curve configuration $E$ such that the classes of the components of $E$ form a rational basis of Pic $Z$, and one is given an identification of the curve configuration $E$ with the graph $\Gamma$.

Pinkham's Theorem in the current situation may be found in [7, (8.1)] and [2, (2.2.3)], yielding

**Theorem 1.1 ([2, (2.3.1)])** *Let* $(X,0)$ *be a singularity as above with a* $\mathbb{Q}$*HD smoothing, and resolution dual graph* $\Gamma$*. Then there exists a one-to-one correspondence between* $\mathbb{Q}$*HD smoothing components of* $(X,0)$ *and* $\Gamma$ *surfaces* $(Z,E)$ *up to isomorphism.*

Examples of $\Gamma$ surfaces are made as follows: take a specific curve configuration $D\subset\mathbb{P}^2$, blow up several times, obtaining $\pi:Z\to\mathbb{P}^2$, with $\pi^{-1}(D)$ consisting of a curve $E$ of type $\Gamma$ plus some $-1$ curves. If $E$ spans Pic$(Z)$ rationally, one has a $\Gamma$ surface. Given the location of the $-1$ curves in relation to the components of $E$, one can reverse the process and blow back down.

For each $\Gamma$, Fowler makes very judicious choices of $D$ and the points to blow up, resulting in either one or two *Basic Models* $(Z,E)$. The models for $\mathcal{W}$, $\mathcal{N}$, and

$\mathcal{M}$ are listed on the first page of Table A.2, where small circles and light lines indicate the location of three $-1$ curves which allow the entire graph to be blown down, to four lines in general position. Basic Models for the other graphs are more complicated and found in Fowler's thesis [2]. The curve configuration $D$ will be unique up to projective equivalence. One may get two models for the same $\Gamma$ and $D$ by blowing up in different ways (sometimes complex conjugate points). The goal is to prove

*Conjecture 1.1* Every $\Gamma$ surface $(Z, E)$ is a Basic Model.

In [2], Fowler proves most of this Conjecture; his nearly complete result, explained below in Sect. 6 states:

**Theorem 1.2 ([2])** *Suppose a $\Gamma$ surface $(Z, E)$ has self-isotropic subgroup which is basic. Then $(Z, E)$ is a Basic Model.*

Fowler also proves that the "basic self-isotropic subgroup" condition is automatically satisfied in all cases except for some $\Gamma$ of type $\mathcal{W}, \mathcal{N}$, or $\mathcal{M}$.

**Corollary 1.1 ([2])** *For $\Gamma$ not of type $\mathcal{W}, \mathcal{N}, \mathcal{M}$, every $\Gamma$ surface is a Basic Model. In particular, for each valency 4 example, there is a unique cross-ratio for which the corresponding singularity has a $\mathbb{Q}$HD smoothing.*

The new contribution of the current paper is to handle the remaining cases.

**Theorem 1.3** *Every $\Gamma$ surface of type $\mathcal{W}, \mathcal{N}, \mathcal{M}$ is a Basic Model.*

The Basic Models for types $\mathcal{W}, \mathcal{N}$, and $\mathcal{M}$ start with four lines in general position, for which the fundamental group of the complement is abelian. Therefore the Milnor fibre of a $\mathbb{Q}$HD smoothing of a singularity of this type has abelian fundamental group (hence is easily computable from $\Gamma$). More generally, we can conclude

**Theorem 1.4** *Let $M$ be the Milnor fibre of a $\mathbb{Q}$HD smoothing of a singularity of type $\Gamma$.*

1. *If $\Gamma$ is of type $\mathcal{W}, \mathcal{N}$, or $\mathcal{M}$, then $\pi_1(M)$ is abelian.*
2. *If $\Gamma$ is of type $\mathcal{A}^4$, $\mathcal{B}^4$, or $\mathcal{C}^4$, then $\pi_1(M)$ is metacyclic, as described in [10].*

In a not-yet-published manuscript by Enrique Artal and the author, it is proved that the fundamental group of a QHD Milnor fibre is abelian in cases $\mathcal{C}_2^3$ and $\mathcal{C}_3^3$, and non-abelian in case $\mathcal{B}_2^3$.

Once we know the number of $\mathbb{Q}$HD smoothings from the main theorem, we can conclude that the explicit examples from the quotient construction (for types $\mathcal{W}, \mathcal{N}$, $\mathcal{A}^4, \mathcal{B}^4, \mathcal{C}^4$) give a complete list of smoothing components in those cases. This also gives the only way to compute the metacyclic fundamental group.

In Sect. 6, we give more details on Fowler's method and list the number of $\mathbb{Q}$HD smoothing components for each $\Gamma$.

The isomorphism type of a $\Gamma$ surface of type $\mathcal{W}, \mathcal{N}$, or $\mathcal{M}$ is determined by the location of the three extra $-1$ curves that are attached to $E$. For special values of

$p, q, r$ for which the graph $\Gamma$ has a symmetry, there could be a second location of the $-1$'s, leading to a different $\Gamma$ surface (which, one recalls, comes equipped with a specific identification with the graph). In light of our new result, we find

**Theorem 1.5 ([2])** *Consider $\mathbb{Q}$HD smoothing components for type $\mathcal{W}, \mathcal{N}$, and $\mathcal{M}$.*

1. *There are two components for $\mathcal{W}(p, p, p)$, $\mathcal{N}(q+2, q, 0)$, and $\mathcal{M}(r+1, q, r)$*
2. *In all other cases, there is a unique $\mathbb{Q}$HD smoothing component.*

(Actually, [2] neglected to mention the exceptional $\mathcal{N}$ case, but it fits in easily with his work.)

Our method, already used in [1], is to blow up and down the given $\Gamma$ surface $(Z, E)$ so that one obtains a surface $(Z', E')$ with central curve of self-intersection $+1$, from which a blowing-down map to $\mathbb{P}^2$ can be constructed. We analyze the possible singularities of the image of $E'$ and the blowing up needed to reach back to $Z'$, leading to location of all possible sets of essential $-1$ curves on $Z$ needed for blowing down. All solutions will be Basic Models.

## 2 Locations of −1 Curves

Suppose $\Gamma$ is a graph of smooth rational curves $E = \Sigma E_i$:

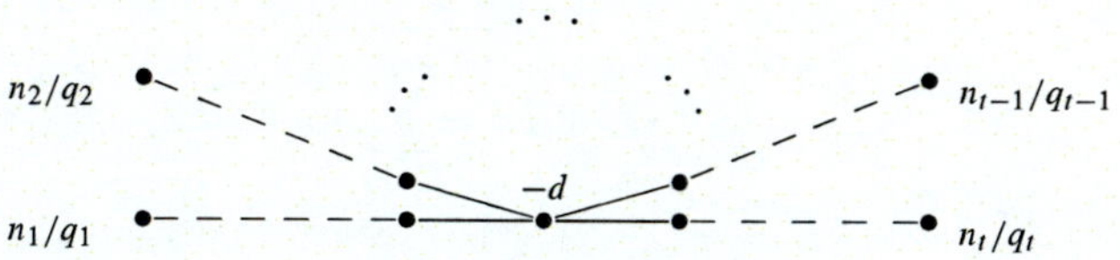


Here, the continued fraction expansion $n/q = b_1 - 1/b_2 - \cdots - 1/b_s$ represents a string of rational curves emanating from the center:

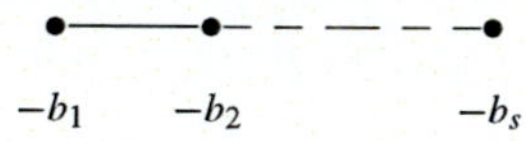


(1)

(We shall assume that $t \geq 3$.) It is well-known that

$$\det \Gamma = \pm n_1 n_2 \cdots n_t (d - \sum_{i=1}^{t} (q_i/n_i)). \tag{2}$$

As long as $\det \Gamma \neq 0$, one can solve the equations

$$K \cdot E_i + E_i \cdot E_i = -2,$$

and write

$$K = \Sigma k_i E_i, \quad k_i \in \mathbb{Q}.$$

Recall a negative-definite $\Gamma$ arises from the resolution of a weighted-homogeneous surface singularity. In Section 2 of [11], the $k_i$ are computed in this case; but only non-degeneracy of $\Gamma$ was used, so the same formulas apply.

The formulas are expressed in terms of the two invariants

$$e = d - \sum_{i=1}^{t}(q_i/n_i)$$

$$\chi = t - 2 - \sum_{i=1}^{t}(1/n_i) = -2 + \sum_{i=1}^{t}(1 - 1/n_i).$$

Since $e \neq 0$ by (2), we can define $\beta = \chi/e$.

As in [11], we consider the rational cycle $-(K+E)$. For a cyclic quotient chain as in (1), let $F_1, \cdots, F_s$ denote the curves. Define the rational cycle $e_i$ by the property $e_i(F_j) = -\delta_{ij}$; it is effective (i.e., has strictly positive coefficients). Then consider the cycle $Y = \beta e_1 - e_s$ (even if $s = 1$.) Denote by $Y_k$ the corresponding cycle for the $k$th string corresponding to $n_k/q_k$ in the graph of $\Gamma$, where $1 \leq k \leq t$. Denoting by $E_0$ the central curve of $\Gamma$, Proposition 2.3 of [11] yields

$$-(K + E) = \sum_{k=1}^{t} Y_k + \beta E_0. \tag{3}$$

**Lemma 2.1** *Assume $\Gamma$ is one of the graphs in Table A.2.*

1. *$\chi \geq 0$, and $\chi = 0$ exactly for the log-canonical singularities $\mathcal{W}(0,0,0)$, $\mathcal{N}(0,0,0)$, $\mathcal{M}(0,0,0)$.*
2. *$e < 0$.*
3. *$\beta < 0$ in all cases except the three log-canonicals above, in which case it is 0.*
4. *$|\beta| < 1$.*

***Proof*** The first statement is a simple check (a sum of three reciprocals of integers is rarely at least 1). For the second, one need only consider the cases when $d = 1$. But all of those examples have 2 chains of $-2$ curves emanating from the central curve; such a chain has $q = n - 1$, so $q/n \geq 1/2$. The third means checking that $\chi < |e|$, or

$$t - 2 - \sum_{i=1}^{t}(1/n_i) < \sum_{i=1}^{t}(q_i/n_i) - d.$$

This statement turns out to be equivalent to that in [11, Lemma 2.4]; but in any case, it is an exercise. (We quickly note that $d = -1$ in case $\mathcal{W}$; for $\mathcal{N}$, we have $d = 0$ and some $q_i = n_i - 1$; for type $\mathcal{M}$, two strings have $q_i = n_i - 1$.)

**Proposition 2.1** *Suppose $(Z, E)$ is a surface of type $\Gamma$, where $\Gamma$ is a graph in Table A.2.*

1. *The canonical divisor of Z is*

$$K = \sum k_i E_i,$$

*where for all $i$ $-1 \leq k_i < 0$.*
2. *$k_i = -1$ only for the log-canonical cases $\mathcal{W}(0,0,0)$, $\mathcal{N}(0,0,0)$, $\mathcal{M}(0,0,0)$, and then only at the central curve.*

***Proof*** Since the divisors $E_i$ span Pic $Z$ rationally, one can write the canonical divisor as $\sum k_i E_i$. These coefficients can therefore be computed as above just from the graph. In terms of the divisor $-(K + E)$, the claim is that its coefficients $-k_i - 1$ are between $-1$ and 0, and equal 0 only at the center for the 3 special cases. Lemma 2.1 (3) verifies this assertion for the central curve.

It remains to show that the coefficients of $Y_k$ are strictly between $-1$ and 0. Writing $Y = \beta e_1 - e_s$, all $e_i$ have strictly positive coefficients; as $\beta \leq 0$, the coefficients of $Y$ are strictly negative.

Next, writing $F = \sum F_j$, we claim that

$$(F + Y) \cdot F_j \leq 0, \text{ all } j;$$

this implies $F + Y$ has strictly positive coefficients, so the coefficients of $Y$ are bigger than $-1$ (one could not have $F = -Y$). For $j = 1$, the term in question is $1 - b_1 - \beta < 2 - b_1 \leq 0$ (it does not matter if $s = 1$). An easier argument handles the other cases.

**Corollary 2.1** *Let $(Z, E)$ be a surface of type $\Gamma$ as above, and $C \subset Z$ an irreducible curve with $C \cdot C < 0$ and not a component of $E$. Then*

1. *$C$ is a smooth rational curve with $C \cdot C = -1$*
2. *If $C \cdot E_i = 1$ for some $i$, then there is another $E_j$ with $C \cdot E_j > 0$.*

***Proof*** $C \cdot E_j \geq 0$ for all $j$, and is positive for at least one $j$ because $E$ supports an ample divisor. By Proposition 2.1, we have $K \cdot C < 0$, so the usual adjunction formula yields that $C$ is a smooth rational $-1$ curve, and $C \cdot K = -1$. In particular, $\sum(-k_j)C \cdot E_j = 1$. So, the second statement will follow once we exclude that for one of the three log-canonicals, there is a $-1$ curve which intersects the central curve transversally but does not intersect any other curve. But in each of those cases, adding such a $-1$ curve to $E$ would give a non-degenerate curve configuration, so that its class could not be a rational combination of the components of $E$.

We can paraphrase the last result by saying there are no "free" $-1$ curves, intersecting $E$ only once.

## 3 How to Find Sets of $-1$ Curves

If $(Z, E)$ is a $\Gamma$ surface, where $\Gamma$ is of type $\mathcal{W}, \mathcal{N}$, or $\mathcal{M}$, we wish to show that it is a Basic Model. That means, $Z$ contains a set of three $-1$ curves which allow one to blow down to $\mathbb{P}^2$; the basic cases identify possible locations of the curves relative to $E$. The blow downs give a projectively rigid configuration in $\mathbb{P}^2$ (4 lines in general position), from which the uniqueness of the $\Gamma$ surface follows.

The method (initially analogous to the one used in [1]) is to blow up and down around the central curve of $E$ to produce $(Z', E')$, on which the new central curve $E'_0$ is a smooth rational curve of self-intersection $+1$. The complete linear system associated to such a curve gives a birational map $\Phi : Z' \to \mathbb{P}^2$ which is an isomorphism in a neighborhood of $E'_0$. It is the analysis of this map which will produce $-1$ curves first on $Z'$ and then on $Z$. In each case, it will follow from the construction that one has an isomorphism of $Z' - E'_0$ with some open set in $Z$; we can conclude (as in Corollary 2.1) that there are no "free" $-1$ curves on $Z'$, and a curve which is not a component of $E'$ has self-intersection $\geq -1$.

Here is how we proceed:

We note $\Phi(E'_0) \equiv L$ is a line. Each curve $C$ in $E'$ adjacent to $E'_0$ is smooth (there are usually 3 such), and $\Phi(C)$ is a (possibly singular) rational plane curve of degree $d = C \cdot E'_0 > 0$. The behavior of these image curves near $L$ is the same as it was on $Z'$, and the key will be to figure out their intersections away from $L$. $\Phi(E')$ will have at most three singular points (away from $L$), and all possible configurations need to be considered.

The construction of $Z'$ shows that the components of $E'$ span Pic$(Z')$ rationally, so no curves are disjoint from $E'$, and $\Phi$ is a sequence of blowing-up points over $\Phi(E')$ away from $L$.

We note $\Phi^{-1}(\Phi(E'))$ consists of $E'$ and (usually) three $-1$ curves, which is the same as for the basic cases (i.e., Basic Models). For, the number of blow-downs given by $\Phi$ depends only on $K^2_{Z'}$, which is computed from $\Gamma'$, so is the same as in the basic case. New curves added to $\Phi^{-1}(\Phi(E'))$ have negative self-intersection, so are $-1$ curves.

Emanating from each adjacent curve $C$ is a chain (possibly empty) $\mathcal{C}(C)$ of rational curves, frequently with a long tail of $-2$ curves; it is disjoint from $E'_0$, so $\Phi$ sends it to a point $\Phi(\mathcal{C}(C)) \in \Phi(C)$. Being a smooth point of $\mathbb{P}^2$, its inverse image is a "blow-downable configuration". It contains $\mathcal{C}(C)$, at least one $-1$ curve, and any other chains $\mathcal{C}(\tilde{C})$ with the same image under $\Phi$. The inverse image of a singular point of $\Phi(E')$ (of course, away from the line $L$) is either a union of chains and $-1$'s, or a single $-1$ curve intersecting only adjacent curves. Since there are at most three new $-1$'s added in the inverse image of $\Phi(E')$, there are at most three singular points.

**Proposition 3.1** *$\Phi(\mathcal{C}(C))$ is a singular point of $\Phi(E')$. More generally, to go from $\mathbb{P}^2$ to $Z'$, one blows up only singular points of the inverse images of $\Phi(E')$.*

***Proof*** Suppose a smooth point on a curve in the blow-up process is blown-up further. Then the inverse image in $\Phi^{-1}(\Phi(E'))$ contains a smooth curve $C$ plus a blow-downable configuration attached transversally at a point of $C$. This configuration has a $-1$ curve. If it were at an end, this would be a free curve, a contradiction. If not, it would be an interior curve, and removing it would leave a bunch of curves disjoint from $E'$. This also is a contradiction.

To unravel $\Phi$, one first examines the possible intersections of the images of the adjacent curves, noting that there are at most three singular points. In each case, $\Phi$ must factor via the minimal resolution of the singular points of $\Phi(E')$; one gathers information about $\Phi^{-1}(\Phi(E'))$, such as possible valency of curves, or whether the $-1$ curves must intersect $E'$ transversally.

Possible blow-downable configurations on $Z'$ are formed by putting together chains and $-1$'s. There are limits to the location of $-1$ curves.

*Remark 3.1* The following two configurations are not negative-definite:

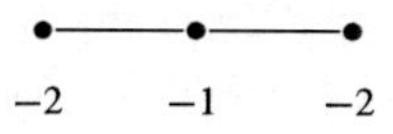


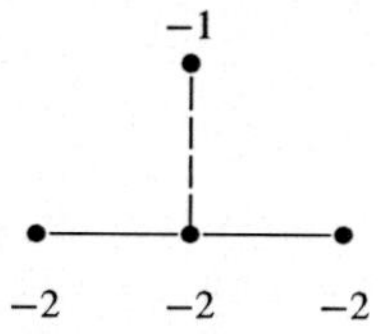


The first example implies that two different chains cannot be connected at $-2$'s. The second implies that a $-1$ curve could intersect a chain of $-2$ curves only at one of its ends. But a connection at the beginning of a $-2$ chain (next to the adjacent curve) has consequences.

*Remark 3.2* The configuration

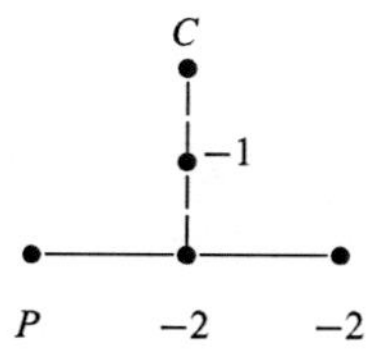


will, when blown-down, produce two curves which do not intersect transversally.

Consequently, if a chain emanating from the $P$ adjacent curve begins with two $-2$ curves, a $-1$ curve intersecting the first of these cannot intersect another curve, unless the final curve in $\mathbb{P}^2$ has a non-transversal intersection. Variants of this situation will arise as well.

It is not a priori clear that the full inverse image of $\Phi(E')$ has normal crossings; while a $-1$ intersects transversally in a blow-downable configuration, it could in principle attach non-transversally to an adjacent curve or curves.

Finally, we introduce a notational convenience. In the various blown-up spaces between $Z'$ and $\mathbb{P}^2$, we shall frequently refer to the image of an adjacent curve $C$ as $C'$, or $C'(s)$ (when the self-intersection at that stage is $s$).

## 4 Type $\Gamma = \mathcal{W}(p, q, r)$

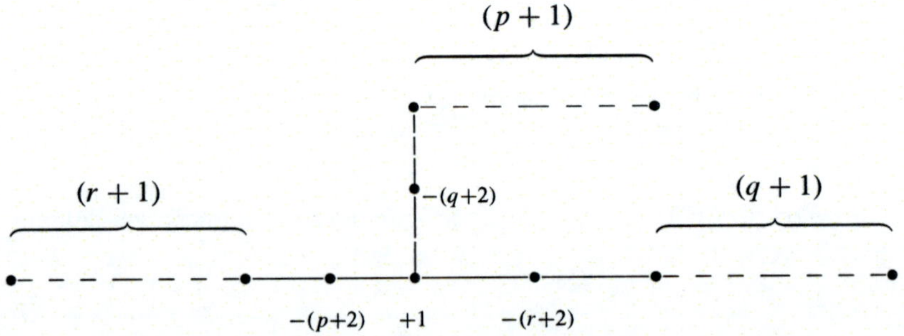


Suppose we are given a surface $Z$ of type $\Gamma = \mathcal{W}(p, q, r)$. The central curve $E_0$ has self-intersection $+1$, so in the above discussion we can set $Z' = Z$. The three curves adjacent to $E_0$ are $P$, $Q$, $R$, with self-intersections respectively $-(p+2)$, $-(q+2)$, $-(r+2)$, and with chains consisting solely of $-2$ curves. We will prove the existence of the three rational $-1$ curves which appear in the Basic Model. (For the case $p = q = r$, there is a second choice, by flipping $Q$ and $R$ and their chains). Each $-1$ curve will connect an adjacent curve with the end of a chain associated to a different adjacent curve.

By earlier discussion, $\Phi(P)$, $\Phi(Q)$, and $\Phi(R)$ are lines intersecting $\Phi(E_0) = L$ in distinct points. Thus $\Phi(E)$ either contains three lines through one point, or consists of four lines in general position. But having a triple point would mean there is only one singular point, so that all three chains would be connected by three $-1$'s. Remark 3.1 shows this is impossible. So $\Phi(E)$ has three ordinary double points, and hence $\Phi^{-1}(\Phi(E))$ has normal crossings and only curves of valency two (of course, not counting intersection with $E_0$).

Each chain connects to other curves in $E$ only with a $-1$ attached at one of its ends. That $-1$ cannot connect with another chain (Remark 3.1), so intersects an adjacent curve. The three $-1$'s are distributed among the three chains. By Remark 3.1, the $-1$ curve appended to a chain must intersect at the far end.

If the $-1$ at the end of $\mathcal{C}(P)$ intersects $Q$, then $\Phi(\mathcal{C}(P))$ is the intersection of the lines $\Phi(P)$ and $\Phi(Q)$. Therefore, the intersection point $\Phi(P) \cap \Phi(R)$ must be the image of the chain $\mathcal{C}(R)$, whose $-1$ curve attachment at the end must intersect $P$. Consequently, $\Phi(Q) \cap \Phi(R)$ comes from the chain $\mathcal{C}(Q)$, with a $-1$ attached to $R$. Blowing-down $E$ completely in this way, collapsing first the 3 $-1$'s and then the adjacent $-2$'s (which have become $-1$'s), one sees that this can happen only if $p = q = r$.

On the other hand, if the $-1$ curve emanating from the end of $\mathcal{C}(P)$ intersects $R$, then the same analysis shows the resulting placement of two $-1$'s as before always blows down exactly to 4 lines in general position. Thus, one has a unique location of the $-1$'s (seen in Table A.2), except in case $p = q = r$, in which case there is a second possibility. These are exactly the Basic Models for $\mathcal{W}$.

## 5 Type $\Gamma = \mathcal{N}(p, q, r)$, $p > 0$

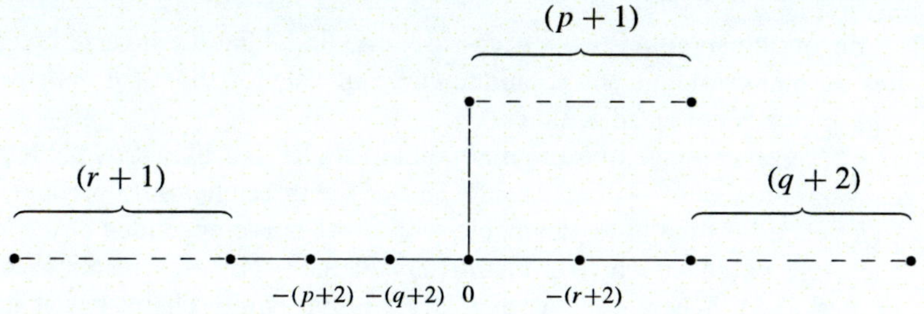


We consider initially the case $p > 0$. Proceed to a new $(Z', E')$ as follows: First, blow-up any point on the central curve not on one of the three adjacent curves. This makes the central curve a $-1$ curve, with four curves emanating from it, and adds a new curve $F$. Now blow-down the old central curve and the one above it in the graph above, yielding:

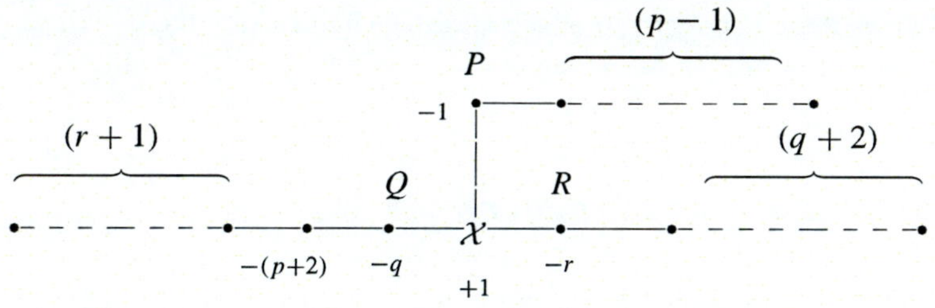


The curve $F$ has now become a $+1$ central curve $E'_0$, intersecting transversally a $-1$ curve $P$, from which a chain of $(p-1)$ $-2$ curves emerge. The two other original adjacent curves are still adjacent, but their self-intersections are now $-q$ and $-r$; we call the new ones $Q$ and $R$. But now $Q$ and $R$ are simply tangent to each other and to $E'_0$, and $P$, $Q$, $R$ all intersect it at the same point of the central curve. (We use the symbol $\mathcal{X}$ as a reminder that the intersections are not transversal.) The usual comparison with $Z$ shows that $Z'$ has no free $-1$'s, and the only negative curves off $E'$ are $-1$'s.

Proceeding as above, one constructs $\Phi : Z' \to \mathbb{P}^2$. Then $\Phi(E')$ consists of the line $\Phi(E'_0) = L$, two smooth conics $\Phi(Q)$ and $\Phi(R)$ intersecting each other and $L$ simply tangentially at a point of $L$, and a line $\Phi(P)$ intersecting transversally at that point. By the usual argument, there are three additional $-1$'s in $\Phi^{-1}(\Phi(E'))$. Here are the possibilities for the other intersections of the images of the three adjacent curves:

Case I The two conics intersect tangentially at one other point, and the line passes through it (one singular point).

Case II The two conics intersect tangentially at one other point, and the line intersects each conic at a different point (three singular points).

Case III The two conics intersect transversally at two other points, and the line passes through one of these points (two singular points).

Resolving singularities in each case, one finds that the inverse image of $\Phi(E')$ has normal crossings and a unique curve of valency three (of course, away from $E'_0$), which is not an adjacent curve.

The inverse image of a singular point of $\Phi(E')$ is a blow-downable graph which is a combination of $-1$ curves and some of the three chains. In particular, $\mathcal{C}(P)$ and $\mathcal{C}(R)$ each become blow-downable with a $-1$ curve appended at the end; further, each one could attach to a chain only at the $-(p+2)$ location of $\mathcal{C}(Q)$ (via Remark 3.1). When $p = 1$, then $\mathcal{C}(P)$ is the empty chain; but at least one $-1$ curve must still emerge from $P$, since $\Phi(P)$ intersects the other curves.

## 5.1 Case I for $\mathcal{N}(p, q, r)$, $p > 0$

Case I does not occur. Since one cannot have a valency four curve, a simple check shows there is no way to attach all three chains using three $-1$'s to get one blow-downable configuration (even in case $p = 1$ when $\mathcal{C}(P)$ is empty).

## 5.2 Case II for $\mathcal{N}(p, q, r)$, $p > 0$

In Case II, the conics $\Phi(Q)$ and $\Phi(R)$ are tangent away from $L$ and the line $\Phi(P)$ intersects each once, so there are three singular points. That means $\mathcal{C}(Q)$ must become blow-downable either with the addition of a single $-1$ curve, or with a single $-1$ joining it and another chain. One computes that adding a single $-1$ to $\mathcal{C}(Q)$ can make it blow-downable only if $p = r$ and the $-1$ is attached to the $-2$ curve adjacent to the $-(p+2)$. That $-1$ curve must intersect one of the other adjacent curves. But as in Remark 3.2, blowing down would give the image of that adjacent curve a worse than simple tangency with $\Phi(Q)$. This is a contradiction.

The only other option is to attach $\mathcal{C}(R)$ with a $-1$ adjoined at the $-(p+2)$ entry of $\mathcal{C}(Q)$; this blows down exactly when $p = q+2$. However, if $r > 0$ one sees that $\Phi(Q)$ and $\Phi(R)$ will have a higher order of tangency; this is ruled out.

So, consider the special situation $p = q+2$ and $r = 0$. We show there is a unique way to find three $-1$'s which blow-down this $E'$. The graph is

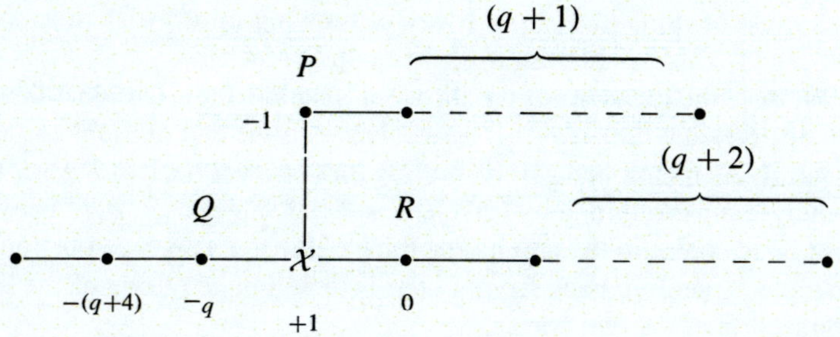


The above discussion states that the $-(q+4)$ curve has valency three in $\Phi^{-1}(\Phi(E'))$ and connects with a $-1$ curve from one of the ends of $\mathcal{C}(R)$.

Here is the minimal blow-up of $\Phi(E')$ which separates the line and two conics:

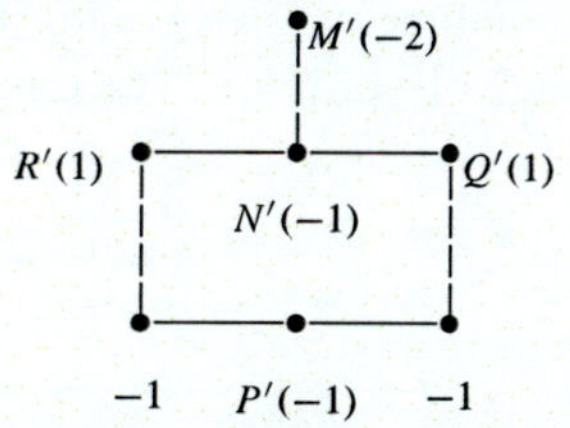


Recall $P'(-1)$, $Q'(1)$, $R'(1)$ are the images of the adjacent curves on partial blow-ups plus their self-intersections there. New curves $M'$ and $N'$ have been named. We specify what is needed in order to blow-up further to get to $E'$. Since $P$ has degree $-1$, there can be no further blow-ups along the bottom line. One cannot blow-up between $N'(-1)$ and $M'(-2)$, as an $M'(-3)$ (with degree $\leq -3$) would eventually become the $-(q+4)$ curve, but not adjacent to $Q$. $N'(-1)$ must be blown up somewhere, else the inverse image of a singular point would be a single $-1$ and $-2$, but not intersecting $P$. Therefore, $N'$ is the valency three curve which will eventually become the $-(q+4)$ curve. But that curve is adjacent to $Q$, so the only place $N'(-1)$ can be blown-up is at the intersection with $R'(1)$. After one blow-up, one reaches

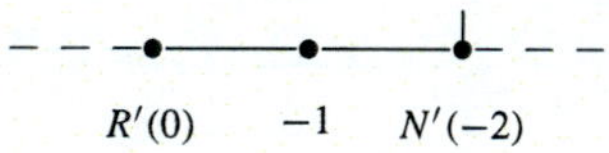

As $R$ has self-intersection 0, the only further blowing-up takes place between the $-1$'s and the curve on the right. After $q+2$ of such blow-ups, one has $R'(0)$ followed by $(q+2)$ $-2$'s, followed by a $-1$, followed by $N'(-(q+4))$. This completes most of the blow-up to $E'$. It remains only to complete the portion

This must be done in the usual way of blowing up $(q+1)$ times between the $-1$ and $Q'(1)$. This completes the blowing up to reach $E'$.

We note the locations of the three $-1$'s which allow the blow-down: between the end of $\mathcal{C}(R)$ and the $-(q+4)$ curve; between the end of $\mathcal{C}(P)$ and $Q$; and between $P$ and $R$. As mentioned before, pulling these curves back to $Z$ gives two of the $-1$'s seen by using the symmetry (given the special values of $p, q, r$) between the top and right hand chains in the graph. The third of the $-1$'s on $Z$ can be found in the above example by pulling back from $Z'$ the inverse image of one of the lines through an intersection of the two conics.

## 5.3 Case III for $\mathcal{N}(p, q, r)$, $p > 0$

There are two singular points; the two conics meet in two distinct points, and the line $\Phi(P)$ passes through one of those. The minimal resolution is

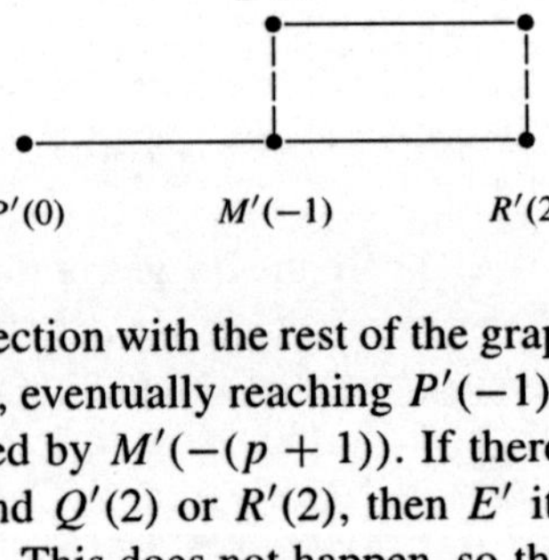


As $P'(0)$ has but one connection with the rest of the graph, as before one must blow up between it and $M'(-1)$, eventually reaching $P'(-1)$ followed by $(p-1)$ $-2$'s, followed by a $-1$, followed by $M'(-(p+1))$. If there were no further blow-ups between $M'(-(p+1))$ and $Q'(2)$ or $R'(2)$, then $E'$ itself would contain a curve intersecting both $R$ and $Q$. This does not happen, so the $M'$ curve must be blown-up at least once more and become $M'(-(p+2))$, the $-(p+2)$ curve in $\mathcal{C}(Q)$ adjacent to $Q$.

Thus, one has $P$ followed by $\mathcal{C}(P)$ (even if empty) followed by $-1$ attached to the $-(p+2)$ curve. Since $Q$ is adjacent to that curve, in the above diagram there is no blowing-up between them; to reach $Q$ of degree $-q$, one blows up repeatedly along the top row. This yields $Q$ followed by a $-1$ followed by $(q+2)$ $-2$'s followed by $R'(2)$. In other words, $Q$ is attached via a $-1$ with the end of $\mathcal{C}(R)$, accounting for the intersection point of the two conics not involving the line. The only way for $\mathcal{C}(Q)$ to attach is via a $-1$ at its end intersecting $R$.

This is exactly the placement of the three $-1$'s on $Z'$ as would happen in the basic case. Passing from $Z'$ back to $Z$, the $-1$'s and their relative location stays the same. We recover a Basic Model.

## 5.4 Type $\Gamma = \mathcal{N}(0, q, r)$

In this situation, we proceed from $(Z, E)$ to $(Z', E')$ exactly as before, first by blowing up to add one curve, and then blowing down two curves. The difference is that there is now no longer an adjacent curve $P$; rather, there are just two adjacent curves $Q$ and $R$, each simply tangent to each other and to the central curve $E'_0$. In this case, the map $\Phi$ sends $E'$ to the line $L$ and the two conics $\Phi(Q)$ and $\Phi(R)$, and two $-1$'s are needed to make the blow-down.

As $p = 0$, the chains $\mathcal{C}(Q)$ and $\mathcal{C}(R)$ consist solely of $-2$-curves, so cannot be connected via a $-1$. They are individually blow-downable by the usual addition of a $-1$ curve at the beginning or end of the chain. So, $\Phi(E')$ has two singular points, hence the conics intersect transversally. If either of these $-1$'s occurred at the beginning of a chain, then blowing down the chain would give a tangency or worse. Thus, the $-1$'s are at the far ends of the chains, and blowing down each chain gives one of the two intersection points of the conics $\Phi(Q)$ and $\Phi(R)$. Thus, there is a unique blow-down, so the basic case is the only one.

If one pulls these two $-1$'s from $Z'$ back to $Z$, one might ask where is the third $-1$ needed for blow-down. This can be found by pulling back to $Z'$ and then $Z$ one of the lines connecting the central point of $L$ with an intersection point of the two quadrics. Again, this is a Basic Model.

# 6 Type $\Gamma = \mathcal{M}(p, q, r)$

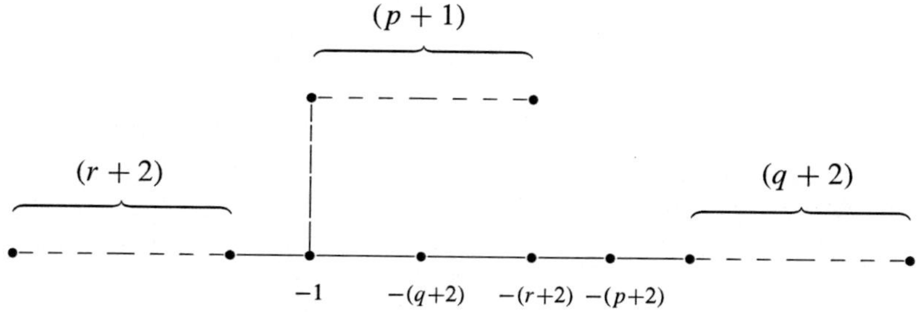


We refer to the three strings emanating from the central $-1$ curve in the diagram as the $r$, $p$, and $q$ directions. Start with the assumption that $p, r > 0$. To form the desired $(Z', E')$, blow-down the central $-1$ curve and the first two curves in the $r$ direction. Then the first curve in the $p$ direction is the new $E_0'$, of self-intersection $+1$, and $E'$ is

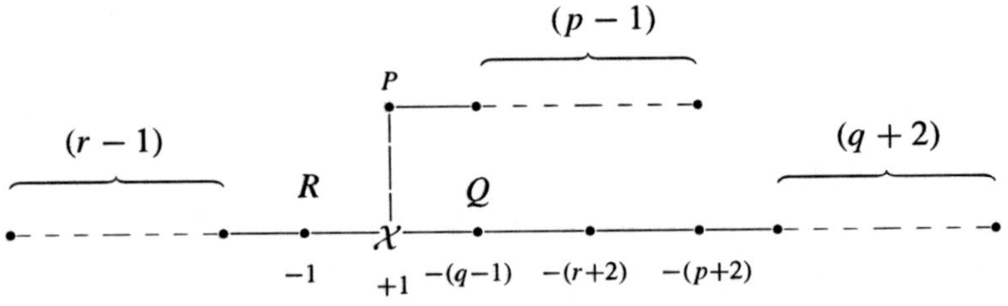


The new central configuration consists of a $+1$ curve $E_0'$ and three adjacent curves $P, Q, R$, but their intersections are no longer transversal. $Q$ is a $-(q-1)$ curve with a tangency of order 3 with $E_0'$ at a point, and $R$ intersects $E_0'$ at that point, transversally to both $E_0'$ and $Q$. Finally, $P$ intersects $E_0'$ transversally at a different point. Note that again there are no free $-1$ curves on $Z'$, given this property on $Z$.

The new map $\Phi : Z' \to \mathbb{P}^2$ arising from $E_0'$ makes $\Phi(E_0') = L$ a line, $\Phi(Q)$ a rational cubic curve with triple tangency at a point of $L$, $\Phi(R)$ a line through that central point transversal to both $L$ and the cubic, and $\Phi(P)$ a line transversal to $L$ intersecting at a different point.

A rational cubic curve is either nodal or cuspidal, and each type is unique up to projective equivalence. Each has a unique flex point, i.e., smooth point whose tangent line intersects with multiplicity 3. A calculation shows that a line through that flex point cannot be tangent to the curve at a smooth point. As a result, here are the only possible configurations of $\Phi(E')$ with at most three singular points:

Case $I_c$ (resp. $I_n$) Cubic is cuspidal (resp. nodal), $\Phi(R)$ passes through the singular point, $\Phi(P)$ has multiplicity three at that point (one singular point for $\Phi(E')$)

Case $II_c$ (resp. $II_n$) Cubic is cuspidal (resp. nodal), $\Phi(R)$ passes through the singular point, $\Phi(P)$ passes through the singular point plus another point of the cubic (two singular points)

Case $III_c$ (resp. $III_n$) Cubic is cuspidal (resp. nodal), $\Phi(R)$ intersects the cubic in two distinct smooth points, $\Phi(P)$ passes through one of those two points and the singular point (three singular points)

We rule out all but Case $I_n$ using the interplay between requirements of the graph $E'$ and resolution of the singularities of $\Phi(E')$. One cannot dismiss a priori the occurrence of a non-transversal intersection of a $-1$ curve with another curve.

## 6.1 Case $III$ for $\mathcal{M}(p, q, r)$, $p, r > 0$

In Case $III$, $\Phi(E')$ would have an ordinary triple point involving all three curves. The inverse image of this singular point would be a single $-1$ curve with valency (at least) three, intersecting all adjacent curves or their chains. We show this cannot happen. Blowing up the three singular points yields the curve configuration

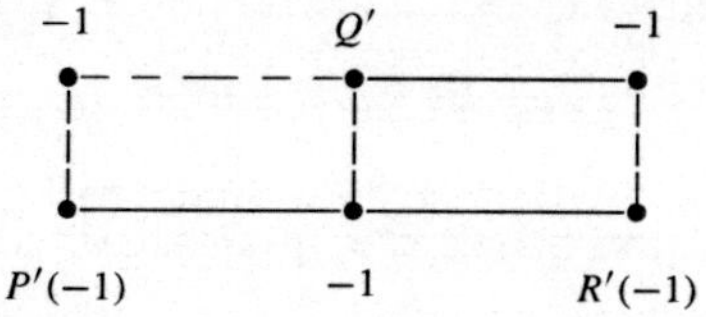


Here as before $P'(-1)$ indicates the image of $P$, with self-intersection $-1$. The dotted connection on the top row between a $-1$ and $Q'$ indicates that the intersection of the curves is not transversal; $Q'$ intersects tangentially in the cusp case, and in two points in the nodal. Since $P$ has self-intersection $-2$, to reach $Z'$ one needs to blow-up the intersection of $P'(-1)$ with exactly one of its neighboring $-1$'s. If one blows up at the $-1$ above, that would make $Q'$ intersect non-transversally with a $-2$ curve, which must be resolved. That further resolution would introduce a fourth $-1$ curve disjoint from the other three; this is a contradiction.

So, one would have to blow up between $P'(-1)$ and the $-1$ on its right, converting that valency three curve to a $-2$, hence no longer eligible to be the $-1$ of valency three or more. The only other possible way to get a trivalent $-1$ would be to blow up the non-transversal intersection between $Q'$ and the $-1$ on its left. In the case of a node, this would result in a new trivalent curve, but it would be a $-2$. In the case of a cusp, the only way to get a trivalent curve would be to blow-up twice, in which case the original $-1$ intersecting $Q'$ would become a $-3$. However, $P$ does not intersect a curve of degree $\leq -3$ (only a $-2$ or $-1$ is allowed). This completes the argument.

## 6.2 Case $II$ for $\mathcal{M}(p, q, r)$, $p, r > 0$

$\Phi(E')$ has two intersection points, with minimal resolution

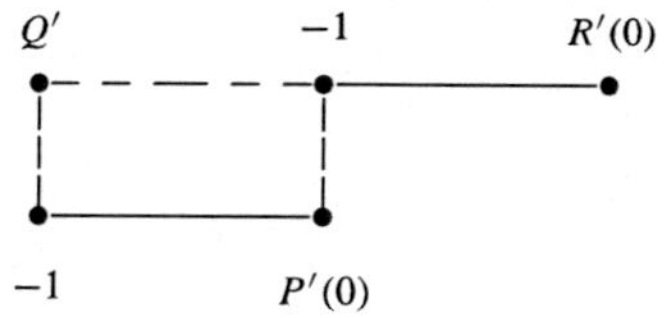

The dotted line again refers to the non-transverse intersections due to the cusp or node. Since $R$ has self-intersection $-1$, one must blow up the point between $R'(0)$ and its neighbor, converting it to a $-2$ and obtaining a new $-1$. But one cannot have a non-transverse intersection between $Q'$ and a $-2$, so that must be blown up, converting the $-2$ to a $-3$ and inserting a $-1$ connecting it to $Q'$ (in both the node and cusp cases). That makes three disjoint $-1$'s. But $P'(0)$ now intersects a $-3$ curve, yet $P$ does not; so, that intersection point must be blown up, giving a fourth $-1$, which is not allowed.

## 6.3 *Case I for* $\mathcal{M}(p, q, r)$, $p, r > 0$

Here $\Phi(E')$ has one singular point. Blowing it up gives a $-1$ curve we shall call $M'(-1)$. It has a simple intersection with $R'(0)$, which intersects no other curves. Since $R$ has self-intersection $-1$, the only way to achieve that is to blow-up the intersection, giving

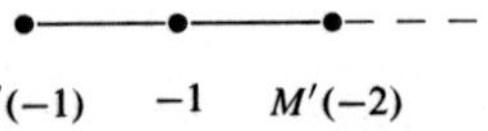


Further, the only way to achieve the string $C(R)$ with $(r-1)$ $-2$ curves is to repeatedly blow up $-1$'s away from $R'(-1)$:

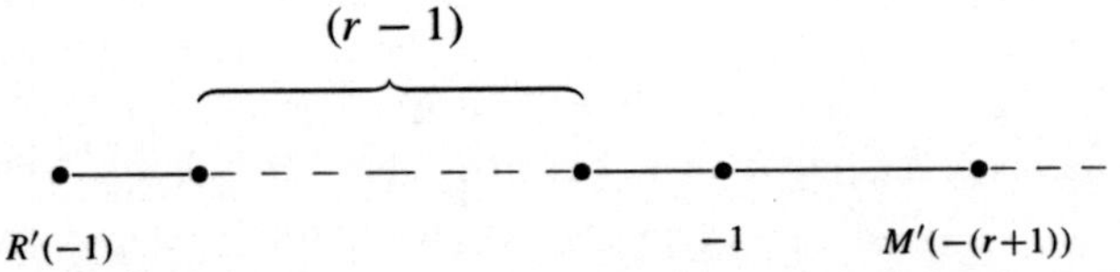


Consider next the other intersections of $M'$. The cubic curve has become $Q'(5)$.

In the cuspidal case, $M'(-(r+1))$ intersects $Q'(5)$ tangentially at a point through which $P'(0)$ passes transversally. Blowing up that point gives a new $-1$ curve $N'(-1)$. It has a simple intersection with $P'(-1)$, which intersects no other curves. Since $P$ has self-intersection $-2$, as above one has to blow-up this intersection point. Continuing as before, one gets part of the graph around $N'$ as

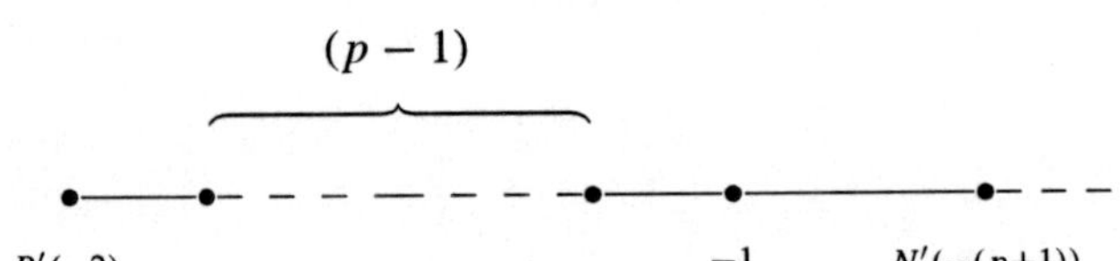

There is as well a point where $N'$, $M'$, and $Q'$ intersect transversally. Since all these curves will appear in $E'$, this point must be blown-up. Putting everything together gives the graph

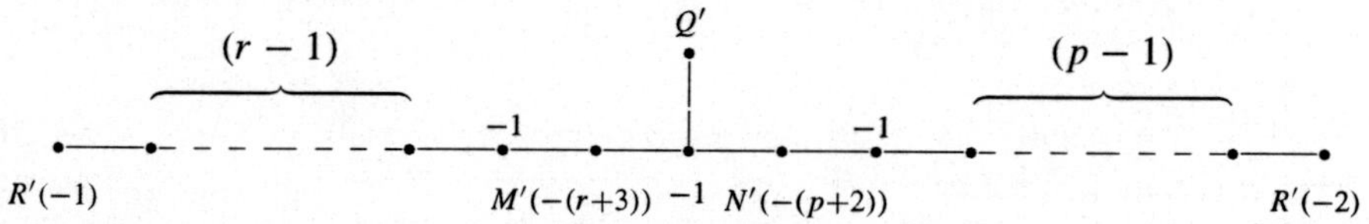


But this configuration cannot be completed to produce $E'$ plus three $-1$ curves. For instance, the $-1$ appended to the chain $\mathcal{C}(R)$ would intersect a curve whose valency remains two, hence an end of a chain. But the self-intersection of that curve is $\leq -3$, an impossibility. The cuspidal case is eliminated.

We are down to the nodal case. $M'(-(r+1))$ intersects $Q'(5)$ transversally in two points, through one of which $P'(0)$ passes with a third tangent direction. Following the same procedure as in the cuspidal case, one reaches the graph

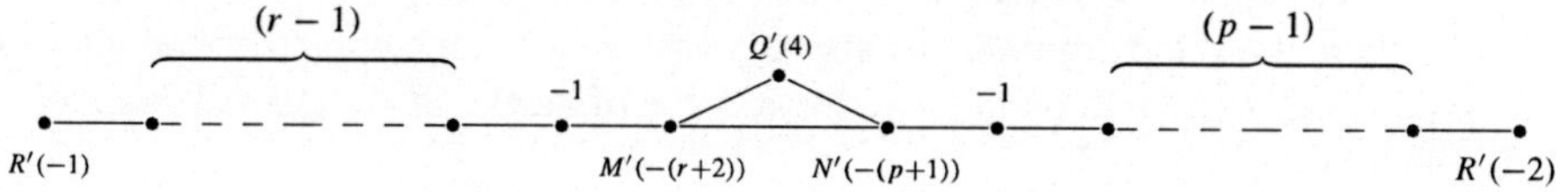


Now, $Q$ has self-intersection $-(q-1)$, so $q+3$ further blow-ups are needed next to $Q'(4)$.

Suppose the first blow-up is between it and $M'(-(r+2))$. Then the new $-1$ is the third one in the diagram, so all future blow-ups of $Q'$ must be adjacent to a $-1$. The final position between $M'$ and $Q'$ is therefore

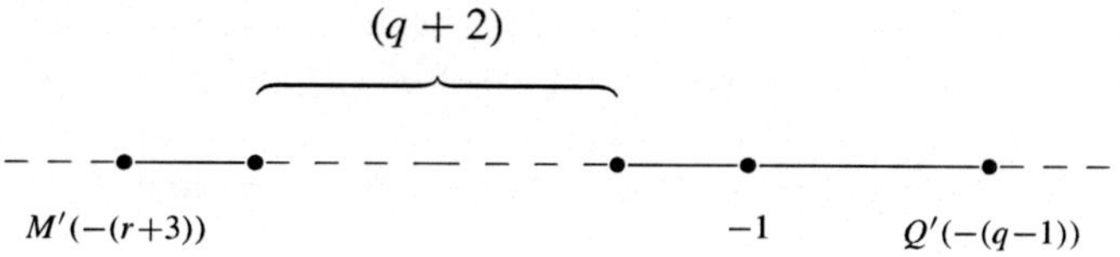


Note that $N'(-(p+1))$ still intersects $M'$ and $Q'$. This fits with the original $E'$ exactly when $r+2=p+1$ and $r+3=p+2$, i.e., $p=r+1$. In that situation, the three desired $-1$'s connect the ends of the three chains as follows: $\mathcal{C}(P)$ to the $-(r+2)$ curve; $\mathcal{C}(R)$ to the $-(p+2)$ curve; and $\mathcal{C}(Q)$ to $Q$. Pulling these $-1$'s back to $Z$ gives the Basic Model for the special value $p=r+1$, when the graph has an obvious symmetry.

If the first blow-up takes place between $Q'(4)$ and $N'(-(p+1))$, then the same procedure as above gives $E'$ for all values of $p, q, r$. Here, the $-1$ locations with the ends of chains are: $\mathcal{C}(P)$ to the $-(p+2)$ curve; $\mathcal{C}(R)$ to the $-(r+2)$ curve; and $\mathcal{C}(Q)$ to $Q$. Pulling back to $Z$, one recovers the $-1$'s for the Basic Model, for all values $p, q, r$ (assuming still that $p, r > 0$).

## 6.4 Type $\Gamma = \mathcal{M}(0, q, r), r \geq 1$

We construct $Z'$ as above, noting that the one curve in the $p$ direction is the new central curve. The diagram is as before, with only two curves adjacent to the central $E_0'$:

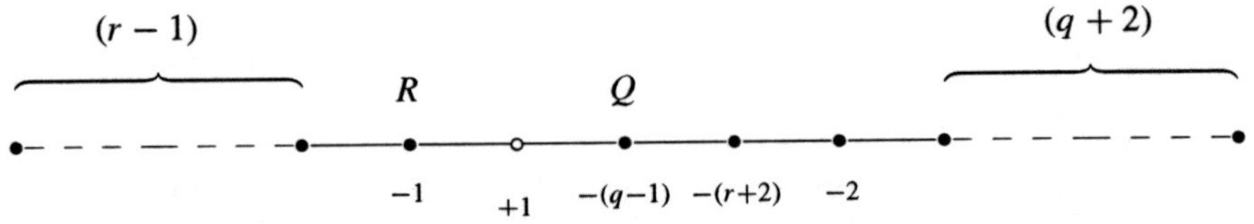


As before, $Q$ is a $-(q-1)$ curve with a tangency of order three with $E_0'$ at a point, and $R$ intersects $E_0'$ at that point, transversally. The basic case requires two $-1$'s to blow down, so the same should be true in general.

If the line $\Phi(R)$ did not pass through the singular point of the cubic $\Phi(Q)$, it would intersect it in two distinct points, giving three singular points on $\Phi(E')$. This cannot happen, so $\Phi(R)$ passes through the singular point. If the singular point is a cusp, the same argument as above produces the same contradiction: the $-1$ at the end of $\mathcal{C}(R)$ would be intersecting a curve of valency two and self-intersection less than or equal to $-3$.

So, the singular point is a node, and one gets a picture as before which gives a $-1$ at the end of $\mathcal{C}(R)$:

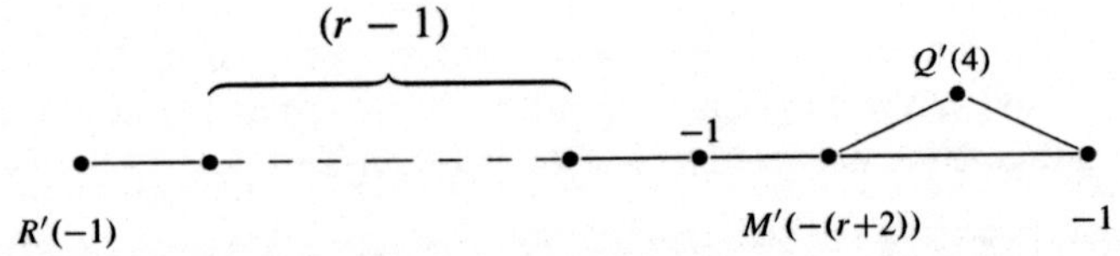


As before, there is a unique way to complete to $E'$, with $-1$ curves appended to the end of chains as follows: $\mathcal{C}(R)$ at the $-(r+2)$ curve, and $\mathcal{C}(Q)$ at $Q$. Pulling back to Z, and adding on the pull-back of the line on $Z'$ through the node and the central point of $L$, gives the Basic Model.

## 6.5 Type $\Gamma = \mathcal{M}(p, q, 0), p \geq 1$

Again, we have the same $Z'$, but $E'$ is now

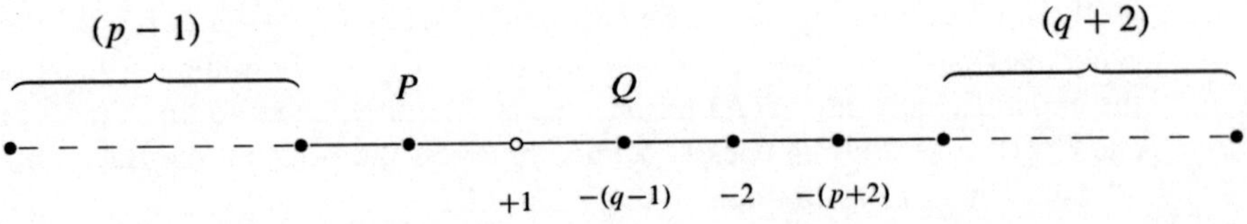

Again $Q$ intersects $E'_0$ with multiplicity three, while $P$ intersects it transversally at a different point. As in the basic case, two $-1$'s are needed to blow-down.

We rule out that $\Phi(E')$ could have two singular points, for which $\Phi(P)$ intersects the cubic at the singular point plus one other. Note that $\mathcal{C}(Q)$ can be made blow-downable with the addition of a single $-1$ only in case $p = q + 2$, and the $-1$ is added to the $-2$ to the right of the $-(p+2)$. In that case, one would have a $-2$ curve of valency three. A brief check of the resolution of $\Phi(E')$ shows that a valency three curve arises only when all non-transversality (from the cusp or node) has been resolved; thus, $\Phi^{-1}(\Phi(E'))$ has normal crossings. Therefore the appended $-1$ curve to $\mathcal{C}(Q)$ intersects $P$ or $Q$ transversally. But blowing-down produces a higher order tangency between two curves, as in Remark 2.3. So this case is eliminated. In addition, one cannot combine $\mathcal{C}(Q)$ and $\mathcal{C}(P)$ with one $-1$ into a blow-downable configuration.

That leaves the case of one singular point of $\Phi(E')$, with $\Phi(P)$ intersecting there with multiplicity three. The same argument as in the original situation shows one cannot have a cuspidal cubic, else $\mathcal{C}(P)$ (or just $P$, if $p = 1$) would have a $-1$ intersecting a valency two curve with self-intersection $\leq -3$. So, one must have a nodal cubic, for which $\Phi(P)$ is tangent to one of the branches of the node. The same partial resolution as above is

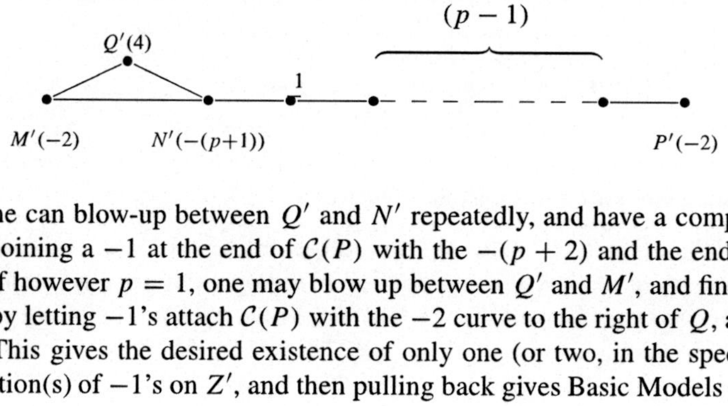


Again, one can blow-up between $Q'$ and $N'$ repeatedly, and have a completion of $E'$ by adjoining a $-1$ at the end of $\mathcal{C}(P)$ with the $-(p+2)$ and the end of $\mathcal{C}(Q)$ with $Q$. If however $p = 1$, one may blow up between $Q'$ and $M'$, and find another solution by letting $-1$'s attach $\mathcal{C}(P)$ with the $-2$ curve to the right of $Q$, and $\mathcal{C}(Q)$ with $Q$. This gives the desired existence of only one (or two, in the special case) configuration(s) of $-1$'s on $Z'$, and then pulling back gives Basic Models on $Z$.

## 6.6 Type $\Gamma = \mathcal{M}(0, q, 0)$

One uses the same $Z'$. There is now one chain, $\mathcal{C}(Q)$, which consists solely of $-2$ curves. One makes it blow-downable only by letting a $-1$ connect $Q$ itself with one or the other end of the chain. It is easy to check that this is possible only if the $-1$ is at the far end of the chain, so there is a unique way to blow-down. So, one must be in the case of the Basic Model.

## 7 Self-Isotropic Subgroups and Fowler's Method

If $L$ is a non-degenerate lattice, the dual $L^* = \mathrm{Hom}(L, \mathbb{Z})$ admits a non-degenerate pairing into $\mathbb{Q}$. Thus, the finite *discriminant group* $D(L) = L^*/L$ admits a non-degenerate *discriminant pairing* into $\mathbb{Q}/\mathbb{Z}$. Overlattices $L \subset M$ of the same rank correspond to isotropic subgroups $\bar{M} = M/L$ of the discriminant group. If $M$ is unimodular, then $\bar{M}$ is self-isotropic. The importance of these notions in smoothing surface singularities may be found in [5, Section 2].

If $\Gamma$ is one of the dual graphs listed in the Bhupal-Stipsicz Theorem, it gives rise to a lattice and discriminant group $D(\Gamma)$. On a $\Gamma$ surface $(Z, E)$, the lattice $\mathbb{E}(\Gamma) = \oplus_i \mathbb{Z}[E_i]$ spanned by the divisor classes in Pic$(Z)$ comes with an identification with the lattice of $\Gamma$.

**Definition 7.1** Let $(Z, E)$ be a $\Gamma$ surface. Then the *self-isotropic group of* $(Z, E)$ is the subgroup $I$ of $D(\Gamma)$ associated with the unimodular overlattice $\mathbb{E}(\Gamma) \subset Pic(Z) \subset \mathbb{E}(\Gamma)^*$.

Fowler studied in [2] the important map

$$\xi : \{\text{Isomorphism classes of } \Gamma \text{ surfaces } (Z, E)\} \to \{\text{Self-isotropic subgroups of } D(\Gamma)\}$$

### 7.1 Fowler's Approach

**Definition 7.2** A self-isotropic subgroup $I \subset D(\Gamma)$ is called *basic* if it is associated with a Basic Model $\Gamma$ surface.

Whenever $D(\Gamma)$ has only one self-isotropic subgroup (as happens most of the time), then of course being "basic" is not an extra condition. On the other hand, there are examples of $\Gamma$ of types $\mathcal{W}, \mathcal{N}, \mathcal{M}$ which have non-basic self-isotropic subgroups (hence the need for the Theorem in this paper). One of Fowler's main theorems is

**Theorem 7.1 ([2])** *Let* $(Z, E)$ *be a* $\Gamma$ *surface whose self-isotropic subgroup is basic. Then* $(Z, E)$ *is basic.*

If a $\Gamma$ surface $(Z, E)$ has basic self-isotropic subgroup $I$, the goal is to prove that it is itself basic. This is accomplished by locating in a precise location $-1$ curves on $Z$ which allow one to blow down in a unique way. The assumption that $I$ is basic implies that for each potential $-1$ curve, there is a line bundle $\mathcal{L}$ with the correct intersection properties with all $E_i$. If $\mathcal{L}$ has a section giving an irreducible curve, it will be the sought after $-1$.

Fowler achieves this for every potential $-1$ curve, via a case by case look at all the types of $\Gamma$, of course assuming that the self-isotropic subgroup is basic.

Fowler starts with general considerations about $K$ similar to those in Sect. 1. He concludes (using Riemann–Roch) that $\mathcal{L}$ has a non-zero section, so is represented by an effective divisor

$$L = \sum n_i E_i + \sum m_j F_j.$$

Here the $F_j$ are irreducible curves not among the $E_i$. One needs to show that all the $n_i$ are 0 as well as all but one of the $m_j$.

The method involves locating in each case various exceptional nef divisors $N$. Then

$$N \cdot L = \sum n_i (N \cdot E_i) + \sum m_j (N \cdot F_j) \geq 0.$$

By $L$'s intersection properties, one can frequently arrange that the product $N \cdot L = 0$. In this case, if $N \cdot E_i > 0$, then necessarily $n_i = 0$. Note that if $F_j$ is not a $-1$ curve, then it is nef (Corollary 2.1).

For instance, in case $\mathcal{W}$, suppose one wishes to prove the existence of a $-1$ curve connecting the end of the chain $\mathcal{C}(P)$ with $R$. Let $L$ be the divisor above representing the potential curve, and choose first $N = E_0$, the central curve. Then $L \cdot N = 0$, while $N$ dots to 1 with the central curve and the three adjacent curves. We conclude that the corresponding four coefficients $n_i$ in the expansion of $L$ equal 0. If we choose as nef divisor $N = (p+2)E_0 + P$, then again $L \cdot N = 0$, so the coefficient of the neighbor of $P$ is 0 as well.

Fowler develops very efficient methods for all $\Gamma$ to show that each potential $-1$ curve actually does exist. This allows careful analysis of the blow-down.

## 7.2 Number of $\mathbb{Q}$HD Smoothing Components

Combining the main result Theorem 1.3 of this paper with Fowler's results, here is the final count of $\mathbb{Q}$HD smoothing components for weighted homogeneous singularities:

1. Two components for $\mathcal{W}(p, p, p)$, $\mathcal{N}(q+2, q, 0)$, and $\mathcal{M}(r+1, q, r)$, with two different self-isotropic subgroups in each case.
2. A unique component for all other $\mathcal{W}$, $\mathcal{N}$, and $\mathcal{M}$.
3. A unique component for type $\mathcal{B}^3_2$ and $\mathcal{C}^3_3$.
4. Two components for type $\mathcal{C}^3_2$, with the same self-isotropic subgroup.
5. Two components for type $\mathcal{A}^4$ with two different isotropic subgroups in each case.
6. Two components for types $\mathcal{B}^4$ and $\mathcal{C}^4(p)$, $p > 0$, with one self-isotropic subgroup in each case.
7. A unique component for type $\mathcal{C}^4(0)$.

As mentioned before, Fowler shows the existence of two components is a consequence either of a symmetry in the graph $\Gamma$ or of complex conjugation in the blowing-up process.

# Appendix

See Tables A.1 and A.2.

**Table A.1** Graphs in the families $\mathcal{W}, \mathcal{N}, \mathcal{M}, \mathcal{B}_2^3, \mathcal{C}_2^3, \mathcal{C}_3^3, \mathcal{A}^4, \mathcal{B}^4$, and $\mathcal{C}^4$

| Family | Graph |
|---|---|
| $\mathcal{W}(p,q,r)$ | 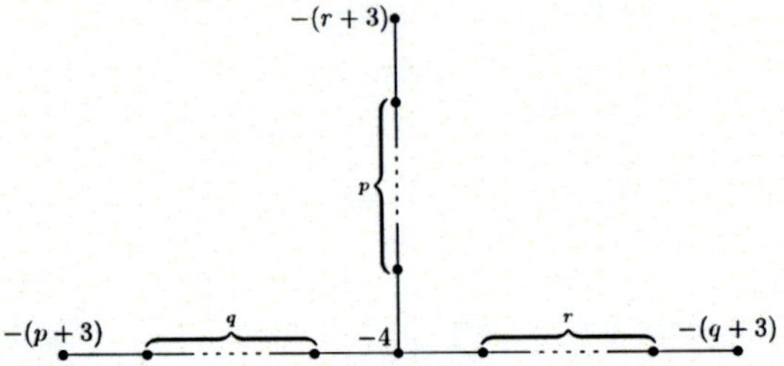 |
| $\mathcal{N}(0,q,r)$ | 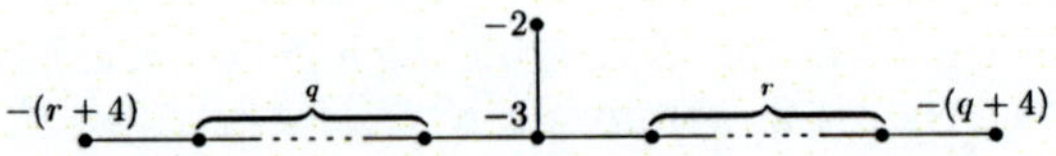 |
| $\mathcal{N}(p,q,r)$ $p \geq 1$ | 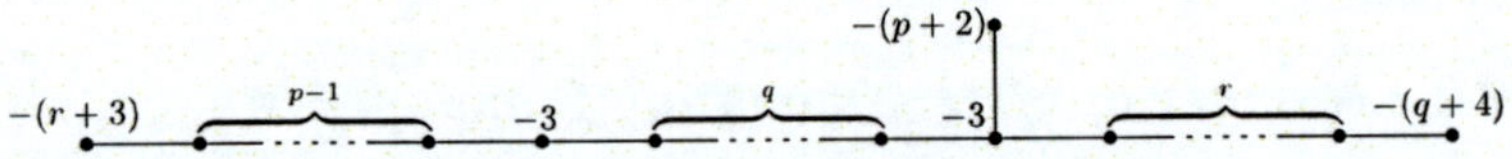 |
| $\mathcal{M}(0,q,0)$ | 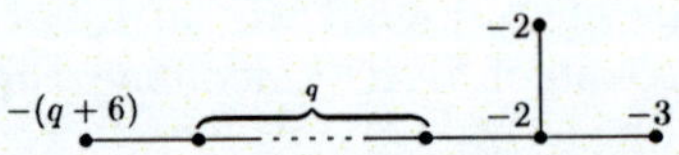 |
| $\mathcal{M}(0,q,r)$ $r \geq 1$ | 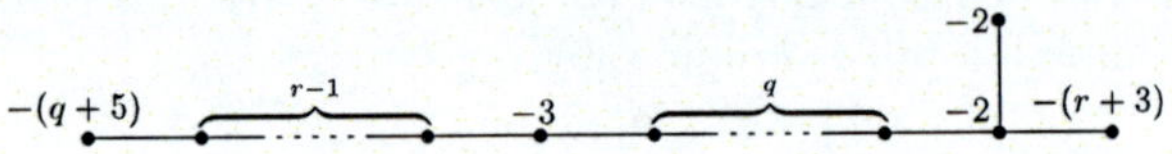 |
| $\mathcal{M}(p,q,0)$ $p \geq 1$ | 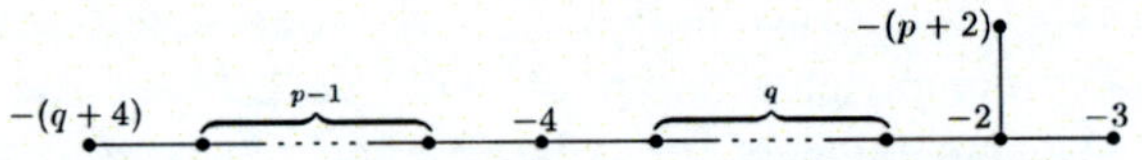 |

Continued on next page

**Table A.1** (Continued)

| Family | Graph |
|---|---|
| $\mathcal{M}(p,q,r)$ <br> $p, r \geq 1$ | 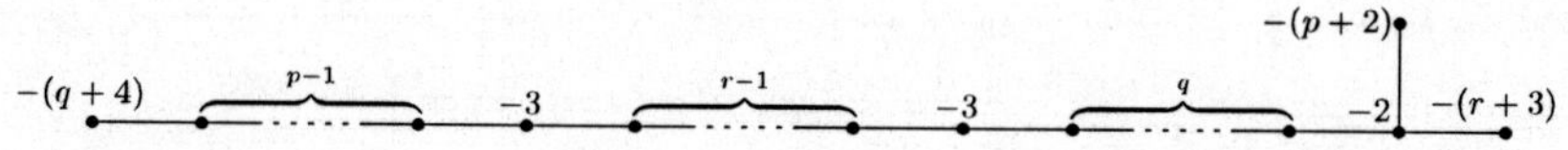  |
| $\mathcal{B}^3_2(p)$ | 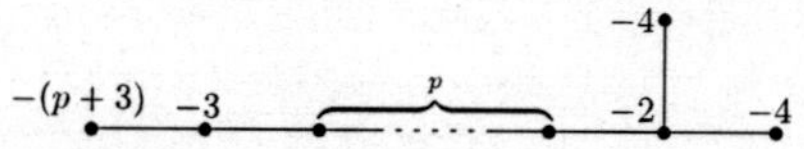  |
| $\mathcal{C}^3_2(p)$ | 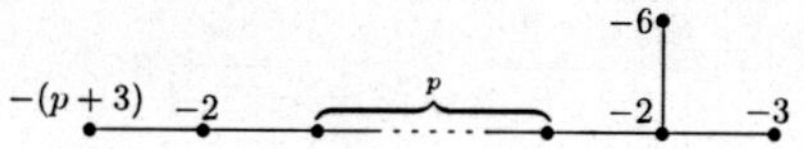  |
| $\mathcal{C}^3_3(p)$ | 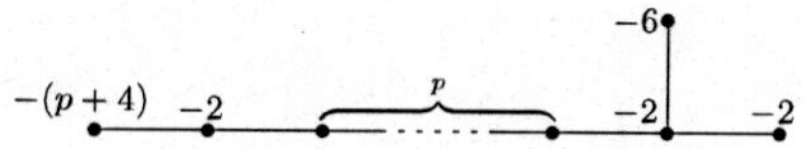  |
| $\mathcal{A}^4(p)$ | 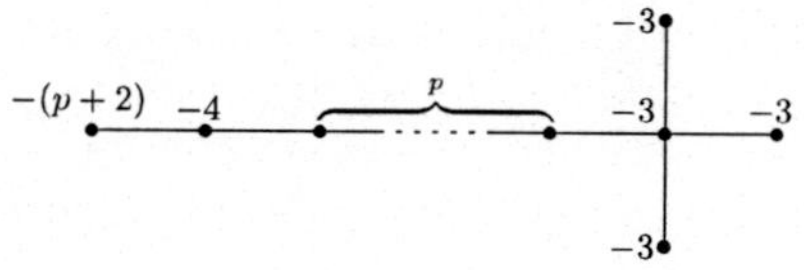  |
| $\mathcal{B}^4(p)$ | 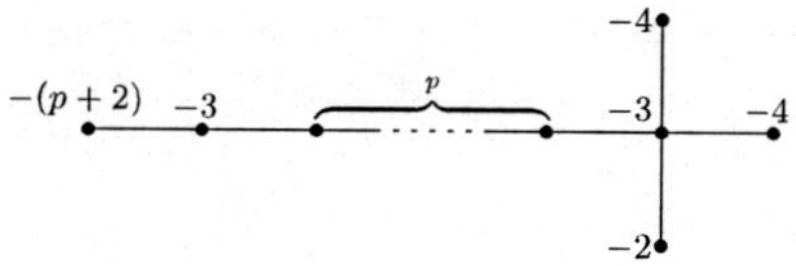  |
| $\mathcal{C}^4(p)$ | 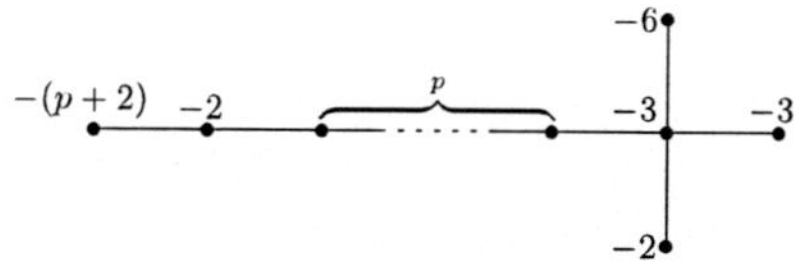  |

**Table A.2** The dual graphs to the graphs in the families $\mathcal{W}, \mathcal{N}, \mathcal{M}, \mathcal{B}_2^3, \mathcal{C}_2^3, \mathcal{C}_3^3, \mathcal{A}^4, \mathcal{B}^4$, and $\mathcal{C}^4$

| Family | Dual graph |
|---|---|

$\mathcal{W}(p, q, r)$

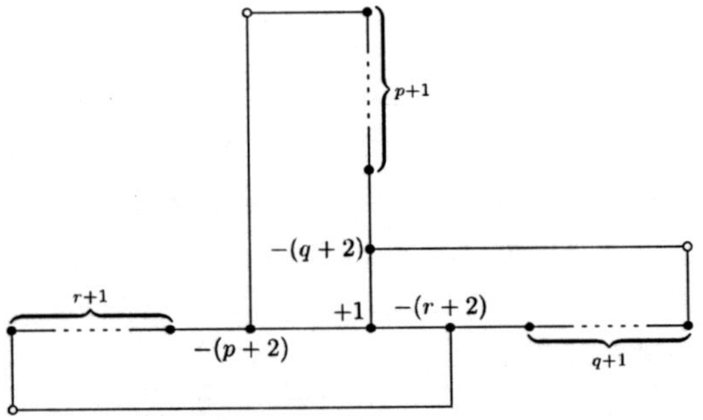


$\mathcal{N}(p, q, r)$

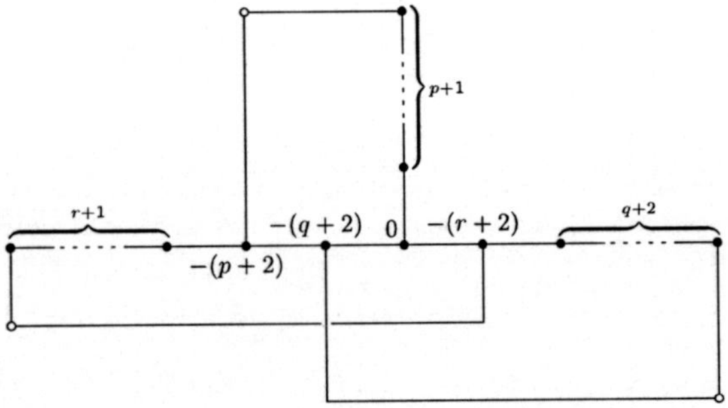


$\mathcal{M}(p, q, r)$

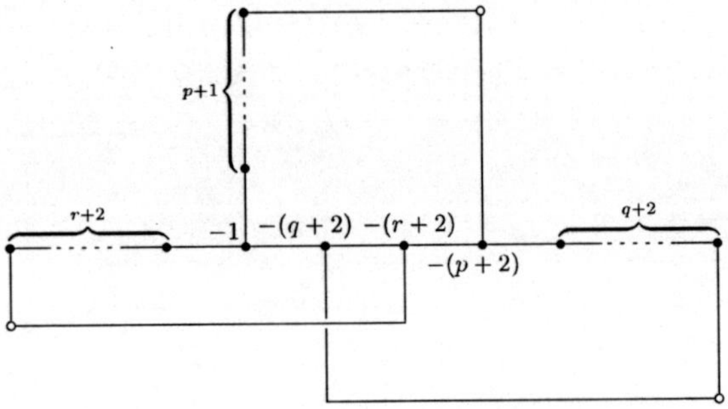


Continued on next page

**Table A.2** (Continued)

| Family | Dual graph |
|---|---|
| $\mathcal{B}_2^3(p)$ | 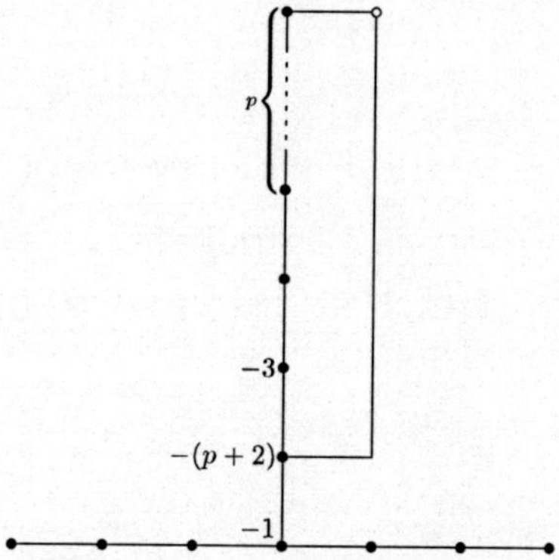  |
| $\mathcal{C}_2^3(p)$ | 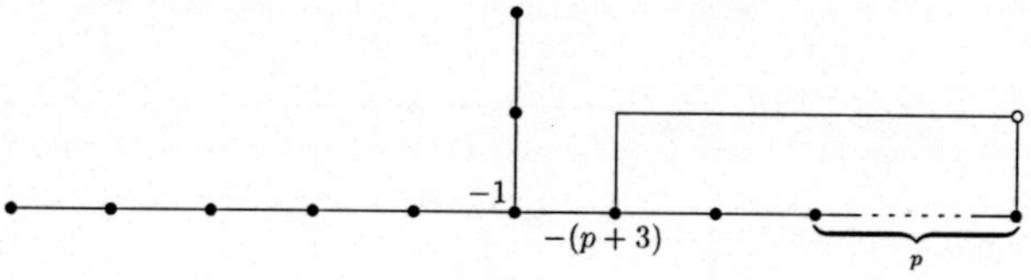  |
| $\mathcal{C}_3^3(p)$ | 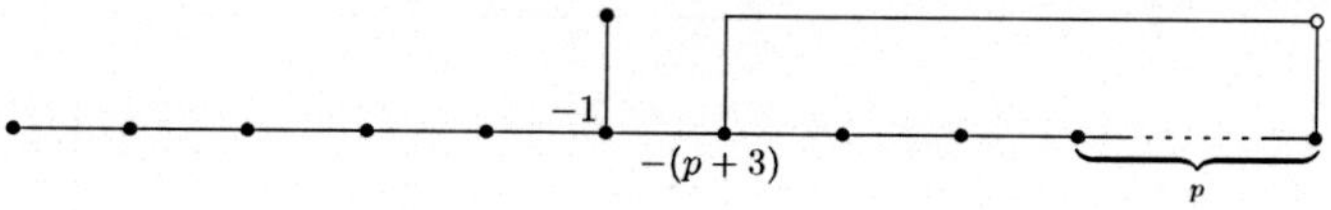  |
| $\mathcal{A}^4(p)$ | 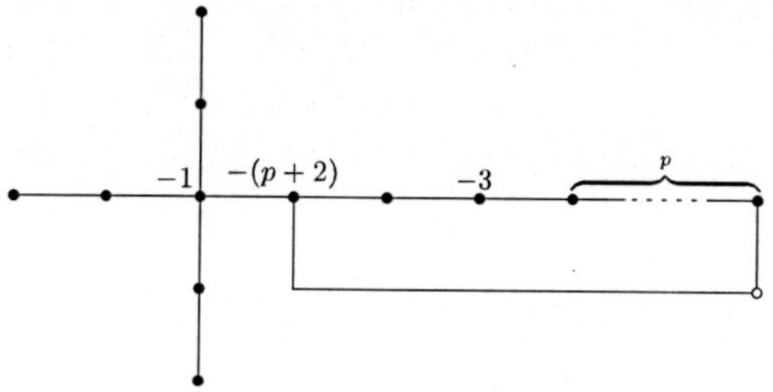  |

Continued on next page

**Table A.2** (Continued)

| Family | Dual graph |
|---|---|
| $\mathcal{B}^4(p)$ | 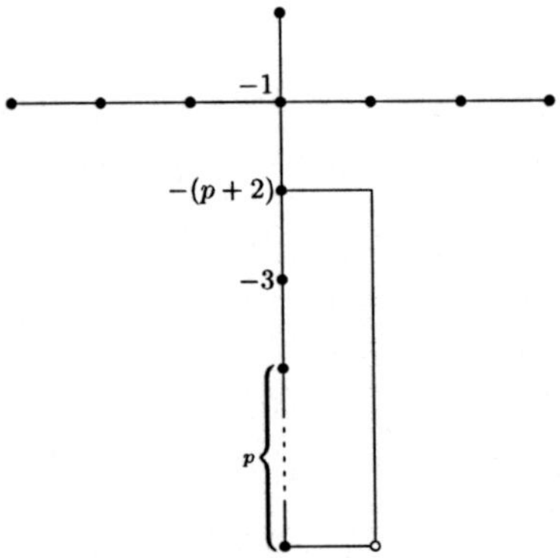 |
| $\mathcal{C}^4(p)$ | 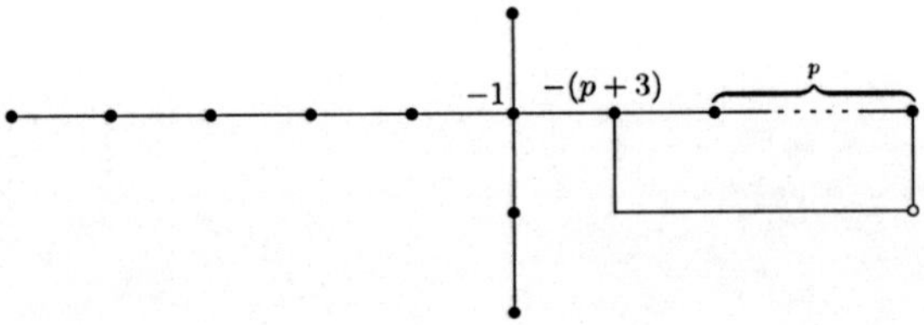 |

## References


1. Bhupal, M., Stipsicz, A.: Weighted homogeneous singularities and rational homology disk smoothings. Am. J. Math. **133**, 1259–1297 (2011)
2. Fowler, J.: Rational homology disk smoothing components of weighted homogeneous surface singularities. Ph.D. Thesis, University of North Carolina (2013). https://cdr.lib.unc.edu/concern/dissertations/b2773w797
3. Greuel, G.-M., Steenbrink, J.H.M.: On the topology of smoothable singularities. Proc. Symp. Pure Math. **40** (part I), 535–545 (1983)
4. Laufer, H.B.: Taut two-dimensional singularities. Math. Ann. **205**, 131–164 (1973)
5. Looijenga, E., Wahl, J.: Quadratic functions and smoothing surface singularities. Topology **25**, 261–291 (1986)
6. Pinkham, H.: Deformations of normal surface singularities with $\mathbb{C}^*$ action. Math. Ann. **232**, 65–84 (1978)
7. Stipsicz, A., Szabó, Z., Wahl, J.: Rational blowdowns and smoothings of surface singularities. J. Topol. **1**, 477–517 (2008)
8. Wahl, J.: Elliptic deformations of minimally elliptic singularities. Math. Ann. **253**, 241–262 (1980)
9. Wahl, J.: Smoothings of normal surface singularities. Topology **20**, 219–246 (1981)
10. Wahl, J.: On rational homology disk smoothings of valency 4 surface singularities. Geom. Topol. **15**, 1125–1156 (2011)
11. Wahl, J.: Log-terminal smoothing of graded normal surface singularities. Michigan Math. J. **62**, 475–489 (2013)

# On Tjurina Transform and Resolution of Determinantal Singularities

**Helge Møller Pedersen**

**Abstract** Determinantal singularities are an important class of singularities, generalizing complete intersections, which recently have seen a large amount of interest. They are defined as preimage of $M^t_{m,n}$ the sets of matrices of rank less than $t$. The rank stratification of $M^t_{m,n}$ gives rise to some interesting structures on determinantal singularities. In this article we will focus on one of these, namely the *Tjurina transform*. We will show some properties of it, and discuss how it can or cannot be used to find resolutions of determinantal singularities.

**Keywords** Resolution of singularities · Determinantal singularities · Nash transformation · Tjurina transformation

**Subject Classifications** 14B05, 32S05, 32S45

## 1 Introduction

Hypersurface singularities have in general been the starting point of singularity theory. They have some very good properties, one of the most important is the existence of the Milnor fibration [8]. The Milnor fibration makes it possible to define the Milnor number $\mu$, which is a very important invariant. So a goal in singularity theory is to find more general families of singularities, for which it is possible to define the Milnor number. A classical example of a generalization, for which the Milnor number can be defined, is the isolated complete intersections. Determinantal singularities are a generalization of complete intersections. They are defined as the preimage of the set of $m \times n$ matrices of rank less than $t$ under certain holomorphic maps. They have seen a lot of interest lately, including several different ways to define the Milnor number of certain classes of determinantal varieties by Ruas and

H. M. Pedersen (✉)
Departamento de Matemática, Universidade Federal do Ceará, Fortaleza, CE, Brazil
e-mail: helge@mat.ufc.br



J. Fernández de Bobadilla et al. (eds.), *Singularities and Their Interaction with Geometry and Low Dimensional Topology*, Trends in Mathematics,
https://doi.org/10.1007/978-3-030-61958-9_13

Pereira [10], Damon and Pike [2] and Nuño Ballesteros et al. [9]. Moreover, Ebeling and Gusein-Zade defined the index of a 1-form [3], and their deformation theory has also been studied by Gaffney and Rangachev in [6].

In this article we study other aspects of determinantal singularities, not directly related to deformation theory, namely, transformations and resolutions. They played a very important role in [3], and the Tjurina transform, which will be one of our main subjects, was also studied for the case Cohen-Macaulay codimension 2 by Frühbis-Krüger and Zach in [5].

We first recall the Tjurina transform, Tjurina transpose transform and Nash transform for the model determinantal singularity in Sect. 3 as done in [3]. This will be our starting point for introducing the transformations for general determinantal singularities. We also explore how these transformations are related and how they are not, and give a description of their homotopy type. We introduce the Tjurina transform (and its transpose) for general determinantal singularities in Sect. 4, give some general properties, for example that the Tjurina transforms of most complete intersections are themselves complete intersections, and give some methods to find the Tjurina transform. In Sect. 5 we show that under some general assumptions the Tjurina transform or its dual is a complete intersection. This means that Tjurina transform cannot be used to provide resolutions in general, but in Sect. 6 we illustrate that by changing the determinantal type of the Tjurina transform of certain hypersurface singularities, we can continue the process of taking Tjurina transform, and in the end reach a resolution. Section 2 introduces determinantal singularities and notions of transformations used throughout the article.

## 2 Preliminaries

In this section we give the basic definitions and properties of determinantal varieties/singularities, and transformations we will need. We will in general follow the notation used in [3].

### 2.1 Determinantal Singularities

Let $M_{m,n}$ be the set of $m \times n$ matrices over $\mathbb{C}$. Then we define the *model determinantal variety of type* $(m, n, t)$, denoted by $M^t_{m,n}$, for $1 \leq t \leq \min\{n, m\}$ to be the subset of $M_{m,n}$ consisting of matrices $A$ of $\operatorname{rank}(A) < t$. $M^t_{m,n}$ has a natural structure of an irreducible algebraic variety, with defining equations given by requiring that the $t \times t$ minors have to vanish. The dimension of $M^t_{m,n}$ is $mn - (m - t + 1)(n - t + 1)$. The model determinantal variety is often called generic determinantal variety as for example in [10].

The singular set of $M_{m,n}^t$ is $M_{m,n}^{t-1}$ and the decomposition of $M_{m,n}^t = \bigcup_{i=1}^t (M_{m,n}^i \setminus M_{m,n}^{i-1})$, where $M_{m,n}^0 := \emptyset$, is a Whitney stratification.

Let $F : U \subseteq \mathbb{C}^N \to M_{m,n}$ be a map with holomorphic entries. $X := F^{-1}(M_{m,n}^t)$ is a *determinantal variety of type* $(m, n, t)$ if $\operatorname{codim}(X) = \operatorname{codim}(M_{m,n}^t) = (m - t + 1)(n - t + 1)$. $X$ has the structure of an analytic variety, with equations defined by the vanishing of the $t \times t$ minors of the matrix $F(x)$. We call this a variety as is custom in the treatment of determinantal singularities and singularity theory in general, even though $X$ need not to be reduced or irreducible. The question if $X$ is reduced or not is not important, since we will always consider $X$ as a subset of $\mathbb{C}^N$ (later a subset of a complex manifold) equipped with the classical topology. In general in this article by a variety we mean a subset of $\mathbb{C}^N$ or a complex analytic manifold given locally as the zero set of a set of analytic equations with appropriate compatibility conditions. This means that we do not distinguish between a set given by non reduced equations, and the same set given by their reduced equations. We also do not make any assumptions on irreducibility. This is because we are interested in studying the classical topology of these sets which does not see whether the equations are reduced or not. Also even if we start with a reduced and irreducible equation, then many of the constructions we will make from them will not give reduced or irreducible equations.

The singular set of $X$ includes $F^{-1}(M_{m,n}^{t-1})$. We make a decomposition $X = \bigcup_{i=1}^t X_i$, where $X_i := F^{-1}(M_{m,n}^i \setminus M_{m,n}^{i-1})$. Notice that even if $X = F^{-1}(M_{m,n}^t)$ is an irreducible determinantal singularity which is given by reduced equations then $Y := F^{-1}(M_{m,n}^s)$ for $s < t$ might not be irreducible, might not be a determinantal singularity or might be given by non reduced equations.

When we talk about the determinantal variety $X$, we do not just consider $X$ as a variety in $\mathbb{C}^N$ but also the map $F : U \subseteq \mathbb{C}^N \to M_{m,n}$ used to define the variety. We will as is customary not include $F$ in the notation and just write $X$, but one has to remember that the determinantal singularity also includes the map $F$. We will therefore also only consider two determinantal varieties $X$ and $X'$ equal if they are given by the same map.

We define *determinantal singularities* as germs of determinantal varieties, i.e. a germ of a space $(X, 0)$ defined as the preimage if of $M_{m,n}^t$ under a germ of a holomorphic map $F : (U, 0) \subseteq (\mathbb{C}^N, 0) \to M_{m,n}$.

Let $\mathrm{GL}_n(\mathcal{O}_N)$ be the group of invertible $n \times n$ matrices with entries in $\mathcal{O}_N$ the sheaf of germs of holomorphic functions on $\mathbb{C}^N$. Let $\mathcal{H} := \mathrm{GL}_m(\mathcal{O}_N) \times \mathrm{GL}_n(\mathcal{O}_N)$ and $\mathcal{R}$ the group of analytic isomorphisms of $(\mathbb{C}^n, 0)$. Then the group $\mathcal{R} \times \mathcal{H}$ acts on map-germs $F : (U, 0) \subseteq (\mathbb{C}^N, 0) \to M_{m,n}$ by composition in the source and multiplication on the left and on the right in the target. We say that two determinantal singularities $(X, 0)$ and $(Y, 0)$ are equivalent (or G-equivalent), if their defining maps are in the same orbit of this action. This in particular implies that $(X, 0)$ is isomorphic to $(Y, 0)$ as germs of varieties.

If $F$ is transverse to the stratum $M_{m,n}^i \setminus M_{m,n}^{i-1}$ at $F(x)$, then the singularity at $x$ only depends upon $\operatorname{rank}(F(x))$. We therefore call such a point *essentially nonsingular*. This naturally leads to the next definition.

**Definition 2.1** Let $(X, 0)$ be a determinantal singularity defined by the map-germ $F$. Then $(X, 0)$ is an *essentially isolated determinantal singularity* (or EIDS for short) if there exists a neighbourhood $U \subset \mathbb{C}^N$ of the origin such that all points $x \in U \setminus \{0\}$ are essentially nonsingular.

An EIDS needs of course not be smooth, but the singularities away from $\{0\}$ are controlled, i.e. they only depend on the strata they belong to. An example of an EIDS is any isolated complete intersection given the type of a $(1, m, 1)$ (or $(m, 1, 1)$) determinantal singularity.

If $(X, 0)$ is a determinantal singularity of type $(m, n, t)$ given by $F: U \subseteq \mathbb{C}^N \to M_{m,n}$ satisfying $F(0) \neq 0$ and $s := \operatorname{rank} F(0)$, then one can find another map $F': U' \subseteq \mathbb{C}^N \to M_{m-s,n-s}$ with $F'(0) = 0$ such that $F'$ gives $(X, 0)$ the structure of a determinantal singularity of type $(m-s, n-s, t-s)$ where $U' \subseteq U$ are open neighbourhoods of the origin. This can be done by action on $F$ by $\mathcal{H}$ to be of the form $\left(\begin{array}{c|c} \mathrm{id}_s & 0 \\ \hline 0 & F' \end{array}\right)$ in a neighbourhood of 0.

## 2.2 *Transformations*

**Definition 2.2** Let $X$ be a variety and $V \subset X$ a closed subvariety, then a *transformation of* $(X, V)$ is a variety $\widetilde{X}$ together with a proper surjective analytic morphism $\pi: \widetilde{X} \to X$, such that $\pi: \pi^{-1}(X \setminus V) \to X \setminus V$ is an isomorphism and $\overline{\pi^{-1}(X \setminus V)} = \widetilde{X}$.

Here closure is the topological closure in the classical topology. The last requirement ensures that $\dim(\pi^{-1}(V)) < \dim(X)$.

This definition is sometimes also called a modification, but since we in this paper work with the Tjurina transform, we will use the word transform.

A resolution of $(X, SingX)$ is then just a transformation where $\widetilde{X}$ is smooth. We want to compare the different transformations, so we define a map between transformations as follows.

**Definition 2.3** Let $f: T_1 \to T_2$ be a map between two different transformations of the same space and subspace $\pi_i: (T_i, E_i) \to (X, V)$. Then we call $f$ a map of transformations if $\pi_1 = \pi_2 \circ f$. We call a map of transformation $f$ an analytic morphism of transformations if it is an analytic morphism and an isomorphism of transformations, if it is an isomorphism of varieties.

# 3 Resolutions of the Model Determinantal Varieties

In [3] the authors introduce three different natural ways to resolve the model determinantal variety $M^t_{m,n}$. The first is the same as the Tjurina transform of $(M^t_{m,n}, M^{t-1}_{m,n})$ which was introduced by Tjurina in [12], and also used [13] and

[5]. Kempf also introduced the same transformation in his thesis [7] under the name *canonical desingularization*, and under that name it is for example used by Eisenbud in [4]. The Tjurina transform is defined as the following variety in $M_{m,n} \times \mathrm{Gr}(n-t+1,n)$:

$$\begin{aligned}\mathrm{Tjur}(M^t_{m,n}) :&= \{(A, W) \in M_{m,n} \times \mathrm{Gr}(n-t+1,n) \mid A(W) = 0\}\\ &= \{(A, W) \in M_{m,n} \times \mathrm{Gr}(n-t+1,n) \mid W \subseteq \ker(A)\}\end{aligned}$$

by considering $A \in M^t_{m,n}$ as a linear map $A\colon \mathbb{C}^n \to \mathbb{C}^m$. It is shown in [1], that this is a smooth variety. Let $\pi\colon \mathrm{Tjur}(M^t_{m,n}) \to M^t_{m,n}$ be the restriction of the projection to the first factor. Then over the regular part of $M^t_{m,n}$ we have that the map $A \to (A, \ker A)$ is an inverse to $\pi$, hence $\pi\colon \mathrm{Tjur}(M^t_{m,n}) \to M^t_{m,n}$ is a resolution. Corollary 3.3 in [5] shows that their definition gives the same as this one, their proof also works for general $n$, $m$ and $t$.

The second resolution is as the Tjurina, but considering $A \in M^t_{m,n}$ as a linear map $A\colon \mathbb{C}^m \to \mathbb{C}^n$. This is of course the map given by the transpose of $A$, so we get the following:

$$\begin{aligned}\mathrm{Tjur}^T(M^t_{m,n}) :&= \{(A, W) \in M_{m,n} \times \mathrm{Gr}(m-t+1,m) \mid A^T(W) = 0\}\\ &= \{(A, W) \in M_{m,n} \times \mathrm{Gr}(m-t+1,m) \mid W \subseteq \ker(A^T)\}\end{aligned}$$

It is clear from the definition that this is also a smooth variety, the same proof as in the case of Tjurina transform works. If one chooses a Hermitian inner product on $\mathbb{C}^m$, then one gets that the relation $W \subseteq \ker(A^T)$ is equivalent to the relation $\mathrm{Im}(A) \subseteq \bar{W}^{\perp}$ where $V^{\perp}$ is the orthogonal complement with respect to the Hermitian inner product and $\bar{W}$ is the image of $W$ under the real linear isomorphism given by complex conjugation. The choice of Hermitian inner product also gives an isomorphism of real algebraic varieties between $\mathrm{Gr}(m-t+1,m)$ and $\mathrm{Gr}(t-1,m)$ defined by sending $V$ to $V^{\perp}$. Hence composing this with the real algebraic isomorphism induced on $\mathrm{Gr}(m-t+1,m)$ by complex conjugation gives an real isomorphism of $\mathrm{Gr}(m-t+1,m)$ and $\mathrm{Gr}(t-1,m)$ defined by sending $W$ to $\bar{W}^{\perp}$. Using this we get that this transform is also real isomorphic to:

$$\mathrm{Tjur}^T(M^t_{m,n}) \cong \{(A, V) \in M_{m,n} \times \mathrm{Gr}(t-1,m) \mid \mathrm{Im}(A) \subseteq V\}. \tag{1}$$

This resolution is called the *dual canonical resolution* in [7].

The third resolution considered by Ebeling and Gusein-Zade is the Nash transform of $M^t_{m,n}$. In section 1 of [3] they show how to get the Nash transform which can be stated as the following proposition:

**Proposition 3.1** *For a model determinantal variety the Nash transform is isomorphic to the following variety:*

$$\{(A, W_1, W_2) \in M_{m,n} \times \mathrm{Gr}(n-t+1,n) \times \mathrm{Gr}(t-1,m) \mid \ker(A) \supseteq W_1 \text{ and } \mathrm{Im}(A) \subseteq W_2\}.$$

It is only a sketch of a proof to this proposition that is given in [3], and we will below show that the two different spaces in the proposition are homeomorphic (remember we are using the classical topology, and not the Zariski topology).

***Proof of Homeomorphism*** In [1] they show that for $A \in M^t_{m,n} \setminus M^{t-1}_{m,n}$, that is the regular points, $T_A M^t_{m,n} = \{B \in M_{m,n} \mid B\big(\ker(A)\big) \subseteq \mathrm{Im}(A)\}$. Consider the map $\alpha\colon \mathrm{Gr}(n-t+1, n) \times \mathrm{Gr}(t-1, m) \to \mathrm{Gr}(d^t_{m,n}, mn)$, where $d^t_{m,n} := mn - (m-t+1)(n-t+1) = \dim(M^t_{m,n})$, given by $\alpha(V, W) := \{B \in M_{m,n} \mid B(V) \subseteq W\}$. It is clear that $\alpha(V, W)$ is a linear subspace of $M_{m,n}$. To find the dimension of $\alpha(V, W)$ we will use the basis $B_{ij}$ of $M_{m,n}$ defined given a basis $v_j$ of $\mathbb{C}^n$ and a basis $w_i$ of $\mathbb{C}^m$ as $B_{ij}(v_j) = w_i$ and $\mathrm{rank}(B_{ij}) = 1$. We choose a basis of $\mathbb{C}^n$ such that $V = \mathrm{Span}\{v_1, \ldots, v_{n-t+1}\}$ and a basis of $\mathbb{C}^m$ such that the $W = \mathrm{Span}\{w_1, \ldots, w_{t-1}\}$. Then $\alpha(V, W)$ is spanned by the $B_{ij}$'s that send one of the first $n-t+1$ basis vectors of $C^n$ to one of the first $t-1$ basis vectors of $\mathbb{C}^m$, and the $B_{ij}$'s that send one of the last $t-1$ basis vectors of $\mathbb{C}^n$ to any basis vector of $\mathbb{C}^m$. This implies that $\dim \alpha(V, W) = (n-t+1)(t-1) + (t-1)m = d^t_{m.n}$. Hence $\alpha(V, W) \in \mathrm{Gr}(d^t_{m,n}, mn)$.

We will first show that $\alpha$ is injective. Assume that there exist two pairs $(W_1, W_2)$ and $(V_1, V_2)$ such that $\alpha(W_1, W_2) = \alpha(V_1, V_2)$. Assume that $W_1 \neq V_1$, let $v_1 \in V_1$ and $v_1 \notin W_1$, since $\dim(W_1) = \dim(V_1)$ such a $v_1$ exists, and choose $v_2 \notin V_2$. Define the linear map $B$ as the map of rank 1 with $B(v_1) := v_2$. Then $B(W_1) = \{0\} \subseteq W_2$ and hence $B \in \alpha(W_1, W_2)$, but $B(V_1) = \mathrm{Span}\{v_2\} \nsubseteq V_2$, so $B \notin \alpha(V_1, V_2)$ and we have a contradiction. Assume now that there exist pairs $(W_1, W_2)$ and $(W_1, V_2)$ such that $\alpha(W_1, W_2) = \alpha(W_1, V_2)$. Assume that $W_2 \neq V_2$, choose $v_1 \in W_1$ and choose $v_2 \in V_2$ and $v_2 \notin W_2$, since $\dim(W_2) = \dim(V_2)$ such a $v_2$ exists. Define $B$ as the linear map as the map of rank 1 with $B(v_1) := v_2$. Then $B(W_2) = \mathrm{Span}\{v_2\} \subseteq V_2$ so $B \in \alpha(W_1, V_2)$, but $\mathrm{Span}\{v_2\} \nsubseteq W_2$ so $B \notin \alpha(W_1, W_2)$ so we have a contradiction. This shows that $\alpha$ is injective.

Next we will show that $\alpha$ is continuous. Let $(V_i, W_i) \in \mathrm{Gr}(n-t+1, n) \times \mathrm{Gr}(t-1, m)$ be a convergent sequence and let $(V, W) := \lim(V_i, W_i)$. Let $\mathcal{B}_i := \alpha(V_i, W_i)$, and choose a convergent subsequence $\mathcal{B}'_i$ which exists because $\mathrm{Gr}(d^t_{m,n}, mn)$ is compact. Let $\mathcal{B} := \lim \mathcal{B}'_i$, choose $B \in \mathcal{B}$ and $B_i \in \mathcal{B}'_i$ a sequence of matrices converging to $B$. Choose $v \in V$ and $v_i \in V_i$ a sequence converging to $v$, set $w_j := B_j v_j$ for any $j$ where $B_j$ is defined. Now since $B_j$ and $v_j$ converge, $w_j$ converges to $w := Bv$, but $w_j \in W_j$ and hence its limit is in $W$. So for all $v \in V$ and all $B \in \mathcal{B}$ $Bv \in W$, hence $\mathcal{B} \subset \alpha(V, W)$. But since $\dim\big(\mathcal{B}\big) = \dim\big(\alpha(V, W)\big)$ we have that $\mathcal{B} = \alpha(V, W)$. So any convergent subsequence of $\mathcal{B}_i$ converges to $\alpha(V, W)$, this implies that $\mathcal{B}_i$ converges to $\alpha(V, W)$ since $\mathrm{Gr}(d^t_{m,n}, mn)$ is compact. Therefore, $\lim \alpha(V_i, W_i) = \alpha(\lim(V_i, W_i))$ for all convergent sequences, hence $\alpha$ is continuous.

Since $\alpha$ is a continuous map from a compact Hausdorff space to a compact space it is closed, and since it is injective this implies it is an topological embedding (Closed Map Lemma).

Let $\beta\colon (M^t_{m,n} \setminus M^{t-1}_{m,n}) \to M_{m,n} \times \mathrm{Gr}(n-t+1, n) \times \mathrm{Gr}(t-1, m)$ be the map $\beta(A) = \big(A, \ker(A), \mathrm{Im}(A)\big)$. We define the map $\alpha'\colon M_{m,n} \times \mathrm{Gr}(n-t+1, n) \times$

$\mathrm{Gr}(t-1,m) \to M_{m,n} \times \mathrm{Gr}(d^t_{m,n}, mn)$ by $\alpha'(A, V, W) = (A, \alpha(V, W))$. Then $(\alpha' \circ \beta)(A) = (A, \mathcal{B})$, where

$$\mathcal{B} = \alpha\big(\ker(A), \mathrm{Im}(A)\big) = \{B \in M_{m,n} \mid B\big(\ker(A)\big) \subseteq \mathrm{Im}(A)\} = T_A M^t_{m,n}$$

So $\alpha' \circ \beta$ is the same as the Gauss map on the regular part of $M^t_{m,n}$. Then we have that $\mathrm{Nash}(M^t_{m,n}) = \overline{(\alpha' \circ \beta)(M^t_{m,n} \setminus M^{t-1}_{m,n})}$. Since $\alpha$ and hence $\alpha'$ is a closed topological embedding we have $\mathrm{Nash}(M^t_{m,n}) = \alpha'\Big(\overline{\beta(M^t_{m,n} \setminus M^{t-1}_{m,n})}\Big)$. Moreover, since $\alpha'$ is an embedding it follows that $\mathrm{Nash}(M^t_{m,n})$ is homeomorphic to $\overline{\beta(M^t_{m,n} \setminus M^{t-1}_{m,n})}$.

The last part of the proof is determining $\overline{\beta(M^t_{m,n} \setminus M^{t-1}_{m,n})}$. Now $\beta(M^t_{m,n} \setminus M^{t-1}_{m,n}) = \{(A, \ker A, \mathrm{Im}\, A) \in M_{m,n} \times \mathrm{Gr}(n-t+1, n) \times \mathrm{Gr}(t-1, m)\}$ and we want to show that the closure $\mathcal{N}$ is

$$\{(A, V, W) \in M_{m,n} \times \mathrm{Gr}(n-t+1, n) \times \mathrm{Gr}(t-1, m) \mid \ker(A) \supseteq V \text{ and } \mathrm{Im}(A) \subseteq W\}.$$

First assume that $(A, V, W) \in \overline{\beta(M^t_{m,n} \setminus M^{t-1}_{m,n})}$ is not in $\mathcal{N}$. This implies that that there is a $v \in V$ such that $Av \neq 0$ or a $v' \in \mathbb{C}^n$ such that $Av' \notin W$. In the first case let $(A_i, V_i, W_i)$ be a sequence in $\beta(M^t_{m,n} \setminus M^{t-1}_{m,n})$ converging to $(A, V, W)$ and $v_i \in V_i$ a sequence converging to $v$, then $A_i v_i$ converges to $Av$ but $A_i v_i = 0$ so this contradicts $Av \notin \mathcal{N}$. In the second case let $(A'_i, V'_i, W'_i)$ be a sequence in $\beta(M^t_{m,n} \setminus M^{t-1}_{m,n})$ converging to $(A, V, W)$ and $v'_i \in V'_i$ a sequence converging to $v'$, then $A'_i v'_i$ converges to $Av'$ but $A_i v'_i \in W$ and hence $Av' \in W$ since $W$ is closed, this gives a contradiction. Let $(A, V, W) \in \mathcal{N}$ and let $r = \mathrm{rank}\, A$. Now $V \subset \ker A$, so let $V' \subset \mathbb{C}^n$ be a subspace satisfying $V \oplus V' = \ker A$, and $\mathrm{Im}\, A \subset W$ so let $W' \subset \mathbb{C}^m$ be a subspace satisfying $\mathrm{Im}\, A \oplus W' = W$. Let $A'$ be a matrix of rank $t-1-r$, such that $\ker A' \oplus V' = \mathbb{C}^n$ and $\mathrm{Im}\, A' = W'$, such a matrix exists since $\dim V' = \dim W' = t-1-r$. Set $A_i = A + \frac{1}{i}A'$ then $\ker A_i = \ker A \bigcap \ker \frac{1}{i}A' = \ker A \bigcap \ker A' = V$ and $\mathrm{Im}\, A_i = \mathrm{Im}\, A + \mathrm{Im}\, \frac{1}{i}A' = W$. Hence $(A_i, V_i, W_i) := (A_i, V, W)$ is a sequence in $\beta(M^t_{m,n} \setminus M^{t-1}_{m,n})$ converging to $(A, V, W)$, so $\mathcal{N} \subseteq \overline{\beta(M^t_{m,n} \setminus M^{t-1}_{m,n})}$ which finishes the proof.

An important consequence of this is the following:

**Corollary 3.1** $\mathrm{Nash}(M^t_{m,n})$ *is a complex manifold.*

***Proof*** Using the description of $\mathrm{Nash}(M^t_{m,n})$ given in Proposition 3.1 we get that the projection to the two last factors $\mathrm{Gr}(n-t+1, n) \times \mathrm{Gr}(t-1, m)$ gives $\mathrm{Nash}(M^t_{m,n})$ the structure of the total space of a vector bundle over a complex manifold.

It follows from Definition 2.3 and Proposition 3.1, that we have a map of transformations $f\colon \mathrm{Nash}(M^t_{m,n}) \to \mathrm{Tjur}(M^t_{m,n})$ by setting $f(A, V, W) = (A, V)$ and a map of transformations $g\colon \mathrm{Nash}(M^t_{m,n}) \to \mathrm{Tjur}^T(M^t_{m,n})$ by setting $g(A, V, W) =$

$(A, W)$ and using (1). These maps are never isomorphisms, as we will see later when we determine the homotopy type of these spaces.

**Proposition 3.2** *There does not exist a continuous maps of transformations between* $\mathrm{Tjur}(M^t_{m,n})$ *and* $\mathrm{Tjur}^T(M^t_{m,n})$.

***Proof*** We start by using (1) to identify $\mathrm{Tjur}^T(M^t_{m,n})$ with the set $\{(A, W) \in M_{m,n} \times \mathrm{Gr}(t-1, m) \mid \mathrm{Im}(A) \subseteq W\}$. Let $f\colon \mathrm{Tjur}(M^t_{m,n}) \to \mathrm{Tjur}^T(M^t_{m,n})$ be a map of transformations, this implies that over $\pi^{-1}(M^t_{m,n} \setminus M^{t-1}_{m,n})$ we have $f(A, \ker A) = (A, \mathrm{Im}\, A)$. Let $\{x_1, \ldots, x_n\}$ be a basis of $\mathbb{C}^n$ and $\{y_1, \ldots, y_m\}$ be a basis for $\mathbb{C}^m$. Let $A$ be the matrix in these bases of the map $A(a_1x_1 + \cdots + a_nx_n) = a_1y_1 + \cdots + a_{t-2}y_{t-2} + 0y_{t-1} + \cdots + 0y_m$, notice that there is at least 2 zeros at the end since $t \leq m$. Now rank $A = t-2$ and hence $A \in M^{t-1}_{m,n}$. Let $V = \mathrm{Span}\{x_t, \ldots, x_n\}$ then it is clear that $\ker A \supset V$.

We now define two different sequences of matrices $A^1_s$ and $A^2_s$ where $A^i_s \in M^t_{m,n}$. The first is defined as $A^1_s(a_1x_1+\cdots+a_nx_n) := a_1y_1+\cdots+a_{t-2}y_{t-2}+\frac{1}{s}a_{t-1}y_{t-1}+0y_t + \cdots + 0y_m$ and the second is defined as $A^2_s(a_1x_1 + \cdots + a_nx_n) := a_1y_1 + \cdots + a_{t-2}y_{t-2} + 0y_{t-1} + \frac{1}{s}a_{t-1}y_t + 0t_{t+1} + \cdots + 0y_m$. It is clear that $\ker A^i_s = V$ and $\lim_{s\to\infty}(A^i_s, V) = (A, V)$ for $i = 1, 2$. Since $A^i_s \in M^t_{m,n} \setminus M^{t-1}_{m,n}$ we get that $f(A^i_s, V) = (A^i_s, \mathrm{Im}\, A^i_s)$. Let $W_1 := \mathrm{Span}\{y_1, \ldots, y_{t-1}\} = \mathrm{Im}\, A^1_s$ and $W_2 := \mathrm{Span}\{y_1, \ldots, y_{t-2}, y_t\} = \mathrm{Im}\, A^2_s$. If $f$ was continuous, then we would have that $f(A, W) = f(\lim_{s\to\infty}(A^i_s, V)) = \lim_{s\to\infty} f(A^i_s, V) = (A, W_i)$ for $i = 1, 2$. But $W_1 \neq W_2$ hence $f$ cannot be continuous. The argument that there is no continuous map of transformations from $\mathrm{Tjur}^T M^t_{m,n}$ to $\mathrm{Tjur}\, M^t_{m,n}$ is similar.

Next we determine the homotopy type of the transformations, and the above shows that in the case $\mathrm{Tjur}(M^t_{m,n})$ and $\mathrm{Tjur}^T(M^t_{m,n})$ are homotopy equivalent they are not isomorphic as transformations. Even in the case $n = m$ where $\mathrm{Tjur}(M^t_{m,m})$ and $\mathrm{Tjur}^t(M^t_{m,m})$ are isomorphic as real varieties by the isomorphism given by $(A, W) \to (A, \bar{W}^{\perp})$, they are not isomorphic as transformations.

**Proposition 3.3** *Let* $\pi\colon (T(M^t_{m,n}), E) \to (M^t_{m,n}, M^{t-1}_{m,n})$ *be one of the three transformations discussed above. Then* $T(M^t_{m,n})$ *deformation retracts onto* $\pi^{-1}(0)$.

This gives that $\mathrm{Nash}(M^t_{m,n}) \sim \mathrm{Gr}(n-t+1, n) \times \mathrm{Gr}(t-1, m)$, $\mathrm{Tjur}(M^t_{m,n}) \sim \mathrm{Gr}(n-t+1, n)$ and $\mathrm{Tjur}^T(M^t_{m,n}) \sim \mathrm{Gr}(t-1, m)$, where $\sim$ denotes homotopy equivalence.

***Proof*** We will only show this for $\mathrm{Nash}(M^t_{m,n})$. The other proofs are similar. Define $F\colon \mathrm{Nash}(M^t_{m,n}) \times \mathbb{C} \to \mathrm{Nash}(M^t_{m,n})$ as $F(A, V, W, s) = f_s(A, V, W) = (sA, V, W)$, using the identification for the Nash transformation given by Proposition 3.1. The map is well defined since $(sA)(V) = s(A(V)) = 0$ and $\mathrm{Im}(sA) = \mathrm{Im}(A) \subset W$ if $s \neq 0$ and $\mathrm{Im}(sA) = \{0\} \subset W$ if $s = 0$. It is continuous since it is just scalar multiplication. Restrict the map to $s \in [0, 1]$. Then $f_1 = \mathrm{id}$, $f_s|_{\pi^{-1}(0)} = id|_{\pi^{-1}(0)}$ and $f_0(\mathrm{Nash}(M^t_{m,n})) = \pi^{-1}(0)$. Hence $f_s$ is a deformation retraction, and $\mathrm{Nash}(M^t_{m,n})$ deformation retracts onto $\pi^{-1}(0)$.

# 4 Transformations of General Determinantal Singularities

In this section we will introduce the transformations defined above for general determinantal varieties. We start by introducing the Tjurina transform. The Tjurina transform of a determinantal variety has been introduced in several places before for example in [1, 3, 12, 13] and [5]. They in general define the Tjurina transform of a determinantal variety $X$ of type $(m, n, t)$ given by $F: \mathbb{C}^N \to M_{m,n}$ as the fibre product $X \times_F \text{Tjur}(M^t_{m,n})$, which works very well in the cases they consider. But this definition gives the following problem in a more general setting: assume that $\dim(X) \leq (t-1)(n-t+1)$ and let $p: X \times_F \text{Tjur}(M^t_{m,n}) \to X$ be the projection to the first factor. Then $p^{-1}(0) \cong \text{Gr}(n-t+1, n)$, hence the exceptional fibre of $p$ has dimension greater than or equal to the dimension of $X$. This means that the fibre product does not satisfy the conditions to be a transformation given in Definition 2.2.[1] We will give an alternative definition that does not have this problem. It should be said that in [3] and [5] they only consider the Tjurina transformation in situations where this does not happen, and that our definition agrees with theirs in these cases. We will see in Proposition 4.3 when the two definitions agree in general.

**Definition 4.1** Let $X$ be a determinantal variety of type $(m, n, t)$ given by $F: \mathbb{C}^N \to M_{m,n}$ and assume that $\overline{X_t} = X$, define $B: X_t \to \text{Gr}(t-1, n)$ as the map that sends $x$ into the row space of $F(x)$. Then we define the Tjurina transform $\text{Tjur}(X)$ of $X$ as

$$\text{Tjur}(X) := \overline{\left\{ \big(x, W\big) \in X_t \times \text{Gr}(t-1, n) \;\middle|\; W = B(x) \right\}} \subseteq X \times \text{Gr}(t-1, n),$$

where we again use the topological closure in the classical topology, and we define the map $\pi^{Tj}: \text{Tjur}(X) \to X$ as the projection to the first factor.

Remember as always we think of the determinantal variety $X$ as the space $X$ and the map $F$, hence as we just write $X$ for the determinantal variety including the map $F$, we also write $\text{Tjur}(X)$ for the Tjurina transform which of course also depends of the map $F$.

The assumption that $\overline{X_t} = X$ is to avoid cases where there are irreducible components of $X$ that do not give components of $X_t$. If $(X, 0)$ is an EIDS then $(X, 0)$ always satisfies this condition in a neighbourhood of the origin.

It is clear that this satisfies the conditions of Definition 2.2 to be a transformation of $(X, X_{<t})$ where $X_{<t} := \cup_{i=1}^{t-1} X_i = F^{-1}(M^{t-1}_{m,n})$, since $\pi^{Tj}|_{\text{Tjur}(X) \setminus (\pi_{Tj})^{-1}(X_t)}$ is the inverse of $B$, it is surjective because $\overline{X_t} = X$ and proper since all fibres are either points or closed subsets of $\text{Gr}(t-1, n)$ hence compact.

[1]It is of course also possible that $p^{-1}(0)$ is a irreducible component of $X \times_F \text{Tjur}(M^t_{m,n})$ even without the condition on the dimensions, we will discuss this later in Proposition 4.3.

Notice that the choice of a Hermitian inner product on $\mathbb{C}^n$ gives a real linear isomorphism between the complex conjugate of the row space of $F(x)$ and $\ker\big(F(x)\big)$ and a real algebraic isomorphism $\mathrm{Gr}(t-1,n) \cong \mathrm{Gr}(n-t+1,n)$. Hence we get a real analytic isomorphism

$$\mathrm{Tjur}(X) \cong \overline{\left\{\big(x,W\big) \in X_t \times \mathrm{Gr}(n-t+1,n) \mid W = \ker F(x)\right\}} \subseteq X \times \mathrm{Gr}(n-t+1,n).$$

We use the row space in our definition, since it makes calculation easier as we see later.

**Proposition 4.1** *Let $(X,0)$ be a determinantal singularity of type $(m,n,1)$, then $(\mathrm{Tjur}(X),0) = (X,0)$.*

***Proof*** Since $X$ is of type $(m,n,1)$ we have that $\mathrm{Tjur}(X) \subseteq X \times \mathrm{Gr}(0,n) = X$ and $B$ is constant. The result then follows since $\mathrm{Tjur}(X) = \overline{X_1} = X$.

Notice that all determinantal singularities of type $(m,n,1)$ are local complete intersections, and that any local complete intersection can be given as a determinantal singularity of type $(m,n,1)$, in fact by a determinantal singularity of type $(m,1,1)$ of $(1,n,1)$. Hence the Proposition says that given the natural representation of a local complete intersection, then the Tjurina transform do not improve the singularity.

Hypersurfaces singularities can also some times be given as determinantal singularities of type $(m,m,m)$, and we will later see some examples of hypersurfaces of type $(m,m,m)$ for which the Tjurina transform is useful to simplify their singularities.

To study the local properties of the Tjurina transform closer we will use the following matrix charts on $\mathrm{Gr}(t-1,n)$. Let $I \subset \{1,\dots,n\}$ such that $\#I = t-1$. For each such $I = \{i_1,\dots,i_{t-1}\}$ let $a = (a_{ji})$ $j \in 1,\dots,t-1$ and $i \in \{1,\dots,n\} \setminus I$ be a $(t-1) \times (n-t+1)$ matrix with variables $a_{ji} \in \mathbb{C}$. We define a chart of $\mathrm{Gr}(t-1,n)$ by the $(t-1) \times n$ matrix $A_I(a)$ which consists of the columns $C_i$ given as follows:

$$C_i = \begin{pmatrix} a_{1i} \\ \vdots \\ a_{(t-1)i} \end{pmatrix} \text{ if } i \notin I, \text{ and } C_{i_l} = \begin{pmatrix} 0 \\ \vdots \\ 0 \\ 1 \\ 0 \\ \vdots \\ 0 \end{pmatrix} \text{ where the 1 is in the } l^{th} \text{ entry.}$$

If we consider $a \in M_{t-1,n-t+1} = \mathbb{C}^{(t-1)(n-t+1)}$ then we can use $A_I(a)$ to define a map $\tilde{A}_I : \mathbb{C}^{(t-1)(n-t+1)} \to \mathrm{Gr}(t-1,n)$ by sending $a$ to the row space of $A_I(a)$.

$\{\tilde{A}_I\}_I$ is a cover of $\mathrm{Gr}(t-1,n)$ by algebraic maps, and if $U_I = \mathrm{Im}(\tilde{A}_I)$ then the change of coordinates from $\tilde{A}_I^{-1}(U_I \cap U_J)$ to $\tilde{A}_J^{-1}(U_I \cap U_J)$ is given by $A_J^T A_I(a)$.

To see the row space of $F(x)$ in a given chart $A_I$, we construct the following $(m+t-1) \times n$ matrix:

$$\tilde{F}_I^{Tj}(x,a) := \begin{pmatrix} A_I(a) \\ \hline F(x) \end{pmatrix}.$$

Then the row space of $F(x)$ is contained in $\tilde{A}_I(a)$ if and only if $\operatorname{rank} \tilde{F}_I^{Tj}(x,a) = t-1$.

Let $\widetilde{\mathrm{Tjur}}_I(X) := (\tilde{F}_I^{Tj})^{-1}(M^t_{m+t-1,n}) \subset X \times \mathbb{C}^{(t-1)(n-t+1)}$, and $\tilde{\pi}_I^{Tj} : \widetilde{\mathrm{Tjur}}_I(X) \to X$ be the projection to the first factor. Then $\widetilde{\mathrm{Tjur}}_I(X)$ is the restriction of the fibre product $X \times_F \mathrm{Tjur}(M^t_{m,n})$ to the chart on $\mathbb{C}^N \times \mathrm{Gr}(t-1,n)$ given by $I$.

From the above construction we get $\mathrm{Tjur}_I(X) := \mathrm{Tjur}(X) \cap (X \times \mathrm{Im}\,\tilde{A}_I) \subset \widetilde{\mathrm{Tjur}}_I(X)$, but they are not necessarily equal. Notice that $\mathrm{Tjur}_I(X)$ is the restriction of the Tjurina transform $\mathrm{Tjur}(X)$ to the chart on $\mathbb{C}^N \times \mathrm{Gr}(t-1,n)$ given by $I$. $\mathrm{Tjur}_I(X)$ can be thought of as the strict transform of $X$ in $\widetilde{\mathrm{Tjur}}_I(X)$. Moreover, $\widetilde{\mathrm{Tjur}}_I(X)$ is not necessarily a determinantal singularity. We have $(\pi_I^{Tj})^{-1}(X_t) = (\tilde{\pi}_I^{Tj})^{-1}(X_t)$. This implies that $\dim \widetilde{\mathrm{Tjur}}_I(X) = \max\big(\dim X, \dim(\tilde{\pi}_I^{Tj})^{-1}(X_{<t})\big)$. Now $\dim(\tilde{\pi}_I^{Tj})^{-1}(X_{<t})$ is the largest of the dimensions of the pullback of $X_1, \ldots, X_{t-1}$. Hence $(\tilde{\pi}_I^{Tj})^{-1}(X_s) \subset X \times \mathrm{Gr}(t-1,n)$ consists of the pairs $(x,W)$ such that $x \in X_s$ and row space of $F(x)$ is a subset of $W$. We will denote the row space of $F(x)$ by $R_x F$. Since $\operatorname{rank} F(x) = s-1$ we can write all such $W$ as $W = R_x F + W_{F(x)}$ where $W_{F(x)}$ is $(t-s)$-dimensional subspace of the complement of $R_x F \subset \mathbb{C}^N$. Moreover, for any $t-s$ dimensional subspace $V$ in the complement of $R_x F \subset \mathbb{C}^N$ we have $\operatorname{rank}(R_x F + V) = t-1$. Hence we get that $\{W \in \mathrm{Gr}(t-1,n) \mid R_x F \subset W\}$ is isomorphic to $\mathrm{Gr}(t-s, n-s+1)$. So we get that $\dim(\tilde{\pi}_I^{Tj})^{-1}(X_s) = \dim F^{-1}((M^s_{m,n} \setminus M^{s-1}_{m,n}) + \dim \mathrm{Gr}(t-s, n-s+1)$.

The above implies that $\dim \widetilde{\mathrm{Tjur}}_I(X) = \dim \mathrm{Tjur}_I(X) = \dim X$ if and only if

$$\begin{aligned}
\dim(\tilde{\pi}_I^{Tj})^{-1}(X^s) &\le \dim X - \dim \mathrm{Gr}(t-s, n-s+1) \\
&= N - (m-t+1)(n-t+1) - (t-s)(n-t+1) \\
&= N - (m-s+1)(n-t+1)
\end{aligned}$$

for all $s = 1, \ldots, t$. If $X$ has an isolated singularity, this becomes $N \ge m(n-t+1)$.

**Proposition 4.2** *If* $\dim \widetilde{\mathrm{Tjur}}_I(X) = \dim X$ *then* $\widetilde{\mathrm{Tjur}}_I(X)$ *is a determinantal variety.*

***Proof*** We just need to check if $\operatorname{codim}\widetilde{\mathrm{Tjur}}_I(X) = \operatorname{codim} M^t_{m+t-1,n} = (m+t-1-t+1)(n-t+1) = m(n-t+1)$. But $\operatorname{codim}\widetilde{\mathrm{Tjur}}_I(X) = \operatorname{codim}\mathrm{Tjur}_I(X) = \operatorname{codim} X + (t-1)(n-t+1) = (m-t+1)(n-t+1)+(t-1)(n+t-1) = m(n-t+1)$.

In this case we get that $\widetilde{\mathrm{Tjur}}_I(X)$ is a determinantal variety of type $(m+t-1, n, t)$. But rank $\tilde{F}^{Tj}_I(0,0) = t-1$, so one can find another matrix $F'_I(x,a)$ defining $\widetilde{\mathrm{Tjur}}_I(X)$ such that $F'_I(0,0) = 0$ and this is a determinantal variety of type $(m+t-1-(t-1), n-(t-1), t-(t-1)) = (m, n-t+1, 1)$. Since $\operatorname{codim}\widetilde{\mathrm{Tjur}}_I(X) = m(n-t+1)$ we get that $\widetilde{\mathrm{Tjur}}_I(X)$ is a complete intersection. We will later show how to explicitly find $F'_I(x,a)$ also in the case $\dim\widetilde{\mathrm{Tjur}}_I(X) \neq \dim X$.

We can also use this to determine when $\mathrm{Tjur}_I(X)$ and $\widetilde{\mathrm{Tjur}}_I(X)$ are equal. Notice that $\widetilde{\mathrm{Tjur}}_I(X) = \big(X \times_F \mathrm{Tjur}(M^t_{m,n})\big) \cap \big(X \times \operatorname{Im}\tilde{A}_I\big)$, hence the next proposition also answers the question, when is our definition of Tjurina transform the same as the one used by other authors. Remember that we earlier defined $X_s := F^{-1}(M^s_{m,n} \setminus M^{s-1}_{m,n})$.

**Proposition 4.3** *Let $X$ be a determinantal variety. Then $\widetilde{\mathrm{Tjur}}_I(X) = \mathrm{Tjur}_I(X)$ if and only if $\dim X_s < N-(m-s+1)(n-t+1)$ for all $s \in 1,\dots,t-1$. Furthermore, if $(X,0)$ is an EIDS, then $\widetilde{\mathrm{Tjur}}_I(X) = \mathrm{Tjur}_I(X)$ if and only if $\dim X_1 < N - m(n-t+1)$.*

***Proof*** Since $\mathrm{Tjur}(X)$ is a transformation, we have that $\dim \pi^{Tj}(X^{t-1}) < \dim X$. Then the above calculations of the dimensions of the fibres give the inequalities, and we get the only if direction.

So assume that the inequalities are satisfied, this implies $\dim\mathrm{Tjur}_I(X) = \dim\widetilde{\mathrm{Tjur}}_I(X)$ and $\dim(\tilde{\pi}^{Tj}_I)^{-1}(X^{t-1}) < \dim X$. Now $\mathrm{Tjur}_I(X)$ is a union of irreducible components of $\widetilde{\mathrm{Tjur}}_I(X)$, and each irreducible component of $\mathrm{Tjur}_I(X)$ is not a proper subvariety of any irreducible variety of the same dimension, since they are closed. This implies that if $\widetilde{\mathrm{Tjur}}_I(X) \neq \mathrm{Tjur}_I(X)$ then there exists another irreducible component $V \subseteq \widetilde{\mathrm{Tjur}}_I(X)$ not contained in $\mathrm{Tjur}_I(X)$. But since $\widetilde{\mathrm{Tjur}}_I(X)$ is a complete intersection it is equidimensional, and hence $\dim V = \dim\mathrm{Tjur}_I(X)$. Since $(\pi^{Tj}_I)^{-1}(X_t) = (\tilde{\pi}^{Tj}_I)^{-1}(X_t)$ we have that $V \subset (\tilde{\pi}^{Tj}_I)^{-1}(X^{t-1})$, but this is a contradiction since $\dim V > \dim \tilde{\pi}^{Tj}_I(X^{t-1})$.

For the case of EIDS remember that if $(X,0)$ is an EIDS, $1 < s < t$ and $X_s \neq \emptyset$ then $\operatorname{codim} X_s = (m-s+1)(n-s+1)$. Hence the inequality of the first part of the theorem is satisfied for all $1 < s < t$ and therefore one only needs that $\dim X_1 < N-(m-1+1)(n-t+1)$ to get the conclusion $\widetilde{\mathrm{Tjur}}_I(X) = \mathrm{Tjur}_I(X)$.

If $X$ is a determinantal singularity and we assume that $X = \overline{X_t}$ then $\dim X_t = N-(m-t+1)(n-t+1)$. But remember even if $X_s \neq \emptyset$ for some $s < t$ then $\overline{X_s}$ needs not be a determinantal singularity, and hence $\dim X_s$ can be larger than $N-(m-s+1)(n-s+1)$ as the following example shows. Hence the assumptions of Proposition 4.3 need not be satisfied.

*Example 4.1* Let $X := F^{-1}(M^3_{3,3})$ be the determinantal singularity given by the matrix:

$$F(x, y, z) := \begin{pmatrix} x & 0 & z \\ y & x & 0 \\ 0 & y & x \end{pmatrix}.$$

We have that $X$ is the variety given by the equation $x^3 + y^2 z = 0$, which have the $z$-axis as its singular set. $X$ is determinantal since $\operatorname{codim} X = 1 = \operatorname{codim} M^3_{3,3}$. Now $X_3$ is $X$ minus the $z$-axis, $X_2$ is the $z$-axis minus the origin and $X_1$ is the origin. This implies that the $\operatorname{codim} X_2 = 2$ but $\operatorname{codim} M^3_{3,3} = 4$, so $X$ do not satisfy Proposition 4.3.

We now want to give an explicit method to find $F'_I(x, a)$. Let $I = \{i_1, \dots, i_{t-1}\} \subset \{1, \dots, n\}$ as before. Now by adding columns of the form $-a_{ji} C_{i_j}$ to the $i$'th column, for all $i \notin I$ and all $j = 1, \dots, t-1$, we get a matrix which has $t-1$ linearly independent rows $R_{i_j}$ of the form $R_{i_j} = (0, \dots, 0, 1, 0, \dots, 0)$, where the 1 is the $i_j$ entry. To this matrix we then add rows of the form $-f_{li_j}(x) R_{i_j}$ to the $l+t-1$'th row for $l = 1, \dots, m$ and $j = 1, \dots, t-1$. We now have a matrix $\bar{F}_I(x, a)$ consisting of the following columns:

$$\bar{F}_i = \begin{pmatrix} 0 \\ \vdots \\ 0 \\ f_{1i}(x) - \sum_{j=1}^{t-1} a_{ji} f_{1i_j}(x) \\ \vdots \\ f_{mi}(x) - \sum_{j=1}^{t-1} a_{ji} f_{mi_j}(x) \end{pmatrix} \text{ if } i \notin I, \text{ and } \bar{F}_{i_l} = \begin{pmatrix} 0 \\ \vdots \\ 0 \\ 1 \\ 0 \\ \vdots \\ 0 \end{pmatrix},$$

where the 1 in $\bar{F}_{i_l}$ is on the $l^{th}$ entry. The $t \times t$ minors of $\bar{F}_I(x, a)$ still defines $\widetilde{\mathrm{Tjur}}_I(X)$. Notice that we can choose special minors $\Delta_{l,i}$, with $l \in \{1, \dots, m\}$ and $i \notin I$, where each row and each column have a single non zero entry, which is 1 except for the $li$'th entry which is $f_{li}(x) - \sum_{j=1}^{t-1} a_{ji} f_{li_j}(x)$. This implies that $\widetilde{\mathrm{Tjur}}_I(X)$ is defined by the $(n-t+1)m$ equations $f_{li}(x) - \sum_{j=1}^{t-1} a_{ji} f_{li_j}(x) = 0$. Hence it is defined by the $1 \times 1$ minors of the matrix $m \times (n-t+1)$ matrix $F'_I(x, a)$ with columns:

$$F'_i = \begin{pmatrix} f_{1i}(x) - \sum_{j=1}^{t-1} a_{ji} f_{1i_j}(x) \\ \vdots \\ f_{mi}(x) - \sum_{j=1}^{t-1} a_{ji} f_{mi_j}(x) \end{pmatrix} \text{ if } i \notin I.$$

This still does not imply that $\widetilde{\mathrm{Tjur}}_I(X)$ is a determinantal variety, since the codimension might not be right. Even if $\widetilde{\mathrm{Tjur}}_I(X)$ is a determinantal variety, it might have components of maximal dimension which is contained in $(\widetilde{\pi}^{Tj})^{-1}(X_{<t})$ and hence $\widetilde{\mathrm{Tjur}}_I(X) \neq \mathrm{Tjur}_I(X)$, as we will see in the next examples.

*Example 4.2* Let $X$ be the irreducible determinantal variety of type (2, 3, 2) defined by the following matrix

$$F_1(x, y, z, w) := \begin{pmatrix} w^l & y & x \\ z & w & y^k \end{pmatrix},$$

with $k, l > 2$. Then $\widetilde{\mathrm{Tjur}}_I(X)$ is a determinantal variety for all $I$. Let us start by looking in the chart defined by $I = \{1\}$.

$$F'_{\{1\}}(x, y, z, w, a_2, a_3) := \begin{pmatrix} y - a_2 w^l & x - a_3 w^l \\ w - a_2 z & y^k - a_3 z \end{pmatrix}.$$

The equations $y - a_2 w^l = 0$, $x - a_3 w^l = 0$ and $w - a_2 z = 0$ all just give the variables $x$, $y$ and $w$ as functions of $z, a_2$ and $a_3$. Using these equations the last equation becomes $a_2^{k(l+1)} z^{kl} - a_3 z = 0$ which shows that $\widetilde{\mathrm{Tjur}}_I(X)$ has two irreducible components. The first given by $\{x = y = z = w = 0\}$ which is the fibre over the origin. The second irreducible component, which is $\mathrm{Tjur}_I(X)$, is given by the equations $y - a_2 w^l = 0$, $x - a_3 w^l = 0$, $w - a_2 z = 0$ and $a_2^{k(l+1)} z^{k(l-1)} - a_3 = 0$ and is hence smooth.

Now let us look closer on the equations in the chart defined by $I = \{2\}$.

$$F'_{\{2\}}(x, y, z, w, a_1, a_3) := \begin{pmatrix} w^l - a_1 y & x - a_3 y \\ z - a_1 w & y^k - a_3 w \end{pmatrix}.$$

Notice that the equations $x - a_3 y = 0$ and $z - a_1 w = 0$ define $x$ and $z$ as holomorphic functions of the other variables, and give embeddings of a $\mathbb{C}^4$ into $\mathbb{C}^6$. Now if we multiply the equations $y^k - a_3 w = 0$ and $w^l - a_1 y = 0$ we get:

$$\begin{aligned} 0 &= (y^k - a_3 w)(w^l - a_1 y) = y^k w^l - a_1 y^{k+1} - a_3 w^{l+1} + a_1 a_3 y w \\ &= y^k w^l - a_1 a_3 y w - a_1 a_3 y w + a_1 a_3 y w = y w (y^{k-1} w^{l-1} - a_1 a_3). \end{aligned}$$

Hence we see that $\widetilde{\mathrm{Tjur}}_I(X)$ is not irreducible. $y = 0$ and $w = 0$ both define the fibre $(\tilde{\pi}_I^{Tj})^{-1}(0)$ which is two dimensional and therefore cannot be a subset of $\mathrm{Tjur}_I(X)$. Therefore, $\mathrm{Tjur}_I(X)$ is given by the equations $y^{k-1} w^{l-1} - a_1 a_3 = 0$, $w^l - a_1 y = 0$ and $y^k - a_3 w = 0$. Hence it can be given as a determinantal variety of the same type as $X$ given by the matrix

$$\begin{pmatrix} w^{l-1} & y & a_3 \\ a_1 & w & y^{k-1} \end{pmatrix}.$$

The case of the last chart $I = \{3\}$ is similar to that of the first chart, hence in that chart we also have that $\mathrm{Tjur}_I(X)$ is smooth. In all charts we have that $\widetilde{\mathrm{Tjur}}_I(X) \neq \mathrm{Tjur}_I(X)$.

*Example 4.3* Let $X \subset \mathbb{C}^4$ be the determinantal variety of type (3, 2, 2) given by

$$F_2(x, y, z, w) := \begin{pmatrix} w^l & z \\ y & w \\ x & y^k \end{pmatrix},$$

with $k, l > 2$. Then $\widetilde{\mathrm{Tjur}}_I(X)$ is given in the two charts $I = \{1\}, \{2\}$ by the matrices

$$F'_{\{1\}}(x, y, z, w, a_1) := \begin{pmatrix} z - a_1 w^l \\ w - a_1 y \\ y^k - a_1 x \end{pmatrix} \text{ and } F'_{\{2\}}(x, y, z, w, a_2) := \begin{pmatrix} w^l - a_2 z \\ y - a_2 w \\ x - a_2 y^k \end{pmatrix}$$

In this case we see that $\widetilde{\mathrm{Tjur}}_I(X) = \mathrm{Tjur}_I(X)$, and hence the Tjurina transform of $X$ is a complete intersection.

Notice that the underlying varieties in Examples 4.2 and 4.3 are the same, it is just their representations as determinantal varieties which are different. In fact the difference is that $F_1(x, y, z, w) = F_2(x, y, z, w)^T$. In Example 4.2 we get that $\widetilde{\mathrm{Tjur}}_I(X) \neq \mathrm{Tjur}(X)$ and in Example 4.3 that $\widetilde{\mathrm{Tjur}}_I(X) = \mathrm{Tjur}(X)$. This does not contradict Proposition 4.3 since in Example 4.2 the inequality $\dim X_s < N - (m - s + 1)(n - t + 1)$ is not satisfied for $s = 1$ since $N - (m - 1 + 1)(n - t + 1) = 4 - 3(3 - 2 + 1) = 0$. In Example 4.3 the inequality is satisfied and we get that $\widetilde{\mathrm{Tjur}}_I(X) = \mathrm{Tjur}(X)$.

Let us define $\mathrm{Tjur}^T(X)$.

**Definition 4.2** Let $X$ be a determinantal variety of type $(m, n, t)$ given by $F : \mathbb{C}^N \to M_{m,n}$ such that $\overline{X_t} = X$, define $C : X_t \to \mathrm{Gr}(t - 1, m)$ as the map that sends $x$ into the column space of $F(x)$. Then we define $\mathrm{Tjur}^T(X)$ of $X$ as

$$\mathrm{Tjur}^T(X) = \overline{\left\{(x, W) \in X_t \times \mathrm{Gr}(t - 1, m) \mid W = C(x)\right\}} \subseteq X \times \mathrm{Gr}(t - 1, m),$$

and we define the map $\pi^{Tj^T} : \mathrm{Tjur}^T(X) \to X$ as the projection to the first factor.

This definition gives us that $\mathrm{Tjur}^T(X) = \mathrm{Tjur}(X^T)$, where $X^T$ is $X$ but defined as a determinantal singularity by $F^T : \mathbb{C}^N \to M_{n,m}$. This means that we can define $\widetilde{\mathrm{Tjur}}^T_I(X)$ as for $\mathrm{Tjur}(X)$, either by setting $\widetilde{\mathrm{Tjur}}^T_I(X) = \widetilde{\mathrm{Tjur}}_I(X^T)$ or by defining it using $\tilde{F}^T_I(x, a) := \left(F(x) \middle| A^T_I(a)\right)$, where $I$ now is a subset of $1, \ldots, m$.

This immediately gives us the following results.

**Proposition 4.4** $\widetilde{\mathrm{Tjur}}^T_I(X)$ *is a determinantal variety if and only if* $\dim X_s \leq N - (m - t + 1)(n - s + 1)$ *for all* $s \in 1, \ldots, t$.

**Proposition 4.5** *$\widetilde{\mathrm{Tjur}_I^T}(X) = \mathrm{Tjur}_I^T(X)$ if and only if $\dim X_s < N - (m - t + 1)(n - s + 1)$ for all $s \in 1, \ldots, t - 1$.*

Notice that this definition of $\mathrm{Tjur}^T(M^t_{m,n})$ is the same as the one we gave earlier, since the column space of a matrix is the same as its image.

The next example shows that just like the blow-up and the Nash transform, the Tjurina transform of a normal variety needs not be normal and that the dimension of the singular set can increase under the Tjurina transform.

*Example 4.4 (*$\mathrm{Tjur}(X)$ *Might Have Singular Locus of Larger Dimension that* $X$*)* Let $X$ be the hypersurface with an isolated singularity at the origin given by $z^2 - x^4 - x^2y^3 - x^2y^5 - y^8 = 0$. It can be given as a determinantal variety of type $(2, 2, 2)$ by the matrix $\begin{pmatrix} z & x^2+y^3 \\ x^2+y^5 & z \end{pmatrix}$. We get that the Tjurina transform is given by the following matrices

$$F'_{\{1\}}(x, y, z, a_2) = \begin{pmatrix} x^2 + y^3 - a_2 z \\ z - a_2(x^2 + y^5) \end{pmatrix} \text{ and } F'_{\{2\}}(x, y, z, a_1) = \begin{pmatrix} z - a_1(x^2 + y^3) \\ x^2 + y^5 - a_1 z \end{pmatrix}.$$

In the first chart we can, by a change of coordinates, see that we have the hypersurface $x^2 + y^3 - a_2^2(x^2 + y^5) = 0$, which has all of the $a_2$-axis as its singular set. In the same way the second chart gives us the hypersurface $x^2 + y^5 - a_1^2(x^2 + y^3) = 0$, which has the $a_1$-axis as its singular set. Hence $\mathrm{Tjur}(X)$ has singularities of codimension 1, and is, therefore, not normal. It also illustrates that the singular set of $\mathrm{Tjur}(X)$ might have larger dimension than the singular set of $X$.

Notice in general that if $X$ is a determinantal singularity given by a matrix $F$ such that all entries in $F$ have orders $\geq 2$ then the singular set of $\mathrm{Tjur}(X)$ contains the full fibre over the common zero locus of the entries of $F$.

We saw in Sect. 3 that $\mathrm{Nash}(M^t_{m,n}) \cong \mathrm{Tjur}(M^t_{m,n}) \times_{M^t_{m,n}} \mathrm{Tjur}^T(M^t_{m,n})$ where the isomorphism is real algebraic. Is this then true in general? Is $\mathrm{Nash}(X) \cong \mathrm{Tjur}(X) \times_X \mathrm{Tjur}^T(X)$? The answer is unfortunately no as we can see in the following. Let $X$ be the determinantal singularity defined in Example 4.2. There we saw that the exceptional divisor of $\mathrm{Tjur}(X)$ consists of two irreducible components. In Example 4.3 we got that the exceptional divisor of $\mathrm{Tjur}^T(X)$ is a single irreducible curve. Hence the exceptional divisor of $\mathrm{Tjur}(X) \times_X \mathrm{Tjur}^T(X)$ consists of three irreducible curves. But in [12] Tjurina shows that $(X, 0)$ is a minimal surface singularity with the following minimal dual resolution graph.

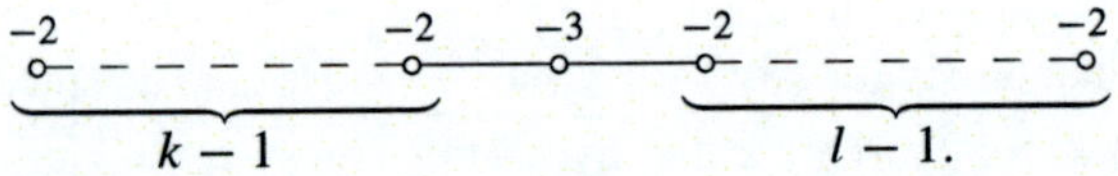


Following the work of Spivakovsky [11] the irreducible components of the exceptional divisor of the normalized Nash transform of a surface singularity correspond to the irreducible components of the exceptional divisor of the minimal resolution intersecting the strict transform of the polar curve of a generic plane projection. By Theorem 5.4 in Chapter III of [11] we find that the polar of a generic plane projection of $X$ intersects the exceptional divisor in two different components. This implies that the exceptional divisor of $\mathrm{Nash}(X)$ has at most two components, since the number of components cannot decrease under normalization. Hence $\mathrm{Nash}(X)$ and $\mathrm{Tjur}(X) \times_X \mathrm{Tjur}^T(X)$ have non isomorphic exceptional divisors, and can, therefore, not be isomorphic as transformations.

## 5 When Is the Tjurina Transform a Complete Intersection

In Lemma 5.2 of their article [5] Frühbis-Krüger and Zach find conditions under which the Tjurina transforms of Cohen-Macaulay codimension 2 singularities in $\mathbb{C}^5$ only have isolated complete intersection singularities. In this section we give some general condition on when the Tjurina transform of an EIDS is a local complete intersection.

If $(X, 0)$ is an EIDS, remember that it means that $F$ is transverse to all strata of $M^t_{m,n}$ in a punctured neighbourhood of the origin, then we get the following result concerning the Tjurina transform.

**Proposition 5.1** *Let $(X, 0) \subset \mathbb{C}^N$ be an EIDS of type $(m, n, t)$, then* $\mathrm{Tjur}(X)$ *is a local complete intersection if $N - m(n - t + 1) > \dim X_1$ and* $\mathrm{Tjur}^T(X)$ *is a local complete intersection if $N - n(m - t + 1) > \dim X_1$.*

***Proof*** To show that $\mathrm{Tjur}(X)$ is a local complete intersection, it is enough to show that $\mathrm{Tjur}_I(X)$ is a complete intersection for all $I$. To do this we show that $\mathrm{Tjur}_I(X) = \widetilde{\mathrm{Tjur}}_I(X)$. Since $(X, 0)$ is EIDS then by Proposition 4.3 we just need that $\dim X_1 < N - m(n - t + 1)$ which follows from the assumption. So $\mathrm{Tjur}_I(X) = \widetilde{\mathrm{Tjur}}_I(X)$ and $\widetilde{\mathrm{Tjur}}_I(X)$ is a complete intersection. Hence $\mathrm{Tjur}(X)$ is a local complete intersection.

The proof for $\mathrm{Tjur}^T(X)$ is similar, just exchange $n$ and $m$.

We are in different situations if $X_1 = \{0\}$ or if $X_1 \neq \{0\}$. Let us first give the following theorem that takes care of the second case.

**Theorem 5.1** *Let $(X, 0)$ be ab EIDS and assume that $X_1 \neq \{0\}$. Then* $\mathrm{Tjur}(X)$ *and* $\mathrm{Tjur}^T(X)$ *are both local complete intersections.*

***Proof*** Assume that $X$ is defined by $F : \mathbb{C}^N \to M_{m,n}$. Since $(X, 0)$ is an EIDS there exist an open neighbourhood of the origin $U$ such that for $x \in (X_1 \setminus \{0\}) \cap U$ we have that $F$ is transverse to the strata $M^1_{m,n}$ at $x$. But this implies that $F$ is a submersion at $x$ because $M^1_{m,n} = \{0\}$. Hence there is an open neighbourhood in $\mathbb{C}^N$ of $(X_1 \setminus \{0\}) \cap U$ on which $F$ is a submersion. Then the Submersion Theorem gives that $(X_1 \setminus \{0\}) \cap U$ is a smooth manifold of dimension $N - mn$. Adding the origin to $(X_1 \setminus \{0\}) \cap U$ does not change the dimension (but might make it singular), hence $\dim X_1 = N - mn$.

Proposition 5.1 is then satisfied for both $\mathrm{Tjur}(X)$ and $\mathrm{Tjur}^T(X)$ since $mn > m(n-t+1)$ and $mn > n(m-t+1)$ for all $1 < t \leq \min\{m, n\}$. If $t = 1$ the result follows from Proposition 4.1.

If $X_1 = \{0\}$ then the equations to determine whether $\mathrm{Tjur}(X)$ is a local complete intersection becomes $\dim X_1 = 0 < N - m(n-t+1)$ or $m(n-t+1) < N$ and likewise $\mathrm{Tjur}^T(X)$ is a local complete intersection if $n(m-t+1) < N$. The assumption on $N$ can be replaced by an assumption on $t$ and the strata of $X$ as seen in the next proposition.

**Proposition 5.2** *Let $(X, 0)$ be an EIDS of type $(m, n, t)$, where $t \geq 3$, $X_1 = \{0\}$ and $X_2 \neq \emptyset$. Then at least one of* $\mathrm{Tjur}(X)$ *and* $\mathrm{Tjur}^T(X)$ *is a local complete intersection.*

***Proof*** First notice that since $t \geq 3$ one of the following two inequalities holds $n - 1 < m(t-2)$ or $m - 1 < n(t-2)$. We will first show that if the first equation holds, then $\mathrm{Tjur}(X)$ is a local complete intersection.

Assume that $n - 1 < m(t-2)$. To show that $\mathrm{Tjur}(X)$ is a complete intersection, we just need to show that $0 < N - m(n-t+1)$. Now $0 < \dim X_2 = N - (m-1)(n-1) = N - mn + m + n - 1 < N - mn + m + m(t-2) = N - m(n-t+1)$ by the assumption $X_2 \neq \emptyset$. So $\mathrm{Tjur}(X)$ is a local complete intersection.

If $m - 1 < n(t-2)$, then the same argument with exchanging $m$ and $n$ shows that $\mathrm{Tjur}^T(X)$ is a local complete intersection.

As we saw in Examples 4.2 and 4.3, this proposition can still hold if $t < 3$, but next we will give an example with $t = 2$ where we have $\mathrm{Tjur}_I(X) \neq \widetilde{\mathrm{Tjur}}_I(X)$ and $\mathrm{Tjur}^T_J(X) \neq \widetilde{\mathrm{Tjur}}^T_J(X)$ for all $I, J$. But in the example, both $\mathrm{Tjur}(X)$ and $\mathrm{Tjur}^T(X)$ are complete local intersections.

*Example 5.1* Let $X \subset \mathbb{C}^3$ be the determinantal variety of type $(3, 2, 2)$ given by

$$F(x, y, z, w) := \begin{pmatrix} z & y & x^{k-3} \\ 0 & x & y \end{pmatrix}.$$

For $k > 4$. Then $\widetilde{\mathrm{Tjur}}_I(X)$ is given in the three charts $I = \{1\}, \{2\}, \{3\}$ as follows. In the first chart the matrix is

$$F'_{\{1\}}(x, y, z, a_2, a_3) := \begin{pmatrix} y - a_2 z & x^{k-3} - a_3 z \\ x & y - a_3 x \end{pmatrix}.$$

We see that $\widetilde{\mathrm{Tjur}}_{\{1\}}(X)$ is the fibre over 0 (given by $x = y = z = 0$) union the $z$-axis (given by $x = y = a_2 = a_3 = 0$), so we get that $\mathrm{Tjur}_{\{1\}}(X)$ is the $z$-axis.

In the second chart we get

$$F'_{\{2\}}(x, y, z, w, a_1, a_3) := \begin{pmatrix} z - a_1 y & x^{k-3} - a_3 y \\ -a_1 x & y - a_3 x \end{pmatrix}.$$

Here we see that $\widetilde{\mathrm{Tjur}}_{\{2\}}(X)$ is the fibre over 0 (given by $x = y = z = 0$) union the curve singularity given by $x^{k-4} - a_3^2 = 0$, $y = a_3 x$ and $a_1 = z = 0$. Hence $\mathrm{Tjur}_{\{2\}}(X)$ is an $A_{k-5}$ plane curve singularity embedded in $\mathbb{C}^5$.

In the last chart we get

$$F'_{\{3\}}(x, y, z, w, a_1, a_2) := \begin{pmatrix} z - a_1 x^{k-3} & y - a_2 x^{k-3} \\ -a_1 y & x - a_2 y \end{pmatrix}.$$

Now we see that $\widetilde{\mathrm{Tjur}}_{\{3\}}(X)$ is the fibre over 0 (given by $x = y = z = 0$) union the curve given by $1 - a_2^2 x^{k-4} = 0$, $y = a_2 x^{k-3}$ and $a_1 = z = 0$. Hence $\mathrm{Tjur}_{\{3\}}(X)$ is a smooth curve in this chart.

So Tjur($X$) is a line disjoint union an $A_{k-5}$ curve, and the fibre over 0 is 2 dimensional.

If we calculate $\widetilde{\mathrm{Tjur}}^T_I(X)$ in the charts {1} and {2}, we get

$$F'_{\{1\}}(x, y, z, a_2) = \begin{pmatrix} -a_2 z \\ x - a_2 y \\ y - a_2 x^{k-3} \end{pmatrix} \text{ and } F'_{\{2\}}(x, y, z, w, a_1) = \begin{pmatrix} z \\ y - a_1 x \\ x^{k-3} - a_1 y \end{pmatrix}.$$

We see that in the first chart we have a line union the fibre over 0 and in the second chart we have an $A_{k-5}$ curve singularity union the fibre over zero.

So in this case we have that Tjur($X$) and $\mathrm{Tjur}^T(X)$ are the same, a line disjoint union an $A_{k-5}$. Notice that in this case Tjur($X$) is also a local complete intersection. Now $X$ is the union of a line $l$ and an $A_{k-3}$ singularity intersecting at the origin. We see that the transformation has separated the line and the singularity, and improved the singularity i.e. what was before an $A_{k-3}$ singularity is now an $A_{k-5}$ singularity.

In Theorem 5.1 we saw that $X_1 \neq \{0\}$ both Tjur($X$) and $\mathrm{Tjur}^T(X)$ are local complete intersections and in Proposition 5.2 we saw that if $t \geq 3$ then one of Tjur($X$) or $\mathrm{Tjur}^T(X)$ is a local complete intersection. The case $t = 1$ is not interesting, because in this case $\mathrm{Tjur}(X) = \mathrm{Tjur}^T(X) = X$ and $X$ is a complete intersection. The next proposition will explain the case $t = 2$.

**Proposition 5.3** *Let* $(X, 0)$ *be an EIDS of type* $(m, n, 2)$ *with* $X_1 = \{0\}$*, then one of* Tjur($X$) *or* $\mathrm{Tjur}^T(X)$ *is a local complete intersection if* $\min(n, m) \leq \dim X$.

***Proof*** To prove that Tjur($X$) is a complete intersection we just need to see that $0 = \dim X_1 < N - m(n - t + 1) = N - m(n - 1)$ by Proposition 5.1. But $(m-t+1)(n-t+1) = (m-1)(n-1) = \operatorname{codim} X$, hence $N = (m-1)(n-1) + \dim X$. Then the inequality becomes $0 < (m - 1)(n - 1) + dim X - m(n - 1)$. Hence $\mathrm{Tjur}(X) = \widetilde{\mathrm{Tjur}}(X)$ and hence a complete intersection if $n - 1 < \dim X$. The case $\mathrm{Tjur}^T(X) = \widetilde{\mathrm{Tjur}^T}(X)$ is gotten by exchanging $n$ and $m$.

These results are only in one direction, because what we really prove is that if the inequalities are satisfied, then $\mathrm{Tjur}(X) = \widetilde{\mathrm{Tjur}}(X)$ or $\mathrm{Tjur}^T(X) = \widetilde{\mathrm{Tjur}^T}(X)$.

But Tjur$(X)$ or Tjur$^T(X)$ can still be local complete intersections, even if this is not true, as we saw in Example 5.1.

## 6 Using Tjurina Transform to Resolve Hypersurface Singularities

In the previous section we saw that very often the Tjurina transform is a complete intersection of type $(m, n, 1)$, which means that one cannot get a resolution by using only the Tjurina transform because of Proposition 4.1. Notice also that in several of the examples Tjur$(X)$ is normal, so using only Tjurina transform and normalizations will also not produce a resolution. In the next example we will look at the case of the $A_n$ surface singularities and see that the Tjurina transform in some cases can be used to achieve a resolution.

*Example 6.1 ($A_n$ Singularities)* In this example we show how different representations of the simple $A_n$ singularity can lead to different Tjurina transforms.

First we can represent $A_n$ as a determinantal singularity of type $(1, 1, 1)$, then the Tjurina transform of $A_n$ is just $A_n$ itself, by Proposition 4.1. But we can also represent $A_n$ as the determinantal singularity of type $(2, 2, 2)$ defined by:

$$F(x, y, z) = \begin{pmatrix} x & z^l \\ z^{n-l+1} & y \end{pmatrix},$$

where $0 < l \leq n$. In this case we get that the Tjurina transform is given by:

$$F'_{\{1\}}(x, y, z, a_2) = \begin{pmatrix} z^l - a_2 x \\ y - a_2 z^{n-l+1} \end{pmatrix} \text{ and } F'_{\{2\}}(x, y, z, a_1) = \begin{pmatrix} x - a_1 z^l \\ z^{n-l+1} - a_1 y \end{pmatrix}.$$

So we see that Tjur$(A_n)$ using these representations has an $A_{l-1}$ and an $A_{n-l}$ singularity, so we have simplified the singularity. It is clear that by writing these new $A_m$ singularities as determinantal singularities of type $(2, 2, 2)$, we can apply the Tjurina transform again to simplify the singularity. By repeatedly doing this we can resolve the $A_n$ singularity.

As we can see in Example 6.1 the Tjurina transform depends not only on the singularity type of $X$ but we can also get different transforms if we have different matrix presentations of the same type.

In the next example we will show how to obtain a resolution through repeated Tjurina transforms changing the determinantal type and matrix presentation. By this we mean that if the Tjurina transform gives us a complete intersection of the form $(m, n, 1)$, which by change of coordinates locally can be seen as a hypersurface, we will then write this hypersurface as a determinantal singularity of type $(t, t, t)$.

*Example 6.2 ($E_7$ Singularity)* The simple surface singularity $E_7$ can be defined by the equation $y^2 + x(x^2 + z^3) = 0$. This can be seen as the determinantal singularity of type $(2, 2, 2)$ given by the following matrix: $\begin{pmatrix} y & x^2+z^3 \\ -x & y \end{pmatrix}$. We then perform the Tjurina transform and get:

$$F'_{\{1\}}(x, y, z, a_2) = \begin{pmatrix} x^2 + z^3 - a_2 y \\ y + a_2 x \end{pmatrix} \text{ and } F'_{\{2\}}(x, y, z, a_1) = \begin{pmatrix} y - a_1(x^2 + z^3) \\ -x - a_1 y \end{pmatrix}.$$

By changing coordinates we see that $F'_{\{1\}}$ is equivalent to the hypersurface $x^2 + z^3 + w^2 x = 0$, which has a singular point at $(0, 0, 0)$, and $F'_{\{2\}}$ is equivalent to the hypersurface $x + v^2(x^2 + z^3) = 0$ which is non singular.

So we will continue working in the first chart, and we will denote this singularity $\mathrm{Tjur}(E_7)$. The exceptional divisor $E_1 = (\pi^{Tj})^{-1}(0)$ is given by $x = z = 0$. We now write $\mathrm{Tjur}(E_7)$ as the matrix $\begin{pmatrix} x & -z^2 \\ z & x+w^2 \end{pmatrix}$ and perform the Tjurina transform.

$$F'_{\{1\}}(x, z, w, a_2) = \begin{pmatrix} -z^2 - a_2 x \\ x + w^2 - a_2 z \end{pmatrix} \text{ and } F'_{\{2\}}(x, z, w, a_1) = \begin{pmatrix} x + a_1 z^2 \\ z - a_1(x + w^2) \end{pmatrix}.$$

The first chart is equivalent to the hypersurface $yw^2 - z^2 - y^2 z = 0$ which has a singularity at $(0, 0, 0)$, and the second chart is equivalent to $v(w^2 - vz^2) - z = 0$ which is smooth. The exceptional divisor consist of two components, the strict transform of the exceptional divisor from before (which we still denote by $E_1$) is given by $z = y = 0$ and a new component $E_2$ given by $x = w = 0$. They intersect each other in the singular point.

We will continue in the first chart and denote this singularity by $\mathrm{Tjur}^2(E_7)$. It can be given by the matrix $\begin{pmatrix} y & -z \\ z & w^2-yz \end{pmatrix}$ as a determinantal singularity of type $(2, 2, 2)$. Its Tjurina transform is given by

$$F'_{\{1\}}(y, z, w, a_2) = \begin{pmatrix} -z - a_2 y \\ w^2 - yz - a_2 z \end{pmatrix} \text{ and } F'_{\{2\}}(y, z, w, a_1) = \begin{pmatrix} y + a_1 z \\ z - a_1(w^2 - yz) \end{pmatrix}.$$

In the first chart we have the hypersurface $xy^2 + w^2 + x^2 y = 0$ which has $(0, 0, 0)$ as its only singular point. The second chart is $z + v(w^2 - vz^2) = 0$ which is smooth. The exceptional divisor consist of $E_1$ given by $z = v = 0$ (so it only exists in the second chart), $E_2$ given by $x = w = 0$ and the new $E_3$ given by $y = w = 0$. $E_1$ and $E_2$ do not meet, but $E_3$ intersects them both, $E_1$ in a smooth point and $E_2$ in the singular point.

We present the singularity $\mathrm{Tjur}^3(E_7)$ as the matrix $\begin{pmatrix} xy & w \\ -w & x+y \end{pmatrix}$. Its Tjurina transform is then given by

$$F'_{\{1\}}(x, y, w, a_2) = \begin{pmatrix} w - a_2 xy \\ x + y + a_2 w \end{pmatrix} \text{ and } F'_{\{2\}}(x, y, w, a_1) = \begin{pmatrix} xy - a_1 w \\ -w - a_1(x + y) \end{pmatrix}.$$

In the first chart we have the hypersurface $x + y + v^2xy = 0$ which is smooth. The second chart gives the hypersurface singularity $xy + z^2(x + y) = 0$, which has a singular point at $(0, 0, 0)$. $E_1$ does not exist in these charts, but intersects $E_3$ in a smooth point in the other charts. $E_2$ is given by $x = z = 0$, $E_3$ is given by $y = z = 0$ and the new $E_4$ is given by $x = y = 0$. $E_2$, $E_3$ and $E_4$ intersect each other in the singular point.

Next we can present the singularity $\text{Tjur}^4(E_7)$ by the matrix $\left(\begin{smallmatrix} x & z(x+y) \\ -z & y \end{smallmatrix}\right)$. Its Tjurina transform is then given by

$$F'_{\{1\}}(x, y, z, a_2) = \begin{pmatrix} z(x + y) - a_2x \\ y + a_2z \end{pmatrix} \text{ and } F'_{\{2\}}(x, y, z, a_1) = \begin{pmatrix} x - a_1z(x + y) \\ -z - a_1y \end{pmatrix}.$$

The first chart gives the hypersurface $zx - wx - wz^2 = 0$ which has a singular point at $(0, 0, 0)$, and the second chart gives $x + v^2y(x + y) = 0$ which is smooth. The exceptional divisor consists of $E_2$ given by $z = v = 0$ so not in the chart that contains the singularity, $E_3$ given by $z = w = 0$, $E_4$ given by $x = w = 0$ and $E_5$ given by $x = z = 0$. $E_2$ intersects $E_5$ in a smooth point, $E_3$, $E_4$ and $E_5$ intersect each other in the singular point, and $E_3$ intersects $E_1$ in a smooth point outside these charts.

We can present $\text{Tjur}^5(E_7)$ by the matrix $\left(\begin{smallmatrix} z & x \\ w & x-wz \end{smallmatrix}\right)$. In this case its Tjurina transform is given by

$$F'_{\{1\}}(x, z, w, a_2) = \begin{pmatrix} x - a_2z \\ x - wz - a_2w \end{pmatrix} \text{ and } F'_{\{2\}}(x, z, w, a_1) = \begin{pmatrix} z - a_1x \\ w - a_1(x - wz) \end{pmatrix}.$$

The first chart gives the hypersurface $yz - wz + yw = 0$ which has a singularity at $(0, 0, 0)$, and the second chart gives the smooth hypersurface $w - vx - v^2wx = 0$. The exceptional divisor consists of $E_1$ and $E_2$ that do not appear in any of these charts, $E_3$ given by $z = v = 0$ (so only appearing in the second chart), $E_4$ given by $w = y = 0$, $E_5$ given by $z = y = 0$ and $E_6$ given by $w = z = 0$. $E_3$ intersects $E_1$ and $E_6$ in different smooth points, $E_2$ intersects $E_5$ in a smooth point, $E_4$, $E_5$ and $E_6$ intersect each other in the singular point.

For $\text{Tjur}^6(E_7)$ we use the matrix $\left(\begin{smallmatrix} y & w \\ z & z+w \end{smallmatrix}\right)$. We get that its Tjurina transform is given by

$$F'_{\{1\}}(y, z, w, a_2) = \begin{pmatrix} w - a_2y \\ z + w - a_2z \end{pmatrix} \text{ and } F'_{\{2\}}(y, z, w, a_1) = \begin{pmatrix} y - a_1w \\ z - a_1(z + w) \end{pmatrix}.$$

The first chart gives the smooth hypersurface $z + xy - xz = 0$, and the second chart gives $z - vz - y = 0$ which is also smooth. So we have reached a resolution of $E_7$. The exceptional divisor consist of $E_1, \ldots, E_7$, where only $E_4 \ldots, E_7$ appear in the last two charts. $E_4$ is given by $y = x - 1 = 0$, $E_5$ is given by $z = v = 0$, $E_6$ is given by $z = x = 0$ and $E_7$ is given by $z = y = 0$. $E_7$ intersects $E_4$, $E_5$ and $E_6$ in three different smooth points, $E_2$ intersects $E_5$ in a smooth point, and $E_3$ intersects

$E_1$ and $E_6$ in two different smooth points. If we represent the exceptional divisor by a dual resolution graph (where vertices represent the curves and edges represent the intersection points) we get:

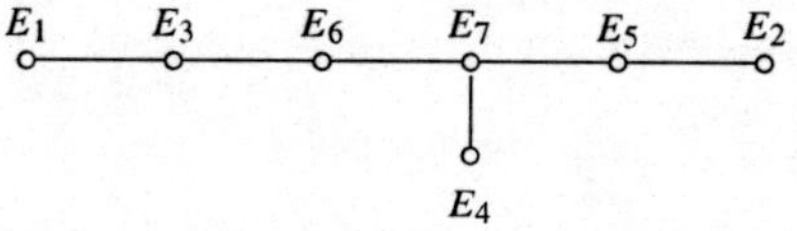


which is indeed the $E_7$ graph.

One can also use this method to produce resolutions of the $D_n$ and $E_6$ singularities, and probably many more. But it is not always possible to use this method. For example the $E_8$ given by $x^2 + y^3 + z^5 = 0$ cannot be written as the determinant of a $2 \times 2$ matrix which is 0 at the origin of $\mathbb{C}^3$, nor can it be written as the determinant of a larger matrix such the value at the origin is 0. If the value at the origin is not zero, then the Tjurina transform does not improve the singularity, it only changes variables.

**Acknowledgments** I wish to thank Maria Ruas for introducing me to the subject of determinantal singularities and for many fruitful conversations during the preparation of this article, and to thank Bárbara Karolline de Lima Pereira who found many mistakes in the earlier version of the article while writing her Master thesis.

The author was supported by FAPESP grant 2015/08026-4.

## References

1. Arbarello, E., Cornalba, M., Griffiths, P.A., Harris, J.: Geometry of algebraic curves. Vol. I. Grundlehren der Mathematischen Wissenschaften [Fundamental Principles of Mathematical Sciences], vol. 267, Springer, New York (1985)
2. Damon, J., Pike, B.: Solvable groups, free divisors and nonisolated matrix singularities II: vanishing topology. Geom. Topol. **18**(2), 911–962 (2014)
3. Ebeling, W., Gusein-Zade, S.M.: On the indices of 1-forms on determinantal singularities. Tr. Mat. Inst. Steklova **267** (2009), no. Osobennosti i Prilozheniya, 119–131
4. Eisenbud, D.: Linear sections of determinantal varieties. Am. J. Math. **110**(3), 541–575 (1988)
5. Frühbis-Krüger, A., Zach, M.: On the vanishing topology of isolated Cohen–Macaulay codimension 2 singularities. To appear in Geometry and Topology. ArXiv e-prints 1501.01915 (2015)
6. Gaffney, T., Rangachev, A.: Pairs of modules and determinantal isolated singularities. ArXiv e-prints 1501.00201 (2015)
7. Kempf, G.R.: The singularities of certain varieties in the Jacobian of a curve. ProQuest LLC, Ann Arbor, MI, 1970, Thesis (Ph.D.)–Columbia University
8. Milnor, J.: Singular points of complex hypersurfaces. Annals of Mathematics Studies, vol. 61. Princeton University Press, Princeton (1968)
9. Nuño-Ballesteros, J.J., Oréfice-Okamoto, B., Tomazella, J.N.: The vanishing Euler characteristic of an isolated determinantal singularity. Israel J. Math. **197**(1), 475–495 (2013)
10. M.A.S. Ruas, Da Silva Pereira, M.: Codimension two determinantal varieties with isolated singularities. Math. Scand. **115**(2), 161–172 (2014)

11. Spivakovsky, M.: Sandwiched singularities and desingularization of surfaces by normalized Nash transformations. Ann. of Math. (2) **131**(3), 411–491 (1990)
12. Tjurina, G.N.: Absolute isolation of rational singularities, and triple rational points. Funkcional. Anal. i Priložen. **2**(4), 70–81 (1968)
13. van Straten, D.: Weakly normal surface singularities and their improvements. Ph.D. thesis, Universiteit Leiden, 1987

# On the Boundary of the Milnor Fiber

**Marcelo Aguilar, Aurélio Menegon, and José Seade**

*To András, in celebration of his first 60th Birthday Anniversary!*

**Abstract** In this work we study the topology of complex non-isolated hypersurface singularities. Inspired by work of Siersma and others, we compare the topology of the link $L_f$ with that of the boundary of the Milnor fiber, $\partial F_f$. We review the three proofs in the literature showing that for functions $\mathbb{C}^3 \to \mathbb{C}$, the manifold $\partial F_f$ is Waldhausen: one by Némethi-Szilárd, another by Michel-Pichon and a more recent one by Fernández de Bobadilla-Menegon. We then consider an arbitrary real analytic space with an isolated singularity and maps on it with an isolated critical value. We study and define for these the concept of vanishing zone for the Milnor fiber, when this exists. We then introduce the concept of vanishing boundary cycles and compare the homology of $L_f$ and that of $\partial F_f$. For holomorphic map germs with a one-dimensional critical set, we give a necessary and sufficient condition to have that $\partial F_f$ and $L_f$ are homologically equivalent.

**Keywords** Boundary Milnor fiber · Vanishing homology · Link

**Subject Classifications** Primary 14J17, 14B05, 32S05, 32S25, 32S30, 32S45; Secondary 14P15, 32C05

M. Aguilar · J. Seade (✉)
Instituto de Matemáticas, Universidad Nacional Autónoma de México, Mexico City, Mexico
e-mail: marcelo@matem.unam.mx; jseade@im.unam.mx

A. Menegon
Universidade Federal da Paraiba, João Pessoa, Brazil
e-mail: aurelio@mat.ufpb.br



J. Fernández de Bobadilla et al. (eds.), *Singularities and Their Interaction with Geometry and Low Dimensional Topology*, Trends in Mathematics,
https://doi.org/10.1007/978-3-030-61958-9_14

# 1 Introduction

Given an analytic map-germ $(\mathbb{R}^m, 0) \xrightarrow{h} (\mathbb{R}^n, 0)$, $m > n$, with a critical value at 0, a fundamental problem is understanding the way the non-critical levels $h^{-1}(t)$ degenerate to the special fiber $V := h^{-1}(0)$. For instance, when $f : (\mathbb{C}^m, \underline{0}) \to (\mathbb{C}, 0)$ is holomorphic, the celebrated fibration theorem of Milnor [17, 34], says that one has a locally trivial fibration:

$$f^{-1}(\mathbb{D}_\eta \setminus \{0\}) \cap \mathbb{B}_\varepsilon \xrightarrow{f} \mathbb{D}_\eta \setminus \{0\}\,,$$

where $\mathbb{B}_\varepsilon$ denotes a sufficiently small ball around the origin in $\mathbb{R}^m$ and $\mathbb{D}_\eta$ is a sufficiently small disc around 0 in $\mathbb{C}$. The set $N(\varepsilon, \eta) := f^{-1}(\mathbb{D}_\eta \setminus \{0\}) \cap \mathbb{B}_\varepsilon$ is usually called a Milnor tube for $f$ and the fibers $F_t := f^{-1}(t) \cap \mathbb{B}_\varepsilon$, $t \neq 0$, are now called the Milnor fibers. Hence one has a family of diffeomorphic manifolds $F_t$ that form a fiber bundle and degenerate to the special fiber $V$. A lot of interesting work has been done studying how this degeneration process $F_t \rightsquigarrow V$ takes place for holomorphic map-germs. This has given rise to a vast literature concerning vanishing cycles and vanishing homology.

This article takes an alternative viewpoint, following our previous work [27] that we briefly explain in what follows. This springs from work by R. Randell [43], D. Siersma [47, 48], F. Michel and A. Pichon [28–30, 32], A. Némethi and Á. Szilárd [39] and J. Fernández de Bobadilla and A. Menegon [13]. For this we recall that a real analytic map-germ $h$ as above has a *link*, which by definition is $L_h := h^{-1}(0) \cap \mathbb{S}_\varepsilon$, the intersection of $V$ with a sufficiently small sphere. The link and its embedding in $\mathbb{S}_\varepsilon$ determine fully the topology of $V$ at 0 and its local embedding in the ambient space (see [34] and [10, Chapter 1, Section 5]).

We shall denote the link by $L_0$ when we want to emphasize that this is the special fiber in the family $\{L_t := \partial F_t\}$ with $t$ in a small disc in $\mathbb{R}^n$.

The link $L_0$ is a real analytic variety and it is non-singular if $h$ has an isolated critical point at 0. In that case, by Ehresmann's fibration lemma, $L_0$ is isotopic to the boundary $L_t$ of the Milnor fiber $F_t$. Otherwise, when $h$ has a non-isolated critical point on $V$, the variety $L_0$ is singular: that is the setting we envisage in this paper.

Given an analytic map-germ $h$ as above, consider a Milnor tube $N(\varepsilon, \eta) := h^{-1}(\mathbb{D}_\eta \setminus \{0\}) \cap \mathbb{B}_\varepsilon$, and let us assume this is a fiber bundle over $\mathbb{D}_\eta \setminus \{0\}$ with projection $h$ (unlike the complex setting, this hypothesis is not always satisfied for real analytic map-germs; see for instance [46] for a thorough discussion about that topic). The fibers $F_t := h^{-1}(t) \cap \mathbb{B}_\varepsilon$, $t \neq 0$, are compact manifolds with boundary $L_t$. While the family $\{F_t\}$ degenerates to the special fiber $V := h^{-1}(0) \cap \mathbb{B}_\varepsilon$, the corresponding family of boundaries $\{L_t\}_{t \neq 0}$, which are smooth compact manifolds, degenerates to the link $L_0$, which is singular.

The purpose of this work is to study and compare the topology of both $L_t$ and $L_0$ by looking at the degeneration process $\{L_t\}_{t \neq 0} \rightsquigarrow L_0$.

We begin this article with a few words about the degeneration of the Milnor fibers to the special fiber, and about the corresponding process as we look at the boundaries (Sect. 2). Then, in Sect. 3, we focus on the case of holomorphic map germs $\mathbb{C}^3 \to \mathbb{C}$ and describe briefly some of the ideas in András Némethi and Ágnes Szilárd's excellent book [39], where it is proved that in this setting, the boundary of the Milnor fiber, which is a 3-manifold, is always Waldhausen. That theorem was announced in 2003 by Françoise Michel and Anne Pichon, providing a proof that worked fine for certain families of singularities. Their complete proof was published in 2016. There is also a third proof by Javier Fernández de Bobadilla and Aurélio Menegon that works more generally: for real analytic map-germs of the form $f\bar{g}$ : $\mathbb{C}^3 \to \mathbb{C}$ with a Milnor fibration. Taking $g$ to be constant one gets the previous assertion. In Sect. 3 we briefly discuss these three viewpoints.

Section 4 is a brief summary of [26], and an extension of it to the case of real analytic map-germs $f : (X, \underline{0}) \to (\mathbb{R}^n, 0)$, $m > n$, where $X$ is an $m$-dimensional real analytic space with an isolated singularity at $\underline{0}$, $f$ has an isolated critical value at 0 and it has a Milnor fibration in a tube. We introduce the notion of a *vanishing zone* for $f$ and for the Milnor fibers. This means a regular neighborhood $W$ of the link $L_\Sigma := \Sigma \cap \mathbb{S}_\varepsilon$ of the singular set of $V = f^{-1}(0)$, with smooth boundary $\partial W$, such that for every regular value $t$ with $\|t\|$ sufficiently small, one has that the boundary of the Milnor fiber $\partial F_t$ meets $\partial W$ transversally, and $\partial F_t \setminus (F_t \cap W)$ is diffeomorphic to $L_f \setminus (L_f \cap W)$, where $L_f := V \cap \mathbb{S}_\varepsilon$ is the link of $V$.

Section 5 is part of [1], a work in progress where we look at the homology of the boundary of the Milnor fiber for holomorphic map germs $\mathbb{C}^{n+1} \to \mathbb{C}$. In analogy with the classical notion of vanishing homology for the Milnor fiber, we introduce the notion of the *vanishing boundary homology*. For this we observe that there is a specialization morphism from the homology of $\partial F_f$ to the homology of the link. We state a theorem from [1] about the homology groups of $\partial F_f$ when the critical set is one-dimensional, which is a special case of a more general theorem. As a corollary we give, for holomorphic map-germs with one-dimensional critical set, a necessary and sufficient condition for having that the boundary of the Milnor fiber and the link are homologically equivalent. We give examples of such cases.

In Sect. 6 we conclude with a couple of remarks. One is for map-germs defined on analytic spaces with arbitrary singular locus. Another is for map-germs with non-isolated critical value. We actually look at an example where the critical values have real codimension 1, so they split the target into several connected components. Yet, the singularities in question are all real analytic isolated complete intersections, and this implies that we have a Milnor fibration over each component. Although the topology of the Milnor fibers varies as we change from one sector to another, they all have boundary isotopic to the link. Notice that if the dimension of the link is even, this implies that all Milnor fibers have the same Euler characteristic. If the dimension of the link is odd, we only have that the Euler characteristic of the Milnor fibers coincides modulo 2.

## 2 From the Non-critical Level to the Special Fiber

The starting point is the classical Milnor's fibration theorem for holomorphic maps, see [34] and [17]. Consider a holomorphic function-germ with a critical point at $\underline{0}$:

$$f : (\mathbb{C}^{n+1}, \underline{0}) \to (\mathbb{C}, 0).$$

Set $V = f^{-1}(0)$ and let $L_f = V \cap \mathbb{S}_\varepsilon$, for $\varepsilon > 0$ sufficiently small, be the link. Given $\varepsilon$, choose $\delta > 0$ small enough with respect to $\varepsilon$, so that every fiber $f^{-1}(t)$ with $|t| \leq \delta$ meets transversally the sphere $\mathbb{S}_\varepsilon$; that such a $\delta$ exists for every Milnor sphere $\mathbb{S}_\varepsilon$ is a consequence of the fact that $f$ has the Thom $a_f$-property, by Hironaka [14].

Set $\mathbb{D}_\delta^* := \mathbb{D}_\delta \setminus \{0\}$, where $\mathbb{D}_\delta$ is the disc of radius $\delta \subset \mathbb{C}$ centered at 0, and consider the *Milnor tube*:

$$N(\varepsilon, \delta) = f^{-1}(\mathbb{D}_\delta^*) \cap \mathbb{B}_\varepsilon .$$

Then part of Milnor's theorem (which follows from Ehresmann's fibration theorem extended to manifolds with boundary) says that we have a locally trivial fibration:

$$f : N(\varepsilon, \delta) \longrightarrow \mathbb{D}_\delta^* .$$

The fibers $F_t$ are the local regular levels of the function. We denote the Milnor fibers by $F_t$ when we emphasize that this corresponds to the value $t$, or by $F_f$ when we look at an arbitrary Milnor fiber of $f$.

So the Milnor fibers are a family of complex Stein manifolds, the local non-critical levels of the function, that degenerate to the special fiber, the critical level $V = F_0$ as $t$ tends to 0. A lot of interesting work, particularly in the isolated singularity case, has been done studying how this degeneration process $F_t \rightsquigarrow V$ takes place (Fig. 1).

When $f$ has an isolated critical point, the fiber $F_t$ is diffeomorphic to a $2n$-ball to which one attaches $\mu = \mu(f)$ handles of middle index $n$, where $\mu$ is the Milnor number of $f$ at $\underline{0}$, by Lê and Perron [19] and Milnor [34]. This number can be computed as the intersection number:

$$\mu = \dim_{\mathbb{C}} \frac{\mathcal{O}_{n+1,\underline{0}}}{\operatorname{Jac} f} ,$$

where $\operatorname{Jac} f$ is the Jacobian ideal, generated by the partial derivatives of $f$. The middle homology group $H_n(F_t)$ is free of rank $\mu$, generated by $\mu$ cycles that "vanish" as $F_t$ degenerates into the special fiber, since $V$ is locally a cone. Therefore these are called *vanishing cycles* and the Milnor number counts how many of these $F_t$ has.

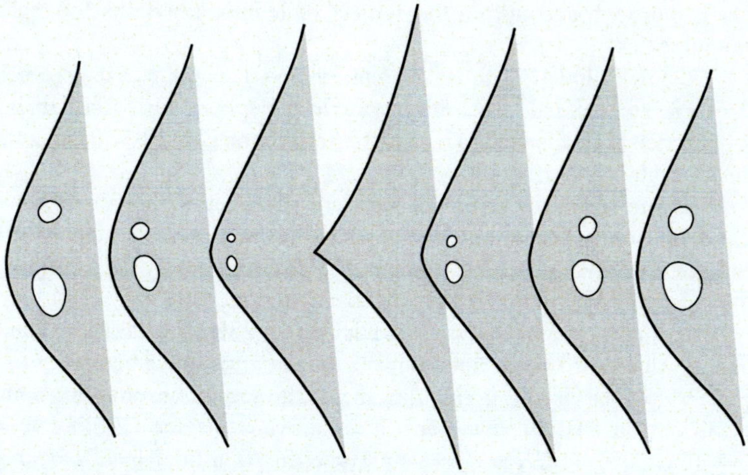

**Fig. 1** The Milnor fibers $F_t$ degenerating to the special fiber $F_0$

The concept of vanishing cycles was first mentioned by Grothendieck in a letter to Serre in 1964 [6, p. 214], where he analyses the difference between the (étale) cohomology of the special fiber and that of the generic fiber in certain families. His theory was developed by Deligne in [9] and has had immense applications.

When $f$, as above, has non-isolated critical points, the fiber $F_t$, being a Stein manifold, has the homotopy type of a CW-complex of middle dimension, by Andreotti and Frankel [2] and Milnor [34], and it is $(n - s - 1)$-connected by Kato and Matsumoto [15], where $s$ is the complex dimension of the singular set of $V = f^{-1}(0)$. In this case, by Massey [23], $F_t$ is diffeomorphic to a $2n$-ball to which one attaches handles of various indices, as indicated by the Lê numbers. The homology groups of $F$ are called groups of vanishing cycles. These have been investigated by several authors; see for instance [23, 24, 48], and [9] for a more general viewpoint. See [44] for a survey on recent results concerning the algebraic computation of vanishing cycles of an algebraic function on a complex quasi-projective variety.

More generally, given an analytic map-germ $(\mathbb{R}^m, 0) \xrightarrow{h} (\mathbb{R}^n, 0)$, $m > n$, with an isolated critical value at 0. Under suitable conditions, one still has a locally trivial fibration as above

$$h^{-1}(\mathbb{D}_\eta \setminus \{0\}) \cap \mathbb{B}_\varepsilon \xrightarrow{h} \mathbb{D}_\eta \setminus \{0\}\,,$$

where $\mathbb{B}_\varepsilon$ denotes a sufficiently small ball around the origin and $\mathbb{D}_\eta$ is a sufficiently small disc around 0 in $\mathbb{R}^n$. The set $N(\varepsilon, \eta) := h^{-1}(\mathbb{D}_\eta \setminus \{0\}) \cap \mathbb{B}_\varepsilon$ is usually called a Milnor tube for $h$, and the fibers $F_t := h^{-1}(t) \cap \mathbb{B}_\varepsilon$, $t \neq 0$, are the Milnor fibers of $h$. A fundamental problem is understanding how the non-critical levels $h^{-1}(t)$ degenerate to the special fiber $V := h^{-1}(0)$. We refer to [46, Section 13] for a thorough account on the subject of Milnor fibrations for real and complex singularities.

We now follow [27] and take an alternative viewpoint to the problem of studying how the non-critical levels degenerate to the special fiber. This springs from work by Randell [43], Siersma [47, 48], Michel-Pichon [28–30, 32], Némethi-Szilárd [39] and Fernández de Bobadilla-Menegon [13].

For this we recall that a real analytic map-germ $h$ as above has a *link*, which by definition is $L_0 := h^{-1}(0) \cap \mathbb{S}_\varepsilon$, the intersection of $V$ with a sufficiently small sphere. The link and its embedding in $\mathbb{S}_\varepsilon$ determine fully the topology of $V$ at 0 and its local embedding in the ambient space (cf. [10, 34]).

The link $L_0$ is real analytic and it is non-singular if $h$ has an isolated critical point at 0. In that case $L_0$ is a smooth manifold, isotopic to the boundary $L_t$ of the Milnor fiber $F_t$. Otherwise $L_0$ is singular: that is the setting we envisage in this paper.

Given an analytic map-germ $h$ as above, consider a Milnor tube $N(\varepsilon, \eta) := h^{-1}(\mathbb{D}_\eta \setminus \{0\}) \cap \mathbb{B}_\varepsilon$, and let us assume this is a fiber bundle over $\mathbb{D}_\eta \setminus \{0\}$ with projection $h$. The fibers $F_t := h^{-1}(t) \cap \mathbb{B}_\varepsilon$, $t \neq 0$, are compact manifolds with boundary $L_t$. While the family $\{F_t\}$ degenerates into the special fiber $F_0 := h^{-1}(t) \cap \mathbb{B}_\varepsilon$, one also has the corresponding family of boundaries $\{L_t\}_{t\neq 0}$ degenerating to the link $L_0$, which may be singular. In the sequel we look at the topology of both $L_t$ and $L_0$, and we study of the degeneration process $\{L_t\}_{t\neq 0} \rightsquigarrow L_0$ for both, real and complex singularities.

This is interesting for two reasons. On the one hand, the boundary of the Milnor fiber, being a smooth manifold, is in many ways easier to handle than the link. Understanding the way $L_t$ degenerates into $L_0$ throws light into the topology of the link, and hence into that of $V$, just as the study of the vanishing cycles on the Milnor fiber throws light into the topology of the special fiber. On the other hand, we can argue conversely: understanding the degeneration $L_t \rightsquigarrow L_0$ allows us to get information about $L_t$ out from $L_0$ itself. For instance, this was the approach followed in [13, 28, 30, 39] to show that in the case of holomorphic map-germs in 3 complex variables, and also for map germs of the form $f\bar{g}$, the boundary $L_t$ is a Waldhausen manifold (see the following section).

## 3 The Case of Complex Surfaces

Now consider a non-constant holomorphic function-germ,

$$f : (\mathbb{C}^3, \underline{0}) \to (\mathbb{C}, 0),$$

with a one-dimensional critical set at $\underline{0}$. Set $V = f^{-1}(0)$ and $\Sigma = \Sigma(f) \subset V$, the singular set. Let $L_f = V \cap \mathbb{S}_\varepsilon$ be the link of $V$ and $L_\Sigma = L_f \cap \Sigma$ the link of $\Sigma$. Notice that $L_\Sigma$ is the singular set of the real analytic three-dimensional variety $L_f$, and $L_\Sigma$ is a disjoint union of circles $S^1$, one for each branch of $V$. We denote by $F$ the Milnor fiber of $f$ and $\partial F$ is its boundary,

R. Randell [43] and D. Siersma [47, 48] determined the homology of the boundary $\partial F$ in several cases, finding, among other things, examples where $\partial F$ is a homology sphere. It was also noticed in [47, 48] that the boundary $\partial F$ consists of two parts, which are compact manifolds glued along their common boundary, which is a disjoint union of tori $S^1 \times S^1$, one for each connected component of $L_\Sigma$. One of these is the portion of $\partial F$ contained within a tubular neighborhood of $L_\Sigma$; the other is the closure of its complement. Following [30] we call these the *vanishing zone* of $\partial F$ and *the trunk*, respectively. The interior of the trunk is diffeomorphic to $L_f \setminus L_\Sigma$. This is discussed with care in the following Sect. 4, so we only say a few words here. This decomposition plays a key role in the work of Némethi-Szilárd [39], Michel-Pichon [30] and Fernández de Bobadilla and Menegon [13]. Those works give three different proofs of the fact that in the setting we now envisage, the boundary $\partial F$ is a Waldhausen manifold. We now say a few words about each of these works.

Recall that a 3-manifold $M$ is a Seifert manifold if it fibers over a surface $S$ with fiber the circle $S^1$ and this fibration is locally trivial away from a finite number of points in $S$. The fibers over those points are called the exceptional fibers. So for instance, every circle bundle over a surface is a Seifert manifold with no exceptional fibers.

A 3-manifold $M$ is a Waldhausen manifold if there exist finitely many tori $S^1 \times S^1$ in $M$, such that cutting $M$ along these torii, the complement is Seifert. And we know from [40] that a 3-manifold $M$ is Waldhausen if and only if it is a graph (or plumbed) manifold.

Plumbing is a construction introduced by Milnor [33] in order to exhibit the existence of exotic spheres. In that construction, the "building blocks", so to say, are $n$-dimensional disc bundles $E$ over compact $n$-manifolds $B$. The boundary $\partial E$ is an $S^{n-1}$ bundle over $B$. Given two of these, say $(E_1, B_1)$ and $(E_2, B_2)$, to perform plumbing on them we choose a small disc $D_i$ in each $B_i$, so that $E_i$ restricted to this disc is a product of two $n$-discs, $D^n \times D_i$. Then we identify the points in $E_1$ and $E_2$ contained in $E_i|_{D_i}$ by identifying a point $(x, y) \in E_1|_{D_1}$ with the point $(y, x) \in E_2|_{D_2}$. What we get is a $2n$-dimensional compact manifold $E_1 \# E_2$ with boundary and with corners. But the corners can be smoothed out in an essentially unique way, up to isotopy.

At the level of the boundaries what we are doing is removing from each $\partial E_i$ the interior of a product $S^{n-1} \times D_i$, and then identifying the boundaries $S^{n-1} \times S_i^{n-1}$ by the map $(x, y) \mapsto (y, x)$.

A plumbed manifold is a manifold obtained by iterating this construction a finite number of times. The relevant case in this section is when $n = 2$, all manifolds and bundles are oriented, and we are plumbing oriented 2-disc bundles over compact 2-manifolds with no boundary. Recall that up to diffeomorphism, every such manifold is classified by its genus. And the oriented 2-disc bundles $E$ over every such manifold $B$ are classified by their Euler class, an integer, which equals the self-intersection number of $B$ in $E$ regarded as the zero-section. Hence in this setting, to every plumbed manifold we can associate a plumbing graph $G$: to each vertex in $G$ we associate a weight $w \in \mathbb{Z}$ and a genus $g \in \mathbb{N}$. This represents a choice of a closed oriented 2-manifold $B$ of genus $g$, and a 2-disc bundle $E$ over $B$ with Euler class $w$. If two vertices are connected by an edge, we do plumbing in the corresponding $E_i$. If we have more than one edge joining two vertices, we repeat this operation several times, choosing disjoint discs.

That is why in singularity theory, in the case $n = 2$, plumbed manifolds are also called graph manifolds. It is not hard to see that every Waldhausen manifold is a graph manifold and viceversa, cf. [40].

Notice that to every graph manifold one associates naturally a symmetric matrix $A = ((E_{i,j}))$, called the intersection matrix. The elements in the diagonal are the weights $w_i$ of the vertices, and the rest of the coefficients are the number of edges connecting the corresponding vertices.

As an example, let $(V, p)$ be a normal complex surface singularity, and let $\pi : \widetilde{V} \to V$ be a good resolution; recall that good means that each irreducible component $S_i$ of the exceptional divisor $E$ is non-singular, all $S_i$ intersect transversally and no three of them intersect. We may now consider its dual graph: to each $S_i$ we associate a vertex $v_i$, with a genus $g_i$ which is the genus of $E_i$ and a weight, the self-intersection number of $S_i$ in $\widetilde{V}$. We then join two vertices $v_i$, $v_j$ by as many edges as the intersection number $S_i \cdot S_j$. We get a plumbing graph. The result of performing plumbing according to this graph is a 4-manifold, homeomorphic to a tubular neighborhood of the exceptional divisor $E$ in $\widetilde{V}$, and its boundary is diffeomorphic to the link $L_f$. Therefore $L_f$ is a graph manifold, and we know from [11, 35] that the corresponding intersection matrix is negative definite. In fact the converse is also true: by Grauert's contractibility criterium, every graph manifold with negative definite intersection matrix, is orientation preserving homeomorphic to the link of a normal complex surface singularity, cf. [40].

Let us consider again a non-constant holomorphic function-germ with a non-isolated critical point at $\underline{0}$,

$$f : (\mathbb{C}^3, \underline{0}) \to (\mathbb{C}, 0).$$

We now say a few words about three different points of view that have been used to establish, among other things, three different proofs of the fact that in this setting, the boundary of the Milnor fiber is a graph manifold.

## 3.1 A Glance on Némethi-Szilárd's Work for Surface Singularities

As pointed out in the introduction to [39], the work by András Némethi and Ágnes Szilárd has its roots in several of the milestones in singularity theory, some of these arising from the rich interplay one has between 3-manifolds and isolated complex surface singularities.

The 3-manifolds that arise as links of normal complex surface singularities, which are the graph manifolds with negative definite intersection matrix, are a particularly interesting class of manifolds, with certain properties that make these manifolds provide a ground for a better understanding of important invariants in low dimensional topology.

In fact, given a normal isolated singularity germ $(V, \underline{0})$, the link $L_f$ can always be regarded as being the boundary of a neighborhood of the special fiber $\pi^{-1}(\underline{0})$ of a resolution $\widetilde{V} \stackrel{\pi}{\to} V$. If $V$ is a hypersurface, or more generally a smoothable singularity, then $L_f$ can also be regarded as being the boundary of the Milnor fiber. One thus has two natural holomorphic fillings of the link: the resolution and the Milnor fiber, and this gives rise to remarkable index-theoretical relations (see for instance [12, 16] and [45, Ch. IV] for the case where $V$ has dimension two). This has been used by Andras in many articles, to produce remarkable results concerning Seiberg-Witten invariants, Floer homology and many other important 3-manifolds invariants, see for instance [36, 37].

Looking at 3-manifolds that are boundaries of Milnor fibers of non-isolated complex singularities defined by an equation $\mathbb{C}^3 \stackrel{f}{\to} \mathbb{C}$ extends the class of manifolds that arise from complex singularities. In this setting the Milnor fiber still provides a holomorphic filling for the natural contact structure on the boundary; but one does not have, *a priori*, the corresponding resolution of the singularity, since the Milnor fiber already is non-singular. A remarkable outcome of the work of Némethi and Szilárd in [39] is that even though these manifolds cannot be in general links of isolated complex singularities, they do appear naturally as links of certain real analytic singularities. Then, resolving these singularities one may regard the boundary of the Milnor fiber as being the boundary of a tubular neighborhood of the resolution, and one gets from this a graph decomposition. In fact the proof in [39] actually provides also an explicit Waldhausen decomposition of it. This is used in [39] to study the topology and geometry of $\partial F$.

The main algorithm in [39] springs from a Iomdin series associated to $f$. In fact, given $f$ as above, let $g$ be a holomorphic function germ in $\mathbb{C}^3$ such that $(f, g)$ defines an isolated complete intersection. Now consider singularities of the form $\{f = |g|^k\}$ for $k > 0$. One has that for large $k$, its link is independent of the choice of $k$, and it turns out to be diffeomorphic to $\partial F$. That means that $\partial F$ appears as the boundary of an arbitrary small neighborhood of a real analytic germ. After resolving this real analytic singularity, the tubular neighbourhood of the exceptional set provides a plumbing representation of $\partial F$.

The way Némethi and Szilárd resolve the above singularities is very interesting, because it not only gives a resolution, but it provides also a lot more information about the singularities in question, and about the boundary $\partial F$. This springs from their earlier work [38], explained in Chapter 6 of their book, where they introduce a decorated graph $\Gamma$ to study hypersurface singularities in three variables with one-dimensional singular locus. That is their main tool in [38] for getting resolution graphs of the singularities in question. Starting with a hypersurface germ $f$ as above, choose $g$ so that $(f, g)$ forms an ICIS. Then the graph $\Gamma$ yields to a resolution of the singularities in the Iomdin series $f + g^k$ for $k$ large. Moreover, the same graph $\Gamma$ also contains enough information to allow them to determine the boundary $\partial F$. This is done by looking at the aforementioned singularities $\{f = |g|^k\}$ for $k$ large.

The work of Némethi and Szilárd was extended by O. Curmi in [7] to functions defined on an arbitrary ambient space. To be precise, Curmi considers the germ $(X, \underline{0})$ of a three-dimensional complex analytic variety, and the germ of a reduced holomorphic function $f : (X, \underline{0}) \to (\mathbb{C}, 0)$ such that its zero locus $V(f)$ contains the singular locus of $X$. Then it is proved that the boundary of the Milnor fiber of $f$ is a graph manifold. This is further refined in [8] where the author gives an algorithm for describing the corresponding plumbing graph of the boundary of the Milnor fiber of Newton non degenerate surface singularities.

## 3.2 *On the Work of Michel-Pichon-Weber*

The references for this subsection are the articles [28–32] by Françoise Michel, Anne Pichon and Claude Weber. In [28] the authors consider a holomorphic map-germ $f : (\mathbb{C}^3, \underline{0}) \to (\mathbb{C}, 0)$ and state the theorem that the boundary $\partial F$ of the Milnor fiber is a graph manifold, with a sketch of the proof (with a gap pointed out in [29]). Complete proofs of this theorem are given in [31, 32] for special families of singularities. Shortly after the appearance of the book by A. Némethi and Á. Szilárd [39], F. Michele and A. Pichon provided a complete proof which is in the spirit of the original method they proposed.

The idea is the following. Firstly they split the boundary $\partial F_f$ in two parts, as already explained, which essentially are the trunk and the vanishing zone; the work above essentially comes from the fact that in these papers the authors do not work with the singular variety $V$ but with its normalization. That the trunk is Waldhausen follows from the classical theory of complex surface singularities, by taking first a normalization of $V$ and then a good resolution. Since the trunk and the vanishing zone are glued along tori, the hard part is showing that the vanishing zone of $\partial F_f$ has a Waldhausen structure compatible with the boundary.

The key point is the use of a "carousel in family". In fact, recall that given a map-germ $g : (\mathbb{C}^2, \underline{0}) \to (\mathbb{C}, 0)$, D. T. Lê developed a remarkable method to construct the corresponding Milnor fiber, known as "the carousel", see for instance [18] or the expository article [20]. When $g$ is irreducible, so it has only one branch, then we know from [3] that its link is an iterated torus knot determined by the Puiseux pairs of $g$. Yet, the Puiseux expansions actually give an additional structure near the singular point, observed by D. T. Lê, that gives rise to what he called the carousel associated to the singularity. This is obtained by considering an auxiliary linear form $\ell$, general enough for $g$, and looking at the distribution of points $\{z_j\}$ in the intersection $\{\ell = t\} \cap \{g(x, y) = 0\}$. Then the carousel arises by a careful study of how the Puiseux pairs describe the points in which the line $\{\ell = t\}$ meets the Milnor fiber, which are distributed regularly around each point $\{z_j\}$, and their distribution is determined iteratively by the Puiseux pairs.

In the setting we envisage here, we want to show that the Milnor fiber of the map-germ $f : (\mathbb{C}^3, \underline{0}) \to (\mathbb{C}, 0)$ has a graph manifold structure within the vanishing zone $W$. This manifold $W$ has a connected component for each component of the singular set $L_\Sigma := \Sigma \cap \mathbb{S}_\varepsilon$ of the link $L_f : V \cap \mathbb{S}_\varepsilon$, where $V = f^{-1}(0)$. Each component of $L_\Sigma$ is a circle. If at each point $z \in L_\Sigma$ we take a small complex two-dimensional disc $H_z$ transversal to $L_\Sigma$ what we get in that disc is the germ of a plane curve. The Milnor fiber of the restriction $f|_{H_z}$ is then described by a carousel. Doing this in a "coherent way" for all points in the corresponding connected component of $L_\Sigma$, we get a family of carousels, parameterized by the circle $S^1$. Hence, in order to construct the part of the Milnor fiber contained in the vanishing zone, we may consider, for each connected component of $L_\Sigma$, a family of carousels parameterized by the circle $S^1$. After taking care of a number of highly non-trivial subtleties, this shows that the boundary of the Milnor fiber of $f$ is a graph manifold.

## 3.3 *On the Work of Fernández de Bobadilla and Menegon*

To finish this section, we say a few words about [13]. This actually considers a more general setting. Let $(X, \underline{0})$ be a three-dimensional complex analytic germ with an isolated singularity at the origin in some $\mathbb{C}^N$, and let $f, g$ be holomorphic function-germs $(X, \underline{0}) \to (\mathbb{C}, 0)$ such that:

- $V(f) := f^{-1}(0)$ and $V(g) := g^{-1}(0)$ have no common irreducible components;
- the real analytic map-germ $f\bar{g} : (X, \underline{0}) \to (\mathbb{C}, 0)$ has an isolated critical value; and
- the real analytic map-germ $f\bar{g} : (X, \underline{0}) \to (\mathbb{C}, 0)$ has a Milnor fibration. That is, we assume that there exists small positive real numbers $\varepsilon > 0$ and $\delta > 0$ with $0 < \delta \ll \varepsilon \ll 1$ such that the restriction

$$(f\bar{g})|_{(f\bar{g})^{-1}(D^*_\delta)\cap X\cap \mathbb{B}_\varepsilon} : (f\bar{g})^{-1}(D^*_\delta) \cap X \cap \mathbb{B}_\varepsilon \longrightarrow D^*_\delta$$

  is the projection of a locally trivial fibration, where $\mathbb{B}_\varepsilon$ and $D^*_\delta$ are a small enough discs around the origin in $\mathbb{C}^N$ and $\mathbb{C}$ respectively.

Then it is proved in [13] that the boundary of the Milnor fibre $\partial F_{f\bar{g}} := (f\bar{g})^{-1}(t) \cap X \cap \mathbb{S}_\varepsilon$, for $t \in D^*_\delta$, is a Waldhausen manifold.

Of course, taking $g$ to be constant one is back in the situation envisaged previously. In other words, this work by Fernández de Bobadilla and Menegon gives a third proof of the theorem that the boundary of the Milnor fiber for holomorphic map germs in three complex variables is a Waldhausen manifold, and that proof works, more generally, for maps $f\bar{g}$ and the ambient space $X$ being any three-dimensional complex space which is non-singular away from a point.

The first step in the proof is based on studying the Milnor fibre of a map-germ of the form $f\bar{g}$ defined on a complex surface with a an isolated singularity, in terms of an embedded resolution of $\{fg = 0\}$. This uses previous work by Pichon and Seade [42] about Milnor fibrations for such maps. Then they proceed to dimension 3. They first split the boundary of the Milnor fibre of $f\bar{g}$ into two parts as before: the trunk and the vanishing zone, which are glued together along a finite union of tori. Most of the work goes for showing that the part of the Milnor fibre inside the vanishing zone is a Waldhausen manifold. They do so by means of a slicing argument, and then using their previous results in dimension 2. A key point is noticing that it is sufficient to decompose the transversal Milnor fibre into pieces which are invariant under the corresponding vertical monodromy, and which decompose the vanishing zone into Waldhausen pieces. This is proved by showing that the vanishing zone can be decomposed into pieces that are either fibre bundles over a circle with fibre a cylinder, or it is a finite unramified covering of a Walhausen manifolds, and therefore it is also Waldhausen.

## 4 The Vanishing Zone

We now consider the germ at $\underline{0} \in \mathbb{R}^N$ of a real analytic variety $X$ such that $X \setminus \{\underline{0}\}$ is a smooth manifold of real dimension $m$. We consider real analytic map-germs

$$f : (X, \underline{0}) \to (\mathbb{R}^n, 0) \quad , \quad m > n > 0,$$

with an isolated critical value at 0, which admit a local Milnor fibration in a tube. That is, there is a Milnor ball $\mathbb{B}_\varepsilon$ for $f$, and $\delta > 0$, depending on $\varepsilon$, such that if we let $N(\varepsilon, \delta)$ be the Milnor tube $N(\varepsilon, \delta) := f^{-1}(\mathbb{D}^*_\delta) \cap \mathbb{B}_\varepsilon$, where $\mathbb{D}^*_\delta$ is a punctured ball in $\mathbb{R}^n$ around 0 of radius $\delta$, then one has a locally trivial fibration

$$f|_{N(\varepsilon,\delta)} : N(\varepsilon, \delta) \to \mathbb{D}^*_\delta \,. \tag{1}$$

This is a stringent condition and yet, there are enough examples to make it an interesting setting. For instance, the above conditions are satisfied in all the following examples:

(a) $f$ has an isolated critical value at $\underline{0}$ and $f^{-1}(0)$ has positive dimension.
(b) $X$ is complex analytic and $f$ is a $\mathbb{C}$-valued holomorphic map.
(c) $X$ is a complex analytic surface with a normal singularity at 0, and $f$ is of the form $f = h\bar{g}$ where $h, g$ are holomorphic with no common branch (see [42]).
(d) $f$ is a polar weighted mixed polynomial $\mathbb{R}^{2m} \to \mathbb{R}^2$.

Statement (a) essentially follows from the classical Ehresmann's fibration theorem, using the implicit function theorem (cf. [4, 5]). Statement (b) follows from Hironaka's theorem in [14], stating that every such map is Thom regular; (c) follows from [42]. For (d) see Oka's paper [41]: the polar action ensures that the critical value must be isolated, while the radial action ensures the transversality condition.

One has the following folklore theorem (see for instance [5, Theorem 2.7] or [26, Proposition 2.5]).

**Theorem 4.1** *Let* $(X, \underline{0})$ *be as above and let* $f : (X, \underline{0}) \to (\mathbb{R}^n, 0), m > n > 0$, *be real analytic with an isolated critical value at* $0 \in \mathbb{R}^n$. *Assume* $f$ *satisfies the following transversality condition: for every Milnor sphere* $\mathbb{S}_\varepsilon$ *there exists* $\delta > 0$, *depending on* $\varepsilon$, *such that each fiber* $f^{-1}(t)$ *with* $|t| \leq \delta$ *intersects transversally the sphere* $\mathbb{S}_\varepsilon$. *Then* $f$ *has a local Milnor-Lê fibration.*

Given $f$ as in this theorem, set $V = f^{-1}(0) \cap X$ and equip $X$ with a Whitney stratification for which $V$ and its singular set $\Sigma$ are union of strata. Let $\mathbb{S}_\varepsilon$ be a sufficiently small Milnor sphere for $f$ so that $L_X := X \cap \mathbb{S}_\varepsilon$ is the link of $X$, and therefore $L_f := V \cap \mathbb{S}_\varepsilon$ is the link of $V$ and $L_\Sigma$ is that of $\Sigma$. We denote by $\partial F_t$ the boundary of a Milnor fiber $F_t$, so:

$$\partial F_t := f^{-1}(t) \cap X \cap \mathbb{S}_\varepsilon ,$$

for $t \in \mathbb{D}_\delta \setminus \{0\}$. Then $\partial F_t$ is a smooth submanifold of $L_X$ that degenerates to the link $L_f$ as $|t|$ goes to 0.

Following the previous discussion, we aim to study and compare the topology of $\partial F_t$ with that of $L_f$. For this we want to define and show the existence of a vanishing zone for $\partial F_t$. We follow [27]:

**Theorem 4.2** *There exists a compact regular neighborhood* $W$ *of* $L_\Sigma$ *in* $L_X$ *such that:*

- *$W$ has smooth boundary $\partial W$ and this boundary intersects $L_f$ transversally;*
- *$W$ has $L_\Sigma$ as a deformation retract;*
- *for every $t$ sufficiently close to 0 we have $\partial F_t \setminus \mathring{W}$ is diffeomorphic to $L_f \setminus \mathring{W}$, where $\mathring{W}$ is the interior.*
- *If the critical set $\Sigma$ of $f$ is either smooth or an isolated singularity, then $W$ can be chosen to be a fiber bundle over $L(\Sigma)$ with fiber a $2(n-k)$-dimensional ball, where $k$ is the dimension of $\Sigma$.*
- *If there is a Whitney stratification of $V$ so that each connected component of $\Sigma \setminus \{0\}$ is a single stratum, then the intersection $W_t := \partial F_t \cap W$ is a topological fiber bundle over $L_\Sigma$, for every $t$ sufficiently near 0.*

The proof is exactly like that of [27, Theorem 2.5] and is left to the reader. Notice that in the last statement in Theorem 4.2 the condition that each connected component of $\Sigma \setminus \{0\}$ be a single stratum is rather stringent. This implies the local topological triviality that gives the fiber bundle structure in that statement.

**Definition 4.3** A *vanishing zone* for $f$ is a regular neighborhood $W$ of $L_\Sigma$ in $L_X$ as in Theorem 4.2.

## 5 The Vanishing Boundary Homology

Consider now an irreducible and reduced complex analytic germ $(X, \underline{0})$ of pure dimension $n+1$ in some $\mathbb{C}^m$, and let $f : (X, \underline{0}) \to (\mathbb{C}, 0)$ be holomorphic. We know from [49] that taking the representative of $X$ to be small enough, $f$ has a unique critical value at 0. And we know from [14] that $f$ has the Thom $a_f$-property, which implies the transversality condition in Theorem 4.1. Hence one has a Milnor fibration in a tube, which is a special case of Lê's fibration theorem in [17], which holds for $X$ with arbitrary singularities. Notice that if $X$ has a non-isolated singularity at $\underline{0}$, then the Milnor fibers of $f$ may not be smooth, due to the singularities of $X$.

Theorem 4.2 hints on looking at relations between the homology of $\partial F_f$ and that of the link $L_f$. In fact one has a specialization morphism (*à la* Verdier, cf. [25, 50]),

$$\mathcal{S} : H_*(\partial F_f) \longrightarrow H_*(L_f) ,$$

by observing that the vanishing zone $W$ extends to a regular neighborhood $\widetilde{W}$ of $L_f$ that contains the boundary of $F_f$ for all Milnor fibers over points sufficiently near 0. Thus one has a morphism $H_*(\partial F_f) \to H_*(\widetilde{W})$ induced by the inclusion. Also, $\widetilde{W}$ has $L_f$ as a deformation retract, so one gets an isomorphism $H_*(\widetilde{W}) \to H_*(L_f)$. The specialization $\mathcal{S}$ is the composition of these two morphisms.

In [1] we show that $\mathcal{S}$ is surjective whenever the germ $(X, \underline{0})$ is an ICIS. It would be interesting to know whether or not $\mathcal{S}$ is always surjective.

We call the kernel of $\mathcal{S}$ *the vanishing boundary homology* of the Milnor fiber, $H_*^{\mathcal{V}}(\partial F_f)$. By Theorem 4.2, $H_*^{\mathcal{V}}(\partial F_f)$ has support in the vanishing zone $W$. Similarly, following [9], the elements in $H_*(\partial F_f)$ can be thought of as being the *nearby boundary cycles*.

In order to study the vanishing and the nearby boundary cycles we prove in [1] the following general theorem:

**Theorem 5.1** *Let $p \colon E \to M$ be a fibration with fiber $F$, where $E$, $M$, and $F$ are CW-complexes of dimension $2n-1$, $2k-1$, and $2(n-k)$ respectively. Assume that $F \simeq \bigvee_\mu S^{n-k}$, $M$ is 0-connected and $2k-1 \le n-k$ $(k > 0)$. Then:*

1. *The Leray-Serre spectral sequence of $p$ collapses to the term $E^2_{*,*}$.*
2. *The induced homomorphism $p_* \colon H_\ell(E; \mathbb{Z}) \to H_\ell(M; \mathbb{Z})$, with $0 \le \ell \le 2k-1$, has the following properties:*
    a. *If $2k-1 < n-k$, then $p_*$ is an isomorphism;*
    b. *If $2k-1 = n-k$, then $p_*$ is an isomorphism when $\ell < 2k-1$ and when $\ell = 2k-1$ it is surjective with kernel isomorphic to $(\mathbb{Z}^\mu)_{\pi_1(M)}$, the group of coinvariants of the action of $\pi_1(M)$ on the homology of the fiber.*

Recall that the group of coinvariants is the quotient of $H_{n-k}(F)$ by the subgroup generated by elements of the form $g \cdot a - a$, with $g \in G$ and $a \in H_{n-k}(F)$.

In particular one has the following theorem; some of the statements in it can also be proved using the Wang sequence as Milnor does in [34, Chapter 8].

**Theorem 5.2** *Let $p \colon E \to S^1$ be a fiber bundle over the circle with fiber $F$ a compact manifold of dimension $2n-2$ ($n \geq 2$), which is homotopically equivalent to a bouquet of $\mu$ $(n-1)$-spheres ($\mu > 0$). Consider the monodromy action of $G \equiv \pi_1(S^1)$ on $H_{n-1}(F) \cong \bigoplus_\mu \mathbb{Z}$. We denote by $H_{n-1}(F)^G$ the group of invariants of the action, i.e., the fixed points, and by $H_{n-1}(F)_G$ the group of coinvariants. Then:*

1. *For all $n \geq 2$ we have:*
   - *$H_q(E) \cong 0$ for all $q \neq 0, 1, n-1, n$, and*
   - *$H_n(E) \cong H_{n-1}(F)^G$ which is a free abelian group.*
2. *For $n = 2$ we have:*
   - *$H_0(E) \cong \mathbb{Z}$ and $H_1(E) \cong H_1(F)_G \oplus \mathbb{Z}$.*
   - rank $H_2(E) =$ rank $H_1(E) - 1$.
   - *$p_*$ is an epimorphism with kernel isomorphic to $H_1(F)_G$.*
3. *For $n > 2$ we have:*
   - *$H_0(E) \cong \mathbb{Z} \cong H_1(E)$ and $H_{n-1}(E) \cong H_{n-1}(F)_G$.*
   - rank $H_n(E) =$ rank $H_{n-1}(E)$.
   - *$p_* \colon H_1(E) \to H_1(S^1)$ is an isomorphism.*

Similar arguments yield:

**Theorem 5.3** *With the hypothesis of Theorem 5.2, let $i : F \hookrightarrow E$ be the inclusion of a fiber in the total space. Then the induced homomorphism $i_*$ in homology satisfies:*

- *In all cases, its kernel is the image of the homomorphism $h_* - Id : H_{n-1}(F) \to H_{n-1}(F)$, where $h_*$ is induced by the monodromy action.*
- *If $n > 2$, $i_*$ is surjective.*
- *If $n = 2$, $i_*$ is never surjective: its image is isomorphic to the group of coinvariants $H_1(F)_G$.*

Now consider a holomorphic function-germ $(\mathbb{C}^{n+1}, \underline{0}) \to (\mathbb{C}, 0)$ with a non-isolated singularity at $\underline{0}$. Notice we know from [27] that in general, the link $L_f$ is not homeomorphic to $\partial F_f$. In fact [27, Theorem 2.8] says that if $n = 2$ and the critical set is one-dimensional, then the boundary of the Milnor fiber is never homeomorphic to the link.

We have the following immediate corollary to the theorems above:

**Corollary 5.4** *Consider a holomorphic function-germ $f : (\mathbb{C}^{n+1}, \underline{0}) \to (\mathbb{C}, 0)$ with a one-dimensional critical set $\Sigma$ at $\underline{0}$. Assume for simplicity that the singular set $\Sigma$ of $V = f^{-1}(0)$ is irreducible. Let $\varepsilon > 0$ be small enough so that $L_f := V \cap \mathbb{S}_\varepsilon$ is the link of $V$ and the circle $L_\Sigma := \Sigma \cap \mathbb{S}_\varepsilon$ is the link of $\Sigma$. Let $z$ be a point in*

*$L_\Sigma$, $F_f^\perp$ a transversal Milnor fiber at z and $h_* : H_{n-1}(F_f^\perp;\mathbb{Z}) \to H_{n-1}(F_f^\perp;\mathbb{Z})$ the corresponding monodromy morphism. Then $L_f$ and the boundary $\partial F_f$ are integrally homologically equivalent if and only if $h_* - Id$ is an isomorphism.*

*Example 5.5* Consider the function $f(x, y, z) = x^a + y^b z$ where $a$ and $b$ are positive integers with $a, b > 1$. Then the boundary of the Milnor fiber is homologically equivalent over the integers to the link. To show this, notice first that its critical locus is $\Sigma = \{x = y = 0\}$. We can take in $\mathbb{C}^3$ balls with corners $(\mathbb{B}^4_{\varepsilon_1} \times \mathbb{B}^2_{\varepsilon_2}) \subset (\mathbb{C}^2 \times \mathbb{C})$. Then $L_\Sigma = \{0\} \times \mathbb{S}^1_{\varepsilon_2}$and the vanishing zone $W = \mathbb{B}^4_{\varepsilon_1} \times \mathbb{S}^1_{\varepsilon_2}$ fibers over $\mathbb{S}^1$ with fiber the ball $B_z := \mathbb{B}^4_{\varepsilon_1} \times \{z\}$. We have

$$W_t = \{(x, y, z) \in \mathbb{B}^4_{\varepsilon_1} \times \mathbb{S}^1_{\varepsilon_2};\ x^a + y^b z = t\},$$

which fibers over $\mathbb{S}^1$ with fiber

$$F_{t,z} := \{(x, y) \in \mathbb{B}^4_{\varepsilon_1};\ x^a + y^b z = t\},$$

for $z \in \mathbb{S}^1_{\varepsilon_2}$. Let us denote by $F^\perp_{t,z}$ the transversal Milnor fiber at a point $z \in L_\Sigma$. One finds (see [1]) that the induced monodromy homomorphism,

$$h_* : H_1(F^\perp_{t,z}) \to H_1(F^\perp_{t,z}),$$

is given by the following $(a-1)(b-1) \times (a-1)(b-1)$-matrix:

$$[h_*] = \begin{pmatrix} M & O & \dots & O \\ O & M & \dots & O \\ \vdots & \vdots & \ddots & \vdots \\ O & O & \dots & M \end{pmatrix},$$

where $O$ is the $(b-1) \times (b-1)$ zero-matrix and $M$ is the $(b-1) \times (b-1)$-matrix given by

$$M = \begin{pmatrix} -1 & 1 & 0 & 0 & \dots & 0 \\ -1 & 0 & 1 & 0 & \dots & 0 \\ -1 & 0 & 0 & 1 & \dots & 0 \\ \vdots & \vdots & \vdots & \vdots & \ddots & \vdots \\ -1 & 0 & 0 & 0 & \dots & 1 \\ -1 & 0 & 0 & 0 & \dots & 0 \end{pmatrix},$$

which appears exactly $(a-1)$-times in the matrix $[h_*]$. Then

$$[h_* - I_*] = \begin{pmatrix} N & O & \dots & O \\ O & N & \dots & O \\ \vdots & \vdots & \ddots & \vdots \\ O & O & \dots & N \end{pmatrix},$$

where $N$ is the $(b-1) \times (b-1)$-matrix given by

$$N = \begin{pmatrix} -2 & 1 & 0 & 0 & \dots & 0 & 0 \\ -1 & -1 & 1 & 0 & \dots & 0 & 0 \\ -1 & 0 & -1 & 1 & \dots & 0 & 0 \\ \vdots & \vdots & \vdots & \vdots & \ddots & \vdots & \\ -1 & 0 & 0 & 0 & \dots & -1 & 1 \\ -1 & 0 & 0 & 0 & \dots & 0 & -1 \end{pmatrix},$$

whose determinant is not zero. Therefore the homomorphism $h_* - I_*$ is an isomorphism, so $L_f$ and the boundary $\partial F_f$ are integrally homologically equivalent.

## 6 Concluding Remarks

We finish this note with a couple of remarks about more general settings:

*Remark 6.1* We may look at germs of holomorphic maps $(X, \underline{0}) \xrightarrow{f} (\mathbb{C}, 0)$ where $X$ is a complex analytic space with arbitrary singularities. In this general setting, if $f$ has an isolated singularity at $\underline{0}$ with respect to some Whitney stratification, by Lê [17] one has a locally trivial fibration

$$f|_{N(\varepsilon,\delta)} : N(\varepsilon, \delta) \to \mathbb{D}^*_\delta,$$

where $N(\varepsilon, \delta) := f^{-1}(\mathbb{D}^*_\delta) \cap \mathbb{B}_\varepsilon$, $\mathbb{B}_\varepsilon \subset \mathbb{R}^m$ is a Milnor ball for $f$ and $\mathbb{D}^*_\delta$ is a punctured ball in $\mathbb{R}^n$ around 0, as before. Thus one has the Milnor fibers $F_t$ degenerating to the special fiber $V = f^{-1}(0) \cap X \cap \mathbb{B}_\varepsilon$ as $t$ tends to 0, and a degeneration of the boundary $\partial F_t$ to the link $L_f$. In this setting one can show the existence of a vanishing zone essentially as in Theorem 4.2. One must replace diffeomorphisms by homeomorphisms, transversality by topological transversality, and Ehresmann's fibration lemma by the first Thom-Mather isotopy lemma.

*Remark 6.2* When considering real analytic map-germs $(\mathbb{R}^m, \underline{0}) \to (\mathbb{R}^n, 0)$, generically the discriminant $\Delta_f$, *i.e.*, the set of critical values, has codimension 1. In that case $\Delta_f$ disconnects the target into several connected components, so if one has a fibration, the base space has several components and the fibers over different components can change. Yet, if the variety $V = f^{-1}(0)$, is a complete intersection of dimension $m - n$ in $\mathbb{R}^m$ with an isolated singularity at a point, say $\underline{0}$, then Theorem 4.1 together with the implicit function theorem imply that we have a Milnor-Lê fibration, which is locally trivial. We still have that the Milnor fibers degenerate to the special fiber. This is obvious when $n = 1$. In that well-studied case we have the right Milnor fibers, and the left Milnor fibers.

For instance, let us look at the following particularly nice example, investigated thoroughly by S. López de Medrano in various papers, e.g. [21, 22]. Consider the map $\mathbb{C}^n \to \mathbb{C}$ defined by:

$$\psi(z) = \sum_{i=1}^{n} \lambda_i |z_i|^2$$

where the $\lambda_i$ are non-zero complex numbers in the Siegel domain. This means that their convex hull contains the origin $0 \in \mathbb{C}$. We further assume that the $\lambda_i$ are generic in the sense that no two of them are in the same line through the origin.

The zero-set $V := V(\psi)$ is a real analytic complete intersection in $\mathbb{C}^n$ defined by two quadrics, the real and the imaginary parts of $\psi$. This has a unique singular point at $\underline{0} \in \mathbb{C}^n$. The topology of the link of $V$ was determined by López de Medrano and it is homeomorphic to a connected sum of products of spheres (depending on the $\lambda_i$). For $n = 3$ the link always is the 3-torus $S^1 \times S^1 \times S^1$.

The critical set of $\psi$ consists of the $n$ coordinate axes. The discriminant (the critical values) are the $n$ half-lines $\mathcal{L}_1, \cdots, \mathcal{L}_n$ determined by the $\lambda_i$. These half-lines split the plane $\mathbb{C}$ into $n$ sectors, and the topology of the fibers of $\psi$ changes as we move from one sector to another. Yet one has that all Milnor fibers have the same boundary, which is diffeomorphic to the link. In fact one has:

**Theorem 6.3** *Let* $f : (\mathbb{R}^m, \underline{0}) \to (\mathbb{R}^n, 0)$ *define a real analytic isolated complete intersection singularity of dimension* $m - n > 0$*. Then all Milnor fibers have diffeomorphic boundaries and these are isotopic to the link* $L_f$*.*

The proof is simple and it is left to the reader.

## References


1. Aguilar, M., Menegon, A., Seade, J.: On the topology of the boundary of the Milnor fiber of non-isolated complex singularities (2021). Preprint
2. Andreotti, A., Frankel, Th.: The Lefschetz theorem on hyperplane sections. Ann. Math. **69**, 713–717 (1959)
3. Brauner, K.: Zur Geometrie der Funktionen zweier komplexen Verändlichen III, IV. Abh. Math. Sem. Hambg. **6**, 8–54 (1928)
4. Cisneros-Molina, J.L., Seade, J., Snoussi, J.: Milnor fibrations and $d$-regularity for real analytic singularities. Int. J. Math. **21**, 419–434 (2010)
5. Cisneros-Molina, J.L., Menegon, A., Seade, J., Snoussi, J.: Fibration theorems à la Milnor for differentiable maps with non-isolated singularities (2017). Preprint, arXiv:2002.07120

6. Colmez, P., Serre, J.-P. (eds.): Correspondance Grothendieck-Serre. Documents Mathématiques (Paris) 2. Société Mathématique de France, Paris (2001)
7. Curmi, O.: Topology of smoothings of non-isolated singularities of complex surfaces. Math. Ann. **377**, 1711–1755 (2020)
8. Curmi, O.: Boundary of the Milnor fiber of a Newton non degenerate surface singularity. Adv. Math. **372**, Article 107281 (2020)
9. Deligne, P., Katz, N.: Groupes de monodromie en géométrie algébrique, SGA7 t. II. LNM, vol. 340. Springer, Berlin (1973)
10. Dimca, A.: Singularities and Topology of Hypersurfaces. Universitext. Springer, New York (1992)
11. du Val, P.: On absolute and non-absolute singularities of algebraic surfaces. Rev. Fac. Sci. Univ. Istanb. (A) **91**, 159–215 (1944)
12. Durfee, A.: The signature of smoothings of complex surface singularities. Math. Ann. **232**, 85–98 (1978)
13. Fernández de Bobadilla, J., Menegon, A.: The boundary of the Milnor fiber of complex and real analytic non-isolated singularities. Geom. Dedicata **173**, 143–162 (2014)
14. Hironaka, H.: Stratification and flatness. In: Real and Complex Singularities (Proc. Ninth Nordic Summer School/NAVF Sympos. Math., Oslo, 1976), pp. 199–265. Sijthoff and Noordhoff, Alphen aan den Rijn (1977)
15. Kato, M., Matsumoto, Y.: On the connectivity of the Milnor fiber of a holomorphic function at a critical point. In: Manifolds Tokyo 1973 (Proc. Internat. Conf., Tokyo, 1973), pp. 131–136. Univ. Tokyo Press, Tokyo (1975)
16. Laufer, H.B.: On $\mu$ for surface singularities. In: "Several Complex Variables". Proc. Symp. Pure Math. AMS, XXX, Part 1, pp. 45–49 (1977)
17. Lê, D.T.: Some remarks on relative monodromy. In: Real and complex singularities (Proc. Ninth Nordic Summer School/NAVF Sympos. Math., Oslo, 1976), pp. 397–403. Sijthoff and Noordhoff, Alphen aan den Rijn (1977)
18. Lê, D.T.: Introduction à la théorie des singularités t. 1 et t. 2. In: Lê, D.T. (ed.) Travaux en cours, pp. 36–37. Hermann Ed., Paris (1988)
19. Lê, D.T., Perron, O.: Sur la fibre de Milnor d'une singularité isolée en dimension complexe trois. C. R. Acad. Sci. Paris Sé r. A-B **289**(2), A115–A118 (1979)
20. Lê, D.T., Nuño-Ballesteros, J.J., Seade, J.: The topology of the Milnor fiber. In: Cisneros, J.L. et al. (eds.) Handbook of geometry and topology of singularities I. Springer, Verlag (2020)
21. López de Medrano, S.: Topology of the intersection of quadrics in $\mathbb{R}^n$. In: Carlsson et al. (eds.) "Algebraic Topology" (Arcata, CA, 1986). Lecture Notes in Math., vol. 1370, pp. 280–292. Springer, Berlin (1989)
22. López de Medrano, S.: Singular intersections of quadrics I. In: Cisneros-Molina, J.L. et al. (eds.) Singularities in Geometry, Topology, Foliations and Dynamics. Trends in Mathematics. Birkhäuser, Basel (2017)
23. Massey, D.B.: Lê Cycles and Hypersurface Singularities. Lecture Notes in Mathematics, vol. 1615. Springer, Berlin (1995)
24. Massey, D.B.: Notes on Perverse Sheaves and Vanishing Cycles (2016). eprint arXiv:math/9908107, version 7
25. Maxim, L.: On the Milnor classes of complex hypersurfaces. In: G. Friedman et al. (eds.) Topology of Stratified Spaces, vol. 58. MSRI Publications, Camb. Univ. Press, Cambridge (2011)
26. Menegon Neto, A., Seade, J.: On the Lê-Milnor fibration for real analytic maps. Math. Nachr. **290**(2–3), 382–392 (2017)
27. Menegon Neto, A., Seade, J.: Vanishing zones and the topology of non-isolated singularities. J. Geom Dedicata **202**(1), 321–335 (2019). https://doi.org/10.1007/s10711-018-0415-5
28. Michel, F., Pichon, A.: On the boundary of the Milnor fiber of non-isolated singularities. Int. Math. Res. Notices **43**, 2305–2311 (2003)
29. Michel, F., Pichon, A.: On the boundary of the Milnor fiber of non-isolated singularities (Erratum). Int. Math. Res. Notices **6**, 309–310 (2004)

30. Michel, F., Pichon, A.: Carousel in family and non-isolated hypersurface singularities in $\mathbb{C}^3$. J. Reine Angew. Math. (Crelles J.) **720**, 1–32 (2016)
31. Michel, F., Pichon, A., Weber, C.: The boundary of the Milnor fiber of Hirzebruch surface singularities. In: Cheniot, D. et al. (eds.) Singularity Theory, pp. 745–760. World Sci. Publ., Hackensack (2007)
32. Michel, F., Pichon, A., Weber, C.: The boundary of the Milnor fiber for some non-isolated singularities of complex surfaces. Osaka J. Math. **46**, 291–316 (2009)
33. Milnor, J.W.: Differentiable structures on spheres. Am. J. Math. **81**, 962–972 (1959)
34. Milnor, J.W.: Singular points of complex hypersurfaces. Ann. Math. Studies, vol. 61. Princeton University Press, Princeton (1968)
35. Mumford, D.: The topology of normal singularities of an algebraic surface and a criterion for simplicity. Publ. Math.. Inst. Hautes Études Sci. **9**, 5–22 (1961)
36. Némethi, A.: Invariants of normal surface singularities. Contemp. Math. **354**, 161–208 (2004)
37. Némethi, A.: Lattice cohomology of normal surface singularities. Publ. Res. Inst. Math. Sci. **44**, 507–543 (2008)
38. Némethi, A., Szilárd, Á.: Resolution graphs of some surface singularities, II. (Generalized Iomdin series). Contemp. Math. **266**, 129–164 (2000)
39. Némethi, A., Szilárd, Á.: Milnor Fiber Boundary of a Non-isolated Surface Singularity. Lecture Notes in Mathematics, vol. 2037. Springer, Berlin (2012)
40. Neumann, W.D.: A calculus for plumbing applied to the topology of complex surface singularities and degenerating complex curves. Trans. Am. Math. Soc. **268**, 299–344 (1981)
41. Oka, M.: Topology of polar weighted homogeneous hypersurfaces. Kodai Math. J. **31**(2), 163–182 (2008)
42. Pichon, A., Seade, J.: Fibered multilinks and singularities $f\bar{g}$. Math. Ann. **342**(3), 487–514 (2008)
43. Randell, R.: On the topology of non-isolated singularities. In: Proceedings 1977 Georgia Topology Conference, pp. 445–473 (1977)
44. Sabbah, C.: Vanishing cycles and their algebraic computation. http://www.cmls.polytechnique.fr/perso/sabbah/livres/sabbah_notredame1305.pdf
45. Seade, J.: On the Topology of Isolated Singularities in Analytic Spaces. Progress in Mathematics, vol. 241. Birkhauser, Basel (2005)
46. Seade, J.: On Milnor's Fibration theorem and its offspring after 50 years. Bull. Am. Math. Soc. **56**, 281–348 (2019)
47. Siersma, D.: Variation mappings on singularities with a 1-dimensional critical locus. Topology **30**, 445–469 (1991)
48. Siersma, D.: The vanishing topology of non isolated singularities. In: New Developments in Singularity Theory, Cambridge 2000, pp. 447–472. Kluwer, Dordrecht (2001)
49. Verdier, J.-L.: Stratifications de Whitney et théorème de Bertini-Sard. Invent. Math. **36**, 295–312 (1976)
50. Verdier, J.-L.: Spécialisation des classes de Chern. Astérisque 82–83, pp. 149–159. Société mathématique de France, Marseille (1981)